A Railroad Atlas of the United States in 1946

VOLUME 1: THE MID-ATLANTIC STATES

PUBLISHING FOR THE WORLD
125 Years
THE JOHNS HOPKINS UNIVERSITY PRESS

CREATING THE NORTH AMERICAN LANDSCAPE

Gregory Conniff, Edward K. Muller, David Schuyler, *Consulting Editors*

George F. Thompson, *Series Founder and Director*

Published in cooperation with the

CENTER FOR AMERICAN PLACES

Santa Fe, New Mexico, and Harrisonburg, Virginia

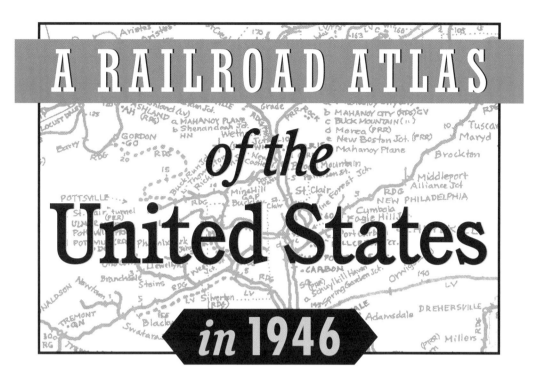

A RAILROAD ATLAS
of the
United States
in 1946

VOLUME 1: THE MID-ATLANTIC STATES

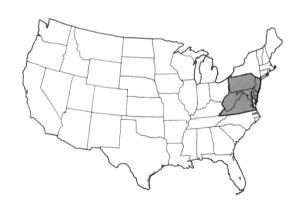

Richard C. Carpenter

THE JOHNS HOPKINS UNIVERSITY PRESS   BALTIMORE AND LONDON

The Johns Hopkins University Press
2715 North Charles Street
Baltimore, Maryland 21218-4363
www. press.jhu.edu

Library of Congress Cataloging-in-Publication Data

Carpenter, Richard C.
   A railroad atlas of the United States in 1946 /
Richard C. Carpenter
      p.   cm. — (Creating the North American landscape)
      Includes bibliographical references and index.
      Contents: v. 1. The Mid-Atlantic states.
      ISBN 0-8018-7331-2 (alk. paper)
      1. Railroads—Middle Atlantic States—Maps.
2. Railroads—Middle Atlantic States—History. I. Title.
II. Series.

G1246.P3 C3 2003
385'.0973'0223—dc21   2002031038 2002040581

A catalog record for this book is available
from the British Library.

Cover photograph taken by Philip M. Carpenter, the author's father, on April 13, 1947, while passing through Bowie, Maryland, on the northeast corridor main line of the then–Pennsylvania Railroad—between Washington, D.C., and Baltimore.

The sketches on pages v, xix, and xxiii were inspired by the photographs of Philip R. Hastings on pages 152, 179, and 138, respectively, of David P. Morgan, *The Mohawk That Refused to Abdicate and Other Tales* (Milwaukee: Kalmbach Publishing Co., 1975).

Contents

Introduction

I was driving across northern Indiana along U.S. 224, east of Huntington, on a trip home from Chicago in 1990. Looking out across the fields, I could see the distinctly visible high fill that had once been the roadbed for the double-track Erie Railroad, by then abandoned. I was following it out of the corner of my eye, as I am wont to do with old railroad lines, when suddenly it disappeared. The fill, which had been cutting a 15-foot-high swath across the verdant fields for miles, was completely gone. I could now look across flat sections of corn and soy as far as my eyes could see. There was absolutely no sign that this farmland had ever been interrupted in any way, never mind that it had watched, for more than one hundred years, the passage of thousands upon thousands of trains, from long, lumbering freights to such passenger trains as the New York–to–Chicago Erie Limited, the Atlantic Express, and the Lake Cities.

As I drove on, I spotted one bit of evidence. It was almost paved over in a farmhouse driveway next to a fully grown cornfield. Four pieces of steel rail sat in the pavement, beginning and ending on each side. These four shining rails, now retired to a tranquil life on a picturesque Indiana farm, were all that remained—in these parts, at least—of the once mighty Erie Railroad. They bore silent witness to the golden age of American railroading, when lines like this criss-crossed the country carrying goods and people and when sublime stretches of summer-evening silence on the prairie were marked by the unmistakable throaty moan of the steam engine's whistle.

I saw many other abandoned railroad rights-of-way as I drove across the historic, complex railscape of northern Indiana, Ohio, and on eastward into Pennsylvania and New Jersey. Somehow it occurred to me that I should not leave un-

recorded or forgotten the existence of these many links in what had once been a United States railroad system totaling some 254,037 route miles. The story of American railroading in its heyday was, and is, a story worth telling. And I became determined to tell it with a clear, easy-to-read atlas.

It seems particularly appropriate to document the railroad network that provided the steel pathway over which President Franklin D. Roosevelt's "arsenal of democracy" had transported the results of its overwhelming war machine. Mr. Roosevelt himself loved to ride the American rails in his private car, the Ferdinand Magellan.

It is my hope that, by producing a graphic record of this transportation network, present and future generations may learn valuable lessons from one of the most glorious episodes of our transportation history. And what better year than 1946, when the impressive system of American railroads emerged from its magnificent performance during World War II to face the challenges of highway and airline competition in the post-war era. This year marked a historic turning point in American rail transportation. In 1946, railroads were still the primary mode of intercity transportation. Indeed, the previous four years of World War II had increased their modal share of passengers and freight, due to gasoline rationing and its severe restrictions on nonessential travel as well as massive wartime troop and materiel shipments. But, with the lifting of wartime restrictions, the steady prewar trend from rail to highway transport and the growing demand for the speed afforded by the airplane would challenge and fundamentally change the American railroad network.

The year 1946 would also mark the beginning of the end of a long period of relative stability in the number of United States railroad companies. For more than a quarter century, between 1920 and 1946, only four national, significant railroad mergers and consolidations occurred. In 1923, the Nickel Plate Road absorbed both the Toledo, St. Louis and Western and the Lake Erie & Western; in 1938, the Duluth, Missabe, and Iron Range absorbed through consolidation the Duluth, Missabe & Northern, Duluth & Iron Range, Interstate Transfer, and Spirit Lake Transfer; in 1939, the Gulf, Mobile & Ohio absorbed through consolidation the Gulf, Mobile & Northern, Mobile & Ohio, and the New Orleans Great Northern; and 1945 saw the Gulf, Mobile & Ohio absorb the Alton through merger.

Only one more merger would take place before 1950, the start of what Frank Wilner, in his 1997 book *Railroad Mergers: History, Analysis, Insight,* has called "the modern merger movement of 1950–1979." In 1947, the Chesapeake & Ohio absorbed the Pere Marquette. This modern merger movement would, by 1990, remove more than 80,000 route miles from the railroad map of the United States. In 1946, there were 137 Class I railroads in the country. By 2001, there were only seven Class I railroad systems in the United States and Canada. Many new "regional" and "shortline" railroads would subsequently be created from those lines not wanted by the merged Class I railroad systems. Some of those would choose to use their former historic "fallen flag" names, such as Wheeling and Lake Erie, Ohio Central, and Wisconsin Central.

In 1946, nearly every rail line, both main and branch, was served by at least one passenger train per day, which also picked up and delivered the mail. The east-west main line of the Pennsylvania Railroad, through the center of the state of Pennsylvania, had some sixty-three daily scheduled passenger trains operating over its four tracks. Often these trains operated with several following separate train sections to accommodate the demand. Nearly all of them had names, some now famous, such as the Broadway Limited between New York City and Chicago and the Spirit of St. Louis between New York City and St. Louis. These particular trains were composed of Pullman cars only.

In 1946, all railroad lines carried freight, both car-load and less-than-car-load, and they moved everything from massive electric generators and giant steel bridge girders on heavy multiwheeled flat cars, to a child's bicycle in a Railway Express car. Nearly every town of any size had its own freight agent, coal unloading trestle, and bulk delivery track. At nearly every junction between two or more railroad companies, specially designated tracks existed to permit the interchange of freight cars. The loaded car interline routing between different railroads could be specifically chosen by the shipper from a variety of published routes, while the return of empty cars usually followed a long-standing "home route" pattern devised by the railroads. If shippers or receivers delayed cars beyond the normal two-day allowance, they were charged "demurrage." And each railroad was required to pay the car owner for every day one of the owner's cars was on line.

More than 1.3 million railroad workers were needed to manage this complex transportation delivery system. In addition to five-person train crews (often with a sixth, or "swing," crewman added in some states when cars exceeded certain limits per train—a holdover from applying freight car brakes by hand!), there were station agents, signal tower operators and signal maintainers, ticket collectors, track laborers, roundhouse and shop workers, yard clerks, and trainmasters. All of these employees, and many more as well, functioned and reported through a military-style chain of command to an all-powerful division superintendent.

My own memories of a typical 1946 railroad station scene may provide the reader with a more meaningful understanding of the feel of American railroading fifty-five years ago. It is important to remember that much of American daily life revolved around the coming and going of trains. If a train was late, people took note. The "train whistle in the night" caused generations of small-town young people to dream of moving to a better life—far away—often in the big city.

Looking down the track, you would see that the rails were jointed, causing the "clickety-clack" sound when trains ran over them. The railroad ties were all made of wood, oozing with creosote preservative on hot summer days. Nowadays, modern trains glide over uninterrupted welded rail, sometimes supported by concrete ties. Overhead bridges were blackened with coal soot, unmistakable accumulated evidence of the passage of countless steam locomotives. These bridges were guarded with "tell-tales," which brushed the heads of brakemen on car roofs, warning them to lie flat, lest they hit the bridge.

Along the edge of the track, parallel with a neat, hand-raked line delineating ballast from cinders, stood a stately procession of telegraph poles. Each pole often supported five or more cross arms, each carrying from eight to ten separate wires. Many of those wires were owned by the Western Union Telegraph Company, and they provided the visual identification mark of a railroad line. Some were for railroad company dispatching and message circuits, others for railroad signal power supply.

And then there were the signals. Many were still of the upper- or lower-quadrant semaphore type, with red or yellow arms and white and black stripes. Many types of buildings—stations, signal towers, freight houses, water tanks, section houses, and tool sheds—most of them wooden, stood along the railroad tracks, each painted in the unique color style of its railroad. Often this was a particular shade of gray, brown, yellow, or green.

At the typical small passenger station, the platform might be wooden, with a yellow stripe painted parallel to the rails to warn waiting passengers to stand back. The station would almost always have a bay window, from which the station agent could easily observe passing trains. There would likely be a train order signal on a high mast near this bay window, with brightly painted mechanical levers controlled by the agent. Train orders would be written on thin paper, nicknamed "flimsies," in either three or five duplicate copies, depending upon the nature of the order. Communication at some stations was still by Morse code telegraph, with its clatter of dots and dashes, and radio was just starting to be used. But mostly, the telephone was in use, with its distinctive headset and mouthpiece on a pantograph-type extension.

If there was a signal tower near the station, the metallic sound of interlocking machine levers being pushed or pulled could also be heard. The imminent approach of a train would be announced within the tower by a distinctive bell, gong, or buzzer, depending upon the direction or origin of the train. At night, colored lanterns—red, yellow, green, and white—stood ready for instant use, but normally they were not visible through a window, so as to not give a false signal to trains. Both stations and towers were ideal places from which to watch trains.

A visitor to a passenger station would see venerable wooden baggage carts, braced with cast iron straps, waiting to be pulled alongside an open baggage car door to allow the transfer of baggage, mail, and express. On the waiting room wall would be colorful posters and advertisements. Near the ticket window would be racks of multicolored railroad public timetables, all free for the taking. Air conditioning was provided by a slow-turning, electric ceiling fan. In winter, warmth was furnished by the pot-bellied stove. Refreshment was available from Chiclets®, gum machines and a porcelain or shiny metal water fountain. Hard wooden waiting room seats were available, from which passengers could listen to the stately ticking of a standard railroad pendulum clock.

Most important of all were the trains. In 1946, nearly everyone watched passenger trains. There were fast passenger "limiteds," often speeding through, non-stop—baggage cars followed by coaches, then the dining car, with perhaps a brief whiff of charcoal smoke from the kitchen stove. Finally came olive-green Pullman cars, with maybe even an open-platform observation car, carrying heavy colored-glass marker lanterns to indicate the end of the train. But the real variety was provided by the long, sometimes slow freight trains—box cars, tank cars, cattle cars, yellow "reefer" (refrigerator) cars, gondola cars, hopper cars, flat cars, and even a few flat cars with highway truck trailers. And there was always a caboose, usually bright red, at the end of the train. Now there is only the winking flash of the "end-of-train device" fastened to the rear coupler of the last freight car.

Especially fascinating was the seemingly infinite variety of railroad names and distinctive railroad company logos and markings—the keystone of the Pennsylvania, the oval of the New York Central, the U.S. Capitol dome of the Baltimore and Ohio, the gracefully interlacing scroll of the New Haven, the black diamond of the Lehigh Valley, the shield of the Union Pacific, the "Heart of Dixie" of the Seaboard Air Line Railway, the "Chessie" cat of the Chesapeake & Ohio and many more. Then there was the unforgettable noise—the rolling, clattering, squealing, and straining of a long freight train.

As you examine the multi-colored maps, read the station names, and trace the course of the active and abandoned lines of this railroad atlas, try to imagine yourself in this railroad age of fifty-seven years ago. Soon a steady procession of railroad mergers, as well as highway and airline competition, would cause most of the 1946 American railroad scene to disappear forever.

MID-ATLANTIC STATES

Volume 1 of the *A Railroad Atlas of the United States in 1946* includes an area with some of the oldest railroads and the most complex network of railroad lines in North America. The Baltimore & Ohio Railroad, the first common carrier railroad in the United States and the first to offer scheduled freight and passenger service to the public, was chartered in 1827 to build a railroad from Baltimore to a suitable point on the Ohio River. The "suitable point" turned out to be Wheeling, West Virginia, to which a 379-mile railroad line was completed on January 1, 1853.

Six states are included in Volume 1: Delaware, Maryland, New Jersey, Pennsylvania, Virginia and West Virginia, as well as the District of Columbia. More than 10 percent of the railroad miles operated in the United States in 1946 were within these six states. Of the 137 Class I railroads operating in the United States in 1946, thirty-five operated within these six Middle Atlantic states. (In 1946, a Class I railroad was one having an annual operating revenue of more than $1 million.) This atlas includes, in addition to the six states listed above, the directly adjacent areas of New York, Ohio, Kentucky, Tennessee, and North Carolina. Of particular interest, New York City and western Long Island are also included.

Following is a brief summary of the major geographic and railroad characteristics of each of the six states in the Mid-Atlantic volume.

Delaware

Delaware is the second smallest state in the union, after Rhode Island, with a total area of 2,370 square miles. It has generally flat terrain, with low hills in the north. The principal railroad city is Wilmington, a station along the Northeast corridor between Washington, D.C., and Boston.

In 1946, three Class I railroads operated in Delaware, with a total of 295 route miles—the Baltimore & Ohio (36 miles), the Pennsylvania (237 miles), and the Reading (22 miles). Major main lines included the Pennsylvania's two-and three-tracked electrified main line from Philadelphia to Washington, D.C., the Baltimore & Ohio's double-tracked Philadelphia-to-Baltimore main line, and the double-tracked Pennsylvania "Del-Mar-Va" main line, from Wilmington to Cape Charles, Virginia.

Among Delaware's railroad engineering features was the grade-separated flyover south of Claymont, Delaware, built in 1904, to allow freight trains on the two center tracks to reach the Shellpot Branch, under the northbound passenger main track. Indeed, the Shellpot Branch, itself, was an example of an early by-pass line, which allowed freight trains to avoid the congestion and the many grade crossings that existed prior to the later elevation of the main line through Wilmington passenger station. The Pennsylvania Railroad constructed many such grade separations and junctions early in the twentieth century.

The only major shop facility in Delaware was the Wilmington Shops, the heavy maintenance point for the electric locomotives and multiple-unit cars of the Pennsylvania Railroad.

District of Columbia

This 70-square-mile district, the capital of the United States, is located within, but is separate from, the states of Maryland and Virginia.

Two Class I railroads own lines that enter the District of Columbia: the Baltimore & Ohio (28 miles) and the Pennsylvania (13 miles). Prior to 1907, each company had a separate passenger terminal in the District. In 1907, the Washington Union Station was opened, and both the Baltimore & Ohio and the Pennsylvania began using this station and closed their separate stations. Station trackage and approaches were operated by the Washington Terminal Company, jointly owned by these two railroads.

The Chesapeake & Ohio, the Southern Railways, and the Richmond, Fredericksburg & Potomac, handling trains of the Atlantic Coast Line and Seaboard Air Line, entered the District from the south via the famous "Long Bridge" across the Potomac River. This bridge, given its strategic and unique status as the only north-south, direct rail connection along the eastern seaboard, was heavily guarded during World War II to prevent sabotage.

Freight lines of the Pennsylvania and the Baltimore & Ohio Railroads used the Pennsylvania main line between Anacostia and Potomac Yard, RF&P's large, hump classification yard in Alexandria, Virginia. Here, cars were classified and exchanged among PRR, B&O, RF&P, C&O, and Southern Railways. In recent years, Potomac Yard has been completely closed and its tracks removed. High real estate value and rail mergers have resulted in through trains operating between more distant classification yards.

Maryland

Maryland ranks forty-first in area among the states, with 12,327 square miles (of which 2,386 are water surface). The generally low elevations surrounding the Chesapeake Bay contrast with the mountains of the western part of the state, both east and west of Cumberland.

In 1828, the Baltimore & Ohio Railroad began to build westward from Baltimore. By 1835, a branch had also been constructed southward to Washington, D.C. The primary objective of the main route west was to reach the Ohio River, thus connecting the Atlantic coast port of Baltimore with this major river, the primary water route of the Midwest. By 1853, the 379-mile rail line had been completed, following the Patapsco and Potomac rivers to Cumberland and beyond, thence over the Appalachian mountains through Grafton and Fairmont to Wheeling and the Ohio River. This strategic connection helped make Baltimore a major world port for both general merchandise and coal.

Four Class I railroads were operating in Maryland in 1946: the B&O (347 miles), the Norfolk and Western (35 miles), the Pennsylvania (607 miles), and the Western Maryland (273 miles), for a total of 1,262 route miles. Other primary main lines included the electrified multiple-track Pennsylvania, from New York to Washington, and the B&O main lines from Philadelphia to Washington and from Washington to Point-of-Rocks, the junction with the original Baltimore-to-Wheeling main line. In addition, the parallel east-west main line of the Western Maryland Railway ran west from Baltimore through Hagerstown and Cumberland to Connellsville, Pennsylvania. Part of the Western Maryland line constituted the so-called "alphabet route" between the Midwest and New England, whereby several smaller connecting railroads competed with the major east-west trunk-line through railroads. Finally, the north-south Shenandoah Valley line of the Norfolk and Western ran from Hagerstown to Roanoke, Virginia. The Western Maryland also connected with the Pennsylvania at Hagerstown and with the Reading at Shippensburg, Pennsylvania. Major classification yards were located at Cumberland and Weverton on the B&O. Major shops were located at Mount Clare in Baltimore, at Cumberland on the B&O and at Hagerstown on the Western Maryland.

New Jersey

New Jersey ranks forty-fifth among the states in size, at 8,224 square miles. It is generally flat, with a few hills to the north and west.

In 1946, eleven Class I railroads operated in New Jersey: the Central Railroad of New Jersey (419 miles), the Delaware Lackawanna & Western (225 miles), the Erie (125 miles), the Lehigh & Hudson River (70 miles), the Lehigh & New England (42 miles), the Lehigh Valley (130 miles), the New York Central (23 miles), the New York, Susquehanna & Western (120 miles), the Pennsylvania (449 miles); Pennsylvania-Reading Seashore Lines (390 miles), and the Reading (55 miles)—a total of 2,048 miles.

New Jersey was the eastern end of seven east-west main and trunk line railroads: the New York Central from the north; the Pennsylvania and the Reading (via the Central Railroad of New Jersey from Bound Brook) from the south; and the Central Railroad of New Jersey, the Delaware, Lackawanna & Western, the Erie and, the Lehigh Valley from the west. Also, B&O passenger trains used the

Reading and the Central Railroad of New Jersey to reach Jersey City.

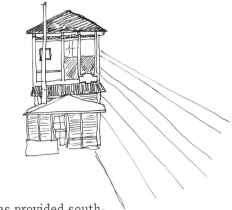

For many years prior to 1946 and continuing to the present day, the extensive suburbs of northern New Jersey have been served by many busy rail commuter lines of the railroads enumerated above. Also, the New Jersey shore area was served by the New York & Long Branch Railroad, which was jointly controlled by the Central Railroad of New Jersey and the Pennsylvania Railroad. Yet another commuter service was provided southeast of Camden and Philadelphia. Here, the Pennsylvania–Reading Seashore Lines, jointly controlled by the two railroads identified in its name, had since 1933 provided rail service to Atlantic City, Ocean City, Cape May, and the Bridgeton-Millville area. Secondary main lines included the Lehigh & Hudson River, a "bridge" line railroad across northwestern New Jersey, providing north-south connecting service between the Central Railroad of New Jersey, the Lehigh Valley, the Pennsylvania, and Reading on the south, and the Delaware, Lackawanna & Western, the Erie, the New York, Susquehanna & Western, the New York, Ontario & Western, the New York Central, and the New Haven on the north, at New Haven Railroad's western gateway at Maybrook, New York. Beyond it, the so-called "Poughkeepsie Bridge route" provided an all-rail route connection to and from New England.

Major New Jersey classification yards included: Secaucus on the Delaware, Lackawanna & Western; Croxton on the Erie; and Oak Island on the Lehigh Valley. Pennsylvania Railroad operated yards at Waverly in Newark, Meadows in Jersey City, and Greenville in Bayonne. All of these railroads interchanged cars with the Long Island and the New Haven railroads via car float ferries across New York Harbor to Brooklyn and the Bronx. In addition, all railroads had piers and floating interchange services for the transfer of goods to and from ocean-going ships and coastwise vessels. Finally, there were two major shops: Elizabethport on the Central Railroad of New Jersey and Kingsland, north of Newark, on the Delaware, Lackawanna & Western.

Pennsylvania

Pennsylvania, the largest of the mid-Atlantic states, ranks thirty-second among the states in size (just above Virginia) and has an area of 45,126 square miles. It is generally hilly and mountainous with major bituminous coal fields in the western half of the state and anthracite coal fields in the eastern, between Reading and Scranton. The mountainous terrain of Pennsylvania has made main line railroad construction difficult. While many railroads operated in Pennsylvania, only three main intercity routes crossed the Allegheny mountain range: the Baltimore

& Ohio's Baltimore-to-Pittsburgh main line and the Pennsylvania's Philadelphia-to-Pittsburgh main line and its Harrisburg-to-Buffalo and Erie, Pennsylvania, lines. Since twenty-two Class I railroads operated in Pennsylvania in 1946, only those of 200 miles or more are listed here: the B&O (1,156 miles), the Central Railroad of New Jersey (230 miles), the Erie (635 miles), the New York Central (800 miles), the Pittsburgh and Lake Erie (212 miles), the Western Maryland (244 miles), the Bessemer & Lake Erie (209 miles), the Delaware, Lackawanna & Western (255 miles), the Lehigh Valley (520 miles), the Pennsylvania (4,017 miles), and the Reading (1,285 miles). The total mileage of these railroad lines is 9,563, and the total of all twenty-two Class I railroads in Pennsylvania in 1946 is 10,473 miles.

In northeastern Pennsylvania, three of these railroads crossed the Pocono Mountains: the Central Railroad of New Jersey, the Lehigh Valley east of Wilkes-Barre, and the Delaware, Lackawanna & Western east of Scranton. The Delaware and Hudson (using Erie trackage) and the Erie crossed the southern Catskills north of Scranton. All of the railroads handled considerable anthracite coal traffic from the mines in northwestern Pennsylvania in 1946, when the primary mode of home and commercial heating was still the coal-fired furnace. As noted in the Maryland description, both the Pennsylvania and the Reading served the Hagerstown/Harrisburg corridor. Some of this traffic flowed north from Harrisburg to the Delaware and Hudson Railroad at Wilkes-Barre, and from there to Canada and Northern New England.

Both the north-south New York–Washington multi-tracked electrified main line of the Pennsylvania and the Reading/B&O routes ran across the southeastern corner of the state, through Philadelphia. The Philadelphia area had an extensive and established commuter area long before 1946, served by both the Pennsylvania and the Reading railroads. As noted in the New Jersey description, some Pennsylvania-Reading Seashore Lines commuter trains crossed the Delaware River and operated directly into Broad Street station, which was also the historic headquarters as well as the Philadelphia terminal station of the Pennsylvania Railroad.

Major railroad engineering landmarks in the state of Pennsylvania included the "Horseshoe Curve" and Gallitzin tunnels of the Pennsylvania Railroad (PRR), west of Altoona; the Sand Patch Tunnel on the B&O south of Johnstown; the reinforced concrete viaducts of the Delaware, Lackawanna & Western at Kingsley and Tunkhannock, northwest of Scranton; Kinzua viaduct on the Erie, north of Mount Jewett in north central Pennsylvania; Starrucca viaduct on the Erie, east of Susquehanna, and south of the New York State line; and, finally, the complex grade-separated junction known as "Zoo Tower" in Philadelphia. Major PRR hump classification yards were at Conway and Pitcairn in Pittsburgh and at Enola (near Harrisburg); Reading's major hump yard was at Rutherford (near Harrisburg); and Central New Jersey's was at Allentown. This last yard was also used by trains of the Reading and the Lehigh & Hudson River Railway. Major shop facilities were to be found at Altoona (PRR), Reading (RDG), Sayre (LV), Susquehanna and Meadville (Erie), Scranton (Delaware, Lackawanna & Western), and Glenwood and DuBois (B&O).

Virginia

Virginia ranks thirty-third among the states, with an area of 42,627 square miles of which 2,365 square miles represent water surface. It is generally hilly and mountainous beyond its tidewater areas.

Some twelve Class I railroads operated in Virginia in 1946. Of these, eight operated 100 or more miles: the Atlantic Coast Line (147 miles), the Chesapeake & Ohio (753 miles), the Clinchfield (107 miles), the Norfolk & Western (2,456 miles), the Richmond, Fredericksburg & Potomac (118 miles), the Seaboard Air Line (178 miles);, the Southern (543 miles), and the Virginian (333 miles). The total mileage of these main railroads was 4,635 miles; the total of all twelve railroads was 4,930 miles. Two major east-west lines crossed Virginia along generally parallel routes. The Chesapeake & Ohio ran from Fort Monroe, near Newport News, to Cincinnati, generally along the James River; the Norfolk & Western ran from Norfolk to Columbus, Ohio, via Roanoke and south of the C&O. The Norfolk & Western was paralleled by the Virginian from Norfolk to Deepwater Bridge, West Virginia, and they ran close together through the Roanoke area.

Several long-distance north-south trunk line railroads traversed the southeastern part of Virginia. The Southern Railway ran from Washington, D.C., to Columbus, Mississippi, via Charlottesville, Lynchburg, and Danville. The Richmond, Fredericksburg & Potomac ran from Richmond to Alexandria, Virginia. South of Richmond, both the Atlantic Coast Line and the Seaboard Air Line had parallel routes from Richmond to Tampa, Florida. Trains of both of these railroads operated via RF&P between Richmond and Washington, D.C. Likewise, the C&O operated over the Southern from Orange to Alexandria, where both the Southern and the C&O operated over RF&P to Washington, D.C. Most of north central Virginia was too rugged to encourage construction of additional east-west main lines. However, along the Shenandoah Valley, the Norfolk & Western operated a major north-south connection, which then turned southwestward toward Knoxville and Chattanooga, Tennessee. The Louisville and Nashville had extensive coal mine branches in the extreme western end of Virginia.

Principal hump classification yards included Roanoke on the Norfolk & Western and the Richmond, Fredericksburg & Potomac's Potomac Yard at Alexandria. This latter yard, across from Washington, D.C., was shared by the B&O, the C&O, the Pennsylvania, and the Southern. Also, the Norfolk & Western and the Virginian had major yards at Norfolk, and the C&O had a major yard at Newport News, for the purpose of transferring coal to ocean-going vessels for shipment. Major shop facilities were located at Roanoke on the Norfolk & Western, and they were still building large steam locomotives for that railroad in 1946.

West Virginia

West Virginia ranks fortieth among the states in area, with 24,170 square miles. The state is completely mountainous, which gives rise to its motto, "Mountain men are always free." More than three quarters of the state constituted one of the largest single coal producing regions in the United States. This immense coal resource was the major reason for the great revenue strength of both the Chesapeake & Ohio and the Norfolk & Western railways.

Nine Class I railroads operated in the state. Those railroads of 100 miles and over included: the Baltimore & Ohio (1,380 miles), the Chesapeake & Ohio (992 miles), the New York Central (258 miles), the Norfolk & Western (1,078 miles), the Virginian (328 miles), and the Western Maryland (323 miles), for a total of 4,359 miles. The total of all nine railroads was 4,521. Primary main lines included the B&O from Baltimore through Grafton to Wheeling, on the Ohio River and the B&O main line west from Grafton to Parkersburg, Cincinnati, and ultimately to St. Louis. The C&O main line ran along the New and Kanawha Rivers to the Ohio River. The Norfolk & Western main line ran along the Tug Fork River and the Big Sandy River to the Ohio River.

The B&O served northern West Virginia coal fields with an extensive network of branch lines, while the C&O, the New York Central, the Norfolk & Western, and the Virginian served the coal deposits of central and southern West Virginia.

The Alphabet Route, described earlier, passed through the "panhandle" of the state, north of Wheeling. Here it used the Pittsburgh & West Virginia Railway between a junction with the Wheeling & Lake Erie in eastern Ohio and the connection with the Western Maryland at Connellsville, Pennsylvania. Also crossing the panhandle was the Pennsylvania main line from Pittsburgh to Columbus, Ohio, Indianapolis, and St. Louis.

The only classification yard of significant size was at Bluefield, on the Norfolk & Western. Major locomotive shops were located at Huntington on the C&O and at Princeton on the electrified western portion of the Virginian.

How to Use This Atlas

FINDING GEOGRAPHICAL AREAS

Railroads within specific geographical areas may be found by consulting the Key Map, which shows state lines, the boundaries of all 177 standard atlas maps, and the locations of detail maps. The boundaries of each of these standard atlas maps represent 30-minute quadrangles between each full degree of latitude and longitude.

Each map has a geographical name and a consecutive number. The name reflects a prominent railroad station within the quadrangle. The number represents the consecutive arrangement of maps from west to east and from north to south. These numbers are also the page numbers for the atlas. The maps in this Middle Atlantic states volume begin with number 1, Erie, Pennsylvania, and end with number 177, Norfolk, Virginia. Areas of complexity are shown on detail maps, which also have a geographical name and a number as well as a letter. The number is the standard atlas map number of the map within which the detail map or maps are located. This number is followed by a letter distinguishing this map as a detail map.

FINDING STATIONS

Particular railroad stations may be found by consulting the alphabetical index. Each index entry gives a station name, followed by the abbreviation for the state in which it is located, followed by the consecutive number of the standard atlas map and detail map on which it is to be found. This number is followed by one of four map quadrant designations, as follows: upper-left quadrant, NW; upper-right quadrant, NE; lower-left quadrant, SW; and lower-right quadrant, SE.

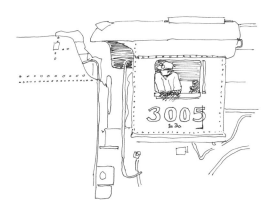

Also included in this index are the names of interlocking stations, both existing and former, coaling stations, track pans, tunnels, and major viaducts and bridges. On every atlas map, all of these names are given in quotation marks to distinguish them from ordinary stations.

RAILROAD COLORS

Each railroad company is identified by one of nine colors: red, yellow, orange, brown, green, light blue, dark blue, purple, and black. Every effort has been made to choose a color for each railroad company that has a historic association with that railroad. For example, red has been used for the Pennsylvania Railroad and the Seaboard Air Line; green represents the Erie Railroad and the Southern Railway System; blue has been used for the Baltimore & Ohio Railroad and the Delaware, Lackawanna and Western Railroad; and purple has been used for the Atlantic Coast Line Railroad and the Western Maryland Railway. However, in order to preserve a graphic color contrast between adjacent railroads, the choice of an unrelated color has sometimes been necessary. Railroads controlled or leased by a major railroad are represented by the same color.

CROSS REFERENCE WITH OTHER MAPS

This railroad atlas can easily be cross-referenced and compared with United States Geological Survey topographic maps, the DeLorme Mapping Company state atlases, and any atlas map that includes latitude and longitude lines, as those lines serve as the basis for this atlas.

The Key Map identifies the names of the USGS United States 1:250,000 scale series, which were used as the base map for the 177 standard atlas maps. These maps, with a scale of one inch equals four miles, were originally produced by the U.S. Army Map Service during the 1940s and 1950s. This map series was the first USGS map product to completely cover the United States, and it has provided a consistent mapping base for this 1946 railroad atlas.

Each standard atlas map measures 30 minutes of latitude and longitude and includes tick marks defining the four possible USGS 15-minute topographic quadrangles that could be contained within. In turn, each of these 15-minute quadrangles may contain as many as four 7.5-minute USGS topographic maps. The United States 15-minute series was the USGS standard from 1910 to 1950, with many early maps dating from 1890 onward. In about 1950, the United States 7.5-minute quadrangle maps, having first been widely produced in the 1940s, became the major product of the USGS. The United States was finally completely covered by these 7.5-minute maps in the 1990s.

Nearly all of the DeLorme Mapping Company state map pages are defined by 30 minutes of north latitude and 26.25 minutes of west longitude.

Finally, nearly all Hammond and Rand McNally state atlas maps include latitude and longitude lines normally drawn to a 30-minute grid.

LIST OF RAILROADS

Each of the railroad companies shown in this atlas are identified by abbreviations or "reporting marks." These are inscribed along each discrete line or branch segment on each map. An alphabetical list of these reporting marks, followed by the railroad name as it existed in 1946 or at the date of abandonment prior to 1946, is provided in the appendix.

MAPPING STYLE

This atlas shows state and county lines as well as major rivers in the interest of geographic orientation. Roads are not shown in the interest of simplicity. Cross-referencing to USGS maps or the DeLorme Mapping Company state atlases is recommended for finding contemporary road network details.

MILE POSTS

All railroad lines have railroad milepost tick marks every five miles. Nearly all of these mileposts have remained unchanged since long before 1946. The place names along the outside of the map neat line indicate the zero milepost location and/or the furthest point of that particular series of mileposts.

SOURCES

Sources used in the compilation of this atlas include a wide range of USGS topographic maps, Rand McNally and Hammond atlases, official railroad company maps, employee timetables, and track charts. Particularly valuable resources included *Moody's Steam Railroads—1946; Poor's Manual of Railroads—1923;* William D. Edson's *Railroad Names;* and many individual railroad histories. A complete list may be found in the "References" section.

ACCURACY

While I have taken every reasonable step to ensure accuracy, there may be some errors or omissions which are my own responsibility. Corrections and additional information are welcomed .

Acknowledgments

In producing this re-creation of the very complex rail network of the Middle Atlantic states, I have benefited from the research and generosity of many individuals—fellow railroad historians, railroad officials and workers, and others.

Particular thanks should be given to the late Henry T. Wilhelm and the late Bruce Coughlin for generosity in sharing their extensive track and signal drawings of U.S. railroads from the 1920s through the 1970s. Also, Ben F. Anthony has always been ready to provide information regarding rail operations and signaling. Finally, particular thanks are given to Robert Gambling for allowing me to use his collection of first and second editions of United States 1:250,000-scale USGS maps.

Among the sources of inspiration for this railroad atlas were the Railway Clearing House's *Railway Junction Diagrams* of pre-grouping British railroads and Jowett's *Railway Atlas of Great Britain and Ireland*. Alan Jowett generously shared with the author advice concerning his methods and graphic style.

Acknowledgment is also due to Mike Walker of Marlow, Bucks., England, who has produced the many volumes of the *Steam Powered Video's Comprehensive Railroad Atlas of North America*, a very detailed contemporary atlas of United States railroads.

Special thanks go to Richard B. Hasselman, retired Senior Vice President, Operations–Conrail, for his advice and encouragement.

Finally, I wish to express special appreciation to George F. Thompson, president of the Center for American Places, for his encouragement and guidance; to Annette Evarts, for her indispensible word processing and dedication; to my son, John, for his editorial advice; to my daughter, Ellin, for her design advice; and to my wife, Mary Jane, for her support and counsel.

THE ATLAS

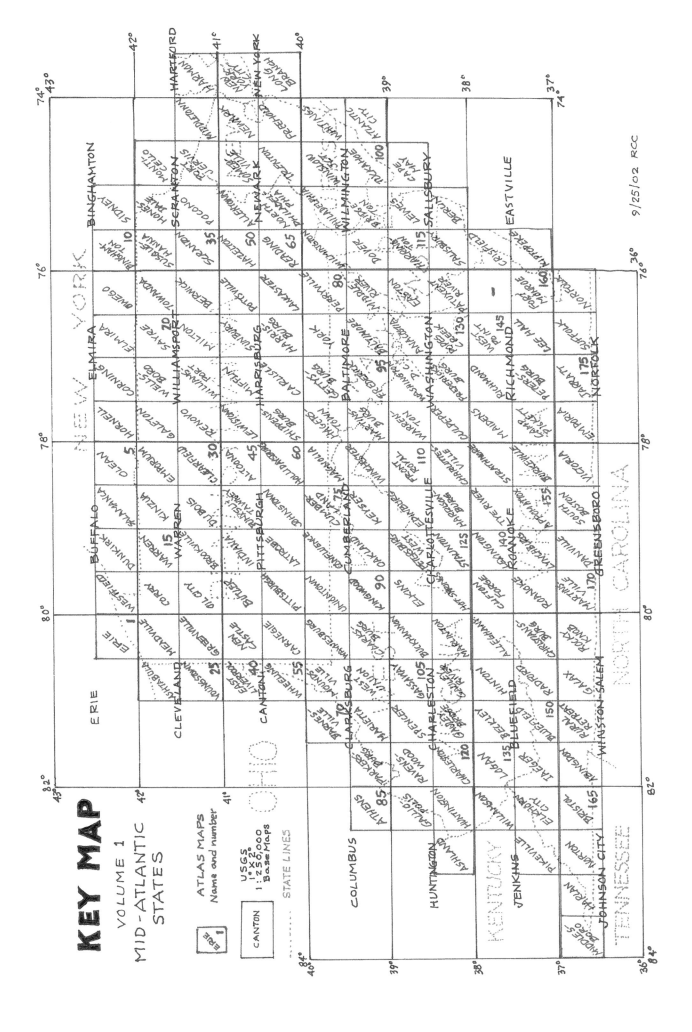

KEY MAP

VOLUME 1
MID-ATLANTIC STATES

ATLAS MAPS
Name and number

USGS
1°X2°
1:250,000
Base Maps

STATE LINES

9/25/02 RCC

SYMBOLS AND ABBREVIATIONS

RAILROAD LINE (In Service)

RAILROAD LINE (Abandoned)

RAILROAD (outline of Yard/Connecting tracks - In Service)

PASSENGER STATION

NON-PASSENGER STATION

VIADUCT/MAJOR BRIDGE (with name)

"Moodna"

TUNNEL (with name)

"Otisville"

TRACK PAN (with name)

"Roelofs"

COALING STATION/MAIN TRACK (with name)

"Yoder"

MILEPOST (every 5 miles)

20

START/END MILEPOST MILEAGE

Buffalo

NYC   1946 OWNER (See Appendix for reporting marks)

PRR (PAA)   1946 OWNER (Associated line - " " ")

ERIE/L&NE   1946 OWNER/(Trackage Rights - " " ")

PRR-NKP   1946 JOINTLY OWNED & OPERATED (" " ")

"VIEW"   INTERLOCKING TOWER (In service, with name)

ALFORD   INTERLOCKING TOWER (Abandoned, with name)

•RN   BLOCK STATION (In service, with name)

LANES CROSSING (R-WH)   INTERLOCKING-REMOTE CONTROLLED (In service, with name and control point name)

••COGAN   BLOCK LIMIT STATION (In service, with name)

RENOVO   CREW CHANGE POINT

NATIONAL BOUNDARY   A   Automatically controlled interlocking

STATE/PROVINCIAL BOUNDARY   D   Draw, lift or swing bridge

COUNTY BOUNDARY

CITY/TOWN BOUNDARY   G   Railroad crossing at grade - Gate

RIVER (With direction of flow)   S   Railroad crossing at grade - Stop sign

CANAL (In Service/Abandoned)   T   Railroad crossing at grade - Target

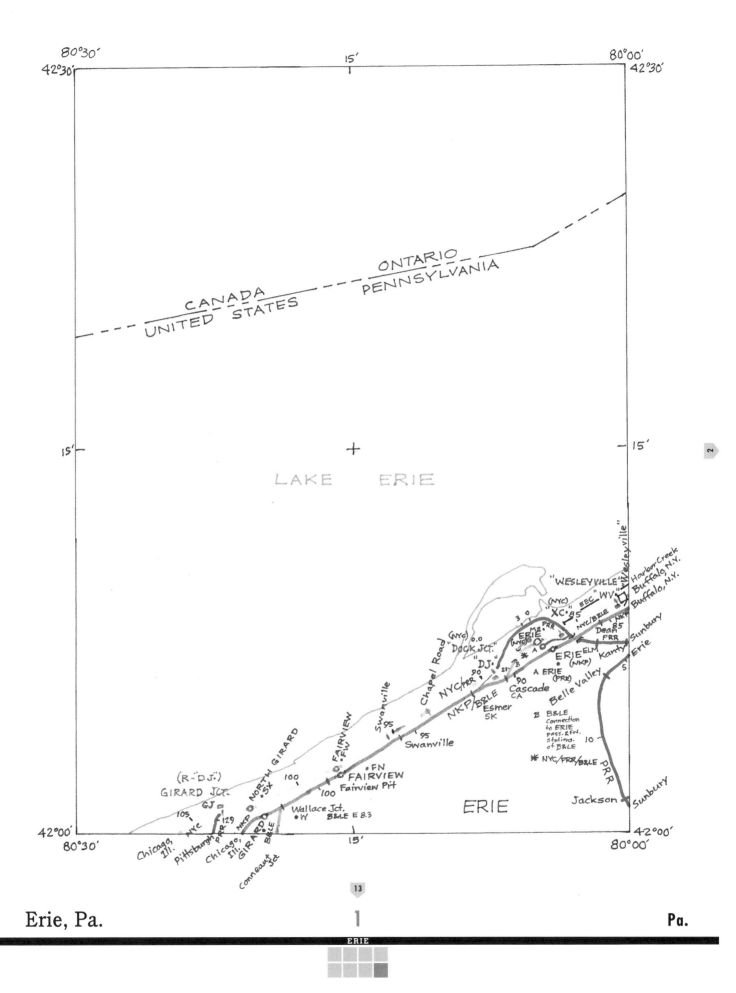

CANADA   ONTARIO
UNITED STATES   PENNSYLVANIA

15'   — 15'

LAKE   ERIE

"WESLEYVILLE"   "Wesleyville"
Harbor Creek
Buffalo, N.Y.
(NYC)   EEC   WV   Buffalo, N.Y.
XC   85   NYC/B&LE
3 0   M.S. PRR   85   NKP
(NYC)   Dean   Sunbury
"Chapel Road" (NYC)   0.0   ERIE   A   PRR   Erie
Dock Jct.   (NYC)   ERIE EL/n
"DJ."   21   B   (NKP)   Kanty   5
NYC/PRR   90   21   A ERIE   Belle Valley
Swanville   90   B&LE   Cascade   (PRR)
NKP/B&LE   CA
95   Esmer
95   SK   B   B&LE
Swanville   Connection
to ERIE
pass. & Fn.
FAIRVIEW   stations.
"FN"   FN   of B&LE   10 —
(R-D.J.)   NORTH GIRARD   FAIRVIEW   * NYC/PRR/B&LE
GIRARD JCT.   .SX   100   Fairview Pit   Jackson   Sunbury
105   GJ   Wallace Jct.   ERIE
NYC   129   • W   B&LE E 8.3
Chicago,   Pittsburgh, PRR   NKP   GIRARD   B&LE   15'
Ill.   Chicago,
Ill.   Conneaut Jct.

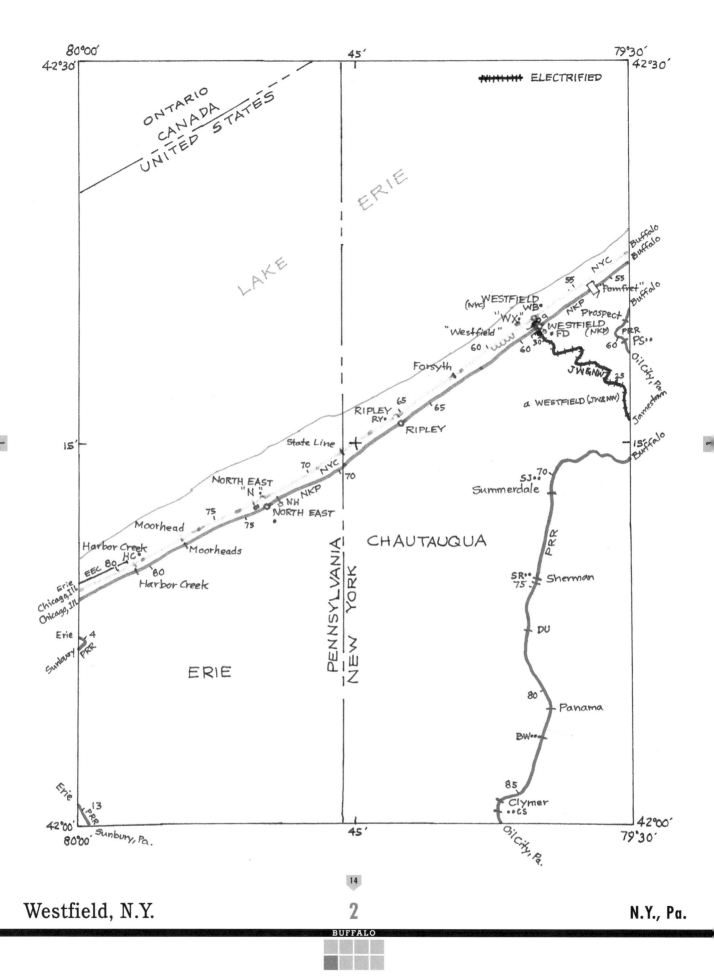

ELECTRIFIED

ONTARIO
CANADA
UNITED STATES

LAKE ERIE

Buffalo
Buffalo
NYC
55   55
"Pomfret"   Buffalo
(NYC) WESTFIELD
WB•   NKP   Prospect
"WX"   •a   PRR
"Westfield"   WESTFIELD (NKP)   PS••
FD   60
60   60 30   Oil City, Pa.
JW&NW   25
Forsyth   a WESTFIELD (JW&NW)   Jamestown
65   15   Buffalo
RIPLEY   65
RY•
RIPLEY
State Line   SJ•• 70
70   Summerdale
NORTH EAST   NKP   70
"N"   NYC   PRR
75   SH   70
Moorhead   NORTH EAST   SR•• Sherman
75   75
Moorheads   DU
Harbor Creek   CHAUTAUQUA
EEC 80   HC•   80   Panama
Harbor Creek   BW••
Erie
Chicago, Ill.
Chicago, Ill.   85
Clymer
Erie   4   •GS
Sunbury   PRR

ERIE

PENNSYLVANIA
NEW YORK

Erie   13
PRR
42°00'   Sunbury, Pa.
80°00'   45'   Oil City, Pa.   79°30'

80°00'   45'   79°30'
42°30'   42°30'
15'   15'
42°00'   42°00'

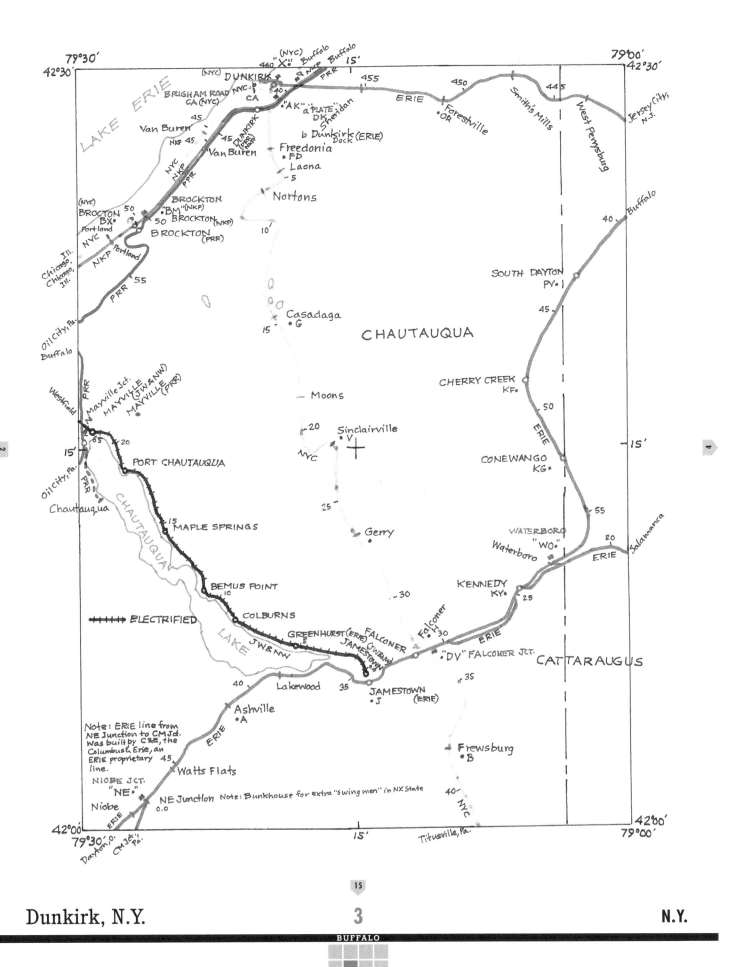

79°30'
42°30'
(NYC)
"X" Buffalo
460 NKP Buffalo
15'
PRR

79°00'
42°30'

LAKE ERIE

(NYC) DUNKIRK
BRIGHAM ROAD
CA (NYC)
NYC b
CA
40

455
ERIE
450
445
Smith's Mills
West Pennsburg
Jersey City, N.J.

"AK" a PLATE
DK
Sheridan

b Dunkirk Dock (ERIE)

Van Buren
NKP 45
45

45
DUNKIRK

Freedonia
FD
Laona
5
Forestville
OR

Van Buren
NYC NKP PPR

Nortons

40 Buffalo

BROCKTON
"BM" (NKP)
50 BROCKTON (NKP)

BROCTON
BX
Portland NYC

50

10'

BROCKTON (PRR)

Casadaga
G
15

CHAUTAUQUA

SOUTH DAYTON
PV

45

NKP Portland
PRR 55

Oil City, Pa.
Buffalo

Moons

CHERRY CREEK
KF

50

ERIE

Westfield
PRR

Mayville Jct.
MAYVILLE (JW&NW)
MAYVILLE (PRR)

63
20

15' Oil City, Pa.
PRR

Chautauqua

PORT CHAUTAUQUA

20
NYC

Sinclairville
V

CONEWANGO
KG

15'

CHAUTAUQUA

15 MAPLE SPRINGS

25

Gerry

55

WATERBORO
"WO"
Waterboro
20
ERIE
Salamanca

BEMUS POINT
10

30

KENNEDY
KY
25

LAKE

+++++ ELECTRIFIED

COLBURNS
JW&NW
GREENHURST (ERIE)
G'w'R&N
JAMESTOWN

Falconer
Falconer H
30

ERIE

"DV" FALCONER JCT.
CATTARAUGUS

40
Lakewood
35

35
JAMESTOWN (ERIE)
J

Ashville
A

Frewsburg
B

Note: ERIE line from
NE Junction to CM Jd.
Was built by C&E, the
Columbus & Erie, an
ERIE proprietary
line.

45
Watts Flats

NIOBE JCT.
"NE"

40
NYC

Niobe
ERIE
NE Junction
0.0
Note: Bunkhouse for extra "swing men" in N.Y. State

42°00'
79°30'
Dayton, O.
CM Jct. Pa.

15'
Titusville, Pa.

79°00'
42°00'

15

Dunkirk, N.Y.   3   N.Y.

BUFFALO

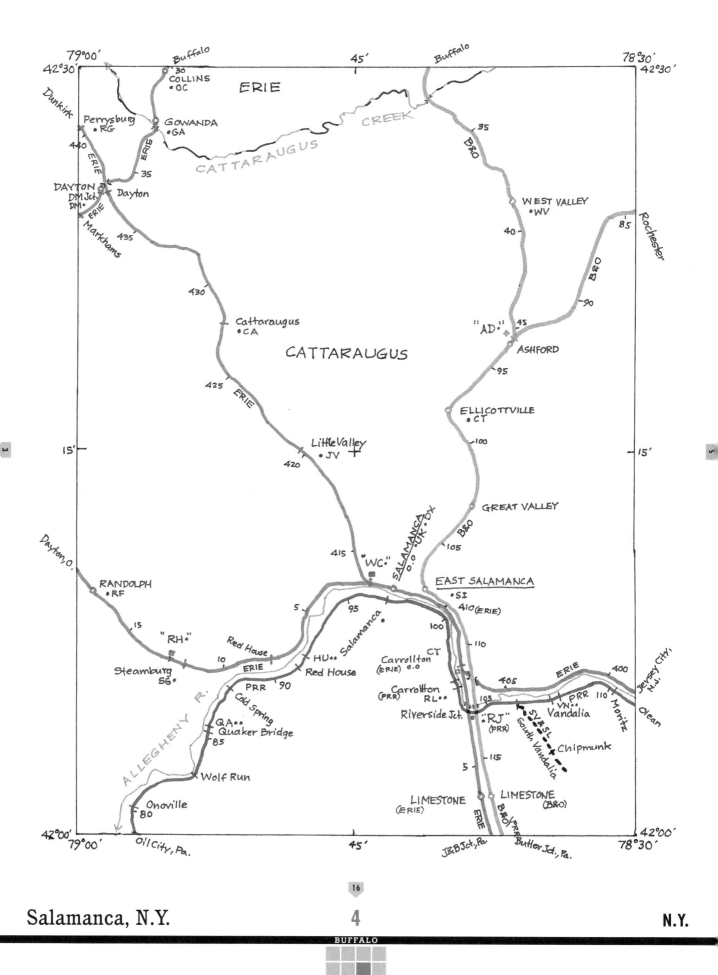

79°00'
42°30'
Buffalo
45'
Buffalo
78°30'
42°30'

30
COLLINS
•OC
ERIE
Dunkirk
Perrysburg
•RG
GOWANDA
•GA
440
ERIE
CATTARAUGUS   CREEK
35
B&O
35
Rochester
DAYTON
DM Jct.
DM•
ERIE
Dayton
WEST VALLEY
•WV
85
Markhams
435'
40
B&O
430
90
Cattaraugus
•CA
"AD"•
45
CATTARAUGUS
ASHFORD
95
425
ERIE
ELLICOTTVILLE
•CT
15'
100
15'
Little Valley
•JV
420
GREAT VALLEY
B&O
105
Dayton, O.
415
"WC."
SALAMANCA
0.0 •UK •DX
EAST SALAMANCA
•SI
RANDOLPH
•RF
95
410(ERIE)
5
100
110
15
Salamanca
•
"RH•"
Red House
CT
405
ERIE
400
Steamburg
SG•
10
HU••
ERIE
Red House
PRR
90
Carrollton
(ERIE) 0.0
Carrollton
RL••
VN••
PRR
110
Jersey City, N.J.
Cold Spring
105
Vandalia
Moritz
QA••
Quaker Bridge
85
Riverside Jct.
"RJ"
(PRR)
SV&SL
South Vandalia
Chipmunk
Olean
Wolf Run
115
5
LIMESTONE
(ERIE)
LIMESTONE
(B&O)
Onoville
80
42°00'
79°00'
Oil City, Pa.
45'
J&BJct., Pa.
ERIE
B&O/PRR
Butler Jct., Pa.
78°30'
42°00'

ERIE

ALLEGHENY R.

16

Salamanca, N.Y.   4   N.Y.

BUFFALO

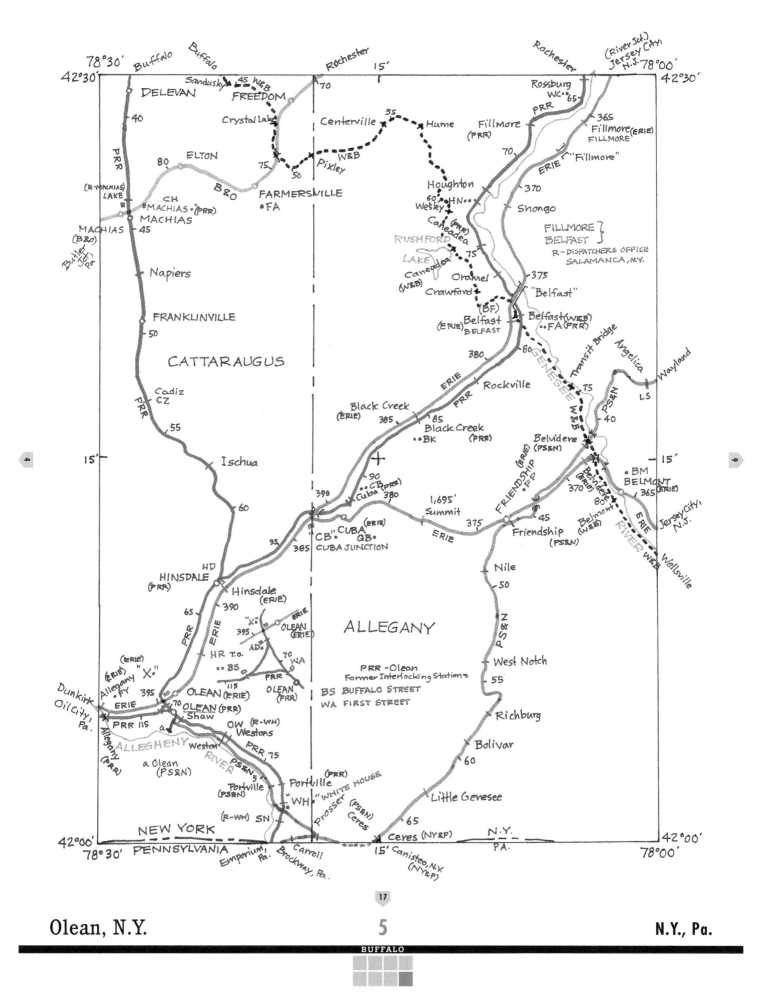

78°30' Buffalo Buffalo Rochester 15' 70
42°30' 42°30'

DELEVAN
Sandusky 45 W&B
FREEDOM
Crystal Lake
Centerville 55 Hume
40
PRR
Rossburg
WC 65
PRR
Fillmore (PRR) 365
Fillmore (ERIE)
FILLMORE
70
ELTON 80
Pixley W&B
75 50
B&O
FARMERSVILLE FA
Houghton
60 HN
Wesley
Caneadea (PRR)
ERIE "Fillmore"
370
Shongo
375

(R—MACHIAS LAKE)
CH "MACHIAS (PRR)
MACHIAS 45
MACHIAS (B&O)
Butler Jct. Pa.

RUSHFORD LAKE
FILLMORE
BELFAST
R—DISPATCHERS OFFICE
SALAMANCA, N.Y.

Napiers
Caneadea (W&B)
Oramel
Crawford
"Belfast"
(BF)
(ERIE) Belfast
BELFAST
Belfast (W&B)
FA (PRR)

FRANKLINVILLE
50

CATTARAUGUS

380
ERIE
PRR Rockville
GENESEE W&B
Transit Bridge
Angelica
Wayland
75
PS&N
40
LS

Cadiz CZ
55

Black Creek (ERIE)
385
85
Black Creek (PRR)
BK
Belvidere (PS&N)
FRIENDSHIP (ERIE) FP
BM
BELMONT
365 ERIE

Ischua
15' 15'
60

390
Cuba CB (PRR)
380
CUBA (ERIE)
QB
CUBA JUNCTION
CB
385
95

1,695' Summit
375
ERIE
Belvidere (ERIE)
370
80
Belmont (W&B)
RIVER W&B
Jersey City, N.J.
Wellsville

Friendship (PS&N)
45

HINSDALE (PRR) HD
Hinsdale (ERIE)
390
"X"
395
OLEAN (ERIE)
HR T.O. AD
70 WA
85
PRR
115
OLEAN (PRR)

ALLEGANY

Nile
50
PS&N
Z
West Notch
55

PRR—Olean
Former Interlocking Stations
BS BUFFALO STREET
WA FIRST STREET

(ERIE) "X"
Allegany FY (ERIE)
395
OLEAN (ERIE)
70
OLEAN (PRR)
Shaw
OW (R—WH)
Westons
Weston
PRR 75
PS&N 5
Portville (PS&N)
(R—WH) SN
Dunkirk
Oil City, Pa.
ERIE
PRR 115
Allegany (PRR)
a Olean (PS&N)
ALLEGHENY RIVER
Portville (PRR)
"WH" "WHITE HOUSE"
Prosser (PS&N) Ceres

Richburg
Bolivar
60
Little Genesee
65

NEW YORK
42°00' 42°00'
78°30' PENNSYLVANIA
Emporium, Pa. Brockway, Pa.
Carroll
15' Canisteo, N.Y. (NY&P)
Ceres (NY&P)
N.Y.
PA.
78°00'

17

BUFFALO

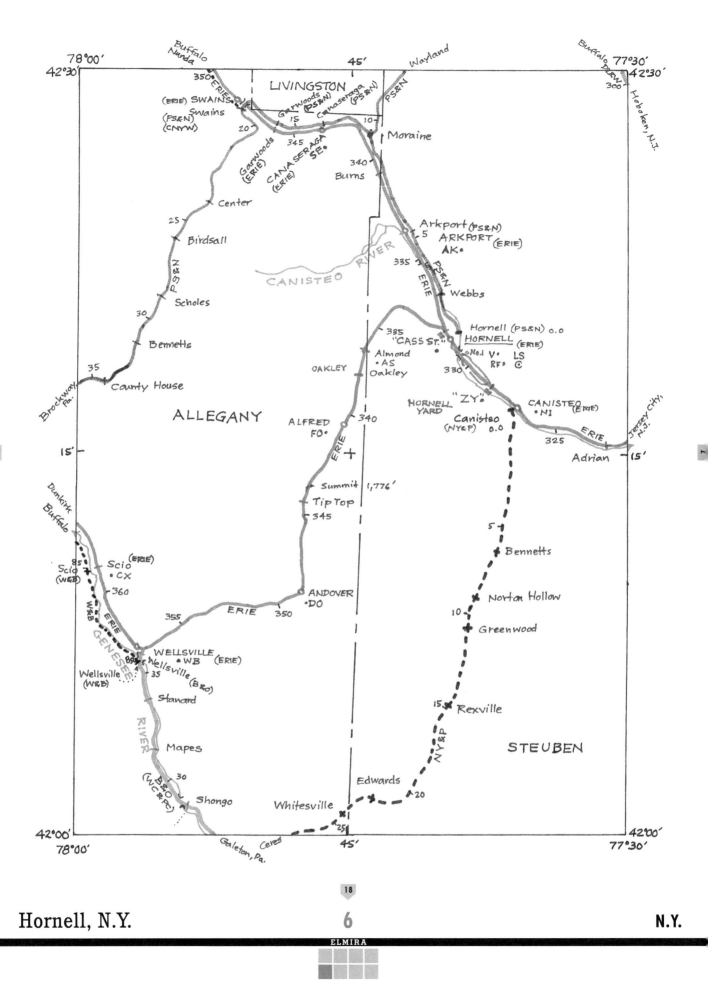

Hornell, N.Y.   6   N.Y.

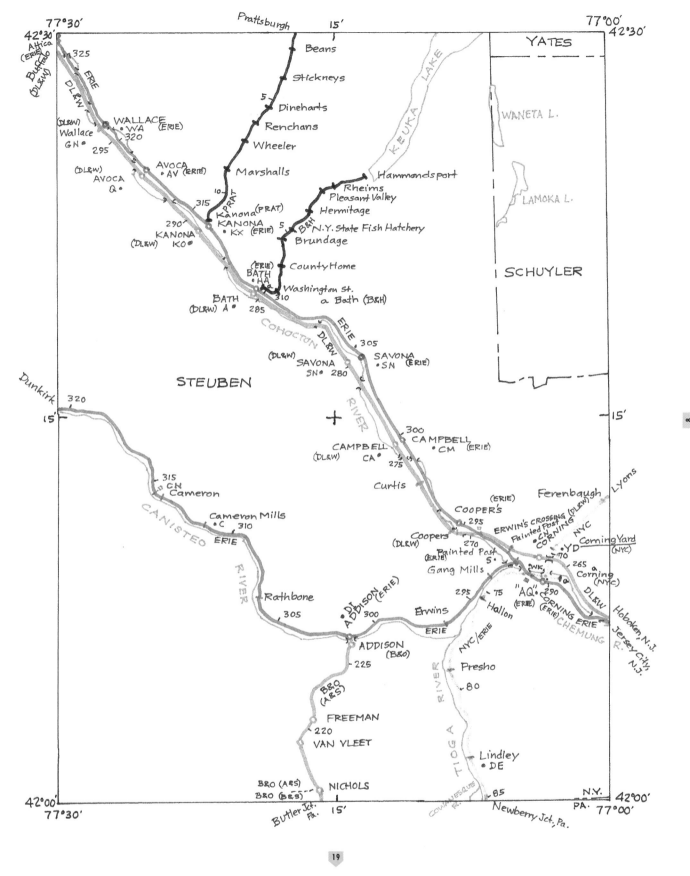

77°30'
42°30'
Athica
(ERIE)
Buffalo
(DL&W)

325

ERIE

DL&W

(DL&W)
Wallace
GN

295

WALLACE
WA
320

(ERIE)

(DL&W)
AVOCA
Q

AVOCA
AV (ERIE)

315

10

K PRAT

290
KANONA
(DL&W) KO

Kanona (PRAT)
KANONA
KX (ERIE)

5

Beans

Sticknesys

5   Dineharts

Renchans

Wheeler

Marshalls

Rheims
Pleasant Valley

Hammondsport

Hermitage

B&H

N.Y. State Fish Hatchery

Brundage

County Home

(ERIE)
BATH
HA

BATH
(DL&W) A

285

310

Washington St.
a Bath (B&H)

COHOCTON

DL&W

ERIE

305

77°00'
42°30'

YATES

KEUKA LAKE

WANETA L.

LAMOKA L.

SCHUYLER

STEUBEN

(DL&W)
SAVONA
SN

SAVONA
SN   280

SAVONA
SN (ERIE)

RIVER

Dunkirk

320

15'

CAMPBELL
(DL&W)

300

CAMPBELL
CA

CAMPBELL
CM (ERIE)

275

15'

315
CN
Cameron

CANISTEO

ERIE

Cameron Mills
C   310

RIVER

Rathbone

305

DI
ADDISON (ERIE)

ADDISON
300

Curtis

COOPER'S

(ERIE)

295

Coopers
(DL&W)

270

(ERIE)
Painted Post
S

Gang Mills

Ferenbaugh

ERWIN'S CROSSING
Painted Post (DL&W)
CN
CORNING

Lyons

NYC

YD

Corning Yard
(NYC)

70

265

Corning
(NYC)

Erwins

ERIE

295

75

Hallon

"AQ"
(ERIE)

WK

290
CORNING ERIE
(ERIE)

DL&W

CHEMUNG

Hoboken, N.J.

Jersey City,
N.J.

ADDISON
(B&O)

225

B&O
(A&S)

FREEMAN
220

VAN VLEET

NYC/ERIE

Presho

80

TIOGA   RIVER

Lindley
DE

N.Y.
PA.   77°00'
42°00'

B&O (A&S)
B&O (B&S)

NICHOLS

85

COHOCTESQUE
R.

Newberry Jct., Pa.

42°00'
77°30'

Butler Jct.
Pa.   15'

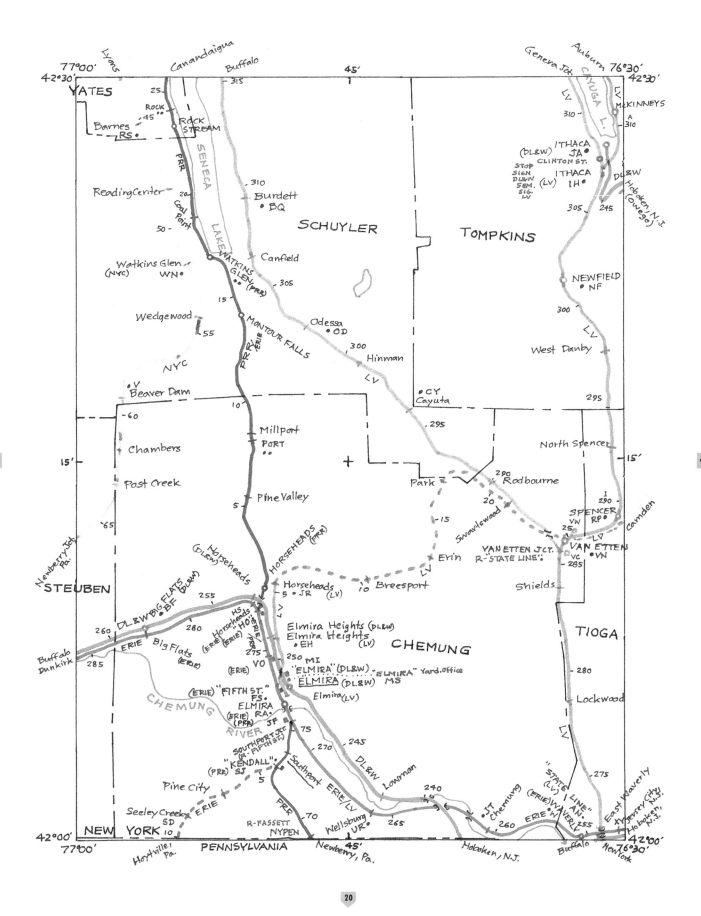

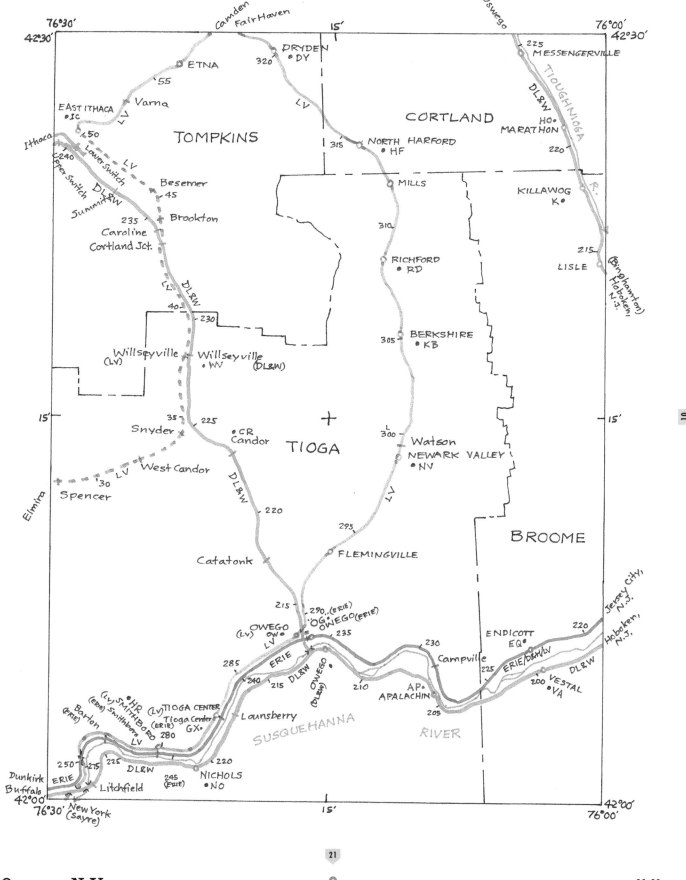

76°30'
42°30'

Camden
Fair Haven

15'

Oswego

76°00'
42°30'

225
MESSENGERVILLE

DRYDEN
• DY

320

ETNA

'55

Varna

TOMPKINS

LV

CORTLAND

TIOUGHNIOGA
DL&W

HO•
MARATHON

EAST ITHACA
• JC

LV

315

NORTH HARFORD
• HF

220

Ithaca

'50

LV

'240
Lower Switch

Upper Switch

DL&W

Summit

Besemer
•45

235

Brookton

Caroline
Cortland Jct.

LV

DL&W

40

230

Willseyville
(LV)

Willseyville
• WV

(DL&W)

MILLS

310

RICHFORD
• RD

KILLAWOG
K •

215

LISLE

R.

(Binghamton)
Hoboken,
N.J.

BERKSHIRE
• KB

305

35

225

Snyder

• CR
Candor

TIOGA

L
300

Watson
NEWARK VALLEY
• NV

15'

15'

Elmira

LV

West Candor

'30

Spencer

DL&W

220

LV

295

BROOME

Catatonk

FLEMINGVILLE

215

290 (ERIE)
'OG•
OWEGO (ERIE)

(LV) OWEGO
OW•
LV

285

240

215

235

ERIE

DL&W
OWEGO
(DL&W)

210

ENDICOTT
EQ•

Campville

230

225

ERIE/DL&W/LV

DL&W

Jersey City,
N.J.

Hoboken,
N.J.

220

200

VESTAL
• VA

• HR
(LV) (LV) TIOGA CENTER
(ERIE) Smithboro

Barton
(ERIE)

SMITHBORO
LV

Tioga Center
(ERIE)
GX•

280

Lounsberry

AP•
APALACHIN

205

SUSQUEHANNA

RIVER

250

275

225

DL&W

220

Dunkirk ERIE
Buffalo
42°00'
76°30'

Litchfield

New York
(Sayre)

245
(ERIE)

NICHOLS
• NO

15'

42°00'
76°00'

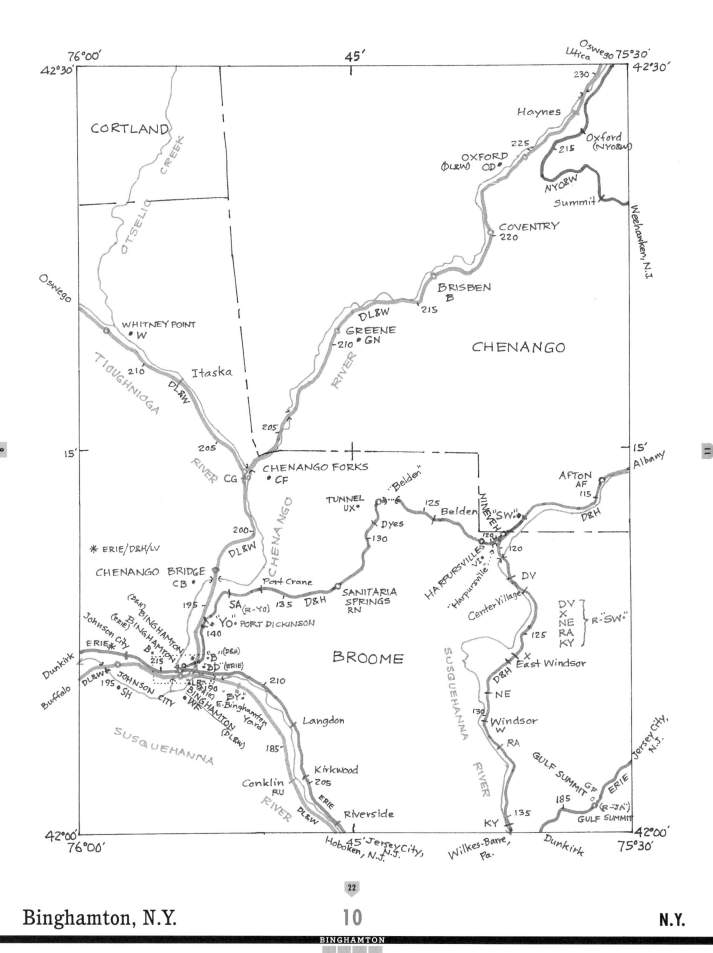

42°30' — 76°00' — 45' — Oswego 75°30' — Utica — 42°30'

230

CORTLAND

Haynes

225

OXFORD
(DL&W) OD

Oxford
(NYO&W)

215

NYO&W

Summit

OTSELIC CREEK

COVENTRY
220

CHENANGO

Oswego

BRISBEN
B

215

WHITNEY POINT
W

DL&W

GREENE
210 GN

TIOUGHNIOGA

RIVER

210

DL&W

Itaska

Weehawken, N.J.

15' — 205

205

15' — Albany

AFTON
AF
115

CHENANGO FORKS
CF

"Belden"

CG

125 Belden

NINEVEH "SW"

D&H

TUNNEL
UX

120

RIVER

Dyes

130

CHENANGO

200

* ERIE/D&H/LV

DL&W

HARPURSVILLE
VI

120

DV

CHENANGO BRIDGE
CB

Port Crane

"Harpursville"

Center Village

DV
X
NE } R·"SW"
RA
KY

195

SA (R-YO) 135 D&H

SANITARIA
SPRINGS
RN

125

SUSQUEHANNA

(D&H) BINGHAMTON
(ERIE) BINGHAMTON

"YO" PORT DICKINSON

140

D&H East Windsor

Johnson City

B
215

"B" (D&H)

BROOME

NE

ERIE*

"BD" (ERIE)

RIVER

Dunkirk

DL&W
195 SH

JOHNSON CITY

"LB" 190 (D&H)
(ERIE) "BY"

130

Windsor
W

Buffalo

WF

BINGHAMTON
(DL&W)

E. Binghamton
Yard

210

RA

Langdon

SUSQUEHANNA

185

GULF SUMMIT
GF

Jersey City, N.J.

Kirkwood
205

ERIE

Conklin
RU

185

(R-"JA")

RIVER

ERIE
DL&W

Riverside

KY 135

GULF SUMMIT

Dunkirk

42°00' — 76°00' — Hoboken, N.J. — 45' Jersey City, N.J. — Wilkes-Barre, Pa. — 42°00' 75°30'

Binghamton, N.Y.

10

N.Y.

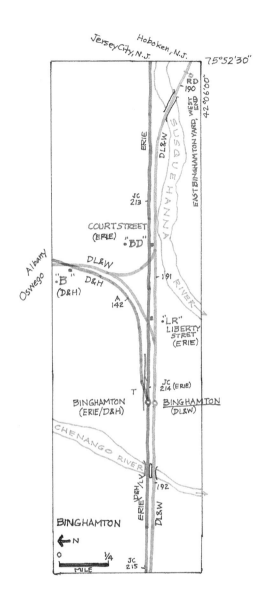

Jersey City, N.J.

Hoboken, N.J.

75°52'30"

RD 190

42°06'00"

ERIE

DL&W

SUSQUEHANNA

EAST BINGHAMTON YARD, WEST END

JC 213

COURTSTREET
(ERIE)

"BD"

Albany
Oswego

DL&W

"B"
(D&H)

D&H

191

RIVER

A 142

"LR"
LIBERTY
STREET
(ERIE)

JC 214 (ERIE)

T

BINGHAMTON
(ERIE/D&H)

BINGHAMTON
(DL&W)

CHENANGO RIVER

ERIE D&H/LV

192

DL&W

BINGHAMTON

← N

0   ¼

MILE

JC 215

10A Binghamton, N.Y.

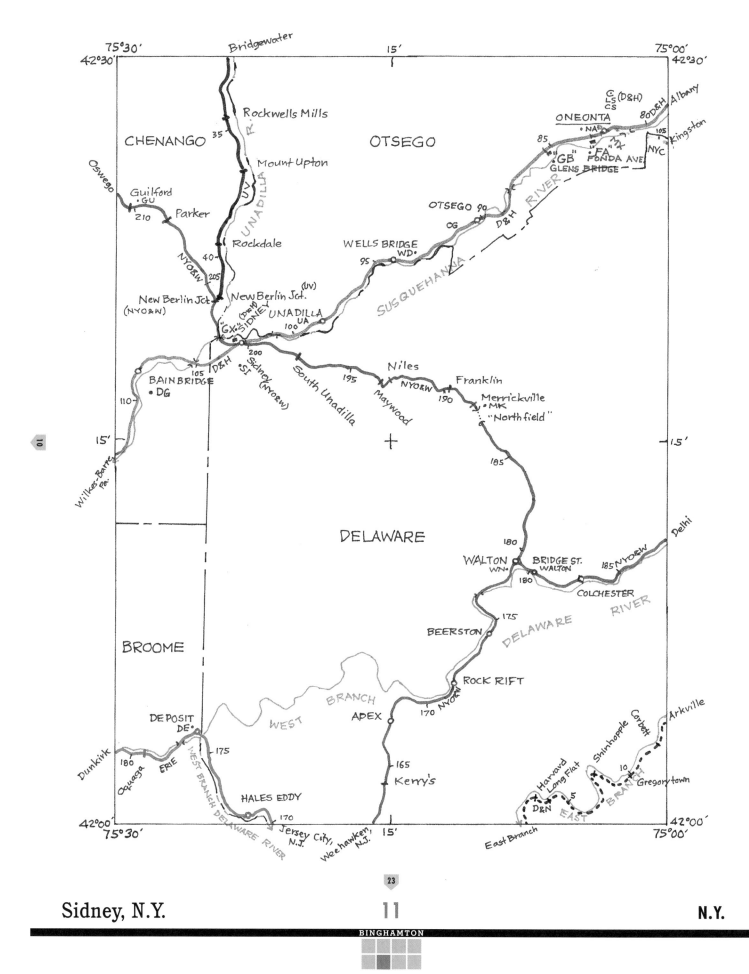

CHENANGO

Rockwells Mills

35

OTSEGO

C (D&H)
LS
CS

ONEONTA

80 D&H   Albany

Kingston

85   NA

"GB"
GLENS BRIDGE

"FA"
FONDA AVE.

105

NYC

Oswego

Guilford
• GU

210   Parker

Mount Upton

UNADILLA
R.

UV

Rockdale

OTSEGO   90

OG

D & H

RIVER

40

205

NYO&W

WELLS BRIDGE
WD•

95

SUSQUEHANNA

New Berlin Jct.
(NYO&W)

New Berlin Jct. (UV)

"GX"• SIDNEY
(D&H)

UNADILLA
UA

100

"SJ"
Sidney (NYO&W)

D&H

105

200

195

Niles

Franklin

River

BAINBRIDGE
• DG

South Unadilla

NYO&W

190

Merrickville
• MK

110

Maywood

"Northfield"

Wilkes-Barre
Pa.

15'   15'

185

DELAWARE

180

Delhi

WALTON
WN•

BRIDGE ST.
WALTON

185 NYO&W

180

COLCHESTER

RIVER

175

DELAWARE

BROOME

BEERSTON

ROCK RIFT

170 NYO&W

Arkville

WEST

BRANCH

APEX

Shinhopple Corbett

DEPOSIT
DE•

175

165

Harvard
Long Flat

10

Gregorytown

Dunkirk

180

Oquaga

ERIE

WEST BRANCH DELAWARE RIVER

Kerry's

5

D&N

EAST

BRANCH

HALES EDDY

170

10

23

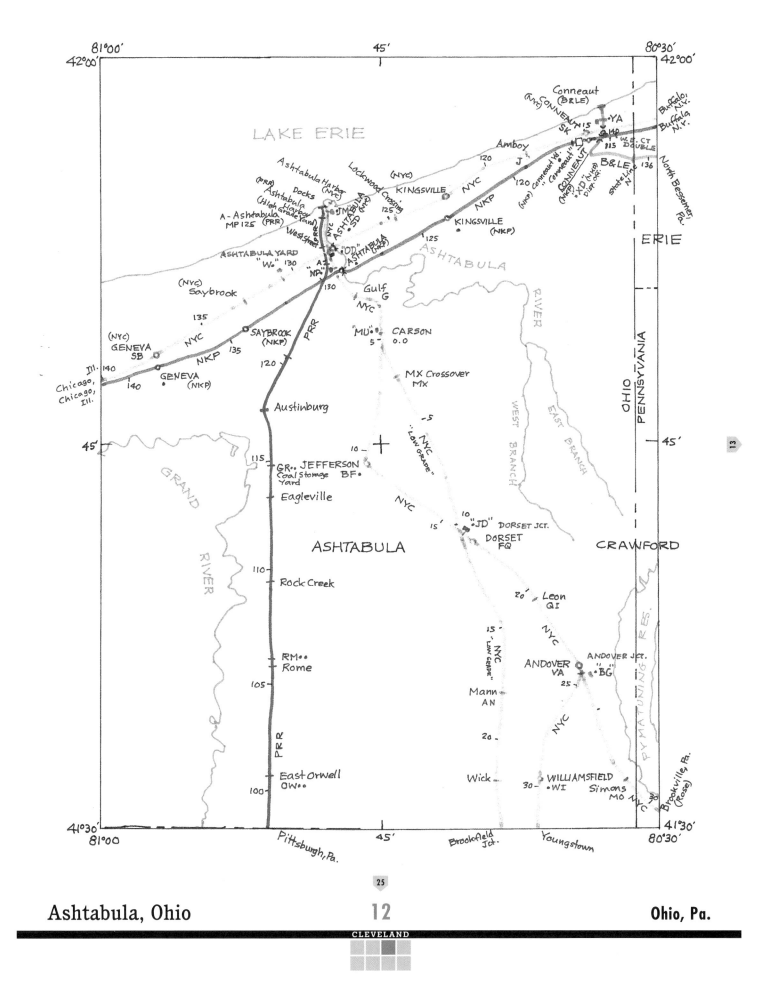

LAKE ERIE

Conneaut
(B&LE)
(NYC) CONNEAUT 15
SK
140
YA
W.E. CT
DOUBLE
Buffalo, N.Y.
Buffalo, N.Y.

Amboy
120
J
120 Conneaut Yd.
(NKP) "Conneaut"
CONNEAUT
"XD" (NKP)
DISP. OFF.
115
136
B&LE
North Bessemer
State Line

Ashtabula Harbor
Lockwood Crossing
(NYC)
KINGSVILLE
(NYC)
NKP
120
NKP

(PRR)
Docks
Ashtabula
Harbor
(High Grade Yard)
A - Ashtabula
MP 125 (PRR)
West Street
ASHTABULA
SD
(NYC)
125
KINGSVILLE
(NKP)
125

ASHTABULA YARD
"W." 130
"OD"
NYC
ASHTABULA
(NKP)
125
ASHTABULA
RIVER

ERIE

"A"
NP
130

Gulf
G
NYC

(NYC)
Saybrook

"MU"
5
CARSON
0.0

135

PRR

MX Crossover
MX

(NYC)
GENEVA
SB
135
SAYBROOK
(NKP)
NKP
120

-5

OHIO
PENNSYLVANIA

45'

Ill. 140
Chicago, Ill.
Chicago, Ill.
140
GENEVA
(NKP)

NYC
"Low Grade"

Austinburg

WEST BRANCH
EAST BRANCH

10 -

GRAND RIVER

115
GR.. JEFFERSON
Coal Storage BF.
Yard
10 -
15'
"JD" DORSET JCT.
DORSET
FQ

CRAWFORD

Eagleville

NYC

ASHTABULA

110
Rock Creek

20'
Leon
QI

PYMATUNING RES.

RM..
Rome
105

15
NYC
"Low Grade"

NYC
ANDOVER
VA
ANDOVER JCT.
"BG"
25

Mann
AN
20

NYC

PRR

East Orwell
OW..
100

Wick
30 -
WILLIAMSFIELD
WI
Simons
MO
30
NYC

Brookville, Pa.
(Rose)

Pittsburgh, Pa.
Brookfield Jct.
Youngstown

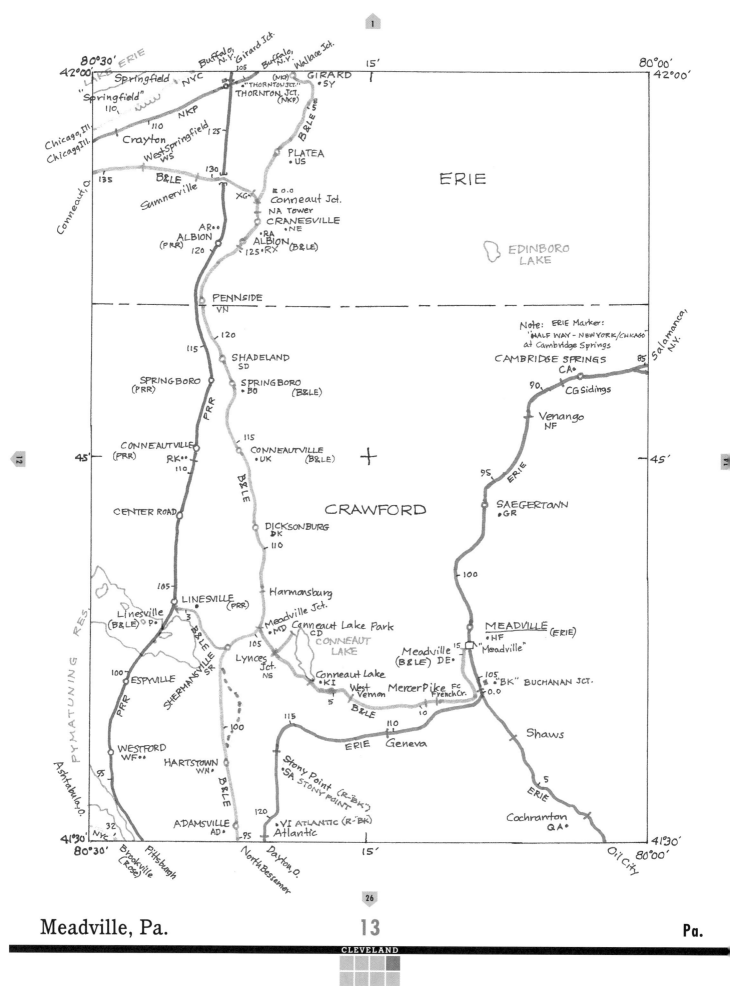

LAKE ERIE

80°30' 42°00'

Buffalo, N.Y. Girard Jct.
Buffalo, N.Y. Wallace Jct.
15'
80°00' 42°00'

Springfield NYC
"Springfield"
110
105
(NKP) GIRARD
"THORNTON JCT." · SY
THORNTON JCT. (NKP)

Chicago, Ill.
Chicago, Ill.
110 NKP
Crayton
West Springfield WS
125
B&LE

ERIE

135 B&LE
130
Sumnerville
XG
E 0.0
Conneaut Jct.
NA Tower
CRANESVILLE · NE
PLATEA · US

EDINBORO LAKE

AR
ALBION (PRR)
120
125
·RA
ALBION (B&LE)
·RX

PENNSIDE VN

Note: ERIE Marker:
"HALF WAY - NEW YORK/CHICAGO"
at Cambridge Springs

80 Salamanca, N.Y.

120
115
SHADELAND SD
CAMBRIDGE SPRINGS
CA·
85

SPRINGBORO (PRR)
SPRINGBORO · BO (B&LE)
90 CG Sidings
Venango NF

PRR
115
CONNEAUTVILLE · UK (B&LE)
+
95 ERIE

CONNEAUTVILLE (PRR)
RK··
110
45'
45'

B&LE
CRAWFORD
SAEGERTOWN · GR

CENTER ROAD
DICKSONBURG DK
110
100

105
LINESVILLE (PRR)
Harmansburg
MEADVILLE (ERIE)
· HF

Linesville (B&LE) P·
3
Meadville Jct.
· MD
CD Conneaut Lake Park
CONNEAUT LAKE
Meadville (B&LE) DE·
15
"Meadville"

PYMATUNING RES.
100 ESPYVILLE
SHERMANSVILLE SR
105
Lynces Jct. NS
Conneaut Lake · KI
West Vernon
5
Mercer Pike Fc
French Cr.
B&LE
10
"BK" BUCHANAN JCT.
0.0
105

PRR
WESTFORD WF··
95
HARTSTOWN WN·
B&LE
100
115
110
ERIE Geneva
Shaws

Ashtabula, O.
Stony Point (R-"BK")
·SA STONY POINT
5 ERIE

32 NYC
ADAMSVILLE AD·
120
95
VI ATLANTIC (R-"BK")
· Atlantic
Cochranton QA·

41°30'
80°30'
Brookville (Rose)
Pittsburgh
North Bessemer
Dayton, O.
15'
80°00' 41°30'
Oil City

Meadville, Pa.

13

Pa.

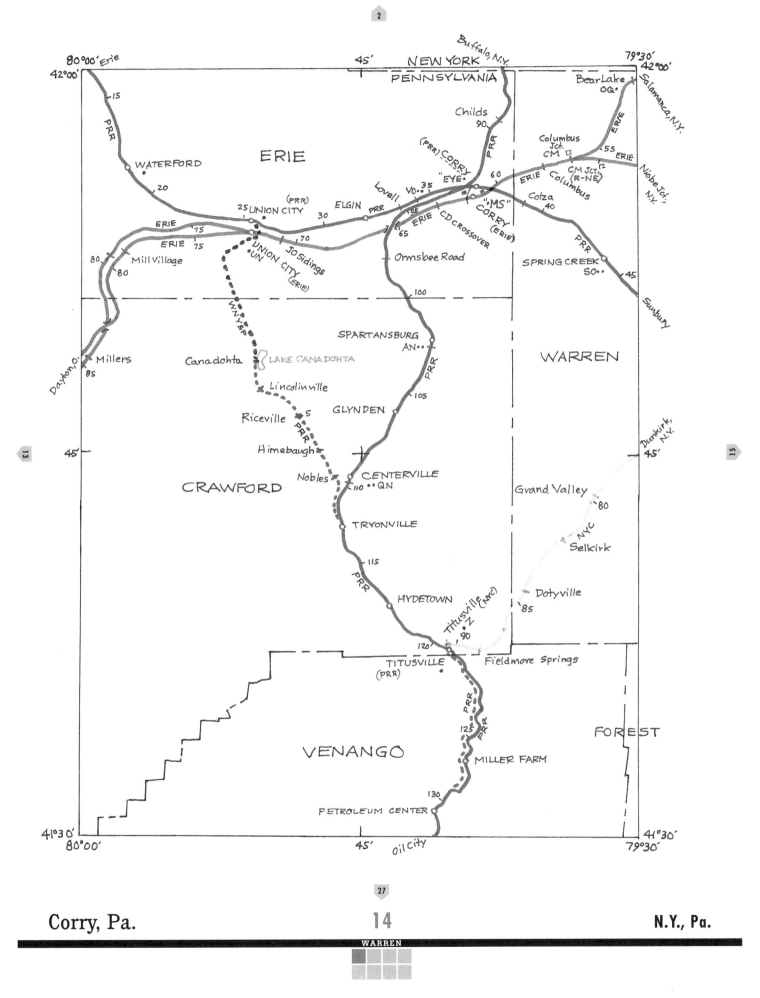

Buffalo, N.Y.

80°00' Erie   45'   NEW YORK   79°30'
42°00'   PENNSYLVANIA   42°00'

15   Bear Lake   Salamanca, N.Y.
                                                   Childs   OQ
PRR   90
                                         ERIE   ERIE
                                                   PRR
                                                              Columbus   55
                          ERIE   (FRR) CORRY   Jct.   ERIE
WATERFORD   "EYE"   60   CM   CM Jct. 2
                                         Lovell VO. 35   ERIE Columbus (R-NE)   Niobe Jct.
20   .35   N.Y.
           (PRR)   ELGIN   PRR   .85   "MS"   Colza
25 UNION CITY   30   PRR   CORRY   40
ERIE   75   ERIE   CD CROSSOVER (ERIE)
ERIE   75   JO Sidings 70   65   PRR
80   UNION CITY •UN   SPRING CREEK
     80   Mill Village   (ERIE)   Ormsbee Road   SO••   45
                                                                          Sunbury
                                              100

                           SPARTANSBURG
Dayton,O. Millers   Canadohta   LAKE CANADOHTA   AN••   PRR   WARREN
85   Dunkirk,
              Lincolinville   PRR   N.Y.
45'   105   45'
          Riceville   5   GLYNDEN
                    PRR
          Himebaugh•   Grand Valley   80
                  Nobles CENTERVILLE   NYC
CRAWFORD   110 •• QN   Selkirk
                   TRYONVILLE
                                                              Dotyville
                     115   85
                  PRR   HYDETOWN   Titusville (NYC)
                                       N
                                       90
                          120   Fieldmore Springs
                    TITUSVILLE   FOREST
                     (PRR)   PRR
                                         PRR
VENANGO   125
                                  MILLER FARM
                            130
               PETROLEUM CENTER
41°30'   41°30'
80°00'   45'   Oil City   79°30'

Corry, Pa.   14   N.Y., Pa.

WARREN

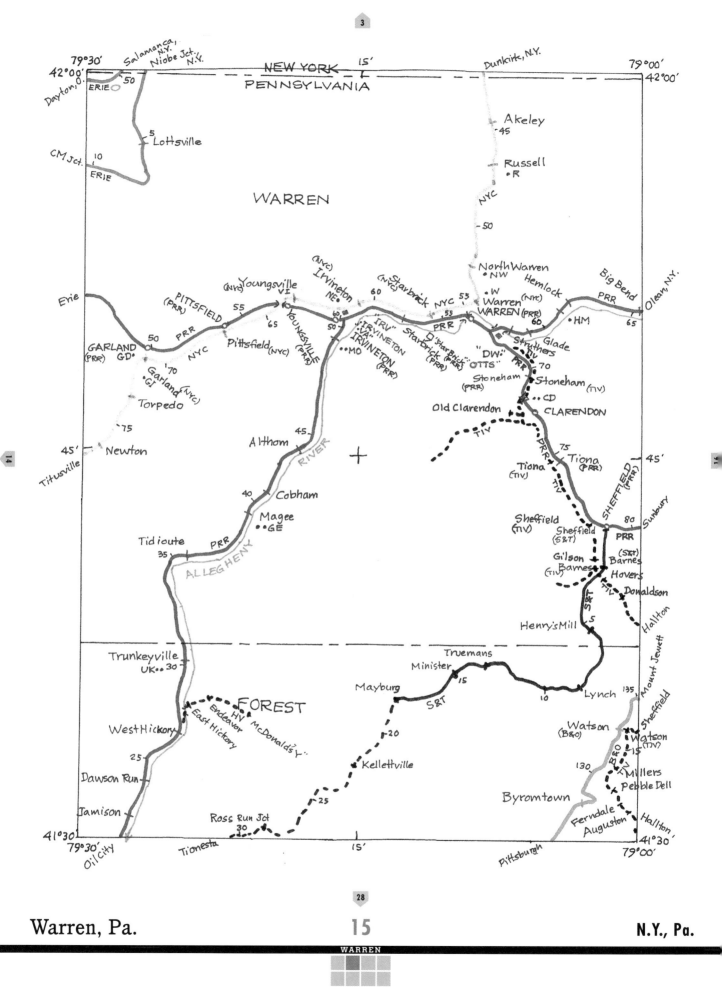

Warren, Pa.　　　　　**15**　　　　　N.Y., Pa.

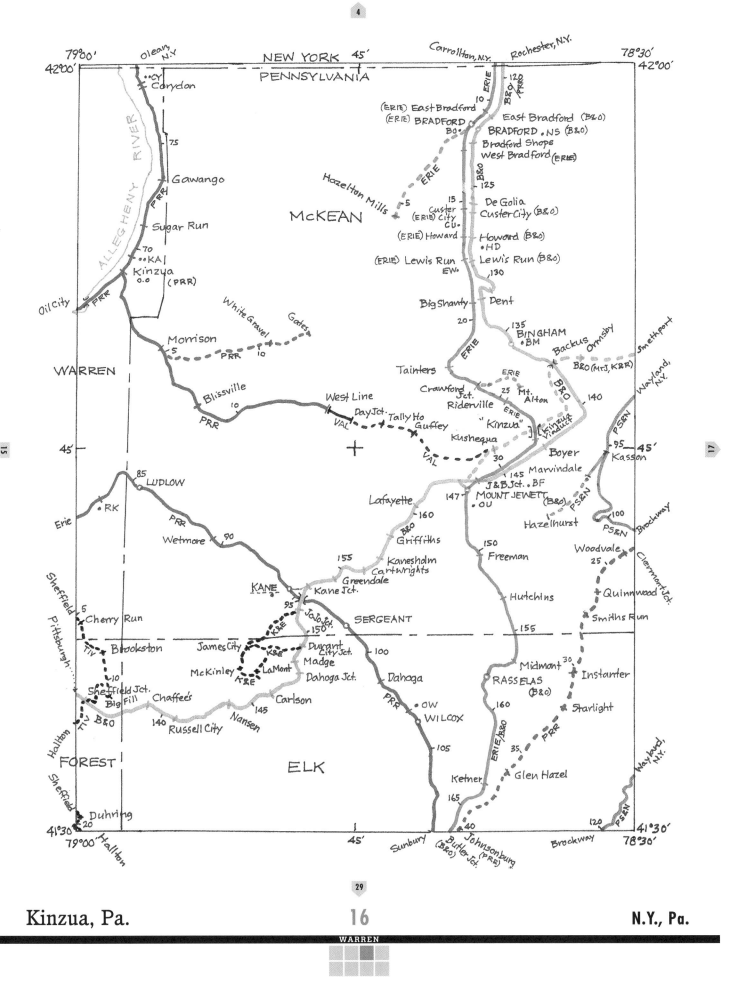

Kinzua, Pa.

16

N.Y., Pa.

WARREN

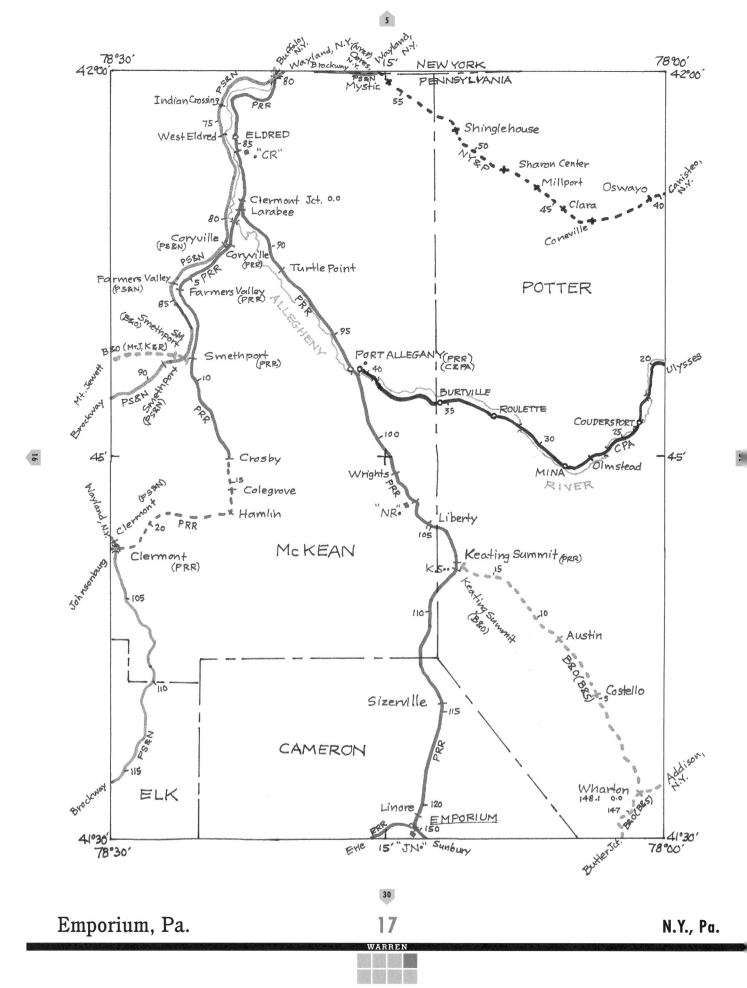

78°30'
42°00'

Buffalo, N.Y.
Wayland, N.Y. (NY&P)
Brockway N.Y. Ceres, N.Y.
Wayland, N.Y.
15'

78°00'
42°00'

NEW YORK
PENNSYLVANIA

PS&N
Indian Crossing
80
PRR
PS&N
Mystic

55

Shinglehouse
NY&P
50

75

West Eldred
ELDRED
85
"CR"

Sharon Center
Millport
45 Clara
Oswayo
40

Canisteo, N.Y.

Clermont Jct. 0.0
Larabee
80

Coneville

POTTER

Coryville
(PS&N)
PS&N
Coryville
(PRR)
90
Turtle Point

PRR
5 PRR
Farmers Valley
(PS&N)
85
Farmers Valley
(PRR)
ALLEGHENY
PRR
95

(B&O)
Smethport
SM
B&O (McJ, K&R)
Smethport
(PRR)
PORT ALLEGANY (PRR)
(C&PA)
40

20
Ulysses

Mt. Jewett
90
PS&N
Smethport
(PS&N)
10
PRR
BURTVILLE
35
ROULETTE
30
COUDERSPORT
25
CPA

Brockway

45'
Crosby
15
Colegrove
100
Wrights
"NR"
PRR
MINA
RIVER
Olmstead

45'

Waylan, N.Y.
(PS&N)
Clermont
20 PRR
Hamlin
Liberty
105

McKEAN

Johnsonburg
Clermont
(PRR)
Keating Summit (PRR)
K.S.
15

105
Keating Summit
(B&O)
10
Austin
B&O (B&S)

110

Sizerville
115

Costello
5

CAMERON
PRR

Addison, N.Y.

115
PS&N
ELK

Linore
120
EMPORIUM

Wharton
148.1 0.0
147
B&O (B&S)

Brockway

41°30'
78°30'

PRR
Erie 15' "JN" Sunbury
150

Butler Jct.

41°30'
78°00'

Emporium, Pa. 17 N.Y., Pa.

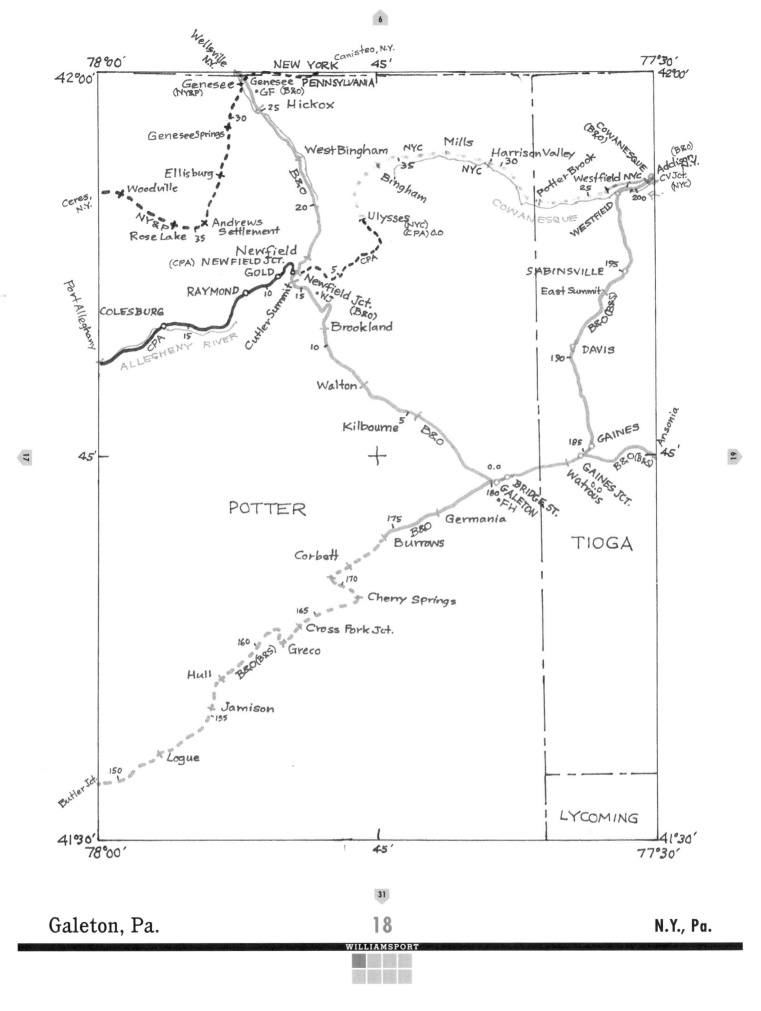

Wellsville N.Y.

78°00'   NEW YORK   Canisteo, N.Y.   45'   77°30'
42°00'   42°00'

Genesee
(NY&P)   • Genesee PENNSYLVANIA
         • GF (B&O)
         25 Hickox

      30
Genesee Springs

                                    NYC   Mills
West Bingham   NYC   Harrison Valley   COWANESQUE
                     35   ,30   (B&O)
Ellisburg   Bingham   COWANESQUE   Addison
      Woodville   NYC   Potter Brook   Westfield NYC   N.Y.
Ceres,   25   CV Jct.
N.Y.   NY&P   Andrews   WESTFIELD R.   200   (NYC)
Rose Lake 35 × Settlement   Ulysses (NYC)
                              (CPA) 0.0   (B&O)

                                                    195
Newfield   SABINSVILLE
(CPA) NEWFIELD JCT.   5   CPA   East Summit
      GOLD   B&O (B&S)
Port   RAYMOND   Newfield Jct.   190 DAVIS
Allegheny COLESBURG   10 15 • WJ (B&O)
              CPA   15   Ansonia
         ALLEGHENY RIVER Cutler Summit Brookland
                                    10   185 GAINES 45'
45'   GAINES JCT. B&O (B&S)
                              Walton ×   Watrous 0.0
                                                              0.0
                        Kilbourne 5   B&O
                              +   0.0
                                          180 BRIDGE ST.
      POTTER   • GALETON
                                          FH
                                    175   Germania   TIOGA
                              Burrows B&O
                        Corbett
                              170
                              Cherry Springs
                        165
                  Cross Fork Jct.
            160 B&O (B&S) Greco
      Hull ×
            Jamison
            155
            Logue
      150
Butler Jct.   LYCOMING

41°30'   41°30'
78°00'   45'   77°30'

Galeton, Pa.   18   N.Y., Pa.

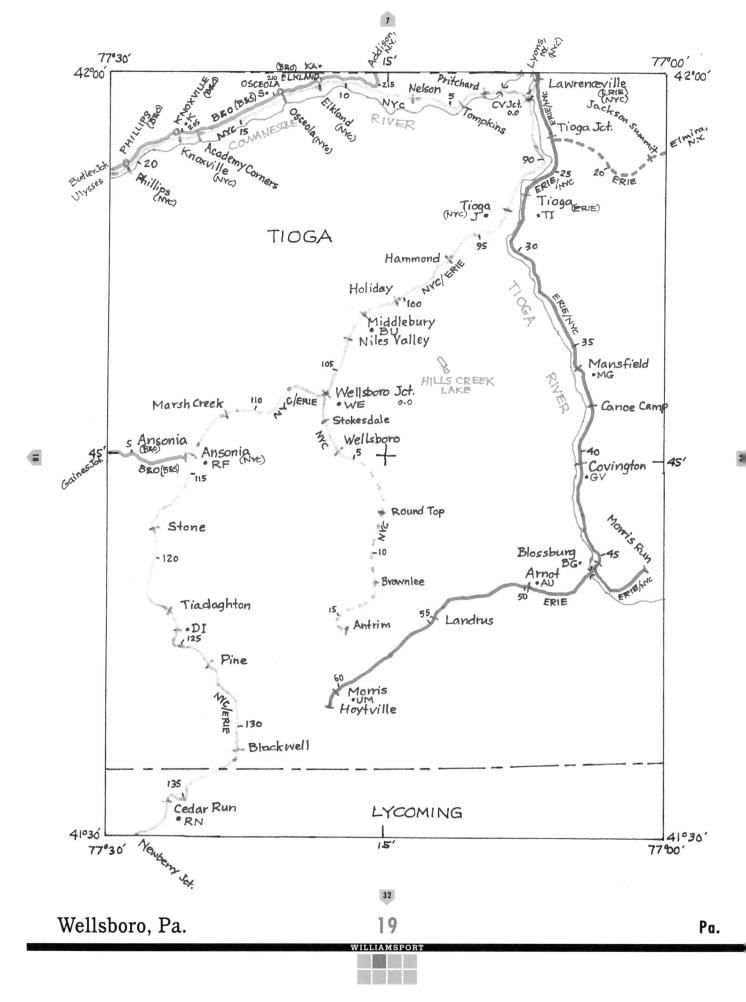

Addison, N.Y.
15'

↑
7

77°30'
42°00'

(B&O) KA•
210 ELKLAND
OSCEOLA
KNOXVILLE (B&O)
(B&O) S.•
PHILLIPS (B&O)
•K
205
B&O (B&S)
NYC 15
Osceola (NYC)
COWANESQUE
Academy Corners
Knoxville (NYC)
Butler Jct.
Ulysses
Phillips (NYC)
20
10
Elkland (NYC)
NYC

Nelson
215
5
Tompkins
CV Jct.
0.0

Lyons, N.Y. (NYC)
Lawrenceville (ERIE)
Jackson Summit (NYC)

77°00'
42°00'

Pritchard
ERIE/NYC
Tioga Jct.
90
25
20 ERIE
ERIE/NYC
Elmira, N.Y.

TIOGA

Tioga (NYC) J•
Tioga (ERIE)
•TI

95
30

Hammond
NYC/ERIE
Holiday
100
TIOGA

ERIE/NYC
35
Mansfield •MG

Middlebury •BU
Niles Valley
105
HILLS CREEK LAKE
RIVER

Marsh Creek
110
Wellsboro Jct. •WE
NYC/ERIE
0.0
Canoe Camp

Stokesdale
NYC
Wellsboro
•5
40
Covington •GV
45'

5 Ansonia (B&O)
Gaines Jct.
Ansonia •RF (NYC)
B&O (B&S)
115

Round Top
NYC

Morris Run
45'

Stone
120
10
Blossburg BG•
45
ERIE/NYC

Brownlee
Arnot •AU
50 ERIE

Tiadaghton •DI
125
15
Antrim
55 Landrus

Pine
NYC/ERIE
130
60 Morris •UM
Hoytville

Blackwell

135
Cedar Run •RN
41°30'
77°30'
Newberry Jct.
LYCOMING
15'
41°30'
77°00'

32

Wellsboro, Pa.   19   Pa.

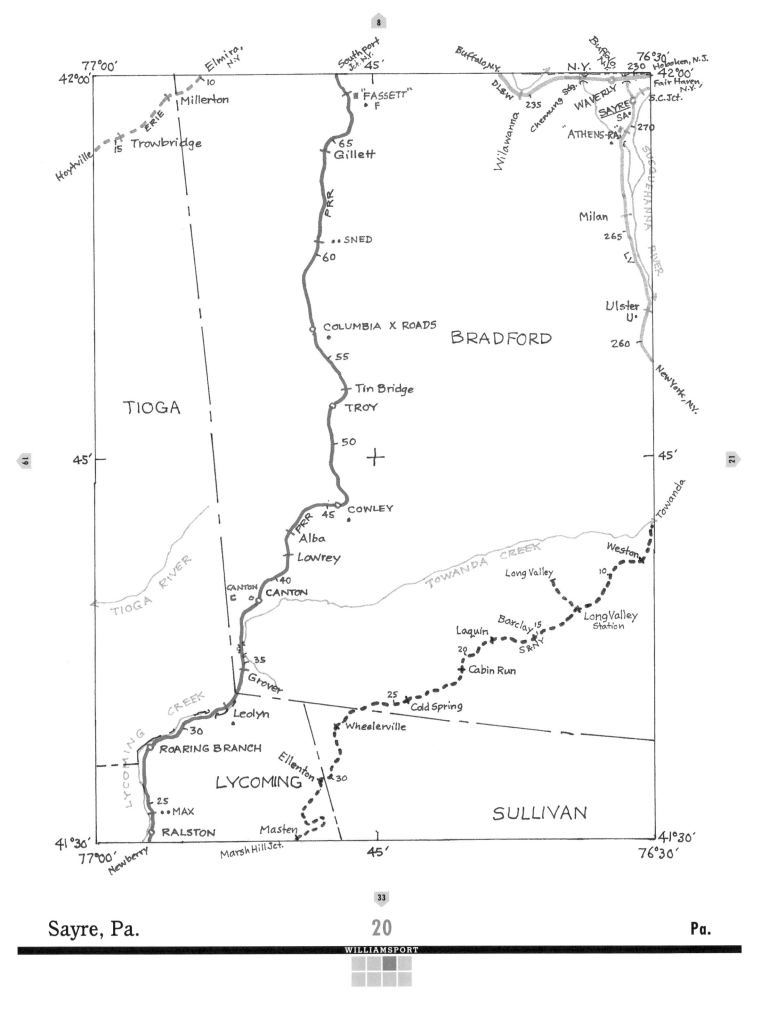

77°00'
42°00'

Elmira, N.Y.
10 Millerton

ERIE
Hoytville
15 Trowbridge

Southport Jct. N.Y.
45'

"FASSETT"
F

Buffalo N.Y.
DL&W

235
Chemung Sdg.

Buffalo N.Y.
230
N.Y.

WAVERLY
SAYRE
SA

76°30'
Hoboken, N.J.
42°00'
Fair Haven, N.Y.
S.C.Jct.

270

ATHENS-RA.

65
Gillett

PRR

Milan
265

SNED
60

Ulster
U.

260

TIOGA

COLUMBIA X ROADS
55

BRADFORD

New York, N.Y.

Tin Bridge
TROY
50

61
45'

45'

Towanda

COWLEY
45
Alba
Lowrey

PRR

Weston

Long Valley
10

21

TOWANDA CREEK

TIOGA RIVER

CANTON
C
40
CANTON

Barclay 15
Laquin
S&NY
Cabin Run
20

Long Valley
Station

35
Grover

LYCOMING CREEK

Leolyn
30
ROARING BRANCH

Cold Spring
25
Wheelerville

LYCOMING

Ellenton
30

SULLIVAN

25 MAX
RALSTON

Masten

41°30'
77°00'
Newberry

Marsh Hill Jct.

45'

41°30'
76°30'

Sayre, Pa.   20   Pa.

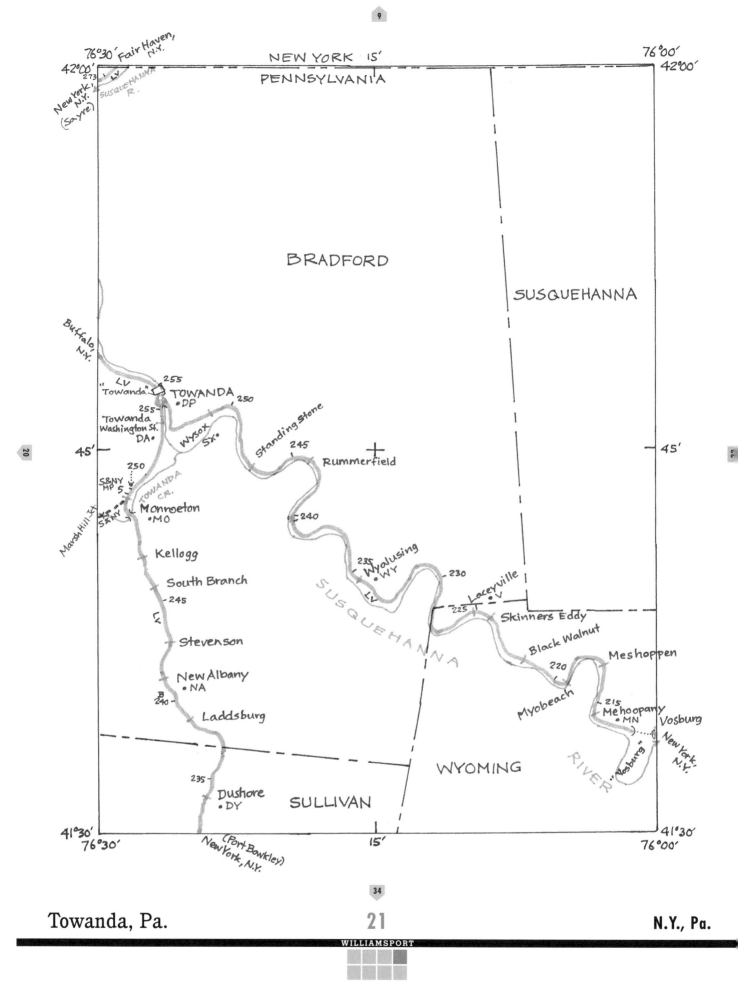

76°30' Fair Haven, N.Y.

NEW YORK 15'

76°00'

42°00'

PENNSYLVANIA

42°00'

New York, N.Y. (Sayre)

273 LV

SUSQUEHANNA R.

BRADFORD

SUSQUEHANNA

Buffalo, N.Y.

LV "Towanda"

255

TOWANDA • DP

250

255

Towanda Washington St. DA•

Wysox SX•

Standing Stone

245

Rummerfield

45'

45'

250

S&NY MP 5

TOWANDA CR.

Monroeton • MO

240

Marsh Hill Jct.

S&NY

Kellogg

Wyalusing • WY

235

Laceyville • V

230

South Branch

245

SUSQUEHANNA

225

Skinners Eddy

LV

Stevenson

Black Walnut

220

Meshoppen

New Albany • NA

Myobeach

215

Mehoopany • MN

Vosburg

B 240

Laddsburg

"Vosburg"

RIVER

New York, N.Y.

WYOMING

235

Dushore • DY

SULLIVAN

15'

41°30'

41°30'

76°30'

(Port Bowkley) New York, N.Y.

76°00'

Towanda, Pa.

N.Y., Pa.

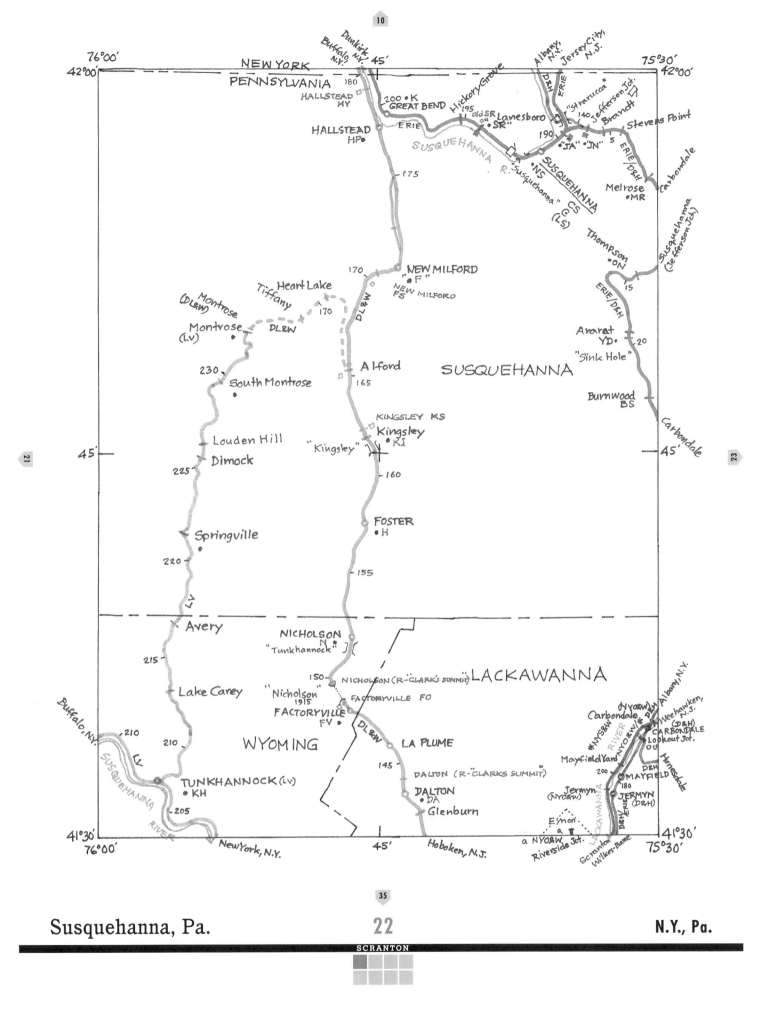

NEW YORK
PENNSYLVANIA

76°00'   45'   75°30'
42°00'   42°00'

Dunkirk, N.Y.
Buffalo, N.Y.
180
HALLSTEAD   200 • K   Albany, N.Y. Jersey City, N.J.
HY   GREAT BEND   D&H ERIE
HALLSTEAD   195   old SR   "Starrucca" Jefferson Jct.
HP •   ERIE   "O" Lanesboro   140 Brandt
                                  • SR   Stevens Point
                                            190   "JA" "JN"   5
SUSQUEHANNA R. "Susquehanna" NS   SUSQUEHANNA   ERIE/D&H   Carbondale
                              CS
                              (LS)   Melrose
                                                               • MR

175   Thompson
                                             • ON   Susquehanna (Jefferson Jct.)

170   NEW MILFORD   ERIE/D&H
Heart Lake "F"   15
Tiffany 170   NEW MILFORD
(DL&W)   FS
Montrose
Montrose   DL&W   SUSQUEHANNA   Ararat
(Lv)   YD •   20
230   "Sink Hole"
South Montrose   Alford
165   Burnwood
DL&W   BS   Carbondale
45'   KINGSLEY KS   45'
Louden Hill   Kingsley
Dimock   "Kingsley"   • KI
225   160

Springville   FOSTER
220   • H
155

Avery   NICHOLSON   LACKAWANNA
                       N
215   "Tunkhannock"   • N
                  150 NICHOLSON (R-"CLARK'S SUMMIT")
Lake Carey   "Nicholson" FACTORYVILLE FO
                   1915   Carbondale NY&W D&H Albany, N.Y.
FACTORYVILLE   *NYS&W   Weehawken, N.J.
WYOMING   FV •   DL&W   LA PLUME   NY&W   (D&H)
210 210   Mayfield Yard Lookout Jct.
         LV   145   DALTON (R-"CLARK'S SUMMIT") 200 D&H Honesdale
Buffalo, N.Y.   180 MAYFIELD
TUNKHANNOCK (Lv)   DALTON   Jermyn
• KH   DA   (NY&W) JERMYN
205   Glenburn   (D&H)
                                                         Eynon D&H ERIE D&H
41°30'   a NY&W
76°00'   NewYork, N.Y.   45'   Hoboken, N.J.   Riverside Jct. Scranton 75°30'
                                                         Wilkes-Barre   41°30'

Susquehanna, Pa.   22   N.Y., Pa.

SCRANTON

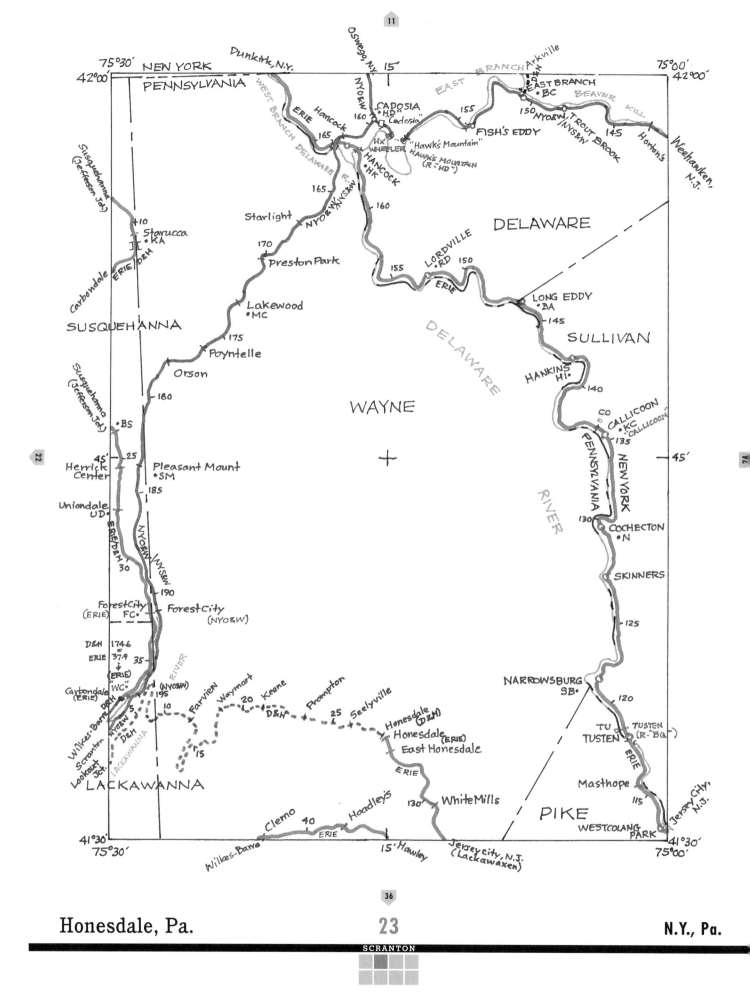

75°30' NEW YORK
42°00'
PENNSYLVANIA

Dunkirk, N.Y.
Oswego, N.Y. 15'
75°00'
42°00'

ERIE
Hancock
NYO&W

CADOSIA
• HD
"Cadosia"
160
165

EAST BRANCH
Arkville
• N
EAST BRANCH
• BC
155
150 NYO&W TROUT BROOK
NYS&W 145
BEAVER KILL
Horton's
Weehawken, N.J.

FISH'S EDDY

HX
WHEELER
"Hawk's Mountain"
HANCOCK
HK
"Hawk's Mountain"
(R "HD")

160

165 NYO&W/NYS&W
165

Starlight
170

Preston Park

Lakewood
• MC
175

DELAWARE

LORDVILLE
RD
150
155
ERIE

LONG EDDY
• BA
145

SULLIVAN

Carbondale ERIE D&H
Susquehanna
(Jefferson Jct.)
+ 10
Starrucca
JI • KA

SUSQUEHANNA

Poyntelle

Orson
180

Susquehanna
(Jefferson Jct.)
• BS
45' 25
Herrick
Center

Uniondale
UD
ERIE/D&H
30
NYO&W/NYS&W

Pleasant Mount
• SM
185

WAYNE
+

HANKINS
HI •
140

CO
CALLICOON
• KC
135
"CALLICOON"

PENNSYLVANIA
NEW YORK
45'

DELAWARE
RIVER

130
COCHECTON
• N

SKINNERS

125

ForestCity
(ERIE)
FC •
Forest City
(NYO&W)
190

D&H 174.6
ERIE 37.9
(ERIE)
"WC"
35

Carbondale
(ERIE)
D&H
Wilkes-Barre
Scranton NYO&W
Lookout D&H
Jct.
195
(NYO&W)
5
RIVER
10
LACKAWANNA

Farview
Waymart
20 Keane
D&H
Prompton
25 Seelyville
15

Honesdale
(D&H)
Honesdale
(ERIE)
East Honesdale

NARROWSBURG
SB •

120

TU
TUSTEN
TUSTEN
(R "BQ")
ERIE

Masthope

115

Jersey City,
N.J.

LACKAWANNA
41°30'
75°30'

Clemo
40
Hoadley's
ERIE

Wilkes-Barre
15' Hawley

130
White Mills

Jersey City, N.J.
(Lackawaxen)

ERIE
PIKE

WESTCOLANG
PARK
75°00'
41°30'

Honesdale, Pa.   **23**   N.Y., Pa.

SCRANTON

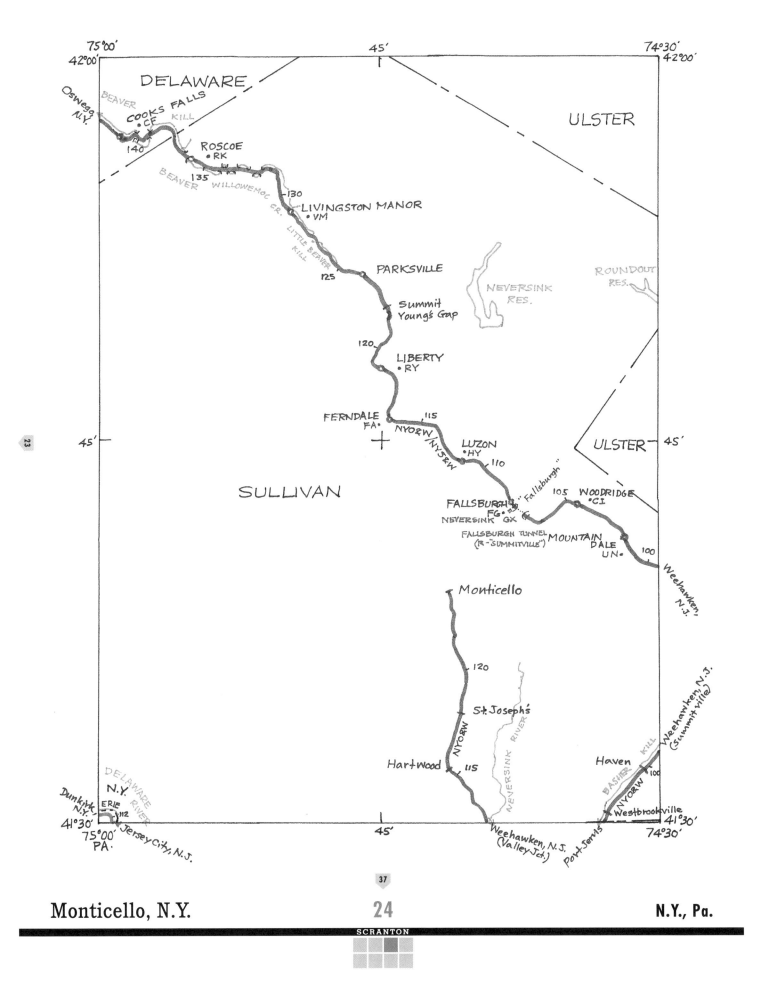

DELAWARE

ULSTER

Oswego N.Y.

BEAVER

COOKS FALLS
CF
KILL
140

BEAVER

ROSCOE
RK
135

WILLOWEMOC CR.

130

LIVINGSTON MANOR
VM

LITTLE BEAVER KILL

PARKSVILLE
125

NEVERSINK RES.

ROUNDOUT RES.

Summit
Young's Gap

120

LIBERTY
RY

FERNDALE
FA
NYO&W
NYS&W
115

LUZON
HY
110

ULSTER 45'

45'

SULLIVAN

"Fallsburgh"
105
WOODRIDGE
CI

FALLSBURGH
FG
NEVERSINK GX

FALLSBURGH TUNNEL
(R-"SUMMITVILLE")

MOUNTAIN DALE
UN

100

Weehawken, N.J.

Monticello

120

St. Joseph's

NYO&W

NEVERSINK RIVER

Weehawken, N.J. (Summitville)

Hartwood

115

Haven
100

BASHER KILL

NYO&W

DELAWARE
N.Y.
RIVER

Dunkirk, N.Y.
ERIE
112

Jersey City, N.J.
PA.

Westbrookville

Weehawken, N.S. (Valley Jct.)

Port Servis

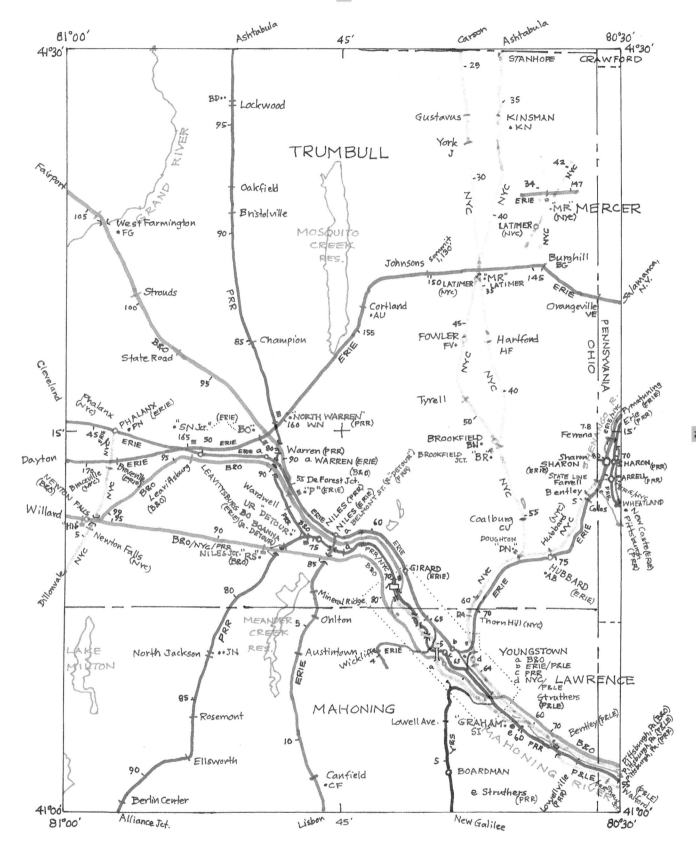

Youngstown, Ohio

Ohio, Pa.

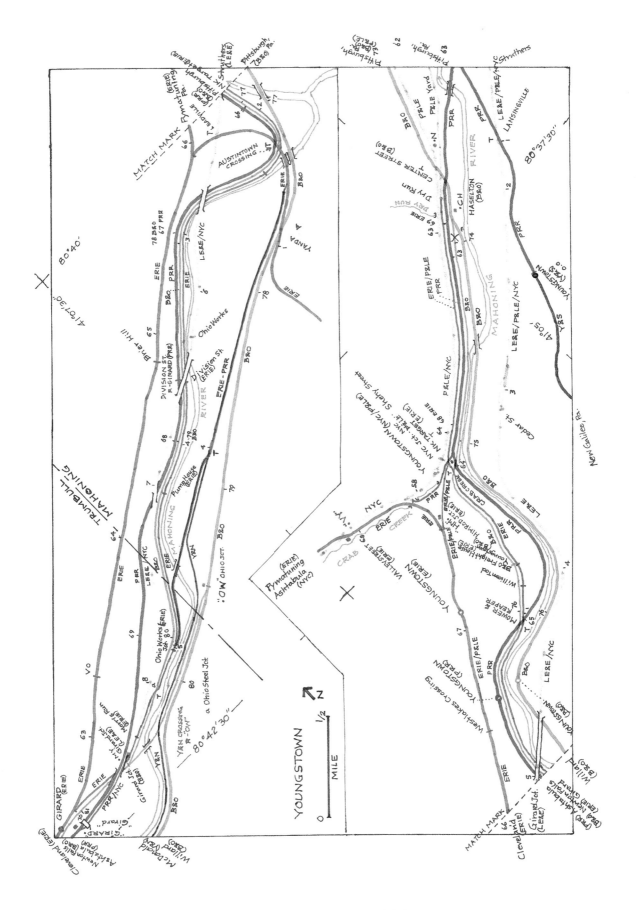

25A Youngstown, Ohio

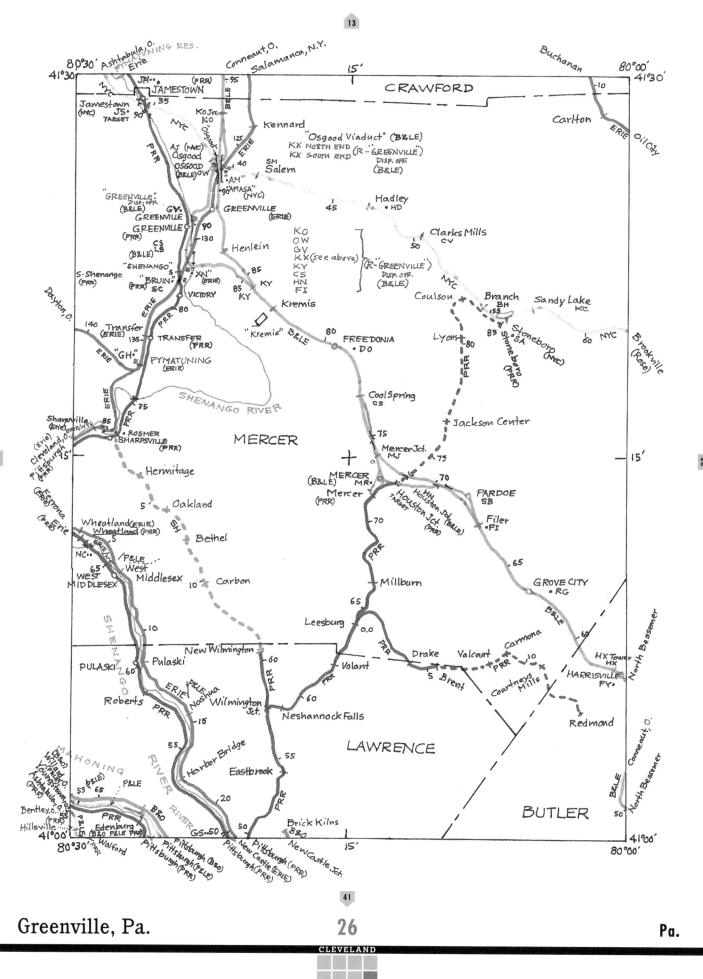

80°30' Ashtabula, O.
41°30'
Conneaut, O.
Salamanca, N.Y.
15'
Buchanan
80°00'
41°30'

CRAWFORD

Carlton

ERIE
Oil City
10

PYMATUNING RES.

Erie

NYC
JH
JAMESTOWN
(PRR)
.95
35
Jamestown
JS
(NYC) TARGET
90
NYC
Ko Jct.
KO
Osgood
125
Kennard

ERIE

"Osgood Viaduct" (B&LE)
KX NORTH END (R-GREENVILLE)
KX SOUTH END DISP. OFF.
(B&LE)

AJ (NYC)
Osgood
OSGOOD
(B&LE) OW
40
SM
Salem

PRR

"AM"
(B&LE) 90 "AMASA"
(NYC)

45
Hadley
• HD

Clarks Mills
CV

"GREENVILLE"
DISP. OFF.
(B&LE)
GV
GREENVILLE
GREENVILLE
(PRR)
90
130
GREENVILLE
(ERIE)

50

KO
OW
GV
KX (see above)
KY
CS
HN
FI

(R-"GREENVILLE")
DISP. OFF.
(B&LE)

CS
(B&LE) LB
Henlein

Coulson
Branch
BH
Sandy Lake
KC

"SHENANGO"
S-Shenango
(PRR)
"BRUIN"
(PRR) SC
"XN"
(ERIE)
85
KY
KY
85
KY
Kremis
155
NYC
85
Stoneboro
SA
Stoneboro
(PRR)
60 NYC

Dayton, O.
140
Transfer
(ERIE)
135
ERIE
TRANSFER
(PRR)
"GH"
VICTORY
80
PRR
"Kremis" B&LE
80
FREEDONIA
• DO
Lyons
80
Branch

Brookville
(Rose)

PYMATUNING
(ERIE)

SHENANGO RIVER

Cool Spring
CS

Jackson Center

75
ERIE
PRR
85
Sharpsville (Erie)
(Erie)
Cleveland,O.
Pittsburgh, (PRR)
75'

ROEMER
SHARPSVILLE
(PRR)

Hermitage

MERCER

75
Mercer Jct.
MJ
75
15'
27

25
Kenyona
(ERIE)
(PRR)

Wheatland (ERIE)
Wheatland (PRR)
5

5
Oakland

Bethel

MERCER
(B&LE) MR
Mercer
(PRR)
HN Houston Jct. (B&LE)
HOUSTON JCT.
(PRR) TARGET
70
PARDOE
SB

Filer
• FI

NC
WEST
MIDDLESEX
65
/P&LE
West
Middlesex

10
Carbon

70
PRR

65

GROVE CITY
• RG

B&LE
60
North Bessemer

SHENANGO
10

New Wilmington
60

Millburn
65
Leesburg
0.0

Drake
Valcourt
Carmona

HX Tower
HX
HARRISVILLE
FY

PULASKI
60
Pulaski
PRR
Volant
PRR
5 Brent
10
PRR
Courtneys Mills

ERIE
P&LE
Noshua
Wilmington
Jct.
60
Neshannock Falls
Redmond

Roberts
PRR
15

55
Harbor Bridge

Eastbrook
55
LAWRENCE

20

Conneaut, O.
B&LE
North Bessemer

BUTLER
50

MAHONING
Willard,O.
(B&O)
Youngstown
(B&O)
Ashtabula O.
(PRR)
55 65
P&LE
Bentley,O.
(PRR) 55
Hillsville
P&LE PRR
41°00'
80°30'
Walford
Edenburg
(B&O P&LE PRR)
PRR
Pittsburgh (B&O)
Pittsburgh (P&LE)
Pittsburgh (PRR)
GS .50
50
Brick Kilns
B&O
New Castle (ERIE)
Pittsburgh (PRR)
New Castle Jct.
15'
41°00'
80°00'

Greenville, Pa.

26

Pa.

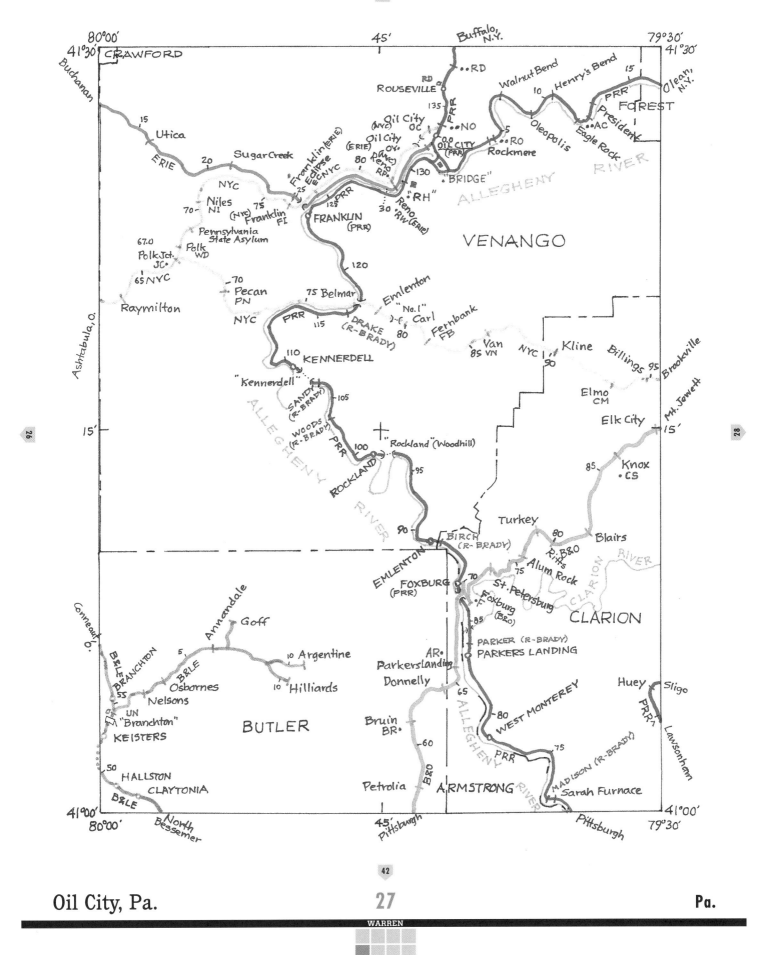

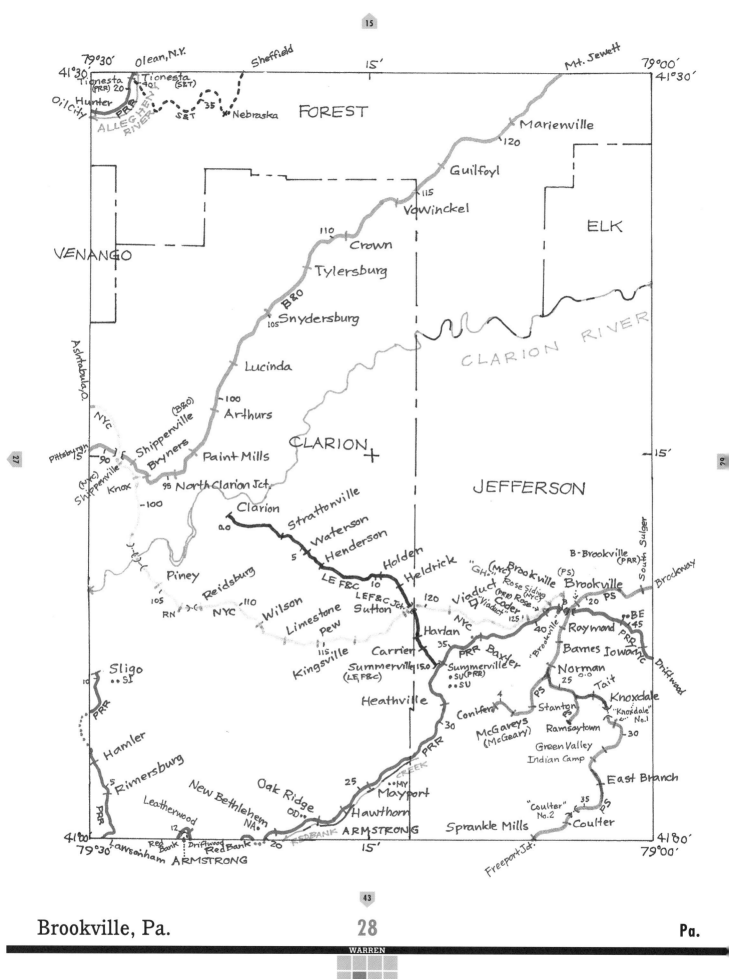

79°30' Olean, N.Y.   Sheffield   15'   Mt. Jewett   79°00'
41°30'   41°30'
Tionesta Tionesta (S&T)   FOREST
(PRR) 20   401
Hunter   Marienville
Oil City   PRR   120
   ALLEGHENY   S&T   Nebraska
        RIVER   Guilfoyl

VENANGO   115
                                                         Vowinckel   ELK

                                          110
                                                Crown
                                          Tylersburg

                                    B&O
                                    105 Snydersburg   CLARION RIVER

                                    Lucinda

Ashtabula,O.   (B&O) 100 Arthurs
NYC   Shippenville
                     Bryners   CLARION
Pittsburgh   Paint Mills   JEFFERSON
5   90
(NYC) Shippenville   95 North Clarion Jct.   15'
    Knox
       100   Clarion
                          Strattonville
                     0.0   Waterson   B-Brookville (PRR)
                              5   Henderson   Holden   "GH" Brookville (PS) Brookville
Piney   Reidsburg   LE F&C 10   Heldrick Rose Siding (NYC)   South Sulger Brockway
     105   NYC 110   Wilson   LEF&C Jct.   120 Viaduct Coder   B 20 PS
RN)-(Reidsburg   Limestone Sutton   125   "Viaduct"   BE
              115 Pew   NYC   Harlan   40 "Brookville" Raymond PRR 45
Sligo   Kingsville Carrier   35 PRR Baxter   Barnes Iowanc Driftwood
10 SI   Summerville 15.0 Summerville   Norman
              PRR   (LE F&C)   SU(PRR)   25   Tait Knoxdale
Hamler   Heathville   SU   PS   "Knoxdale" No.1
              Conifer   McGaveys   Stanton Ramsaytown   30
Rimersburg   30   (McGeary)   PS   Green Valley
     5   New Bethlehem   25   PRR   Indian Camp
PRR   Leatherwood Oak Ridge OD MY Mayport   "Coulter" 35   East Branch
        12   NA   Hawthorn   No.2   25
41°00'   Red Bank Driftwood ARMSTRONG   Sprankle Mills Coulter 41°00'
79°30' Lawsonham RedBank 20   15'   Freeport Jct. 79°00'
        ARMSTRONG   REDBANK CREEK

Brookville, Pa.   28   Pa.

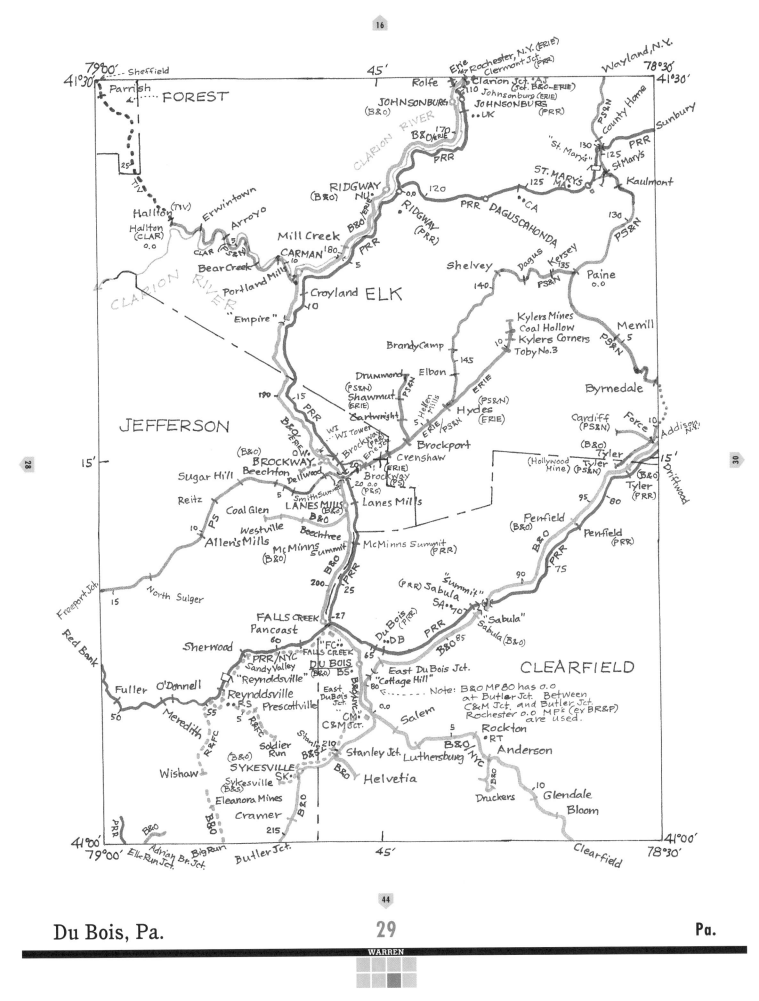

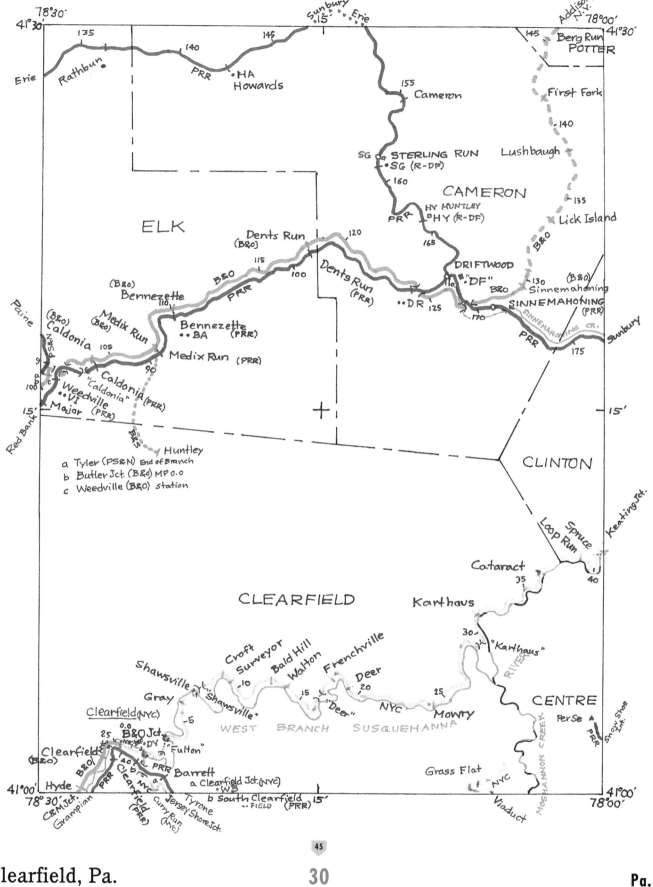

78°30'
41°30'

135
Erie
Rathbun
140
PRR
HA
Howards
145
Sunbury Erie
15'

Addison N.Y. 78°00'
145
Berg Run
POTTER
41°30'

155
Cameron

First Fork

140

ELK

SG SG STERLING RUN
SG (R-DF)
160
PRR
CAMERON
Lushbaugh

135

B&O
Lick Island

Dents Run
(B&O)
120

HY HUNTLEY
HY (R-DF)
165
Dents Run
(PRR)

DRIFTWOOD
"DF"
110
B&O
130
(B&O)
Sinnemahoning

115
B&O
PRR
100
DR 125
SINNEMAHONING (PRR)

(B&O)
Bennezette
110
170
SINNEMAHONING CR.
Sunbury

(B&O)
Caldonia
Medix Run
(B&O)
Bennezette
BA
(PRR)
PRR
175

Pine
(B&O)
105
Medix Run (PRR)
9
90
Red Bank
100
Caldonia
"Caldonia" (PRR)
Weedville
VI
Major (PRR)
15'

15'
CLINTON

B&S
Huntley

a Tyler (PS&N) End of Branch
b Butler Jct. (B&O) MP 0.0
c Weedville (B&O) station

Spruce
Loop Run
Keating Jct.

Cataract
35
40

CLEARFIELD

Karthaus

30
"Karthaus"

Croft
Surveyor
Bald Hill
Walton
Frenchville
Deer
20
RIVER

Shawsville Jct.
10
25

Gray
"Shawsville"
15
"Deer"
NYC
Mowry
CENTRE

Clearfield (NYC)
-5
WEST BRANCH SUSQUEHANNA
PerSe
Snow Shoe Int.
PRR

0.0
25
B&O Jct.
NYC DY "Fulton"
40 b
Grass Flat
MOSHANNON CREEK

Clearfield (B&O)
B&O
PRR
Barrett
a
NYC
WB
NYC

41°00' Hyde
C&M Jct.
Grampian
PRR
Clearfield (PRR)
Curry Run (NYC)
Jersey Shore Jct.
Tyrone
a Clearfield Jct. (NYC)
b South Clearfield FIELD (PRR)
15'
Viaduct
41°00'
78°00'

78°30'

Clearfield, Pa.

Pa.

WARREN

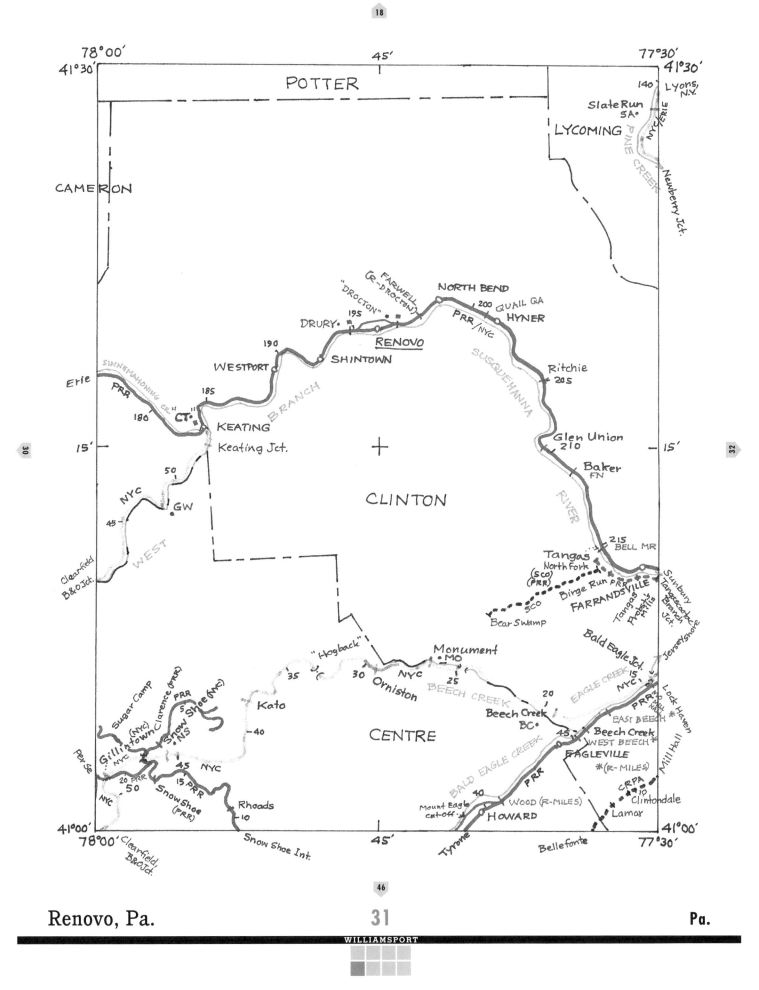

78°00'   45'   77°30'

41°30'   41°30'

POTTER

CAMERON

LYCOMING

140 Lyons, N.Y.

Slate Run SA

NYC ERIE

Newberry Jct.

PINE CREEK

FARWELL (R - DROCTON)

NORTH BEND

"DROCTON"

195

200 QUAIL QA

DRURY

PRR/NYC

HYNER

190

RENOVO

SUSQUEHANNA

Erie

SINNEMAHONING CR.

WESTPORT

SHINTOWN

Ritchie
205

PRR

185

BRANCH

180

CT

15'   KEATING

Glen Union
210

15'

Keating Jct.

Baker
FN

50

CLINTON

NYC

GW

45

RIVER

215 BELL MR

WEST

Tangas
North Fork
(SCO)
(PRR)

Sumbury
Tangascoobre
Branch
Jct.

Clearfield
B&O Jct.

Birge Run PRR

FARRANDSVILLE

Bear Swamp

SCO

Probst's
Mills

Tangas
Branch

Jersey Shore

"Hogback"

Monument
MO

Bald Eagle Jct.

35 JC

30

NYC

25

EAGLE CREEK

15
NYC

Lock Haven

Sugar Camp

Kato

Ornilston

BEECH CREEK

PRR

EAST BEECH

Snow Shoe (NYC)

(NYC) Clarence (PRR)

PRR

40

20

Beech Creek
BC

45

Beech Creek
WEST BEECH

Mill Hall

Gillintown

NS

CENTRE

EAGLEVILLE

*(R-MILES)

Pen Se

NYC

45 NYC

BALD EAGLE CREEK

PRR

CRPA

20 PRR

15 PRR

40

WOOD (R-MILES)

10
Clintondale

50

NYC

Snow Shoe
(PRR)

Rhoads
10

Mount Eagle
cut-off

HOWARD

Lamar

41°00'   41°00'

78°00' Clearfield
B&O Jct.

Snow Shoe Int.

45'

Tyrone

Bellefonte

77°30'

Renovo, Pa.   31   Pa.

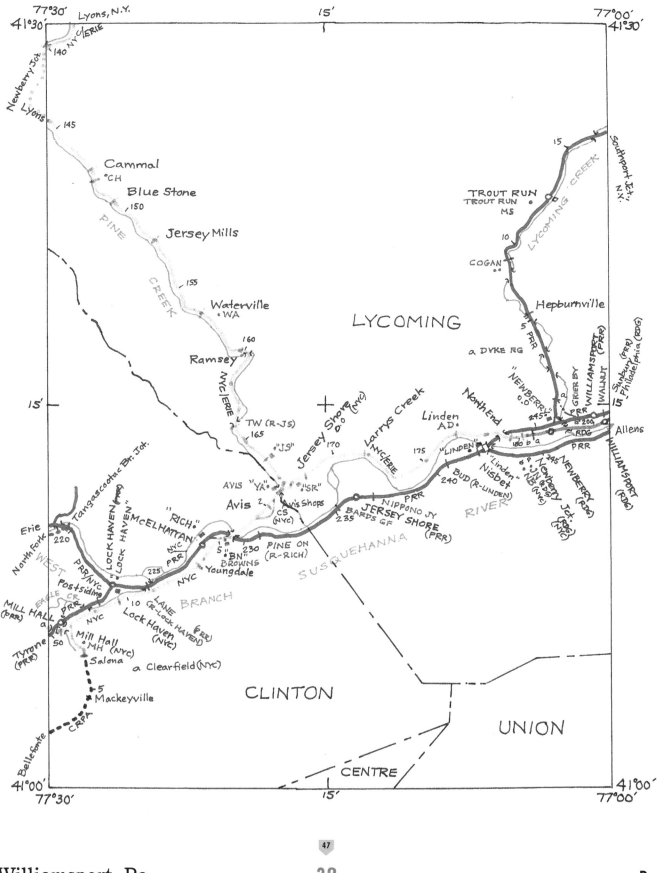

77°30' Lyons, N.Y.
41°30'

Newberry Jct.
140 NYC/ERIE
Lyons
145

Cammal
°CH

Blue Stone
150

Jersey Mills

PINE

155

CREEK

Waterville
•WA

160
76

Ramsey

NYC/ERIE

15'

TW (R-JS)
165

"JS"

15'

77°00'
41°30'

15

Southport Ext.,
N.Y.

LYCOMING CREEK

TROUT RUN
TROUT RUN
MS

10

COGAN

Hepburnville

5 PRR

a DYKE RG

"NEWBERRY"
0.0

GRIER BY

PRR

WILLIAMSPORT (PRR)
WALNUT

Sunbury (PRR)
Philadelphia (RDG)

15

LYCOMING

Jersey Shore (NYC)

Larrys Creek

Linden
AD.

North End

200

245

PRR

RDG

Allens

PRR

WILLIAMSPORT (RDG)

170

NYC/ERIE

175

"LINDEN"

"LINDEN"
Nisbet

Newberry Jct.
(RDG)
JN RDG
NB (NYC)

180 b a

245

Newberry
(NYC)

AVIS "YA"

Avis ²

"SR"

Avis Shops

CS
(NYC)

"Linden"

BUD (R-LINDEN)
240

NIPPONO JY

PRR

JERSEY SHORE
(PRR)

RIVER

"RICH"

McELHATTAN

BARDS GF
235

Erie

North Fork
220

Tangascootac Br. Jct.

LOCK HAVEN (PRR)
"LOCK HAVEN"
NYC
PRR

BN"
230
BROWNS

Youngdale

PINE ON
(R-RICH)

SUSQUEHANNA

WEST

PRR/NYC
225
NYC

Post siding

EAGLE CR.
PRR

'10
LANE
Cr-Lock Haven
(PRR)

BRANCH

MILL HALL
(PRR)
a

NYC

Lock Haven (NYC)

Tyrone
(PRR)

50

Mill Hall
MH (NYC)

Salona

a Clearfield (NYC)

5
Mackeyville

CRPA

Bellefonte

CLINTON

CENTRE

UNION

41°00'
77°30'

15'

77°00'
41°00'

47

31

33

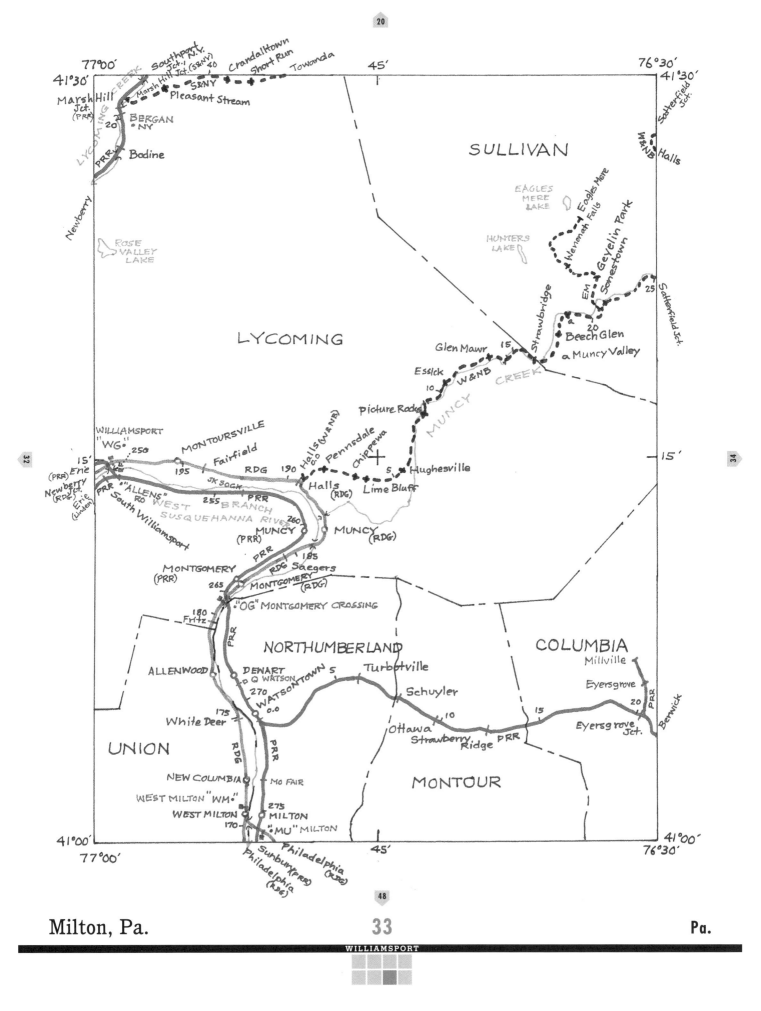

77°00'
41°30'
Southport
Jct., N.Y.
Marsh Hill Jct. (S&NY)
Crandalltown
Short Run
Towonda
45'
76°30'
41°30'
Satterfield Jct.
LYCOMING CREEK
S&NY
40
Pleasant Stream
Marsh Hill Jct. (PRR)
20
BERGAN NY
PRR
W&NB
Halls
Badine

Newberry

ROSE VALLEY LAKE

SULLIVAN

EAGLES MERE LAKE

HUNTERS LAKE

Weronah Falls
Eagles Mere
Geyelin Park
Sonestown
EM
25
Satterfield Jct.

LYCOMING

Strawbridge
a
20
Beech Glen
a Muncy Valley

Glen Mawr
15
Esslck
10
W&NB
MUNCY CREEK

Picture Rocks

WILLIAMSPORT "WG"
250
MONTOURSVILLE
Fairfield
Halls (W&NB)
20
Pennsdale
Chippewa
15'

32

34

15'
Erie (PRR)
195
RDG
190
5
Hughesville
JK SOGK
Halls (RDG)
Lime Bluff
Newberry Jct. (RDG)
Erie (Linden)
PRR
"ALLENS" RO
255
WEST BRANCH
PRR
South Williamsport
SUSQUEHANNA RIVER
260
MUNCY (PRR)
MUNCY (RDG)
PRR
185
MONTGOMERY (PRR)
265
RDG Saegers
MONTGOMERY (RDG)
"OG" MONTGOMERY CROSSING
180
Fritz
PRR
NORTHUMBERLAND
COLUMBIA
Millville
ALLENWOOD
DEWART
Q WATSON
270
WATSONTOWN
0.0
5
Turbotville
Schuyler
10
Eyersgrove
20
PRR
Bernick
175
White Deer
Ottawa
Strawberry Ridge
15
PRR
Eyersgrove Jct.
RDG
PRR
UNION
MONTOUR
NEW COLUMBIA
MO FAIR
WEST MILTON "WM"
WEST MILTON
275
MILTON
170
"MU" MILTON
41°00'
77°00'
Philadelphia Sunbury (PRR)
Philadelphia (RDG)
45'
76°30'
41°00'

Milton, Pa.   33   Pa.

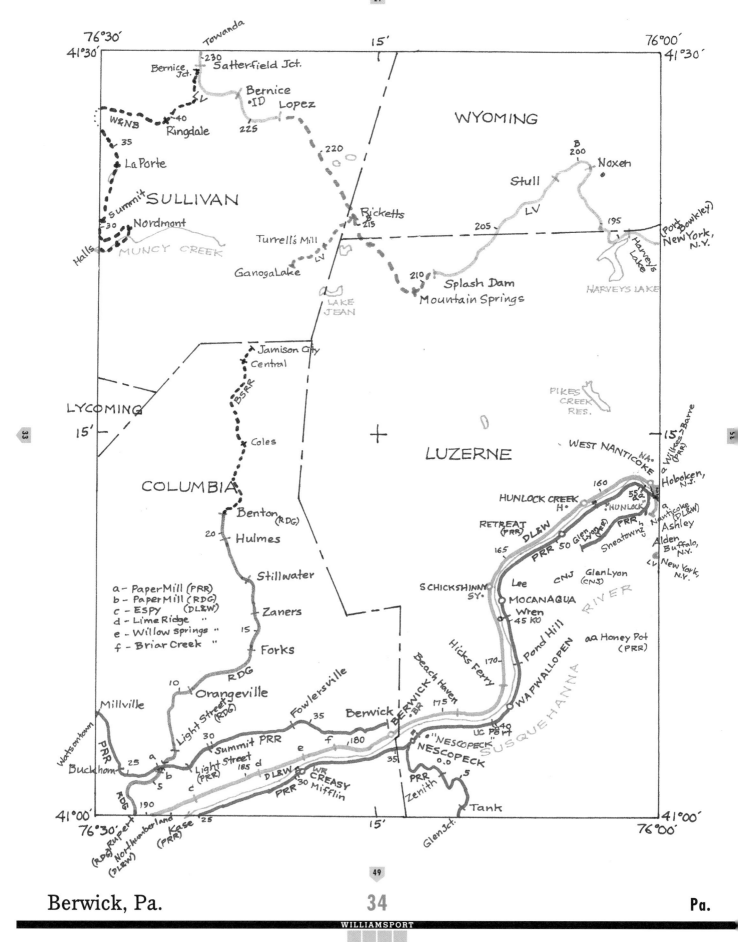

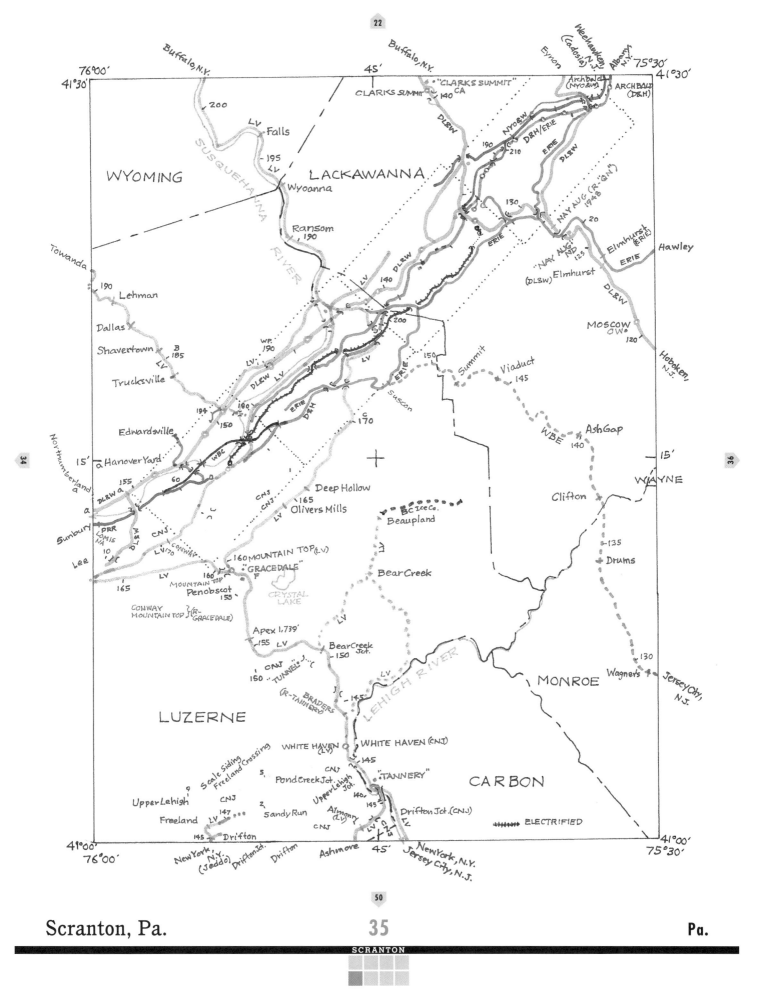

WYOMING

LACKAWANNA

Buffalo, N.Y.

76°00'

41°30'

45'

Buffalo, N.Y.

Weehawken (Cadosia) N.J.

Eynon

Albany, N.Y.

75°30'

41°30'

"CLARKS SUMMIT" CA

CLARKS SUMMIT

140

Archbald (NYO&W)

ARCHBALD (D&H)

200

LV Falls

195

LV

NYO&W

D&H/ERIE

DL&W

190

210

ERIE

DL&W

Wyoanna

130

NAY AUG (R-"GN") 1948

ERIE

Ransom

190

NAY AUG N.D.

125

Elmhurst (ERIE)

Hawley

SUSQUEHANNA RIVER

20

Towanda

190

ERIE

"NAY AUG"

(DL&W) Elmhurst

DL&W

Lehman

LV

140

DL&W

MOSCOW OW

120

Hoboken, N.J.

Dallas

200

Shavertown

B 185

LV

WP. 190

Summit

Viaduct

145

Trucksville

DL&W

LV

LV

ERIE

Suscon

150

WBE

Ash Gap

140

15'

WAYNE

194

160

150

D&H

Clifton

Edwardsville

C

170

Hanover Yard

WBC

Deep Hollow

BC Ice Co.

Beaupland

135

Northumberland

15'

155

60

CNJ

165

Drums

a

DL&W

CNJ

Olivers Mills

Bear Creek

Sunbury

PRR

LOMIS N.A.

CNJ

LV

CONWAY

160 MOUNTAIN TOP (LV)

Lee

10

LV 70

160

"GRACEDALE"

MONROE

130

Wagners

Jersey City, N.J.

165

MOUNTAIN TOP

F

Penobscot

155

LV

CRYSTAL LAKE

CONWAY

MOUNTAIN TOP (R-GRACEDALE)

Apex 1,739'

155 LV

LV

Bear Creek Jct.

150

LUZERNE

CNJ

150

TUNNEL

BRADERS (R-TANNERY)

145

LEHIGH RIVER

LV

WHITE HAVEN (LV)

WHITE HAVEN (CNJ)

145

CARBON

Scale Siding

Freeland Crossing

CNJ

Pond Creek Jct.

"TANNERY"

Upper Lehigh

CNJ

Sandy Run

Upper Lehigh Jct.

147

140

Almonry (LV)

145

Drifton Jct. (CNJ)

ELECTRIFIED

Freeland

LV

CNJ

LV CNJ

LV

145

Drifton

New York, N.Y. (Jeddo)

Drifton Jct.

Drifton

Ashmore

45'

New York, N.Y.

Jersey City, N.J.

41°00'

76°00'

41°00'

75°30'

Scranton, Pa.

35

Pa.

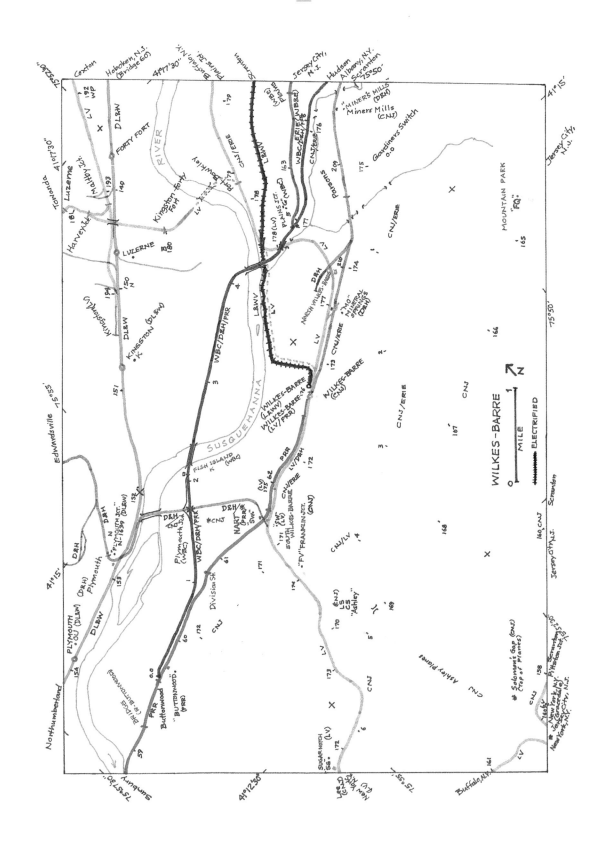

35A Wilkes-Barre, Pa.

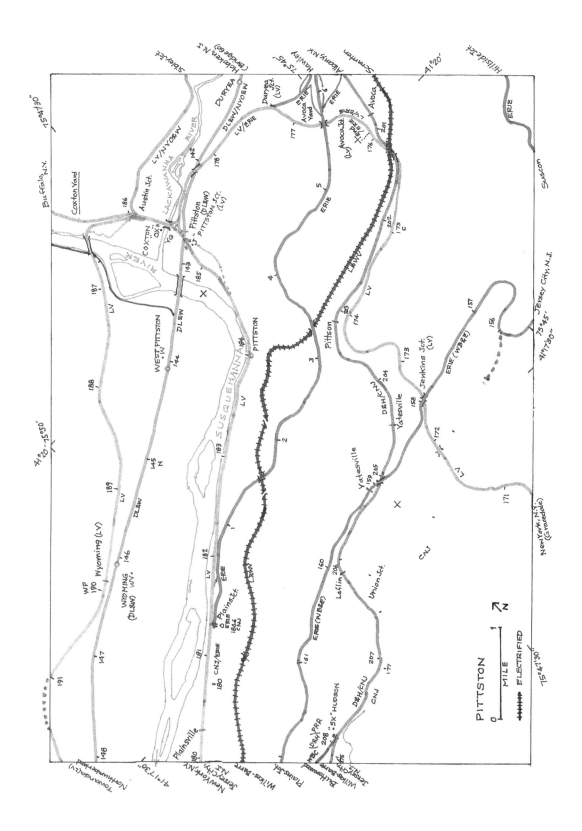

35B Pittston, Pa.

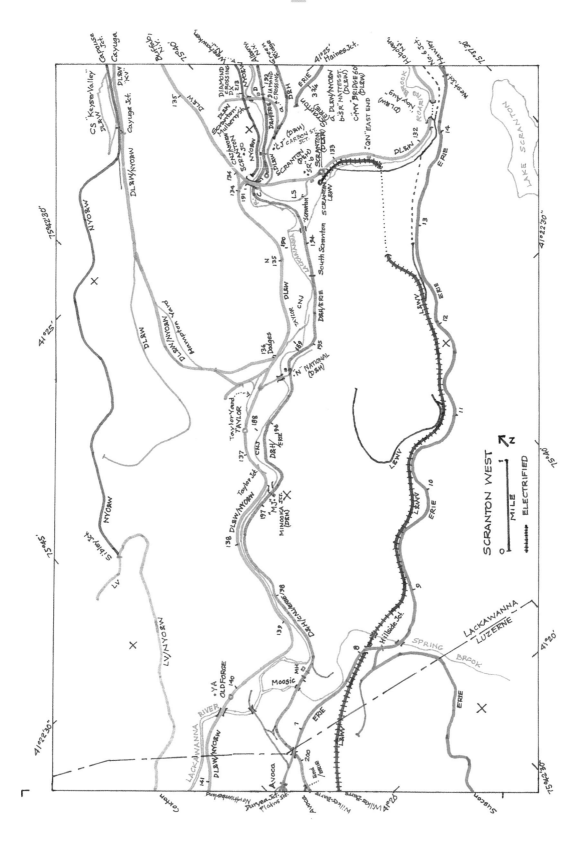

35C Scranton West, Pa.

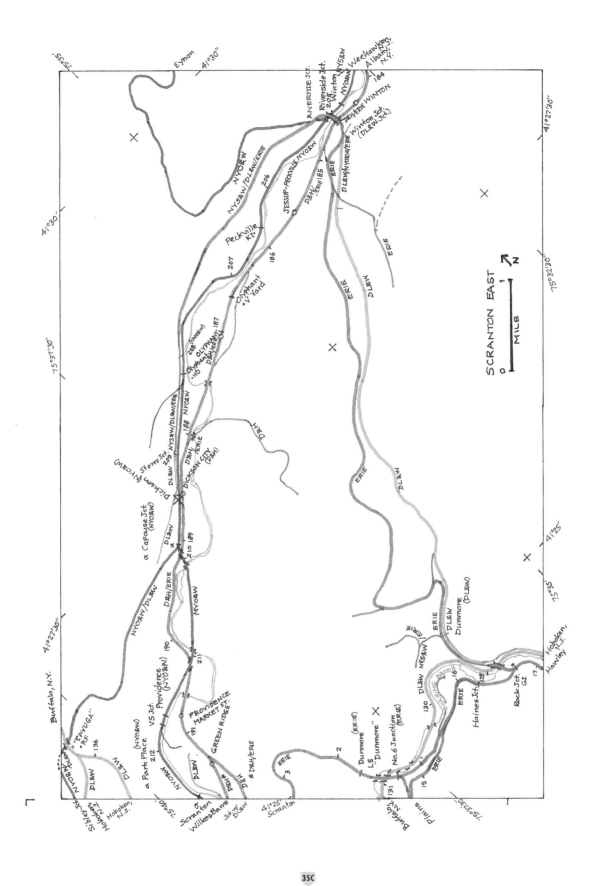

SCRANTON EAST

0   1
MILE

35C

35D Scranton East, Pa.

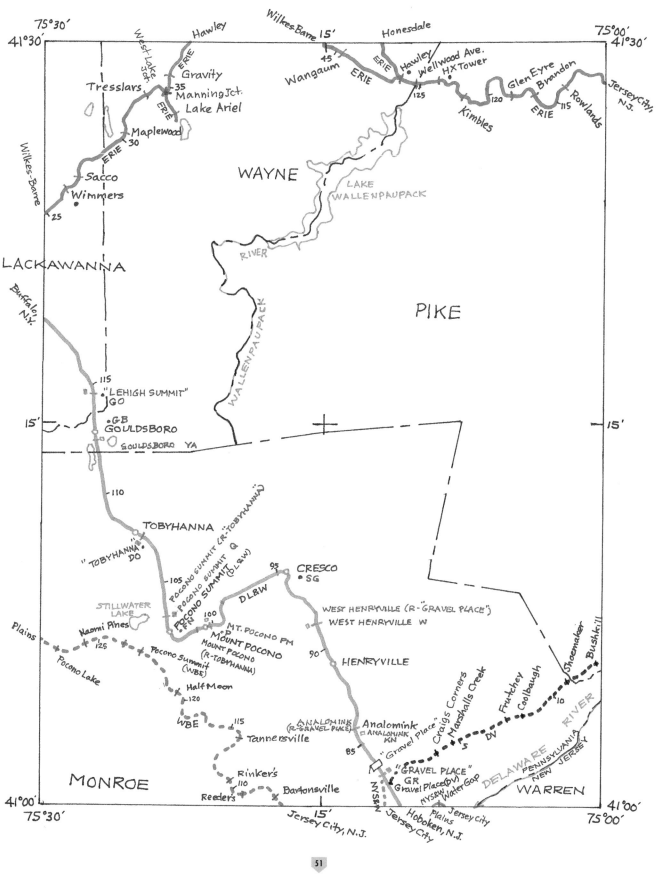

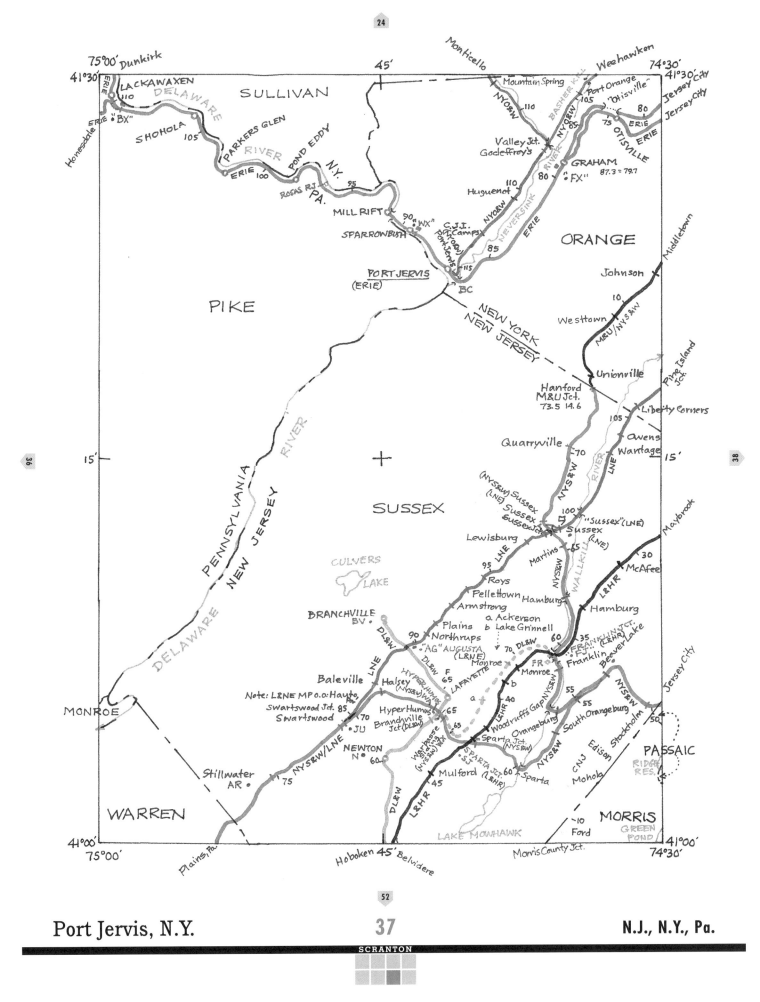

75°00' Dunkirk
41°30'
ERIE 110
LACKAWAXEN
Honesdale ERIE "BX"
SHOHOLA
105
45'
SULLIVAN
DELAWARE
PARKERS GLEN
RIVER
POND EDDY
ERIE 100
ROSAS RJ.
PA.
N.Y.
95
MILL RIFT
90 "WX"
SPARROWBUSH

Monticello
Weehawken
Mountain Spring
NYO&W 110
BASHER KILL
Port Orange
105
"Otisville"
80
Jersey City
85
75
ERIE 80
Jersey City
NYO&W
Valley Jct.
Godeffroy's
OTISVILLE
ERIE
GRAHAM
"FX" 87.3 = 79.7
80
Huguenot
110
NEVERSINK
ERIE
ORANGE
C.J.I.
(Troloy)
Port Jervis
85
PORT JERVIS
(ERIE)
115
BC

PIKE

NEW YORK
NEW JERSEY

Johnson
Westtown
10
M&U/NYS&W
Unionville

Hanford
M&U Jct.
73.5 14.6
Pine Island Jct.
Liberty Corners
105
Owens
Wantage
15'

Middletown

41°30' CITY
Jersey City

74°30'

15'
Quarryville
70
NYS&W
RIVER
LNE

Maybrook

36

DELAWARE RIVER

PENNSYLVANIA

NEW JERSEY

SUSSEX

CULVERS LAKE

(NYS&W) Sussex
(LNE) Sussex
Sussex Jct.
100
"Sussex"(LNE)
Sussex
(LNE)

Lewisburg
95
LNE
Martins
NYS&W
Hamburg
WALLKILL
55
L&HR
30
McAfee

BRANCHVILLE
BV.
DL&W
90
Northrups
"AG" AUGUSTA
(L&NE)
DL&W
F
65
LAFAYETTE
Monroe
70
DL&W
60
Franklin Jct.
FJ. (L&HR)
Franklin
BeaverLake

Roys
Pellettown
Armstrong
Plains
a Ackerson
b Lake Gr'nnell
Hamburg
35
FR
Monroe
NYS&W
55

Baleville
Halsey
HyperHumus
(NYS&W)
Note: L&NE MP 0.0: Hauto,
Swartswood Jct. 85
Swartswood
70
JU
HyperHumus
65
Branchville
Jct.(DL&W)
65
b
a
Woodruffs Gap
40
Sparta Jct.
(NYS&W)
Orangeburg
55
South Orangeburg
Stockholm
50
Jersey City
NYS&W
PASSAIC
RIDGE RES.
CNJ Edison
Mohola

NYS&W/LNE
Stillwater
AR
75
NEWTON
N.
60
War Pass
(NYS&W) WA
Mulford
(L&HR)
45
SPARTA JCT.
S.J.
Sparta Jct. 60
Sparta

DL&W
L&HR
LAKE MOWHAWK
10
Ford
GREEN POND

MONROE

WARREN
41°00'
75°00'
Plains, Pa.
MORRIS
74°30'
41°00'

Hoboken 45' Belvidere
Morris County Jct.

38

Port Jervis, N.Y.   37   N.J., N.Y., Pa.

SCRANTON

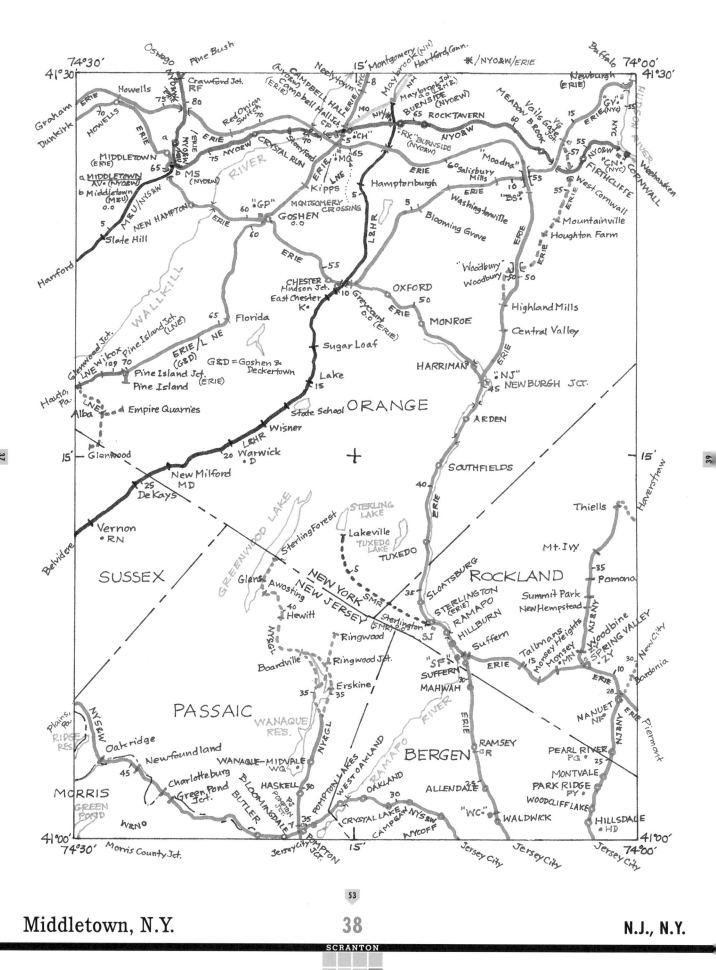

SCRANTON

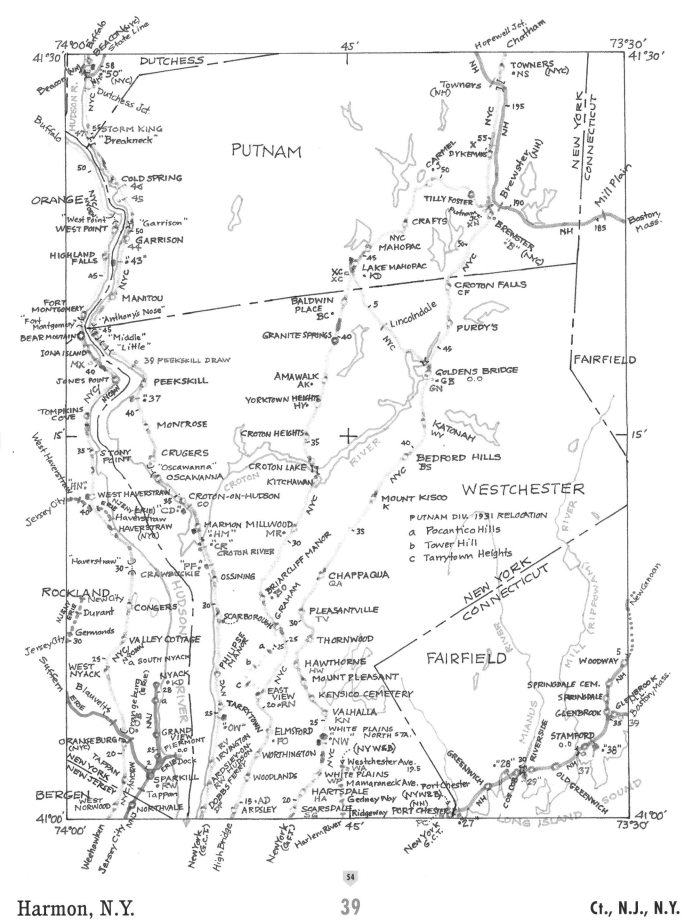

74°00' Buffalo BEACON(NYC) State Line
41°30' Beacon(NYC)
DUTCHESS
58 NY"50"(NYC)
HUDSON R.
Dutchess Jct.
Buffalo
47
55 STORM KING "Breakneck"
PUTNAM
50
COLD SPRING
46
45
ORANGE
"West Point"
WEST POINT
50 "Garrison"
44 GARRISON
HIGHLAND FALLS
"43"
45
MANITOU
FORT MONTGOMERY
"Fort Montgomery"
"Anthony's Nose"
45
BEAR MOUNTAIN
"Middle" "Little"
IONA ISLAND
MX
40
39 PEEKSKILL DRAW
JONES POINT
PEEKSKILL
TOMPKINS COVE
"37"
40
15'
MONTROSE
35 STONY POINT
CRUGERS
"Oscawanna"
OSCAWANNA
"HN"
CROTON-ON-HUDSON CO
WEST HAVERSTRAW
35 NJ-NY ERIE "CD"
ERIE
40 Haverstraw
HAVERSTRAW (NYC)
HARMON MILLWOOD MR
"HM"
"CR"
CROTON RIVER
"Haverstraw"
30
"PF"
CRAWBUCKIE
OSSINING
ROCKLAND
New City
Durant
CONGERS
HUDSON
Germonds
30
VALLEY COTTAGE
SOUTH NYACK
25
WEST NYACK
ERIE
Blauvelts
NYACK KD 28
GRAND VIEW
25 PIERMONT 0.0
ORANGEBURG (NYC)
NEW YORK
TAPPAN
NEW JERSEY
BERGEN
WEST NORWOOD
NORTHVALE
41°00'
74°00'
Weehawken
Jersey City
New York (G.C.T.)
High Bridge
New York (G.C.T.)
Harlem River

Towners (NH)
TOWNERS (NYC) NS
195
NYC
NH
CARMEL DYKEMANS
50
55
NH
Brewster (NH)
190 Putnam Jct. XN
BREWSTER "B"(NYC)
NH 185
TILLY FOSTER
CRAFTS
NYC MAHOPAC
XC XC
45 LAKE MAHOPAC KD
50
CROTON FALLS CF
BALDWIN PLACE BC
5
Lincolndale
PURDY'S
GRANITE SPRINGS 40
45
AMAWALK AK
YORKTOWN HEIGHTS HY
GOLDENS BRIDGE GB 0.0
GN
CROTON HEIGHTS
35
KATONAH WY
BEDFORD HILLS BS
CROTON LAKE
KITCHAWAN
WESTCHESTER
MOUNT KISCO K
40
NYC
PUTNAM DIV. 1931 RELOCATION
a Pocantico Hills
b Tower Hill
c Tarrytown Heights
35
BRIARCLIFF MANOR
CHAPPAQUA QA
GRAHAM 30
PLEASANTVILLE TV
SCARBOROUGH
30
THORNWOOD
25
PHILIPSE MANOR
a 25
b
HAWTHORNE HW
MOUNT PLEASANT
c
EAST VIEW 20 RN
KENSICO CEMETERY
TARRYTOWN
"OW"
25
VALHALLA KN
ELMSFORD FO
WHITE PLAINS NORTH STA.
"NW" (NYW&B)
WORTHINGTON WA
Westchester Ave. 19.5
WOODLANDS
WHITE PLAINS WP
Mamaroneck Ave.
HARTSDALE HA
Gedney Way
SCARSDALE SG
Ridgeway
PORT CHESTER
15 AD
ARDSLEY
20
PC "27"
New York G.C.T.

Hopewell Jct. Chatham
73°30'
41°30'
NEW YORK CONNECTICUT
Mill Plain
Boston, Mass.
FAIRFIELD
15'
NEW YORK CONNECTICUT
MILL (RIPPOWAM) RIVER
New Canaan
FAIRFIELD
WOODWAY 5
SPRINGDALE CEM.
SPRINGDALE WA
GLENBROOK NH
GLENBROOK Boston, Mass.
35 39
STAMFORD 0.0
MIANUS
"38"
GREENWICH
"28" 30 RIVERSIDE
CC CB "29" 37
NH OLD GREENWICH
PORT CHESTER (NYW&B) (NH)
LONG ISLAND SOUND
41°00'
73°30'

38

54

Harmon, N.Y.

39

Ct., N.J., N.Y.

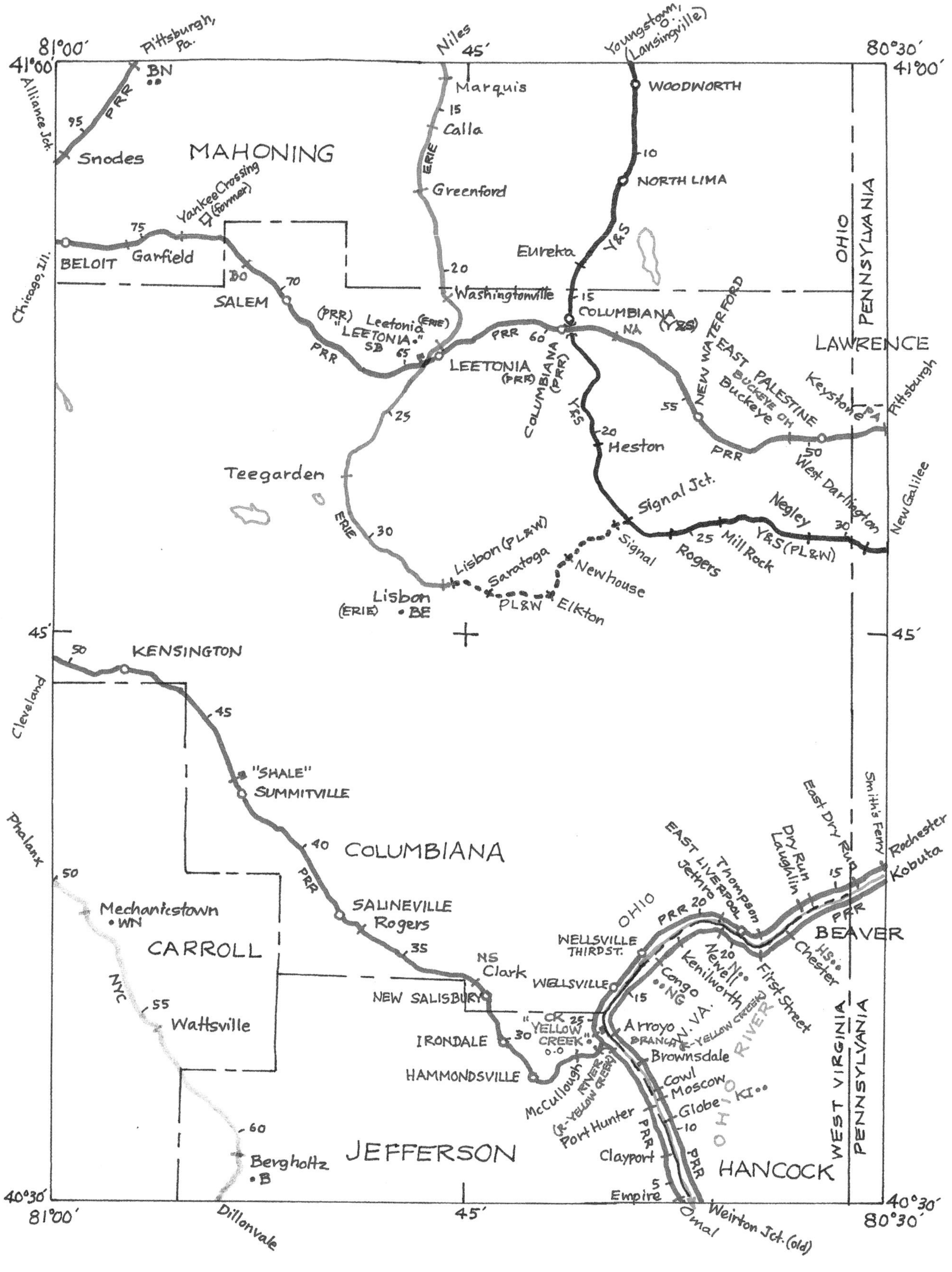

81°00'
41°00'
Pittsburgh, Pa.
BN
Niles
45'
80°30'
41°00'
Youngstown, (Lansingville)
WOODWORTH

Alliance Jct.
95
PRR
Snodes

MAHONING

Marquis
15
Calla

10
NORTH LIMA

Y&S

OHIO
PENNSYLVANIA

Greenford

Eureka

Yankee Crossing (former)

Chicago, Ill.
75
Garfield
BELOIT

20

Washingtonville
15
COLUMBIANA
NA
(Y&S)

LAWRENCE

SALEM
BO
70

(PRR) Leetonia (ERIE)
"LEETONIA."
SB
65
PRR
60

COLUMBIANA (PRR)
COLUMBIANA (PRR)

EAST PALESTINE
NEW WATERFORD
BUCKEYE OH
Buckeye

Keystone
Pittsburgh

LEETONIA (PRR)

25

Y&S
55
PRR

50
West Darlington

New Galilee

Teegarden

30
ERIE

Heston
20

Signal Jct.

Negley
30

Y&S (PL&W)
25
Mill Rock

Lisbon (PL&W)
Saratoga
Newhouse
Signal
Rogers

Lisbon (ERIE)
BE
PL&W
Elkton

45'
45'

41

45'
50
KENSINGTON

Cleveland
45

Phalanx
50

"SHALE"
SUMMITVILLE

COLUMBIANA
40

Mechanicstown
WN

CARROLL

PRR

SALINEVILLE
Rogers

NYC
55
Wattsville

35

NS
Clark

NEW SALISBURY

East Dry Run
Smith's Ferry
Rochester
Kobuta

EAST LIVERPOOL
Jethro
Thompson
Laughlin
Dry Run

15
PRR

OHIO
WELLSVILLE THIRD ST.
PRR 20

NII
NeNell
Kenilworth
Congo
NG
First Street
Chester
HS

BEAVER

WELLSVILLE
15

Arroyo
N. VA.
Brownsdale

"CR 25
"YELLOW CREEK"
BRANCH-YELLOW CREEK
RIVER-YELLOW CREEK
OHIO RIVER
KI

IRONDALE
30

Cowl
Moscow

HAMMONDSVILLE

McCullough
(R-YELLOW CREEK)
Port Hunter
Globe
10
PRR

WEST VIRGINIA
PENNSYLVANIA

60

Berghottz
B

JEFFERSON

Clayport

HANCOCK

40°30'
81°00'
Dillonvale
45'

Empire
5
PRR
O'mal
Weirton Jct. (old)
80°30'
40°30'

East Liverpool, Ohio   40   Ohio, Pa., W.Va.

CANTON

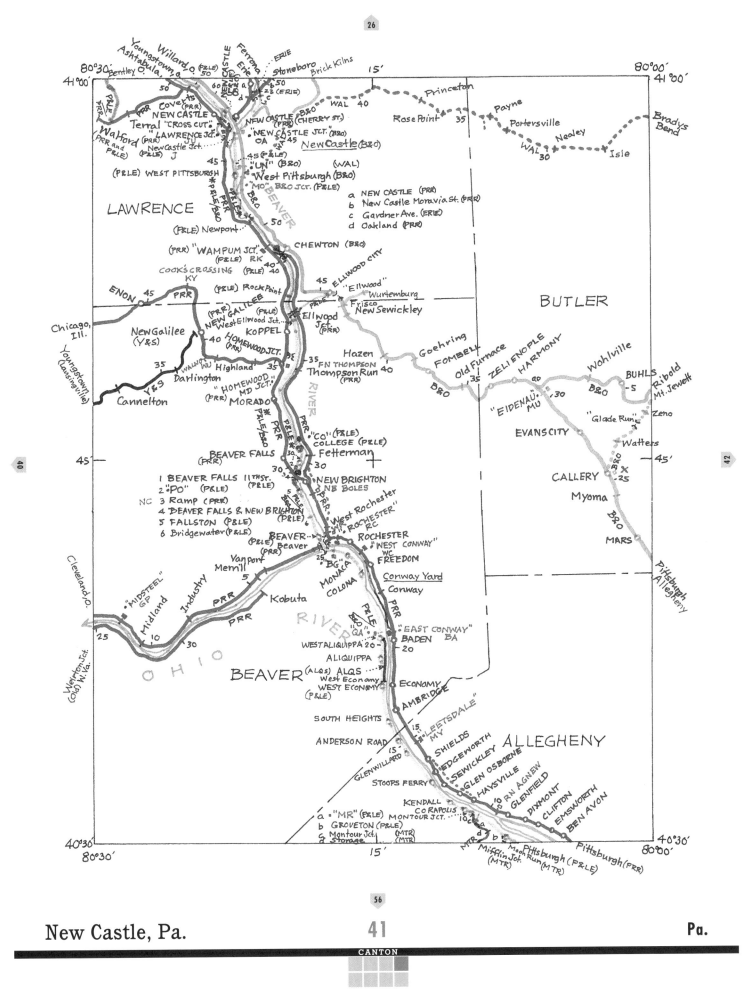

Willard O. (P&LE) NEW CASTLE Ferrona Erie ERIE
Youngstown O. 50 60 50 Stoneboro Brick Kilns
Ashtabula, Bentley O. 50 23 (ERIE)
80°30' 41°00'
(P&LE) PRR Coverts (PRR) NEW CASTLE B&O (CHERRY ST.) WAL 40 Princeton 41°00' 80°00'
NEW CASTLE NEW CASTLE JCT. (PRR) Rose Point 35 Payne
Terral "CROSS CUT" NEW CASTLE JCT. B&O Portersville Bradys Bend
Walford "LAWRENCE JCT." OA 45 New Castle (B&O) Nealey Isle
(PRR and P&LE) NewCastle Jct. "UN" (B&O) WAL 30
(P&LE) J 45 (P&LE)
(P&LE) WEST PITTSBURGH West Pittsburgh (B&O) (WAL)
"MO" B&O JCT. (P&LE)
LAWRENCE a NEW CASTLE (PRR)
b New Castle Moravia St. (PRR)
c Gardner Ave. (ERIE)
d Oakland (PRR)
(P&LE) Newport 50
(PRR) "WAMPUM JCT." CHEWTON (B&O)
(P&LE) RK 40
COOK'S CROSSING KY (P&LE) 40
ENON 45 PRR (P&LE) Rock Point 45 ELLWOOD CITY "Ellwood" BUTLER
Chicago, Ill. (PRR) NEW GALILEE (P&LE) Wurtemburg
Youngstown (Lansingville) West Ellwood Jct. Ellwood Frisco New Sewickley
New Galilee (Y&S) 40 HOMEWOOD JCT. Jct. (PRR)
35 WU Highland (PRR) 35 Hazen Goehring
WALNUT Darlington FN THOMPSON FOMBELL Old Furnace ZELIENOPLE HARMONY Wohlville
Y&S Thompson Run (PRR) 40 B&O 35 BUHLS 5 Ribold Mt.Jewett
Cannelton "HOMEWOOD 30 "EIDENAU" MU Glade Run" Zeno
(PRR) MD JCT. MORADO EVANS CITY Watters
"CO" (P&LE) 45'
P&LE/B&O COLLEGE (P&LE) CALLERY B&O X 25
BEAVER FALLS 30 Fetterman Myoma
(PRR) 30
1 BEAVER FALLS 11TH ST. (P&LE) NEW BRIGHTON MARS B&O
2 "PO" (P&LE) NB BOLES
NC 3 Ramp (PRR) West Rochester Pittsburgh Allegheny
4 BEAVER FALLS & NEW BRIGHTON (P&LE) ROCHESTER RC
5 FALLSTON (P&LE) ROCHESTER
6 Bridgewater (P&LE) BEAVER "WEST CONWAY" WC
(P&LE) Beaver FREEDOM
(PRR) Conway Yard
Van port "BG" "A" Conway
Merrill 25 MONACA
"MIDSTEEL" GP COLONA "EAST CONWAY" BA
Midland Industry PRR Kobuta WEST ALIQUIPPA 20 BADEN
Cleveland O. 10 30 PRR ALIQUIPPA 20
25 OHIO RIVER BEAVER (ALQS) ALQS
(Weirton Jct.) West Economy ECONOMY
(Old) W.Va. WEST ECONOMY (P&LE) AMBRIDGE
SOUTH HEIGHTS "LEETSDALE" MY
ANDERSON ROAD 15 SHIELDS ALLEGHENY
GLENWILLARD EDGEWORTH
15 SEWICKLEY
STOOPS FERRY GLEN OSBORNE RN AGNEW
KENDALL HAYSVILLE 10 GLENFIELD DIXMONT
a "MR" (P&LE) CORAPOLIS CLIFTON
b GROVETON (P&LE) MONTOUR JCT. 10 EMSWORTH
c Montour Jct. (MTR) BEN AVON
d Storage (MTR)
40°30' MTR Pittsburgh (P&LE) Pittsburgh (PRR) 40°30'
80°30' Mifflin Jct. Moon Run (MTR) 80°00'
15' (MTR)

New Castle, Pa.   **41**   **Pa.**

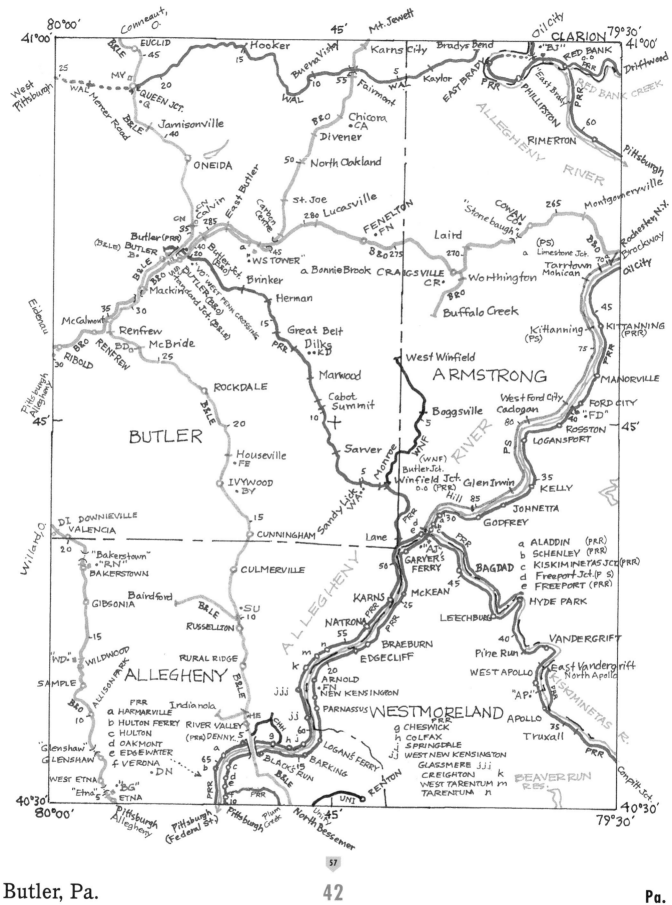

27

West Pittsburgh

Conneaut, O.

80°00' 41°00'
EUCLID •45
BLE
MV 0
WAL Mercer Road
25
QUEEN JCT. •Q
20
BLE
Jamisonville •40
ONEIDA
CN Calvin
CN 35 285
Butler (PRR)
(B&LE) BUTLER
BUTLER B•
BLE
B&O 40 20
35 30
"V" WEST PENN CROSSING
"VO" WEST PENN (B&O)
Standard Jct.(B&LE)
Mackirndard Jct.(B&LE)
McCalmont
35 30
Renfrew
B&O McBride
B&O RENFREW •25
RIBOLD
30
BUTLER

Hooker 45'
15

Buena Vista
10 55 WAL
WAL Fairmont

Karns City Bradys Bend
5 Kaylor
WAL

Mt. Jewett

B&O Chicora
•CA
Divener
50 North Oakland

St. Joe
Carbon 280 Lucasville
Centre
East Butler
45
"WS TOWER" a
Brinker
Herman
15
Great Belt
Dilks
PRR ••KD
Marwood
Cabot
Summit
10

FENELTON
•FN
B&O 275
a BonnieBrook CRAIGSVILLE
CR•
Buffalo Creek
B&O

COWAN
"Stonebaugh" 265
CO.
270
(PS)
a Limestone Jct.
Tarrtown
Mohican
Worthington

Oil City
CLARION 79°30' 41°00'
"BJ"
RED BANK
0.0
"East Brady PRR Driftwood
EAST BRADY PRR
PHILLIPSTON RED BANK CREEK
RIMERTON
60 ALLEGHENY
Montgomeryville RIVER
Rochester,N.Y.
B&O Brockway
70 OIL CITY
KITTANNING
45 (PRR)
75 PRR

Eidenau
Pittsburgh
& Allegheny
41 45'

DI DOWNIEVILLE
VALENCIA
20
"Bakerstown"
"RN"
BAKERSTOWN
Baindford
GIBSONIA
15
"WD" WILDWOOD
SAMPLE ALLISON PARK
B&O
10
PRR Indianola
a HARMARVILLE
b HULTON FERRY
c HULTON
d OAKMONT
e EDGEWATER
f VERONA
•DN
"Glenshaw"
GLENSHAW
WEST ETNA
"Etna" 5 "BG"
40°30' ETNA
80°00'
Pittsburgh
Allegheny Pittsburgh
(Federal St.)

ROCKDALE
BLE
20
Houseville
•FE
IVYWOOD
•BY
15
CUNNINGHAM
CULMERVILLE
BLE •SU
•10
RUSSELLTON
RURAL RIDGE
BLE
"HE" CHH
River Valley
(PRR) DENNY •5
g
65 15
BLACK'S RUN
BLE
PRR PRR
Pittsburgh Plum
Creek
North Bessemer

West Winfield
ARMSTRONG
Boggsville
5
WNF
(WNF)
Butler Jct.
Winfield Jct.
0.0 (PRR)
Sarver
5
Monroe

Lane

Sandy Lick WA.

"AJ"
GARVER'S
FERRY
50
KARNS
McKEAN
NATRONA PRR
55 PRR 25
m n
BRAEBURN
k EDGECLIFF
20
ARNOLD
•FN
NEW KENSINGTON
jjj PARNASSUS
jj LOGAN'S FERRY
h j
g BARKING
RENTON
UNI
45'
Uni. "A"

RIVER
Glen Irwin
Hill 85
JOHNETTA
d a 30 GODFREY
e b
PRR

BAGDAD
45
LEECHBURG

Pine Run
WEST APOLLO
"AP."
APOLLO

West Ford City
Cadogan FORD CITY
80 40 "FD"
ROSSTON
LOGANSPORT
PS
35
KELLY

a ALADDIN (PRR)
b SCHENLEY (PRR)
c KISKIMINETAS JCT(PRR)
d Freeport Jct.(P S)
e FREEPORT (PRR)
HYDE PARK
VANDERGRIFT
40 East Vandergrift
North Apollo
KISKIMINETAS R.
Truxall 35
PRR

WESTMORELAND
g CHESWICK
h COLFAX
j SPRINGDALE
jj WEST NEW KENSINGTON
GLASSMERE jjj
CREIGHTON
WEST TARENTUM m
TARENTUM n
BEAVER RUN
RES.
Compht Jct.
40°30'
79°30'

MANORVILLE
Kittanning
(PS)
45'

ALLEGHENY

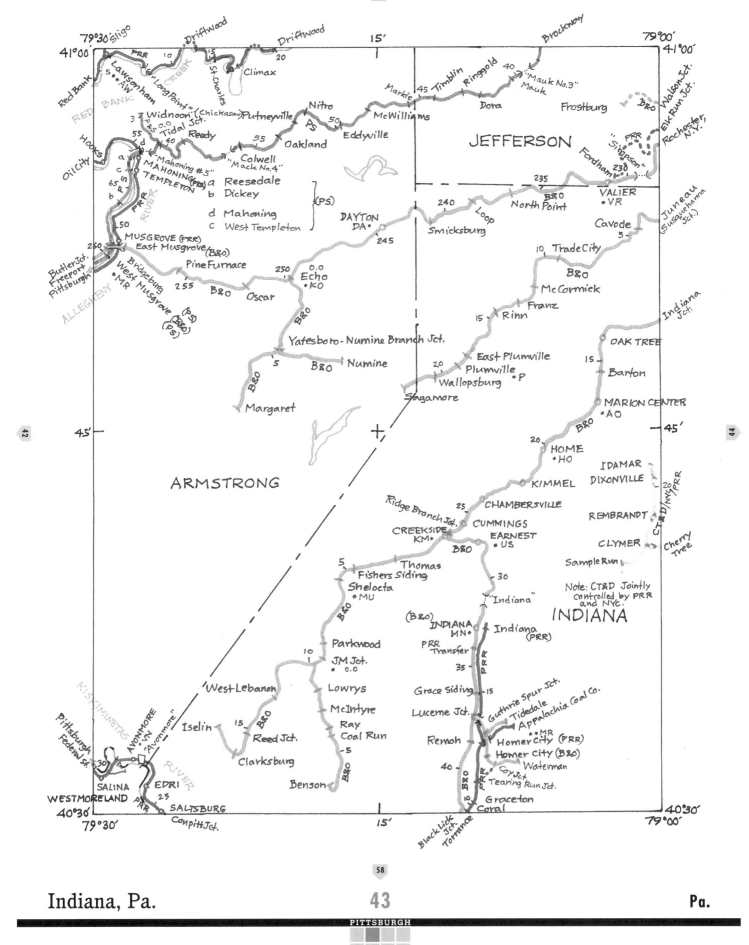

Indiana, Pa.

43

Pa.

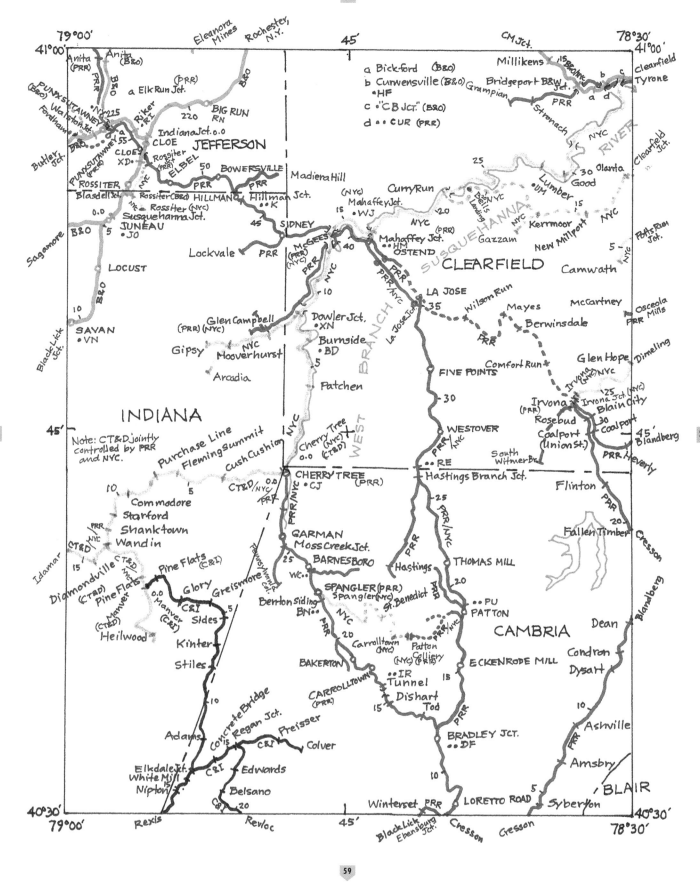

41°00' 79°00'   45'   78°30' 41°00'

Eleanora Mines   Rochester, N.Y.   CM Jct.

Anita (PRR)   Anita (B&O)   Millikens   Beehive   Clearfield
        a Elk Run Jct. (PRR)   a Bickford (B&O)   Bridgeport B&W Jct.   b c
PUNXSUTAWNEY (B&O)   NYC 225   Riker RI   b Curwensville (B&O)   Grampian   a d Tyrone
Walston Jct.   220   BIG RUN RN   • HF   Strenach   PRR
Fordham   Butler Jct.   Indiana Jct. 0.0   c • "C B Jct." (B&O)   Clearfield Jct.
PUNXSUTAWNEY (PRR) 55   CLOE   JEFFERSON   d • CUR (PRR)   NYC   RIVER
        CLOE XD   Rossiter (PRR)   Clearfield Jct.
ROSSITER   ELBEL 50   BOWERSVILLE   25   Lumber 30 Olanta
Blasdell Jct.   Rossiter (B&O)   PRR   PRR   Madiera Hill   CurryRun   B&O Landing   Good 15
0.0   Rossiter (NYC)   HILLMAN   Hillman Jct.   (NYC)   Mahaffey Jct.   NYC   Kerrmoor   NYC
JUNEAU   Susquehanna Jct.   • K   15   • WJ   NYC (PRR)   Potts Run Jct.
B&O 5   • JO   45'   SIDNEY   McGEES   40   Mahaffey Jct.   Gazzam   New Millport   5   NYC
LOCUST   Lockvale   PRR   (PRR)(NYC)   • HM   OSTEND   CLEARFIELD   Camwath
Saganore B&O   PRR   NYC   10   PRR   PRR
Black Lick Jct.   10   SAVAN • VN   Glen Campbell (PRR)(NYC)   Dowler Jct. • XN   La Jose   35   Wilson Run   Mayes   McCartney   Osceola Mills
45'   INDIANA   Gipsy   Hooverhurst NYC   Burnside • BD   La Jose Jct.   Berwinsdale   PRR   Glen Hope   Dimeling
Note: CT&D jointly controlled by PRR and NYC.   Arcadia   5   Patchen   WEST BRANCH   Comfort Run   Irvona NYC   125
Purchase Line   Cherry Tree (NYC)(CT&D)   FIVE POINTS   30   Irvona (PRR)   Irvona Jct. (NYC)   Blain City   30
Fleming Summit   Cush Cushion   0.0   WESTOVER   Rosebud   Coalport   Blandberg   45'
10   5   CT&D/NYC 0.0   CHERRY TREE (PRR)   • RE   South Witmer Br.   Coalport (Union St.)   PRR Beverly
Commodore   CT&D/NYC PRR   • CJ   Hastings Branch Jct.   Flinton
Starford   PRR NYC   GARMAN   25   PAR
Shanktown   Moss Creek Jct.   PRR   Fallen Timber   20
Idamar   CT&D NYC   Wandin   BARNESBORO   Hastings   THOMAS MILL   Cresson
15   CT&D   Pine Flats (C&I)   WC ••   20   Blandberg
Diamondville (CT&D)   Glory   Greismore   SPANGLER (PRR)   St. Benedict   PU PATTON
Pine Flats   Manver (C&I)   Berrton Siding   Spangler (NYC)   CAMBRIA   Dean
Heilwood   C&I Sides   5   BN ••   NYC   •• PU PATTON   Condron   Dysart
Kinter   20   Carrolltown (NYC)   Patton Colliery (NYC)(PRR)   ECKENRODE MILL
Stiles   10   BAKERTON   • IR Tunnel   15   10
Adams   Concrete Bridge   Regan Jct.   Preisser   CARROLLTOWN (PRR)   Dishart Tod   Ashville
Elkdale Jct.   C&I   Edwards   Colver   15   BRADLEY JCT. • DF   PRR   Amsbry
White Mill   C&I   Belsano   10   BLAIR
40°30'   Nipton   C&I 20   Winterset PRR   LORETTO ROAD   Syberton   40°30'
79°00'   Rexis   Revloc   45'   Black Lick Ebensburg Jct.   Cresson   Cresson   78°30'

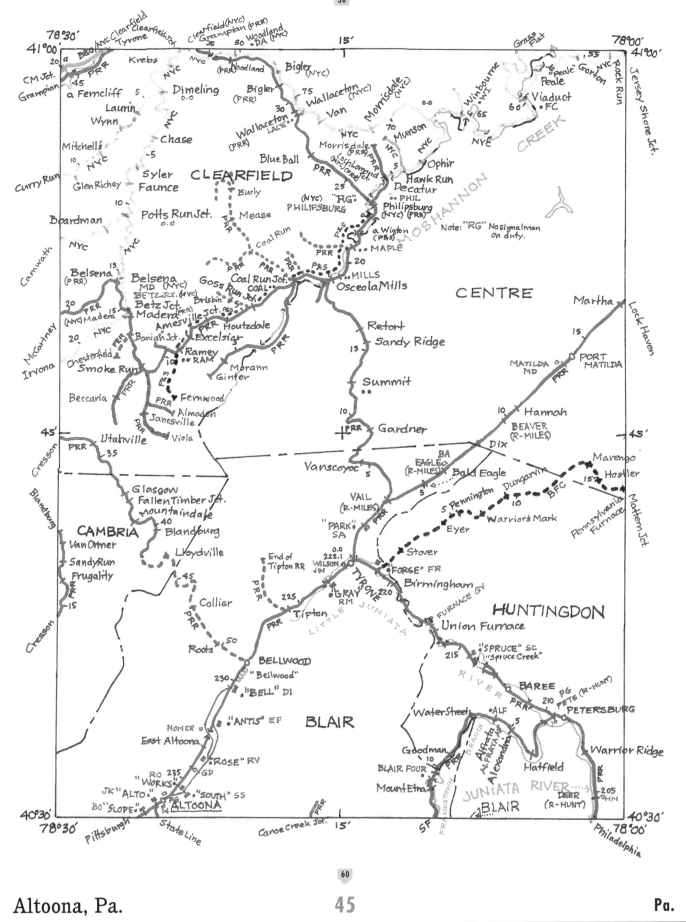

78°30'
41°00'
a B&O NYC Clearfield
Clearfield Jct.
Tyrone
Clearfield (NYC)
Grampian (PRR)
80 Woodland
DA (NYC)
15'
78°00'
41°00'
Grass Flat
55
Jersey Shore Jct.

20 a
CM Jct.
Grampian
45
a Ferncliff
Laurin
Wynn
Krebs
PRR
5
Dimeling
0.0
Bigler
(PRR)
(PRR) Woodland
Bigler
(NYC)
30
Wallaceton
(NYC)
75
Van
Morrisdale
(NYC)
0.0
70
Munson
NYC
NYC
Winbourne
WI
4 65
60
NYC
Peale
"Peale" Gorton
Rock Run
Viaduct
FC

Mitchell's
10 NYC
Chase
5
Blue Ball
Morrisdale
(PRR)
PRR
Loch Lomond
NYC/PRR
5
Ophir
NYC
Hawk Run
Decatur
CREEK

Syler
Faunce
CLEARFIELD
Burly
(NYC) "RG"
PHILIPSBURG
25
PHIL
Philipsburg
(NYC) (PRR)
Note: "RG" No signalman
on duty.
MOSHANNON

Glen Richey
10
Boardman
Potts Run Jct.
0.0
Mease
Coal Run
PRR
P&S
a Wigton
(P&S)
MAPLE

Carwath
NYC
NYC
PRR
PRR
PRR
P&S
20
MILLS

Belsena
(PRR)
15
Belsena
MD (NYC)
BETZ JCT. (NYC)
Goss Run Jct.
COAL
Coal Run Jct.
P&S
Osceola Mills
CENTRE
Martha
Lock Haven

20
McCartney
PRR
15
Madera
NYC
Betz Jct.
(NYC)
Madera
Brisbin
PRR
Amesville Jct.
PRR Houtzdale
Retort
Sandy Ridge
15
MATILDA
MD
0
15
PORT
MATILDA
PRR

Irvona
20
Chesterfield
Smoke Run
PRR
Baniah Jct.
Excelsior
Ramey
RAM
10
Morann
Ginter
Summit
BEAVER
(R-MILES)
10
Hannah

Beccaria
PRR
Fernwood
Almoden
Janesville
10
Viola
Gardner
PRR
Dix
BA
EAGLE
(R-MILES)
Bald Eagle
Marengo
15
Hastler
Pennsylvania Furnace
Mattern Jct.

45'
Cresson
PRR
Utahville
35
Vanscoyoc
5
VAIL
(R-MILES)
5
Pennington
Dungarvin
10
BFC
Warrior's Mark
Eyer
45'

Blandburg
Glasgow
Fallen Timber Jct.
mountaindale
40
Blandburg
"PARK"
SA
PRR
Stover
45'

CAMBRIA
Van Ormer
Sandy Run
Frugality
PRR
15
Lloydville
End of
Tipton RR
WILSON
0.0
222.1
JN
"FORGE" FR
Birmingham
FURNACE GY
HUNTINGDON

Collier
PRR
225
PRR
Tipton
"GRAY"
RM
220
TYRONE
Union Furnace

Cresson
45
Roots
50
PRR
BELLWOOD
LITTLE JUNIATA
"SPRUCE" SC
"Spruce Creek"
215

230
"Bellwood"
"BELL" DI
RIVER
BAREE
PRR
210
PETE (R-HUNT)
PG
PETERSBURG

"ANTIS" EF
BLAIR
Water Street
ALF
5
Warrior Ridge

HOMER
East Altoona
"ROSE" RV
Goodman
10
Alfrata
ALFRATA
Alexandria
Hatfield
PRR

"WORKS"
RO 235
GD
BLAIR FOUR
FRANKSTOWN
BRANCH
JUNIATA RIVER
DEER
(R-HUNT)
205

40°30'
78°30'
Pittsburgh
JK "ALTO"
BO "SLOPE"
"SOUTH" SS
ALTOONA
State Line
Canoe Creek Jct.
15'
Mount Etna
SF
BLAIR
Philadelphia
40°30'
78°00'

Altoona, Pa.

Pa.

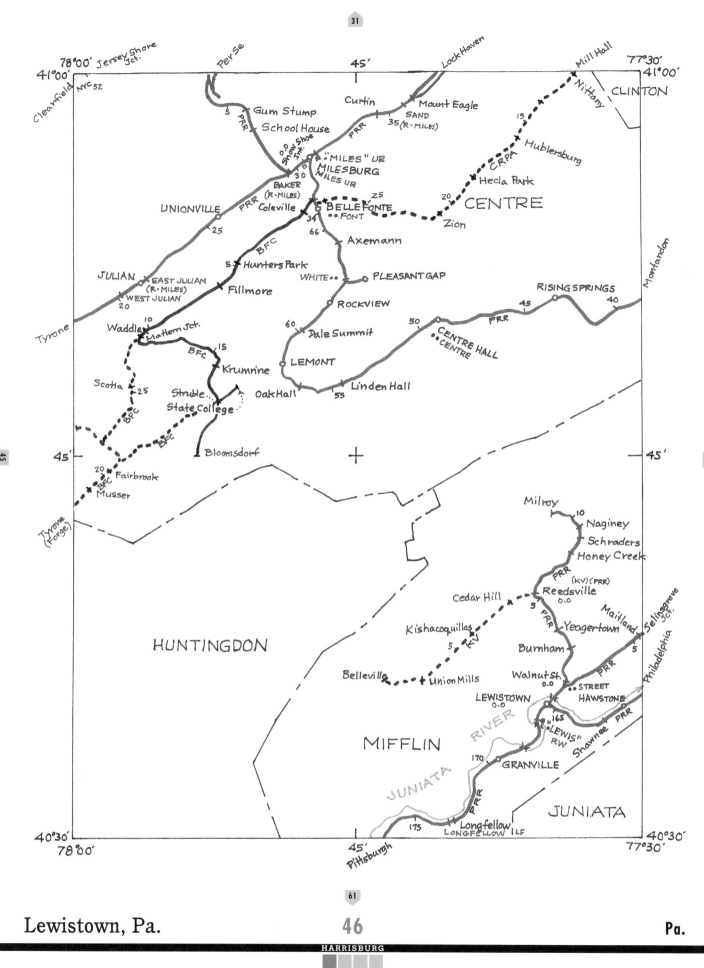

Lewistown, Pa.   46   Pa.

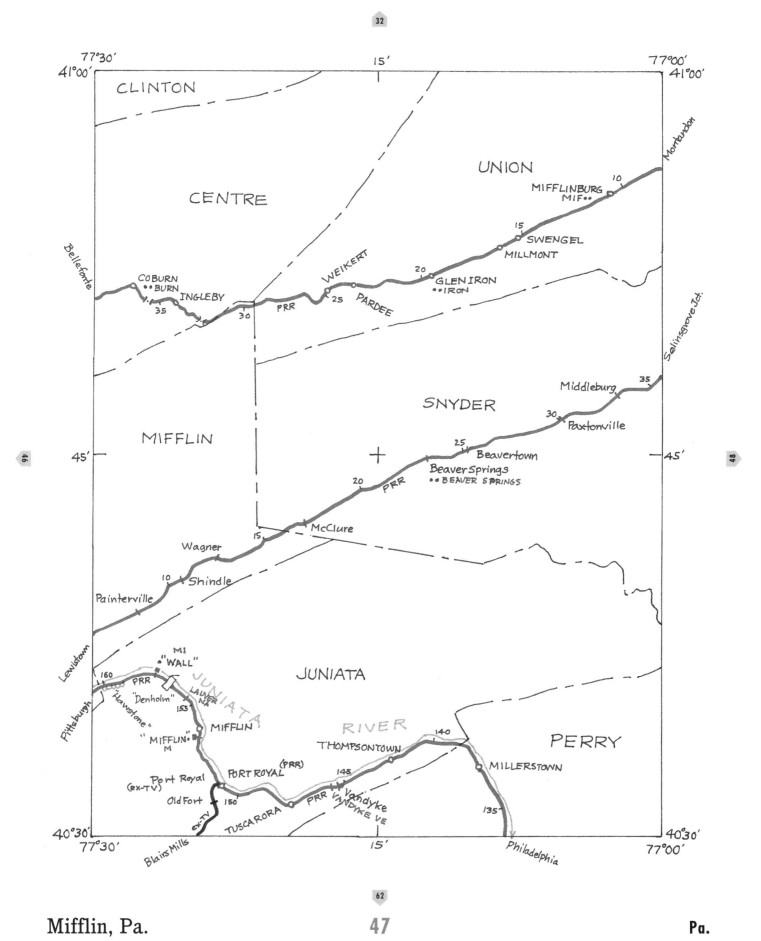

77°30'
41°00'

15'

77°00'
41°00'

CLINTON

UNION

CENTRE

Mortandon

10

MIFFLINBURG

MIF ••

15

SWENGEL

MILLMONT

Bellefonte

COBURN

•• BURN

INGLEBY

35

30

PRR

WEIKERT

25

PARDEE

20

GLEN IRON

•• IRON

Selinsgrove Jct.

35

Middleburg

SNYDER

30

Paxtonville

MIFFLIN

45'

25

Beavertown

Beaver Springs

•• BEAVER SPRINGS

20

PRR

McClure

Wagner

15

Painterville

10

Shindle

45'

Lewistown

M1

"WALL"

JUNIATA

160

PRR

LAUVER

N

"Denholm"

153

"Newstone"

"MIFFLIN"

M

MIFFLIN

JUNIATA

RIVER

140

THOMPSONTOWN

PERRY

Pittsburgh

Port Royal

(ex-TV)

PORT ROYAL (PRR)

145

MILLERSTOWN

Old Fort

150

ex-TV

TUSCARORA

PRR

Vandyke

VANDYKE VE

135

40°30'
77°30'

Blairs Mills

15'

Philadelphia

40°30'
77°00'

Mifflin, Pa.

Pa.

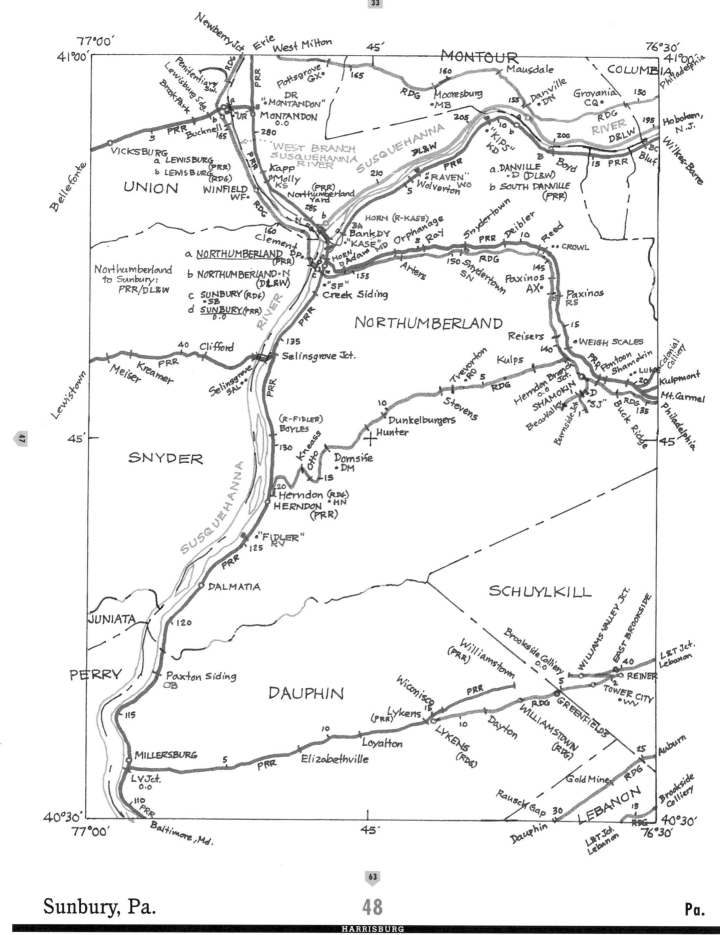

Sunbury, Pa.

48

Pa.

HARRISBURG

SUSQUEHANNA RIVER
Jamison City
Hoboken, N.J.
Wilkes-Barre
15'
Nescopeck
76°00'
41°00'
41°00'
76°30'

Scotch Valley
PRR
LUZERNE

Bloom BM
EAST BLOOMSBURG
190
25
DL&W
Mainville
30
PRR
MountainGrove
10
RockGlen
43
154
LV/PRR
Harleigh Jct.
150
NewYork
"NORCA"
RT
PRR
25
Mainville
Rock Glen
Glader
150
LV
NewYork

194 d
RDG
140
Shumans
RG Glen Jct.
Gowen
Tomhicken
(LV)
Long Run
147
LV/PRR
NewYork

North-
umberland
Sunbury
PRR
145
CA Jct.
"REDPEN"
McAuley
Mifflin
Cross Roads
"Shumans"
Fern Glen
40
GUM RUN
Tomhicken
(PRR)
150

a Rupert • RU (RDG)
b Bloomsburg (RDG)
c BLOOMSBURG • B (DL&W)
d CATAWISSA (DL&W)
e CATAWISSA (PRR)
f Catawissa (RDG)
Beaver
Valley
BY
130
135
Ranicks
Oneida
155
LV
Harwood
Jct.
29

COLUMBIA
Ferndale
125
Brandonville
BN.
Sheppton
115 Girard Manor

NORTHUMBERLAND
Ringtown
RN.
Krebs
120
RDG Port Place
BARRYVILLE
160
BARNESVILLE
105
"UG"
EASTMAHANOY
Jct.
RDG
Philadelphia

Colonial Colliery (RDG)
(PRR)
Centralia
Aristes
Aristes Jct.
Lost Creek
166 162
163 LV "WK"
160
2
105
5
Nescopeck

Luke
PRR
Kulpmont
PRR
Susque-
hanna
Colliery
(PRR)
LV/PRR
Ashland
Girard
ville
175
Packer No.5
170
Nicholas
Mahanoy Tunnel
105
Tamaqua

Luke
Newberry Jct.
Locust
25
Locust
Dale
Jct.
Ashland
upper
LV
110
105
LV/PRR
Vulcan
NEWTON
"Mahanoy
BF."

Sunbury
Fulton
130
RDG
120
Big Mine Run Jct.
GIRARDVILLE
RDG F.A.
Frackville
Head of
Grade
a Mahanoy City (LV) Buck Siding
b MAHANOY CITY (RDG) "CV"
c BUCK MOUNTAIN (")
d Morea (PRR)
e New Boston Jct. (PRR)
f Mahanoy Plane
RDG
Tuscarora
Maryd

RDG a
EXCELSIOR
LOCUST GAP SU.
Locust Summit
LOCUST SUMMIT
125
ASHLAND
(LV)
AH (RDG)
a MAHANOY PLANE
b Shenandoah
HN
Wetherill
Jct.
WH
(RDG)
100
MORRIS
Q (PRR)
Broad Mountain
Brockton

a Enterprise (RDG)
Barry
LOCUST DALE
GORDON
GO
RDG
20
RDG
15
Buck Run Jct.
Thomaston Jct.
Richardson Jct.
New
Castle Col.
s Patterson St. Jct.
St. Clair
Pine Forest Jct.
Middleport
Alliance Jct.
RDG
NEW PHILADELPHIA
5
Cumbola
SCHUYLKILL

POTTSVILLE
a "St. Clair" tunnel (PRR)
b ULMER (PRR)
c Pottsville (PRR)
d POTTSVILLE (RDG)
Phoenix Park
DO
Mine Hill
GAP
Buckley
St.
Clair
60
Eagle Hill Jct.
Port Carbon
"CA" MILLCREEK JCT.
95
PRR

Otto Col.
Llewellyn
Branchdale
Steins
John Veith
Jct.
West
Jct.
West
150
PRR
5
3
"MJ" POTTSVILLE JCT.
CARBON
SC
(PRR)
Schuylkill Haven Jct.
Spring Garden Jct.
Orwigsburg
140
LV

Good Spring Col.
GOOD SPRING
FG
Hazelbrook Jct.
WEST END
DONALDSON
Newtom
TREMONT
QN
Swatara Jct.
LV Silverton
(RDG)
Blackwood (RDG)
(LV)
155
5
RDG
Becks
DALE
Adamsdale
DREHERSVILLE
85
New York
Newberry Jct.
LV

South Good Spring
KEFFERS
Rausch Cr.
RDG
30
RG
Tremont Jct.
CRESSONA
West Cressona
Mine Hill Crossing
(RDG) J
Landingville
85
Millers
Molino
BN
85
Auburn
RDG

Brookside
Colliery
RDG
5
RDG
Lorberry
LORBERRY JCT.
SCHUYLKILL HAVEN
SC
(RDG)
AUBURN
AR
SCHUYLKILL
80
"PN"

Dauphin
Outwood
Beuchler
20
25
NORTH PINE GROVE
NG
PINE GROVE
GO
RDG
Jefferson
Aucheys
Roeders
Summit
Stonemont
Moyers
STONY CREEK
PRR 80
(RDG) NY
PRR PORT CLINTON
Philadelphia

Lebanon
LEBANON
RDG IRVING
SUEDBURG
20
EXMOOR
Auchenbach
Stanhope
15
Rock
RDG 10
Berne
Perry
72
Newberry Jct.
Philadelphia

40°30'
76°30'
15'
BERKS
40°30'
76°00'

Pottsville, Pa.

Pa.

HARRISBURG

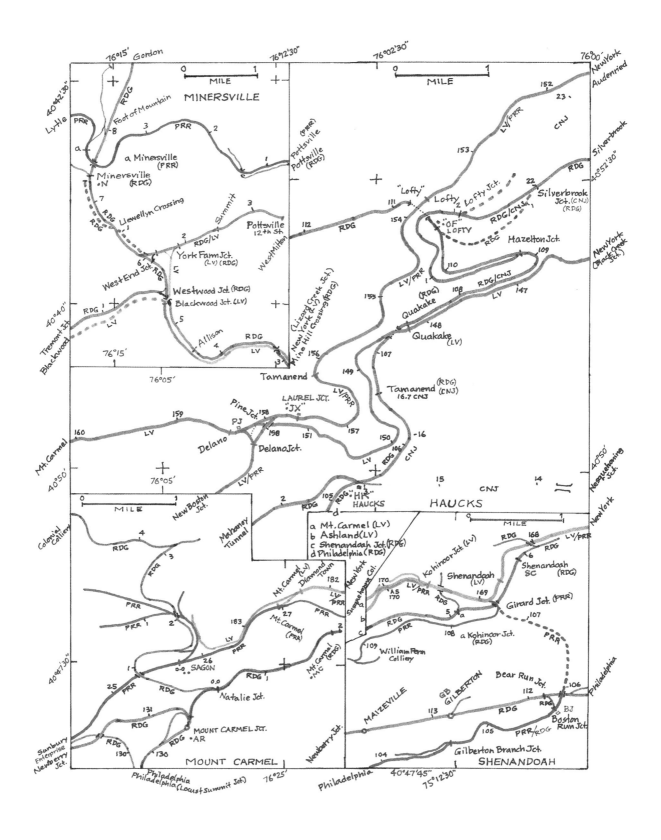

49A Mount Carmel, Pa. **49B** Minersville, Pa.

49C Shenandoah, Pa. **49D** Haucks, Pa.

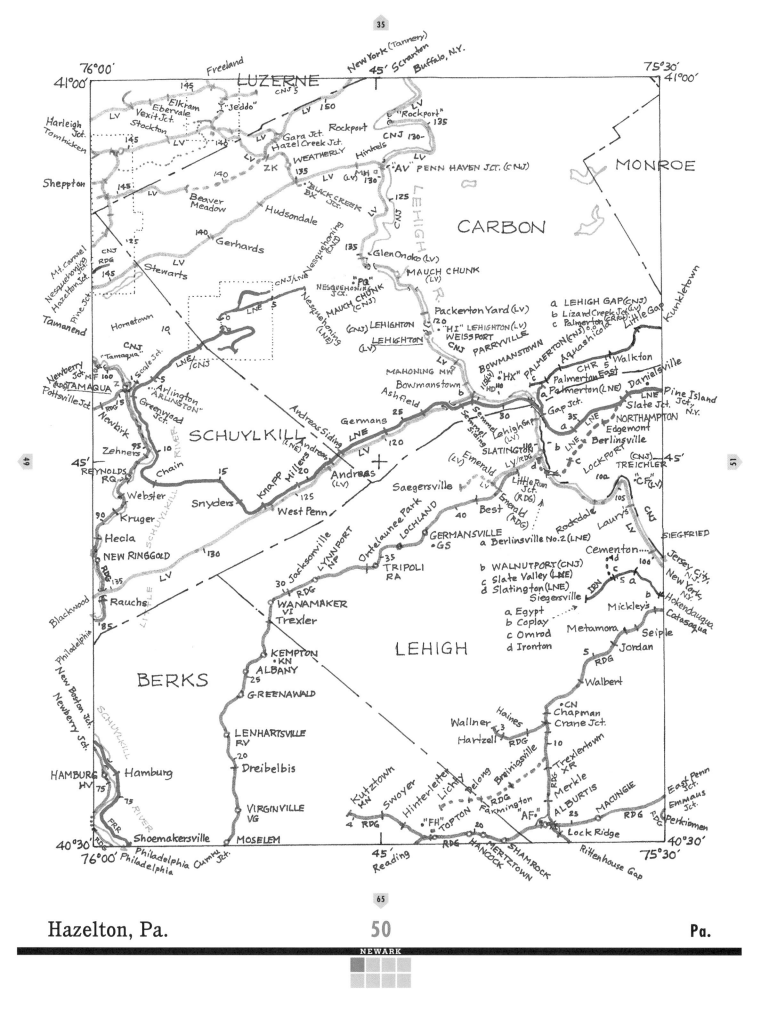

New York (Tannery)
45' Scranton
Buffalo, N.Y.
Freeland
LUZERNE

76°00'
41°00'
75°30'
41°00'

145
CNJ 5
LV 150
Elkram
Ebervale
Vexit Jct.
Stockton
LV
"Jeddo"
LV
"Rockport"
135
LV

Harleigh Jct.
Tombicken
145
LV 140
Gara Jct. Rockport
Hazel Creek Jct.
CNJ 130
CNJ 130

MONROE

Sheppton
145
LV
ZK
135
WEATHERLY
Hinkels
MH
LV
"AV" PENN HAVEN JCT. (CNJ)
(LV) 130

Beaver Meadow
140
BLACK CREEK BX
LV
125
CNJ

Hudsondale

Mt. Carmel
Nesquehoning
Hazelton Jct.
CNJ RDG
145
125
LV
140 Gerhards
BLACK CREEK JCT.
135
GlenOnoko (LV)

CARBON

Stewarts
CNJ
LV
Nesquehoning (CNJ)
MAUCH CHUNK (LV)

Tamanend
Hometown
10
CNJ
LNE 5
"PQ"
NESQUEHONING JCT.
MAUCH CHUNK (CNJ)
Packerton Yard (LV)
Nesquehoning (LNE)
(CNJ) LEHIGHTON
"HI" LEHIGHTON (LV)
WEISSPORT
LEHIGHTON (LV)
CNJ PARRYVILLE
120

a LEHIGH GAP (CNJ)
b Lizard Creek Jct. (LV)
c Palmerton (CRR)
Aquashicola
Little Gap
Kunkletown

Newberry Jct. MF 100
"TAMAQUA"
Z
LNE/CNJ
Scale Jct.
5
"Arlington"
ARLINGTON
Greenwood Jct.
Pottsville Jct.
Newkirk
15
RDG
MAHONING MW
HD 10
"HX"
HD
BOWMANSTOWN
PALMERTON (LV)
Palmerton East
a Palmerton (LNE)
Walkton
CHR 5
Danielsville
Pine Island Jct. N.Y.
LNE
NORTHAMPTON
Edgemont
Berlinsville
TREICHLER

Zehners
95
10
Chain
SCHUYLKILL RIVER
REYNOLDS RQ
45'
Webster
Snyders
KNAPP
Millers
15
125
West Penn
Andreas Siding
SCHUYLKILL
Germans
25
Semmel
Semmel Siding
LehighGap
SLATINGTON
Emerald (LV)
Andreas (LNE)
LV
Andreas (LV)
120
Saegersville
Ashfield
30
35
LNE
100a
Gap Jct.
LV/RDG
Little Run Jct. (RDG)
Emerald (RDG)
Rockdale
Laury's
105
CNJ
SIEGFRIED
"CF" (LV)
45'

Kruger
90
Hecla
NEW RINGGOLD
130
RDG
135
LV
LOCHLAND
Ortelaunee Park
40
Best
Jersey City, N.J.
New York, N.Y.
Hokendauqua
Catasauqua

Rauchs
Blackwood
85
Philadelphia
New Boston Jct.
Newberry Jct.
JacksonVille
30
RDG
LYNNPORT NP
35
TRIPOLI RA
GERMANSVILLE
GS
a Berlinsville No.2 (LNE)
Cementon
d
c
IBN
s a
106
Mickley's
Metamora
Seiple
Jordan

SCHUYLKILL RIVER
WANAMAKER VI
Trexler
b WALNUTPORT (CNJ)
c Slate Valley (LNE)
d Slatington (LNE)
Siegersville
a Egypt
b Coplay
c Omrod
d Ironton
5
RDG

BERKS
KEMPTON
KN
ALBANY
25
GREENAWALD
LEHIGH
Walbert
CN
Chapman
Crane Jct.
Haines
Wallner
Hartzell
3
RDG
10
Trexlertown XR
Merkle
ALBURTIS
MACUNGIE
East Penn Jct.
EMMAUS Jct.

LENHARTSVILLE RV
20
Dreibelbis
Kutztown
KN
Swoyer
Hinterleiter
Lichty
Delong
Breinigsville
"AF."
RDG
25
Lock Ridge
RDG
RDG
Perkiomen

HAMBURG HV
Hamburg
75
75
VIRGINVILLE VG
"FH" TOPTON
20
MERTZTOWN
SHAMROCK
Rittenhouse Gap
East Penn Jct.

40°30'
76°00' Philadelphia
Philadelphia
Catawissa Jct.
PRR
Shoemakersville
MOSELEM
45' Reading
RDG HANCOCK
75°30'
40°30'

Hazelton, Pa.
50
Pa.

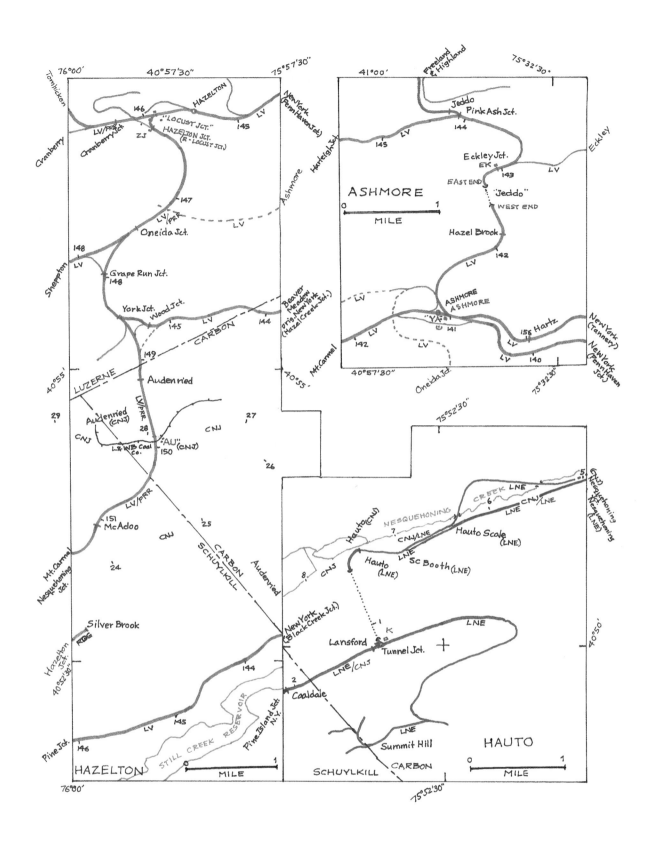

50A Hazelton, Pa. **50B** Ashmore, Pa. **50C** Hauto, Pa.

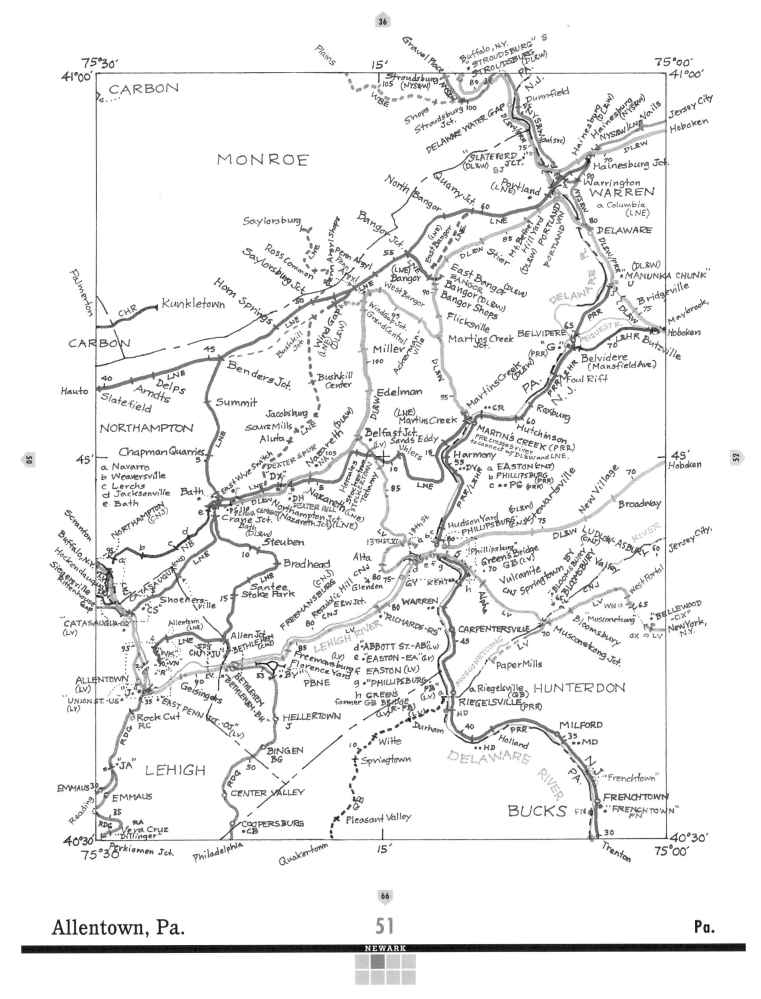

Allentown, Pa.   51   Pa.

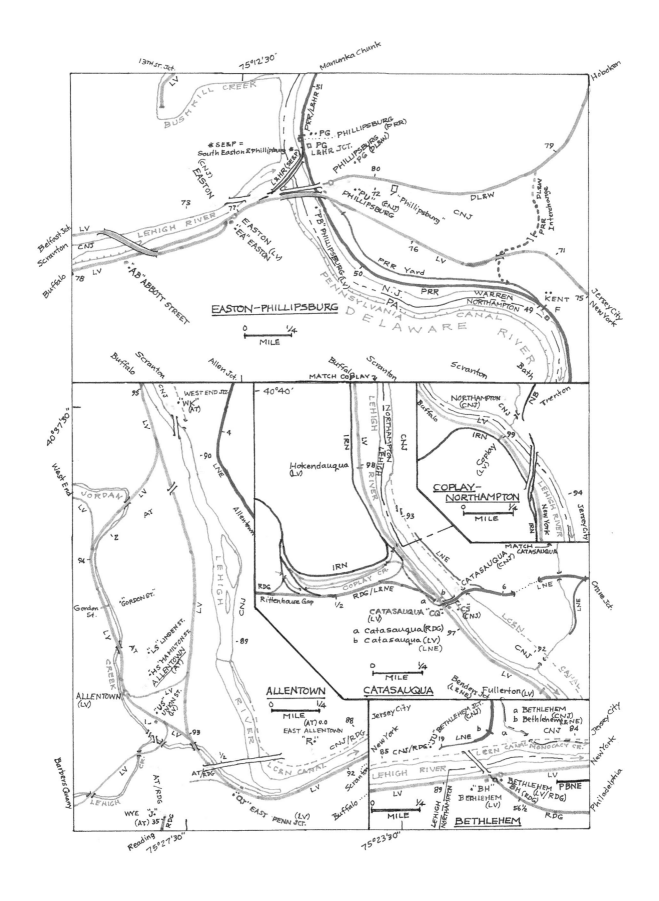

51A Easton, Pa.-Phillipsburg, N.J.   **51B** Allentown, Pa.

51C Catasauqua, Pa.   **51D** Coplay-Northampton, Pa.   **51E** Bethlehem, Pa.

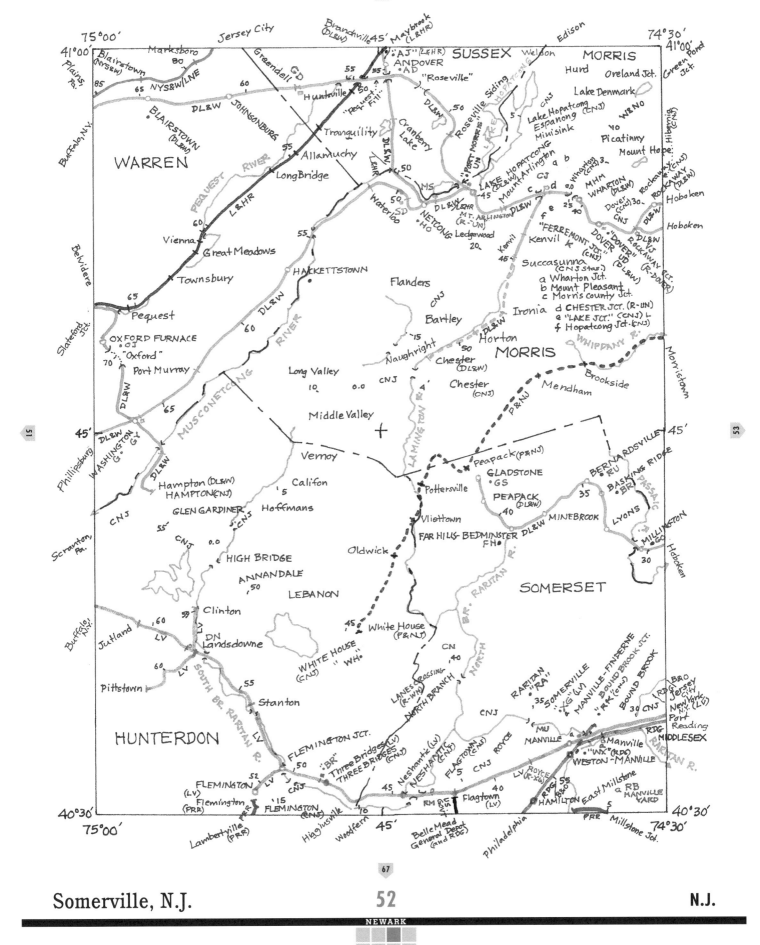

Somerville, N.J.

N.J.

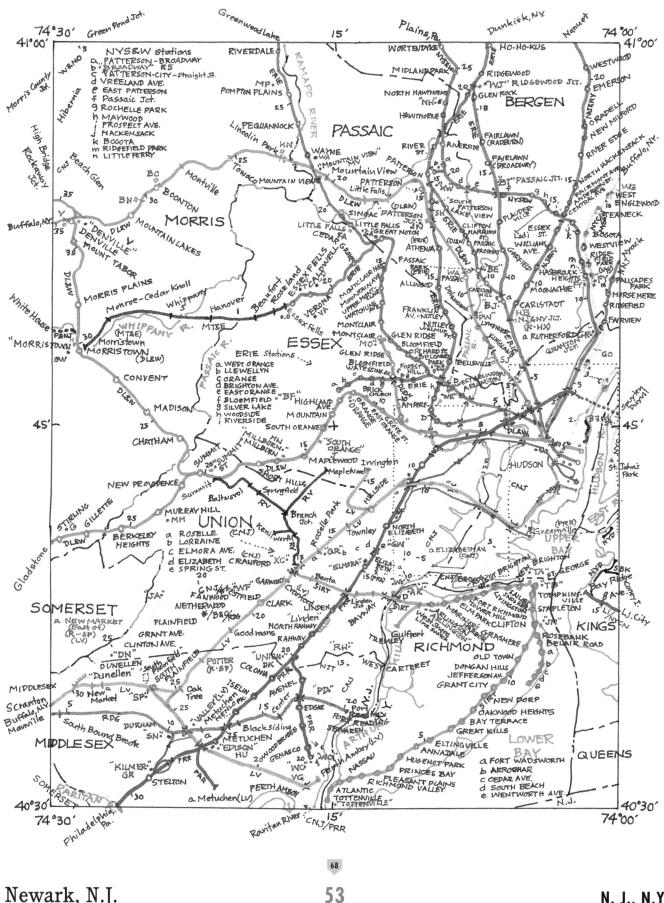

38

68

Newark, N.J.   **53**   **N. J., N.Y.**

NEWARK

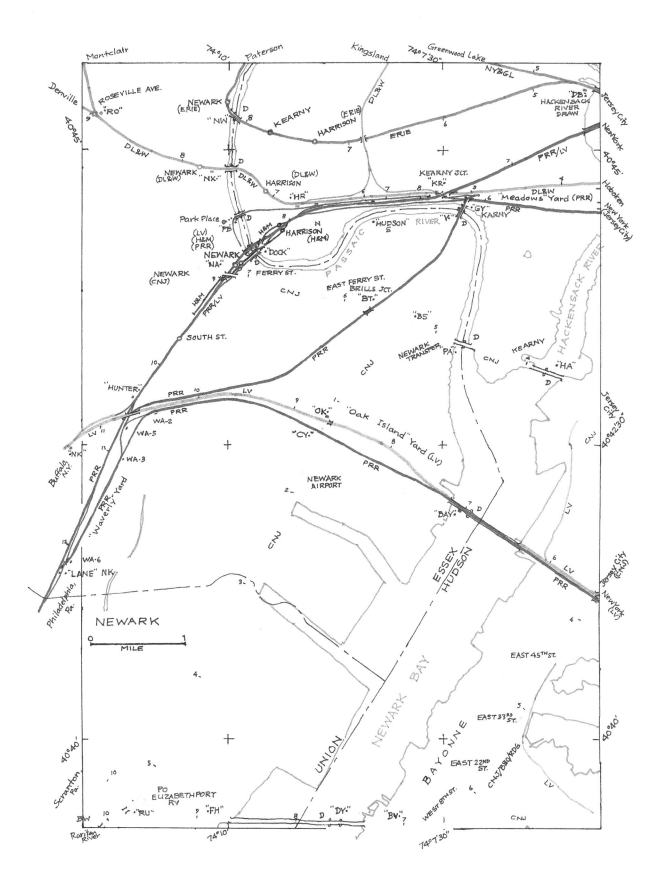

53A Newark, N.J.

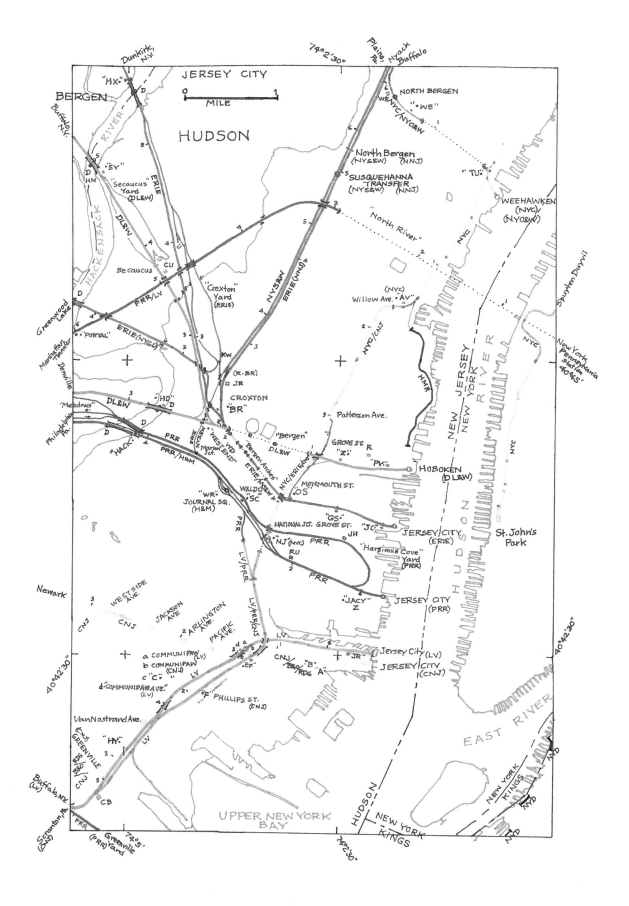

JERSEY CITY

HUDSON

0   MILE   1

BERGEN

"HX"

Dunkirk, N.Y.

HACKENSACK RIVER

Buffalo, N.Y.

ERIE

DL&W

"SY"

HN

Secaucus Yard (DL&W)

NORTH BERGEN

NY&W/NYO&W

"WE"

North Bergen (NYS&W) (NNJ)

SUSQUEHANNA TRANSFER (NYS&W) (NNJ)

"North River"

"TU"

WEEHAWKEN (NYC) (NYO&W)

NYC

Spuyten Duyvil

Secaucus

CU

PRR/LV

"Croxton" Yard (ERIE)

NYS&W

ERIE (NNJ)

(NYC) Willow Ave. "AV"

NYC/CNJ

NEW JERSEY / NEW YORK

NEW YORK Pennsylvania Station 40 45'

NYC

Greenwood Lake

Denville

"PORTAL"

ERIE (NYGL)

KW

(K-BR) JR

"Meadows"

DL&W

"HD" D

CROXTON "BR"

HMR

Paterson Ave.

Philadelphia, Pa.

D

"HACK"

PRR

ERIE NYS&W "WEST END"

Marion Jct.

WD

Bergen Arches ERIE/NYC/NYS&W

"Bergen" DL&W

GROVE ST. "Z" R

HOBOKEN (DL&W)

HUDSON RIVER

"PV"

Manhattan Transfer

PRR/H&M

D

MONMOUTH ST.

NYC/ERIE/CNJ

"OS"

St. John's Park

"WR" JOURNAL SQ. (H&M)

WALDOW "SC"

NATIONAL JCT. GROVE ST.

"GS"

"JC"

JERSEY CITY (ERIE)

PRR

"NJ" (NYC) RU

"Harsimus Cove" Yard (PRR)

Newark

WEST SIDE AVE.

JACKSON AVE.

ARLINGTON AVE.

PACIFIC AVE.

LV/PRR

LV/PRR/CNJ

"JACY" Z

JERSEY CITY (PRR)

CNJ

CNJ

a COMMUNIPAW (LV)
b COMMUNIPAW (CNJ)
c "C"

"COMMUNIPAW AVE." (LV)

LV

CF

"F" PHILLIPS ST. (CNJ)

LV

CNJ/RDG "B" RDG A

"JR"

Jersey City (LV)

JERSEY CITY (CNJ)

VanNostrand Ave.

CNJ/RDG/CNJ GREENVILLE

"HY"

LV

Buffalo, N.Y. (LV)

Scranton, Pa. (CNJ)

CB

PRR

Greenville (PRR) Yard

74 5'

UPPER NEW YORK BAY

HUDSON KINGS

NEW YORK KINGS

EAST RIVER

NEW YORK KINGS

NYD

74 2'30"

40 42'30"

53B Jersey City, N.J.

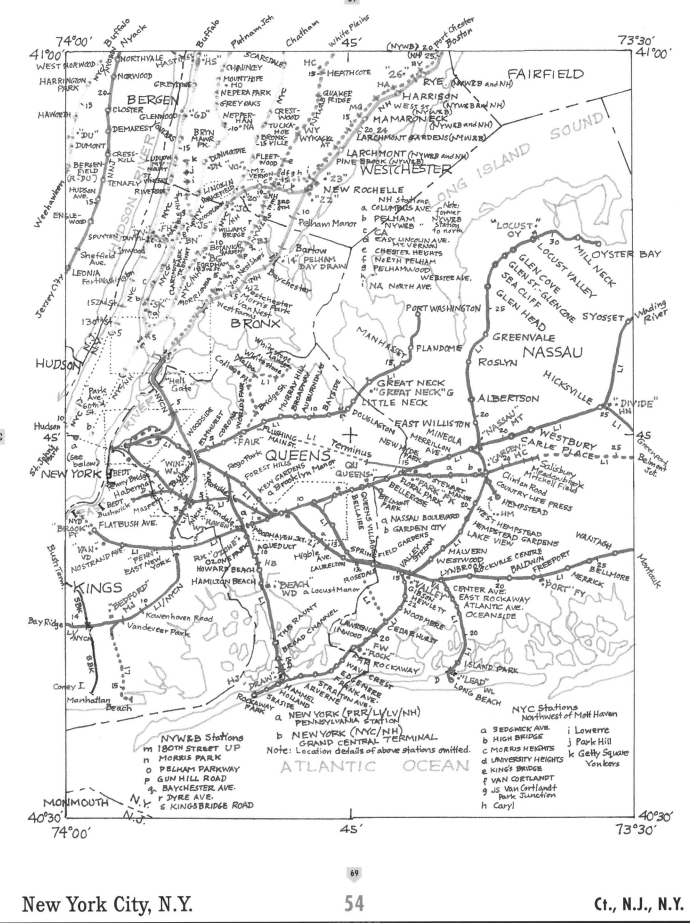

New York City, N.Y.

54

Ct., N.J., N.Y.

NEW YORK

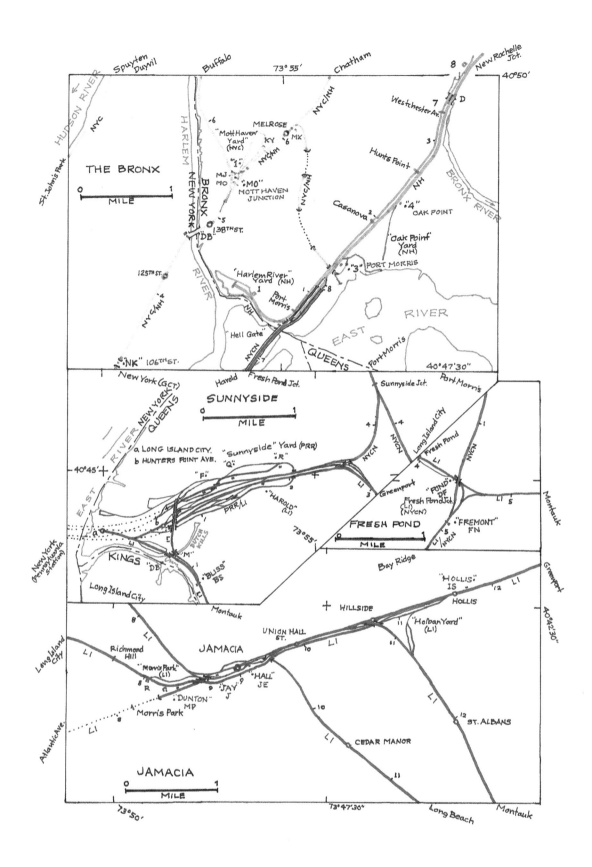

54A The Bronx, N.Y. **54B** Sunnyside, N.Y.

54C Fresh Pond, N.Y. **54D** Jamaica, N.Y.

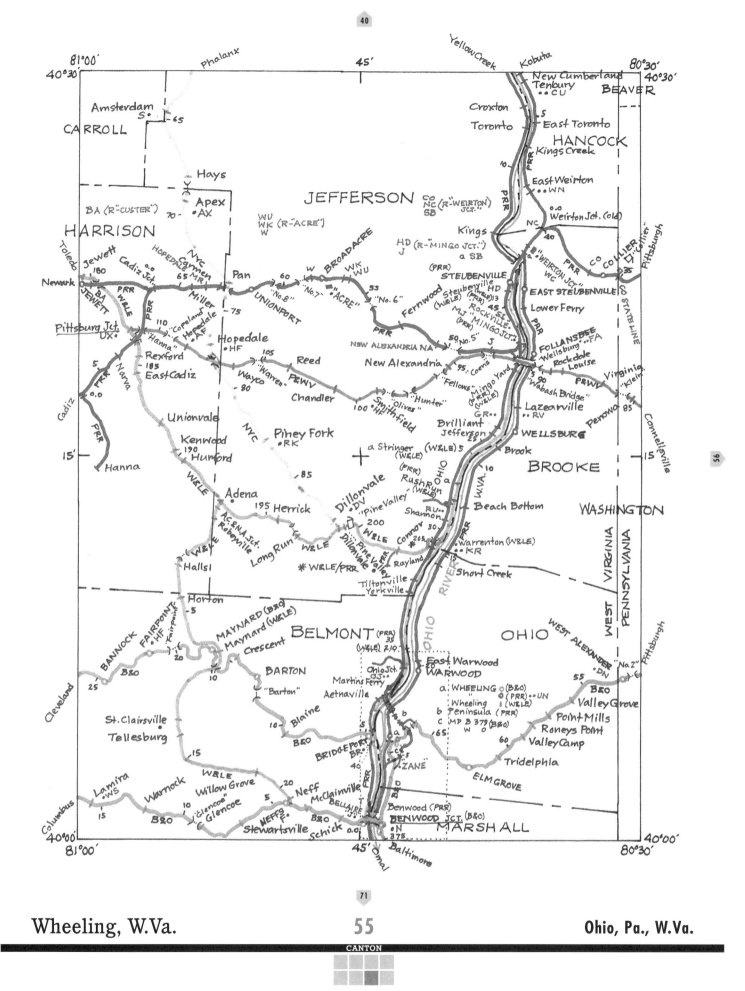

Wheeling, W.Va.   **55**   Ohio, Pa., W.Va.

CANTON

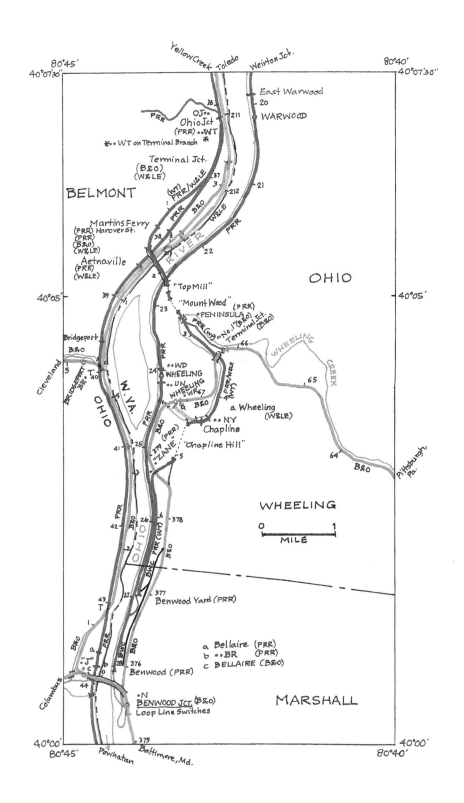

55A Wheeling, W. Va.

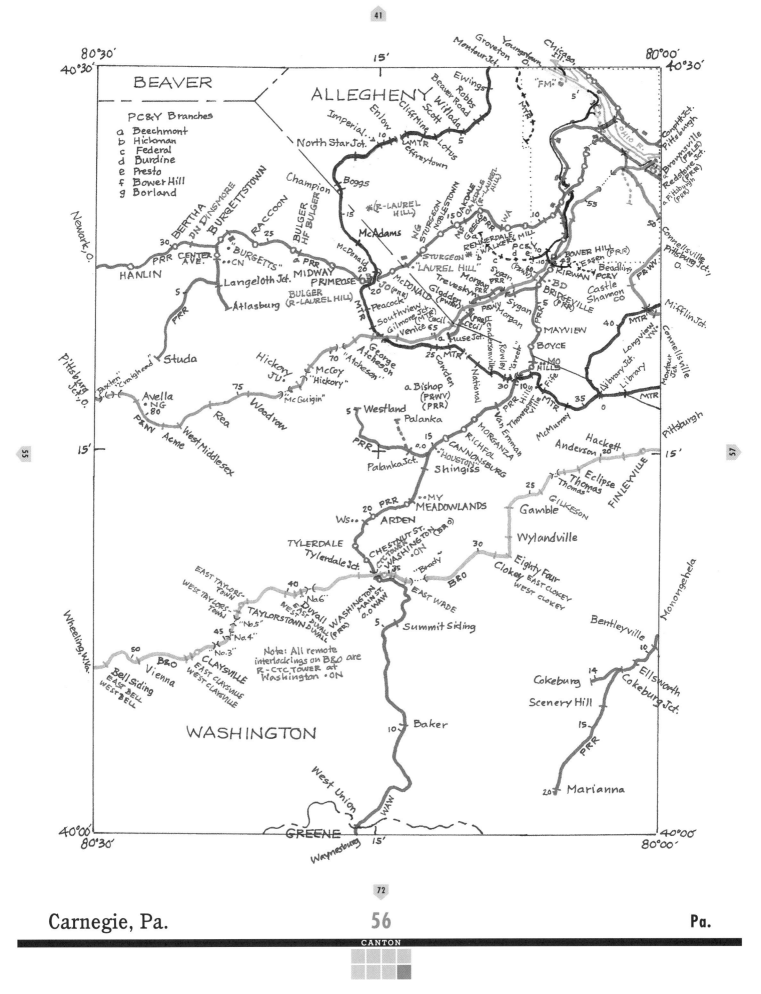

BEAVER

PC&Y Branches
a Beechmont
b Hickman
c Federal
d Burdine
e Presto
f Bower Hill
g Borland

ALLEGHENY

80°30'
40°30'

15'

80°00'
40°30'

Groveton
Montour Jct.
Youngstown
Chicago
Ill.
"FM"
5'
Corupth Jct.
Pittsburgh
Ewings
Rabbs
Beaver Road
Witfada
Scott
Brownsville
(P&LE)
Redstone Jct.
a Pittsburgh
(PRR)
Imperial
Enlow
Cliffmine
Lotus
"MTR"
Jeffreytown
Ohio R.
Connellsville
Pittsburgh
O.
North Star Jct.
10
Boggs
McAdams
*(R-LAUREL
HILL)
15
NG
Sturgeon
Noblestown
OAKDALE
OA-KPAUS
(R-LAUREL
HILL)
WA
50
53
Gregg PRR
RENKERDALE MILL
WALKER'S MILL
PC&Y.O
10
BD
BOWER HILL (PRR)
Essen
Beadling
PC&Y
P&WV
Champion
BULGER
HF BULGER
McDonald
STURGEON
"LAUREL HILL"
b
X PC&Y
KIRWAN
Castle
Shannon
CO
BERTHA
DN DINSMORE
BURGETTSTOWN
RACCOON
25
BULGER
PRR
CN
Langeloth Jct.
MIDWAY
PRIMROSE
20
McDONALD
JO (PRR)
20
Morgan
Treveskyn PRR
Gladden
(P&WV)
Sygan
Morgan
PRR
P&WV
Sygan
BRIDGEVILLE
5 (PRR)
MAYVIEW
Mifflin Jct.
30
CENTER
AVE.
BURGETTS
PRR
5
BULGER
(R-LAUREL HILL)
Peacock
Southview
Gilmore (MTR)
Venice 65
P&WV Morgan
Cecil
Hendersonville
Rowley
"Creek"
BOYCE
MO
HILLS
Longview
VN.
40
MTR
Connellsville
VN.
Newark, O.
HANLIN
PRR
5
Atlasburg
Studa
George
Atcheson
"Atcheson"
70
Muse Jct.
a
Cowden
National
25
MTR
30
Fife
10 15
McMurray
MTR
35
Library Jct.
Library
MTR
Montour
Jct.
Pittsburg
Jct. O.
Baxter
"Craighead"
Hickory
JU.
McCoy
"Hickory"
"McGuigin"
a Bishop
(P&WV)
(PRR)
Van Emman
PRR Hills
Thompsonville
Hackett
Anderson 20
Pittsburgh
15'
Avella
NG
80
75
Woodrow
Rea
Acme
West Middlesex
Westland
Palanka
15
MORGANZA
RICHFOL
0.0
CANNONSBURG
"HOUSTON"
Shingiss
25
Eclipse
Thomas
"Thomas"
FINLEYVILLE
15'
55
P&WV
PRR
Palanka Jct.
20
PRR
MY
MEADOWLANDS
Ws
ARDEN
GILKESON
Gamble
Wylandville
Monongahela
57
TYLERDALE
Tylerdale Jct.
CHESTNUT ST.
CTC TOWER
WASHINGTON
ON
"Brady"
30
Eighty Four
Clokey EAST CLOKEY
WEST CLOKEY
B&O
EAST TAYLORS-
TOWN
WEST TAYLORS-
TOWN
40
"No.6"
E.Duvall
WEST DUVALL
TAYLORSTOWN DUVALL
45
"No.5"
EAST DUVALL
"No.4"
WASHINGTON
MAIN ST.
0.0 WAW
(PRR)
EAST WADE
Summit Siding
Bentleyville
10
Wheeling, W.Va.
50
B&O
Vienna
"No.3"
CLAYSVILLE
EAST CLAYSVILLE
WEST CLAYSVILLE
Note: All remote
interlockings on B&O are
R-CTC. TOWER at
Washington • ON
5
Cokeburg
14
Cokeburg Jct.
Ellsworth
Bell Siding
EAST BELL
WEST BELL
Scenery Hill
15
PRR
WASHINGTON
10
Baker
West Union
WAW
20
Marianna
40°00'
80°30'
GREENE
15'
Waynesburg
40°00'
80°00'

Carnegie, Pa.   56   Pa.

CANTON

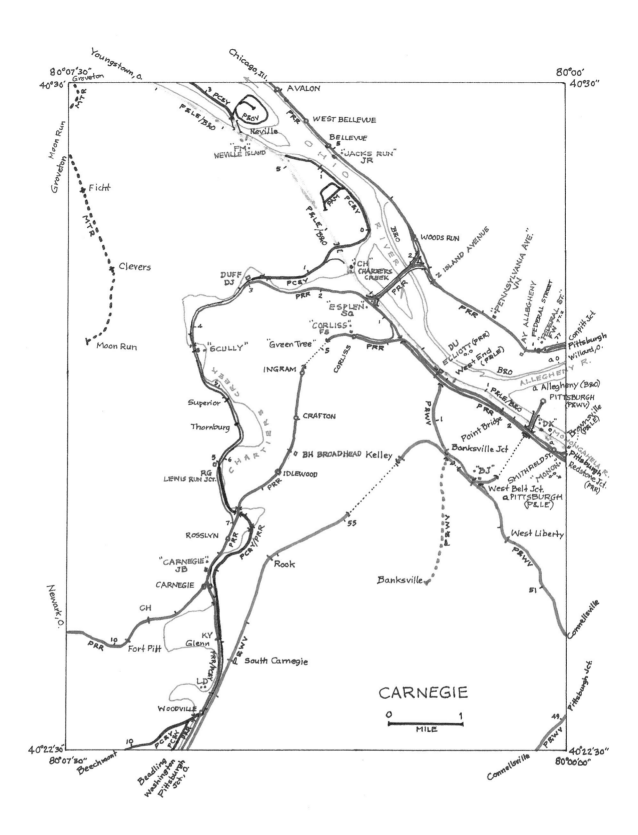

56A Carnegie, Pa.

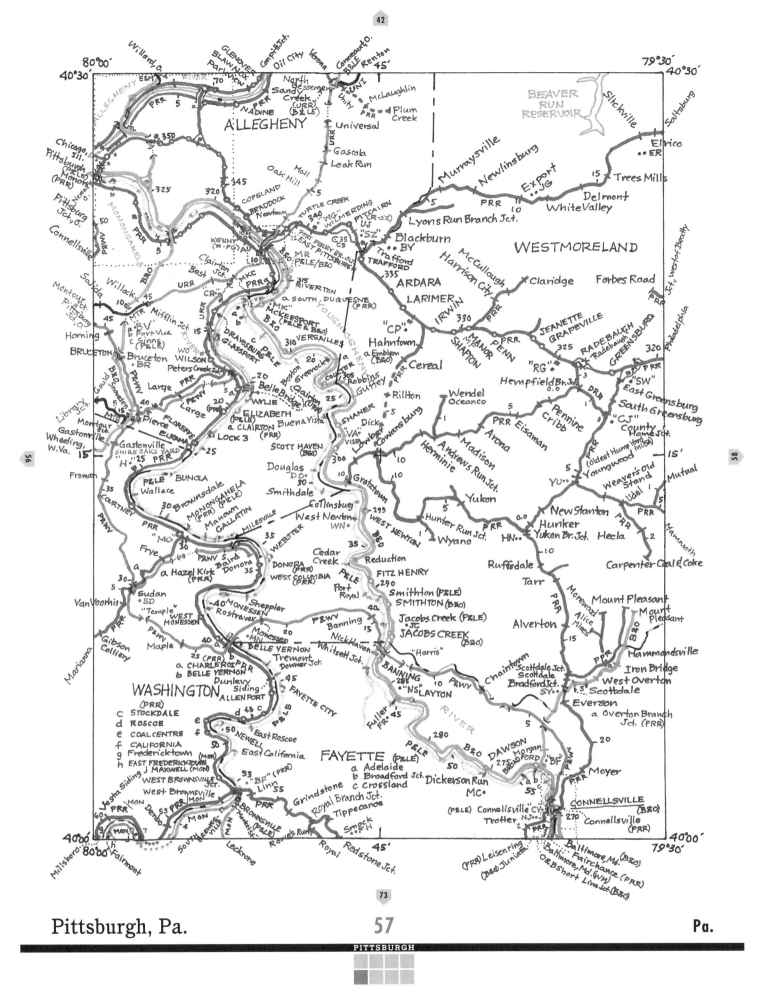

Pittsburgh, Pa.

Pa.

40°30' 80°00'

79°52'30" 40°30'

Willard, O.

ETNA
"BG"
E&M
Pine Creek
B&O
Sharpsburg
ETNA
"UV"
ASPINWALL
Etna
PRR
Conpitt Jct.
4
ALLEGHENY RIVER
(R-UV)
PRR
5
PRR
70
"CZ"
Oil City
54th St.
Coleman
B&O
PRR
VI
2
7
PRR
MILLVALE
(B&O)
NB
43rd St. Yard
BRILLIANT
"Y"
Millvale
(PRR)
36th St. Yard
7S
2
1
Stock
Yards
"FY" 33RD ST. VIADUCT (B&O)
PRR
"CQ"
328
B&O
BEN VENUE
"DV"
EAST LIBERTY
"CM"
28th
St.
PRR
350
ROUP
HOMEWOOD
Federal St.
Chicago,
Ill.
NA
SHADYSIDE
PRR
B&O PRR
b
PRR
Pittsburgh Yards
"BU"
"Schenley"
WILKINSBURG
Allegheny
"UF"
US.
PH
PITTSBURGH (PRR)
SC
SCHENLEY
EDGEWOOD
No.1"
a PITTSBURGH (B&O)
b 11TH ST. (PRR)
WK
WK
(R-CM)
a
FOURTH
AVE.
Pittsburgh
Monon
327
B&O
SWISSVALE
345
"MONON"
(PRR)
22ND
ST.
P&LE
"J&L"
B&O
LAUGHLIN JCT.
"GN"
YJ
SOUTH SIDE
PRR
M&AC
325
Dennison
320 B&O PRR
HAWKINS
30TH ST.
34TH
ST.
(P&LE)
(R-BK)
MARION JCT.
MJ
P&LE
B&O
Philadelphia
BECKS RUN
"OB"
HAZELWOOD
RIVER
URR
URR
PRR
8
Baltimore,
Md.
BECKS RUN
"BK"
(P&LE)
GLENWOOD
GLENWOOD JCT.
HM
MONONGAHELA
HOMESTEAD
(P&LE)
HOMESTEAD (PRR)
MUNHALL
9
a
c
b
Connellsville
(P&LE)
LUCAS
(P&LE)
(R-BK)
"WJ"
"Hays"
5
MESTA (PRR)
West Homestead (P&LE)
a RANKIN (B&O)
b RANKIN (P&LE)
c Green Springs (PRR)
Monongahela Jct.
Redstone Jct. (URR)
(PRR)
Pittsburgh Jct.
50
HAYS
(PRR)
5
6

PITTSBURGH

0   1
MILE

P&WV
Connellsville
Rand
Streets Run
Branch Sw.
B&O

40°22'30" 80°00'

Wheeling, W.Va.

79°52'30" 40°22'30"

56A

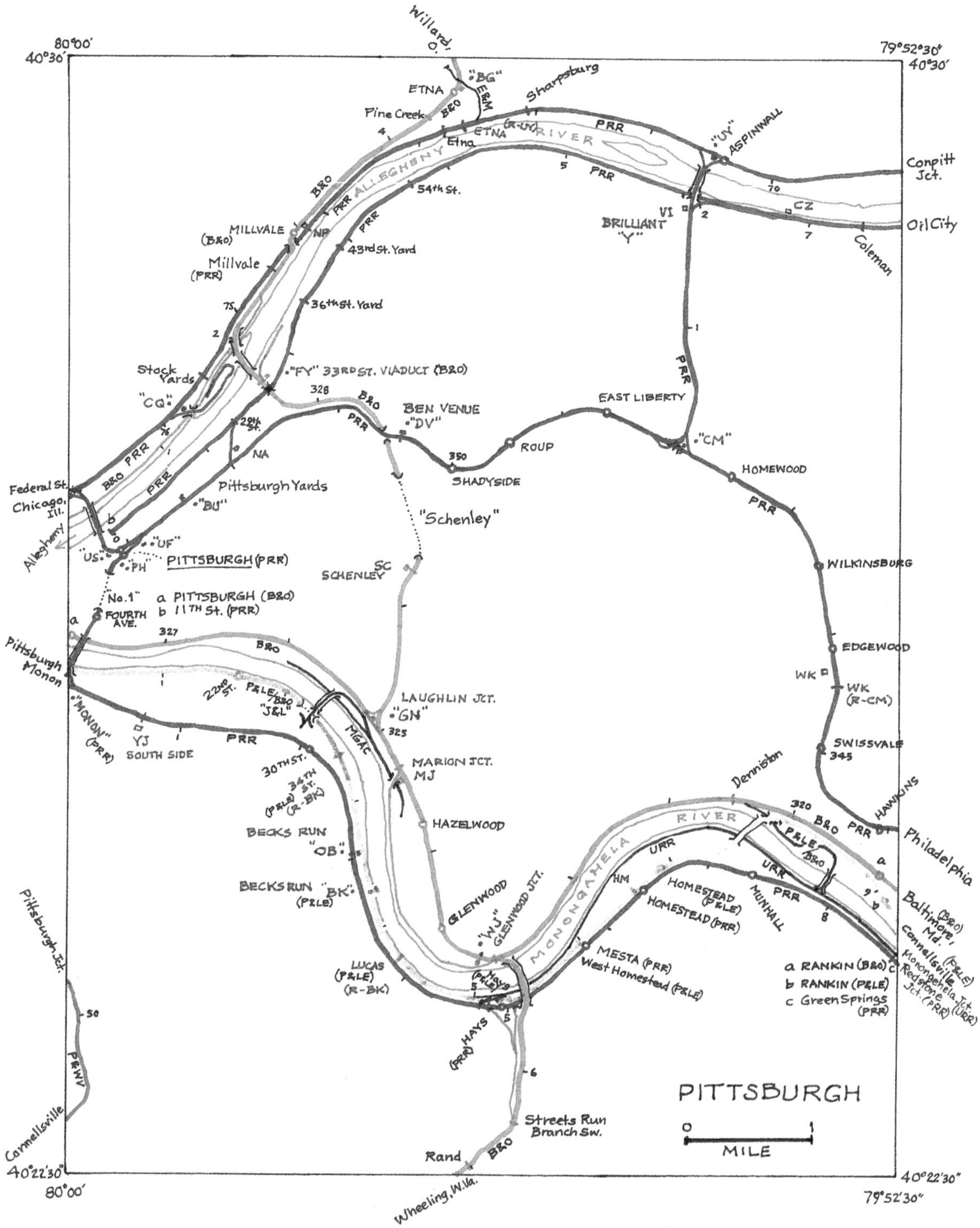

57A Pittsburgh, Pa.

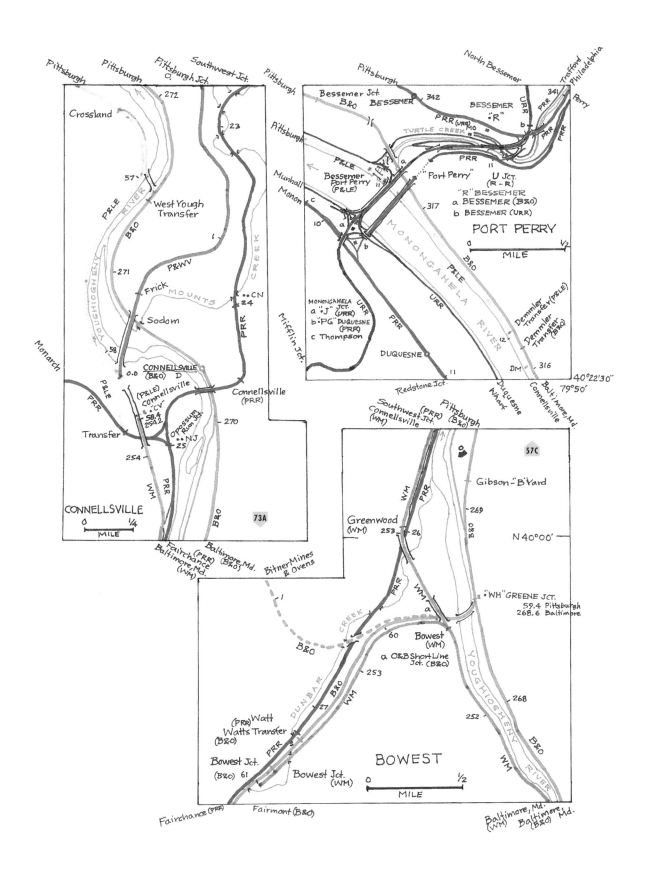

57B Port Perry, Pa. **57C** Connellsville, Pa. **73A** Bowest, Pa.

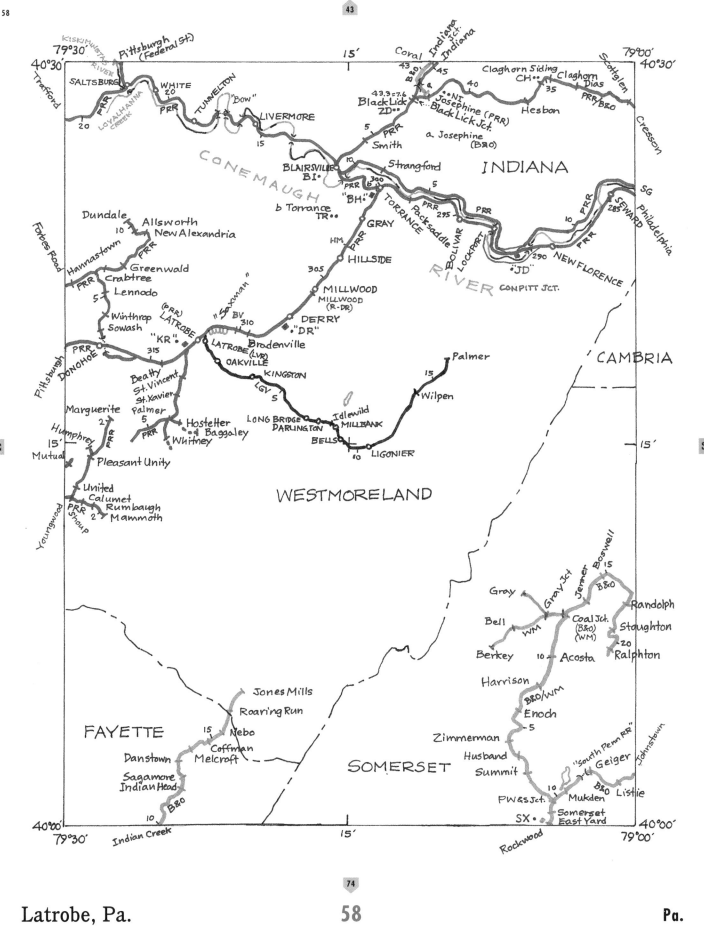

79°00'
40°30'
45'
Cresson
Black Lick
Punxsutawney
Blandburg
78°30'
40°30'

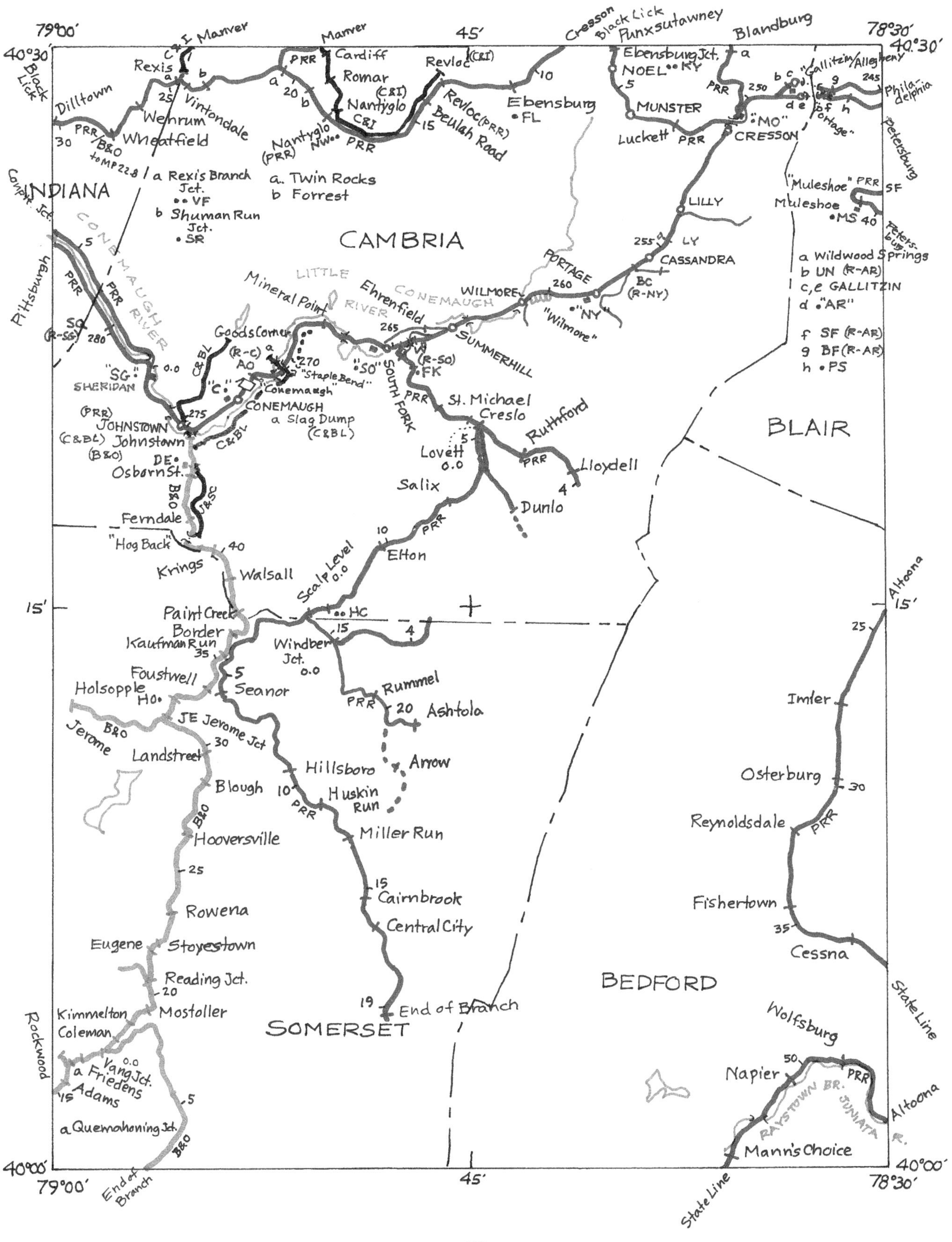

Black Lick
C & I Manver
Manver
Cardiff
Revloc (C&I)
Ebensburg Jct.
Gallitzin/Allegheny
Rexis
C
b
PRR
Romar
(C&I)
Revloc (PRR)
10
NOEL
KY
a
b c
9
245
Dilltown
a
a
20
b
Nantyglo
Beulah Road
5
PRR
250
d
e
"Portage"
Philadelphia
Wehrum
25
Vintondale
Nantyglo
C&I
15
Ebensburg
FL
MUNSTER
"MO"
f
h
Wheatfield
30
PRR/B&O
(PRR)
NW
PRR
Luckett PRR
CRESSON
Petersburg

to MP 228
a Rexis Branch Jct.
a. Twin Rocks
LILLY
"Muleshoe" PRR
SF
INDIANA
Conm Jct.
VF
b. Forrest
255
LY
Muleshoe
MS 40
Petersburg
b Shuman Run Jct.
CASSANDRA
a Wildwood Springs
SR
CAMBRIA
PORTAGE
BC
b UN (R-AR)
5
LITTLE Ehrenfield
260
(R-NY)
c,e GALLITZIN
CONEMAUGH
Mineral Point
RIVER
WILMORE
d "AR"
CONEMAUGH RIVER
Goods Corner
265
"Wilmore"
"NY"
f SF (R-AR)
Pittsburgh
PRR
(R-C)
SUMMERHILL
g BF (R-AR)
SG
280
C&BL
AO
a
270
"SO"
NY
h PS
(R-SG)
0.0
"C"
"Staple Bend"
(R-SO)
"SG"
"Conemaugh"
South Fork
FK
SHERIDAN
275
St. Michael
(PRR)
CONEMAUGH
Creslo
BLAIR
JOHNSTOWN
a Slag Dump
Ruttford
(C&BL) Johnstown
C&BL
(C&BL)
Lovett
PRR
Lloydell
(B&O)
DE
5
0.0
4
Osborn St.
Salix
Dunlo
B&O
Ferndale
10
PRR
"Hog Back"
40
Scalp Level
Elton
Krings
Walsall
0.0
15'
15'
Paint Creek
HC
Altoona
Border
15
4
25
Kaufman Run
35
Windber Jct.
Foustwell
0.0
Holsopple
5 Seanor
Rummel
Imler
Ho
PRR
20 Ashtola
Jerome
JE Jerome Jct.
Osterburg
B&O
30
Hillsboro
Arrow
30
Landstreet
Reynoldsdale
PRR
Blough
10
Huskin
PRR
Run
B&O
PRR
Miller Run
Hooversville
Fishertown
25
15
Rowena
Cairnbrook
35
Eugene Stoyestown
Central City
Cessna
Reading Jct.
BEDFORD
Kimmelton
20
Mostoller
Wolfsburg
Rockwood
Coleman
SOMERSET
19
End of Branch
Napier
50
PRR
a Vang Jct.
0.0
a Friedens
Altoona
15
Adams
5
JUNIATA R.
a Quemahoning Jct.
RAYSTOWN BR.
B&O
40°00'
Mann's Choice
40°00'
79°00'
End of Branch
45'
State Line
78°30'
State Line

Johnstown, Pa.

59

Pa.

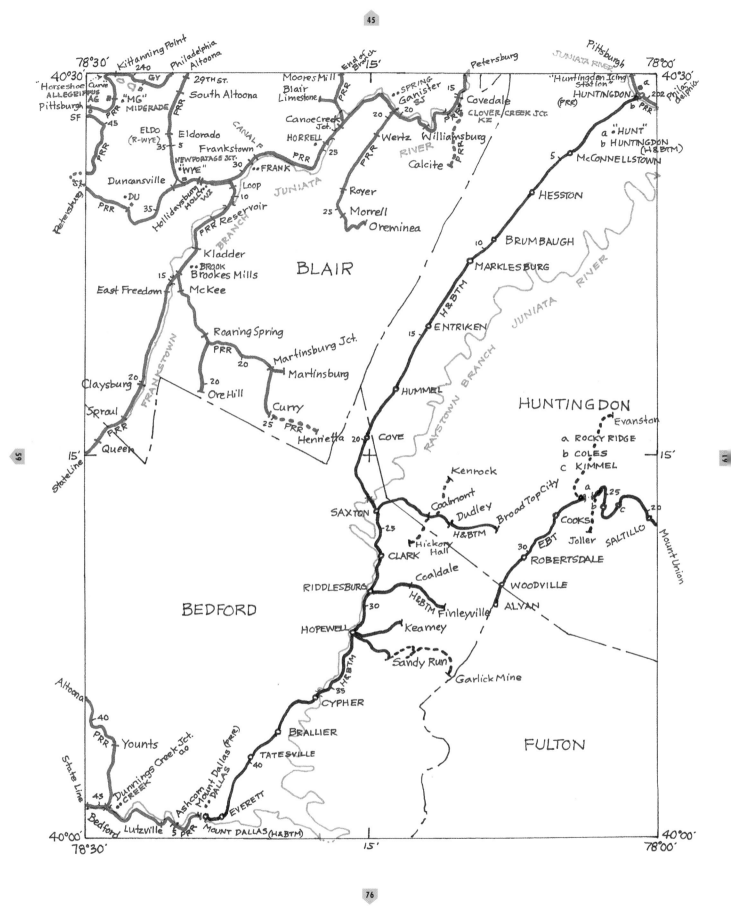

Horseshoe Curve 240 Kittanning Point
78°30' 40°30' Philadelphia Altoona
ALLEGRIPPUS AG 29TH ST.
Pittsburgh SF "MG" MIDGRADE South Altoona
45 ELDO (R-WYE) Eldorado
35 5 Frankstown
NEWPORTAGE JCT. PRR
Duncansville "WYE" 30
DU 35 FRANK
Hollidaysburg HOLW Loop
PRR Reservoir 10
Petersburg 57
Kladder
BROOK
15 Brookes Mills
East Freedom McKee

Moores Mill End of dr Brast 15'
Blair Limestone PRR SPRING
Canoe Creek Jct. Ganister SS
HORRELL 20 20
25 Wertz
Williamsburg
Calcite
Royer
25 Morrell
Oreminea

BLAIR

CANAL P JUNIATA
BRANCH RIVER
PRR

Petersburg
Pittsburgh JUNIATA RIVER
78°00' 40°30'
"Huntingdon Icing Station" 202 Phila-delphia
HUNTINGDON b PRR
(PRR)
a "HUNT"
b HUNTINGDON (H&BTM)
5 McCONNELLSTOWN
HESSTON
10 BRUMBAUGH
MARKLESBURG
H&BTM

Covedale
CLOVER CREEK JCT.
KZ

Roaring Spring
PRR 20
Claysburg 20 Martinsburg Jct.
Sproul 20 Martinsburg
PRR Ore Hill
15' Queen 25 Curry PRR
State Line Henrietta 20 COVE

FRANKSTOWN

15 ENTRIKEN
HUMMEL

HUNTINGDON
Evanston
a ROCKY RIDGE
b COLES
c KIMMEL
15'

RAYSTOWN BRANCH JUNIATA RIVER

Kenrock
Coalmont
Dudley
Broad Top City
SAXTON
25
Hickory Hall
CLARK
RIDDLESBURG
Coaldale
30
HOPEWELL Kearney
H&BTM Finleyville
Sandy Run
Garlick Mine

BEDFORD

a 25 b
c 20
COOKS
Joller SALTILLO
30 EBT
ROBERTSDALE
WOODVILLE
ALVAN
H&BTM
Mount Union

FULTON

Altoona
40 PRR Younts
State Line Dunnings Creek Jct.
45 DUNNINGS CREEK
Bedford Lutzville 5 PRR MOUNT DALLAS (H&BTM)
40°00'
78°30' 15'

Ashcom Mount Dallas (PRR)
DALLAS
35 CYPHER
BRALLIER
TATESVILLE 40
EVERETT

40°00'
78°00'

Hollidaysburg, Pa. 60 Pa.

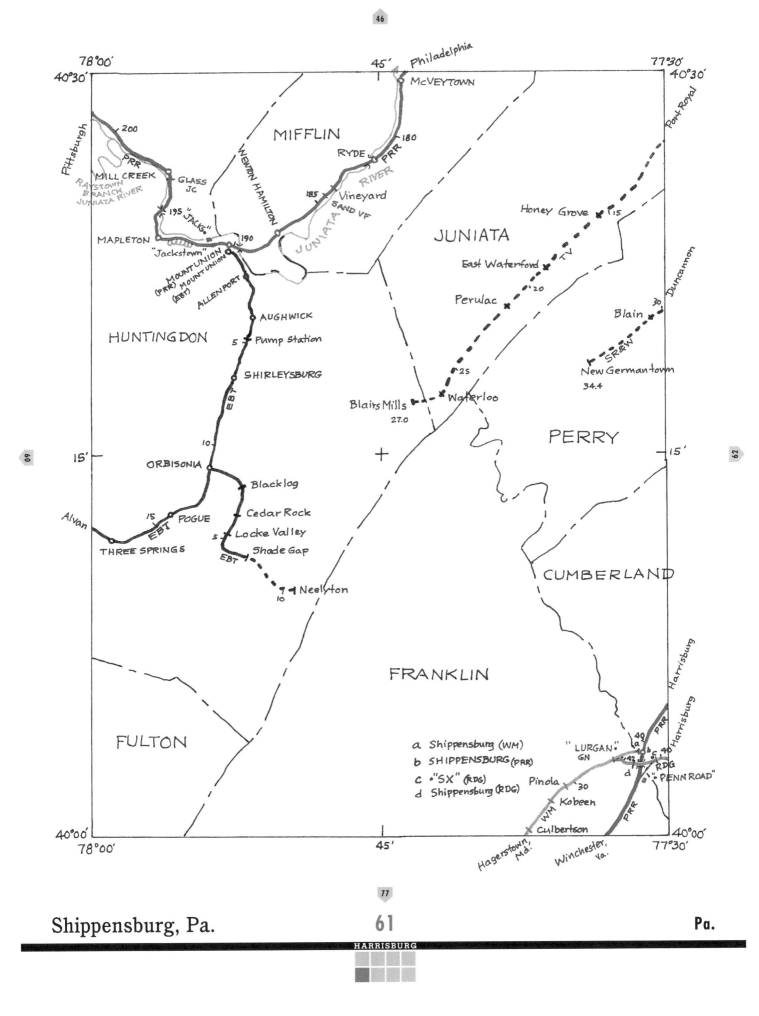

78°00'

40°30'

Pittsburgh

PRR

200

MILL CREEK

RAYSTOWN BRANCH JUNIATA RIVER

GLASS JC

195 "JACKS"

MAPLETON

190

"Jackstown"

MOUNT UNION

(PRR) MOUNT UNION

(EBT)

ALLENPORT

HUNTINGDON

WENTON HAMILTON

MIFFLIN

RYDE

PRR

180

Philadelphia

45'

McVEYTOWN

77°30'

40°30'

185

Vineyard

SAND VF

JUNIATA RIVER

Port Royal

JUNIATA

Honey Grove

15

East Waterford

TV

Perulac

20

Blain

30

Duncannon

New Germantown

SR&W

34.4

AUGHWICK

5

Pump Station

SHIRLEYSBURG

EBT

10

ORBISONIA

Blacklog

15

EBT

POGUE

Cedar Rock

5

Locke Valley

THREE SPRINGS

EBT

Shade Gap

Alvan

Neelyton

10

25

Blairs Mills

27.0

Waterloo

PERRY

15'

+

15'

CUMBERLAND

FRANKLIN

FULTON

a Shippensburg (WM)
b SHIPPENSBURG (PRR)
c •"SX" (RDG)
d Shippensburg (RDG)

"LURGAN"
GN

40

a

b

c

40

42

d

PRR

RDG

Harrisburg

Harrisburg

Pinola

30

Kobeen

WM

"PENN ROAD"

Harrisburg

PRR

Culbertson

40°00'

78°00'

Hagerstown, Md.

Winchester, Va.

45'

77°30'

40°00'

09

60

62

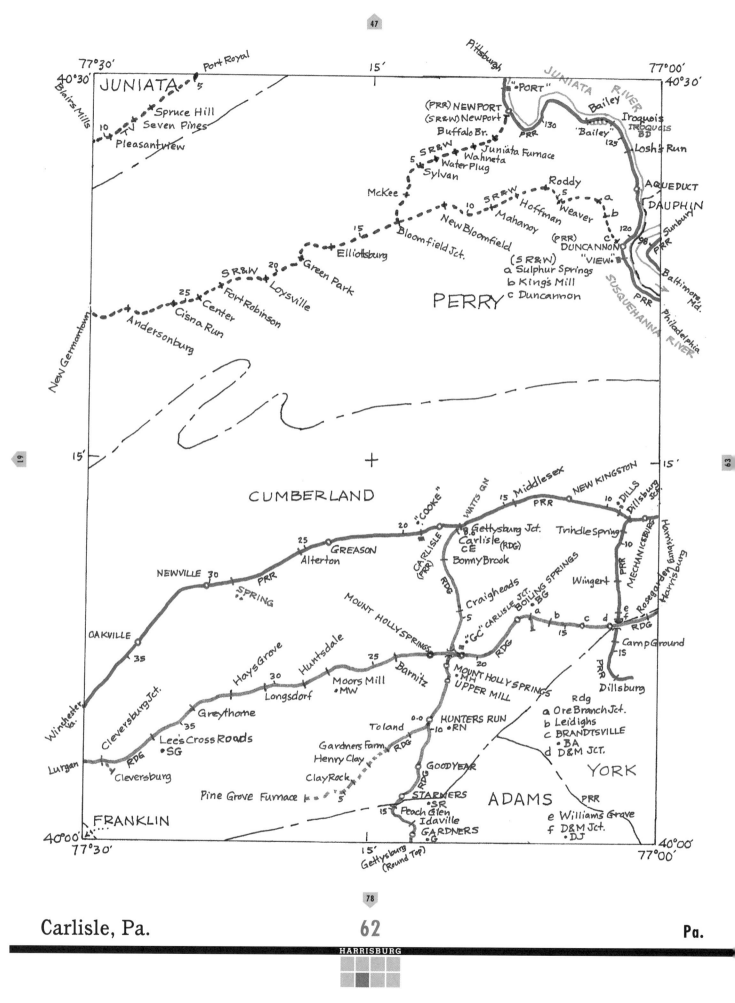

77°30' JUNIATA 15' Pittsburgh JUNIATA RIVER 77°00'
40°30' 40°30'

Port Royal 5
Blairs Mills Spruce Hill
10 TJ Seven Pines
Pleasantview

"PORT"
(PRR) NEWPORT Bailey Iroquois
(SR&W) Newport 130 "Bailey" IROQUOIS BD 125
Buffalo Br. PRR Losh's Run
SR&W Juniata Furnace
5 Wahneta Roddy AQUEDUCT
Water Plug 5 a DAUPHIN
Sylvan Hoffman Weaver b Sunbury
McKee 10 SR&W Mahanoy 120
15 New Bloomfield (PRR) c 98 PRR
Bloomfield Jct. DUNCANNON
Elliottsburg (SR&W) "VIEW"
20 Green Park a Sulphur Springs Baltimore, Md.
SR&W Loysville b King's Mill
25 Fort Robinson c Duncannon PERRY PRR
Center Cisna Run
Andersonburg

SUSQUEHANNA RIVER
Philadelphia

New Germantown

15' 15'

CUMBERLAND

WATTS GN Middlesex NEW KINGSTON DILLS
20 "COOKE" 15 PRR 10 Dillsburg Jct.
25 Gettysburg Jct. Trindle Spring Harrisburg
GREASON Carlisle (RDG) 10
Alterton CE
NEWVILLE 30 Bonny Brook Wingert PRR Harrisburg
PRR Mechanicsburg
SPRING RDG Craigheads Rosegarden +
OAKVILLE 5 "GC" CARLISLE JCT. BOILING SPRINGS a b c d e f RDG
35 MOUNT HOLLY SPRINGS BG Camp Ground
25 Hays Grove Huntsdale Barnitz 15 PRR 15
Cleversburg Jct. Longsdorf Moors Mill MOUNT HOLLY SPRINGS Dillsburg
30 MW MH
Greythorne UPPER MILL
35 Lee's Cross Roads Rdg
SG Toland HUNTERS RUN a Ore Branch Jct.
Lurgan RDG 0.0 10 RN b Leidighs
RDG Gardners Farm RDG YORK c BRANDTSVILLE
Cleversburg Henry Clay GOODYEAR BA
Clay Rock RDG d D&M JCT.
Pine Grove Furnace 5 STARNERS ADAMS PRR
SR e Williams Grove
FRANKLIN 15 Peach Glen f D&M JCT.
Idaville DJ
40°00' GARDNERS 40°00'
77°30' 15' G 77°00'
Gettysburg
(Round Top)

Carlisle, Pa. **62** Pa.

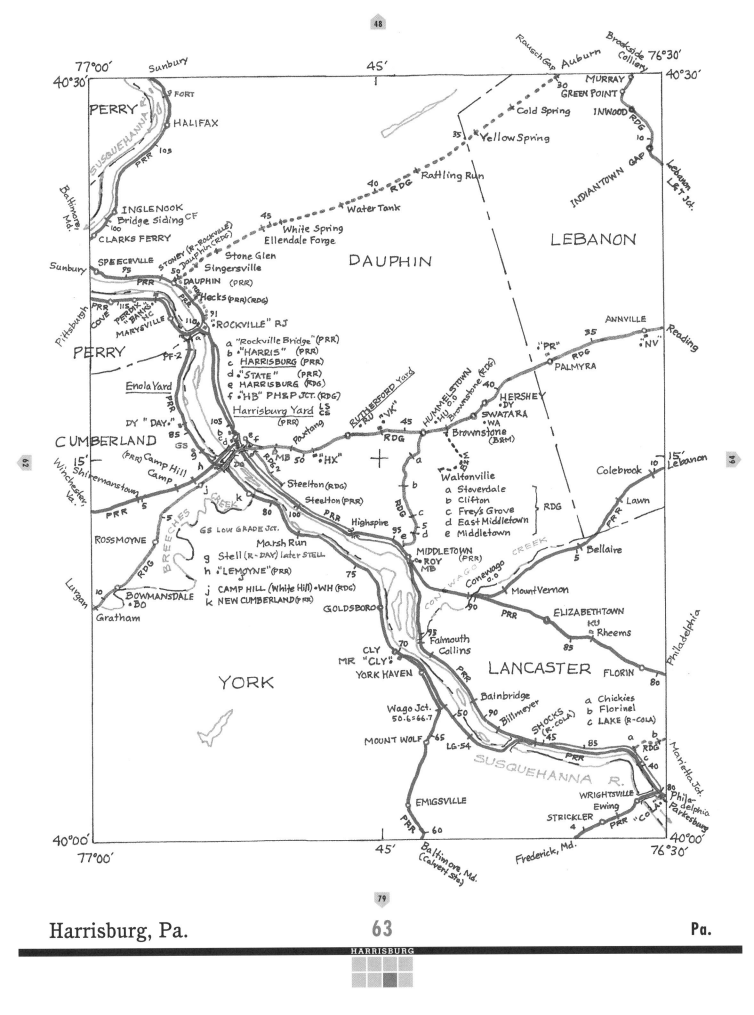

Rousch Gap Auburn
Brookside 76°30'
Colliery
40°30'

77°00'   Sunbury   45'
40°30'   40°30'

PERRY   9 FORT   MURRAY
                                                               GREEN POINT
         HALIFAX   INWOOD   RDG
                                              Cold Spring
                            PRR 105
                                       35   Yellow Spring   10
Baltimore   Lebanon
Md.   L&T Jct.
                                        40   INDIANTOWN GAP
                          45   RDG   Rattling Run
    INGLENOOK   Water Tank   LEBANON
    Bridge Siding CF
         100   45   White Spring
    CLARKS FERRY   Ellendale Forge
Sunbury SPEECEVILLE   DAUPHIN
              95   STONEY (R-ROCKVILLE)   ANNVILLE
                      50 Dauphin (RDG)   35   Reading
              PRR   Stone Glen   "PR"   "NV"
                    Singersville
         PRR "115" DAUPHIN (PRR)   RDG
Pittsburgh PRR COVE Hecks (PRR)(RDG)   PALMYRA
         PERDIX "BANKS" 91   40
         COVE MARYSVILLE HC 110   HUMMELSTOWN (RDG)
PERRY   "ROCKVILLE" RJ   HU 0.0 HERSHEY
              PF-2   a "Rockville Bridge" (PRR)   Brownstone (RDG) DY
                          b "HARRIS"   (PRR)   SWATARA
                          c HARRISBURG (PRR)   WA
    Enola Yard   d "STATE"   (PRR)   Brownstone
                          e HARRISBURG (RDG)   (B&M)
              PRR   f "HB" PH&P JCT. (RDG)
    DY "DAY."   Harrisburg Yard LS   Waltonville   Colebrook 10 15'
CUMBERLAND   105   (PRR)   CS   a Stoverdale   Lebanon
              85   b   RUTHERFORD Yard RU "VK" 45   b Clifton   Lawn
              GS   c d Poxtang   RDG   c Frey's Grove   RDG   PRR
    15'   e f   d East Middletown   5 Bellaire
Shiremanstown (PRR) Camp Hill MB 50 "HX"   e Middletown
Winchester   Camp   h   RM   a
Va.   PRR   5 DO RDG 2   b
                                Steelton (RDG)   CONEWAGO CREEK
ROSSMOYNE   j k CREEK Steelton (PRR)   c   MIDDLETOWN
              RDG   80   100   Highspire RDG 95   (PRR)   5
         BREECHES GS Low Grade Jct.   95 e 5 d ROY MB   ELIZABETHTOWN
                         Marsh Run   e d   Mount Vernon   KU
Lurgan   g Stell (R-DAY) later STELL   90   PRR   85 Rheems
         RDG 10   h "LEMOYNE"(PRR)   FLORIN
    BOWMANSDALE   j CAMP HILL (White Hill) WH (RDG)   Conewago 0.0   80
    BO   k NEW CUMBERLAND (PRR)   75   Philadelphia
Gratham   GOLDSBORO   PRR   a Chickies
                                                                                b Florinel
         YORK   c LAKE (R-COLA)
                                           95 Falmouth   SHOCKS
                                  CLY 70   Collins   (R-COLA) 45   85
                             MR "CLY"   Bainbridge   a b
                                 YORK HAVEN   Billmeyer   RDG
                    Wago Jct.   90   SUSQUEHANNA R. c 40
                    50.6=66.7   50   80
                    MOUNT WOLF 65   LG-54   Philadelphia
                                                                        WRIGHTSVILLE Parkesburg
                          EMIGSVILLE   LANCASTER   Ewing
                                       PRR   STRICKLER   PRR "CO"
                                        60   4
40°00'   40°00'
77°00'   45'   Baltimore Md.   Frederick, Md.   76°30'
                          (Calvert Sta)

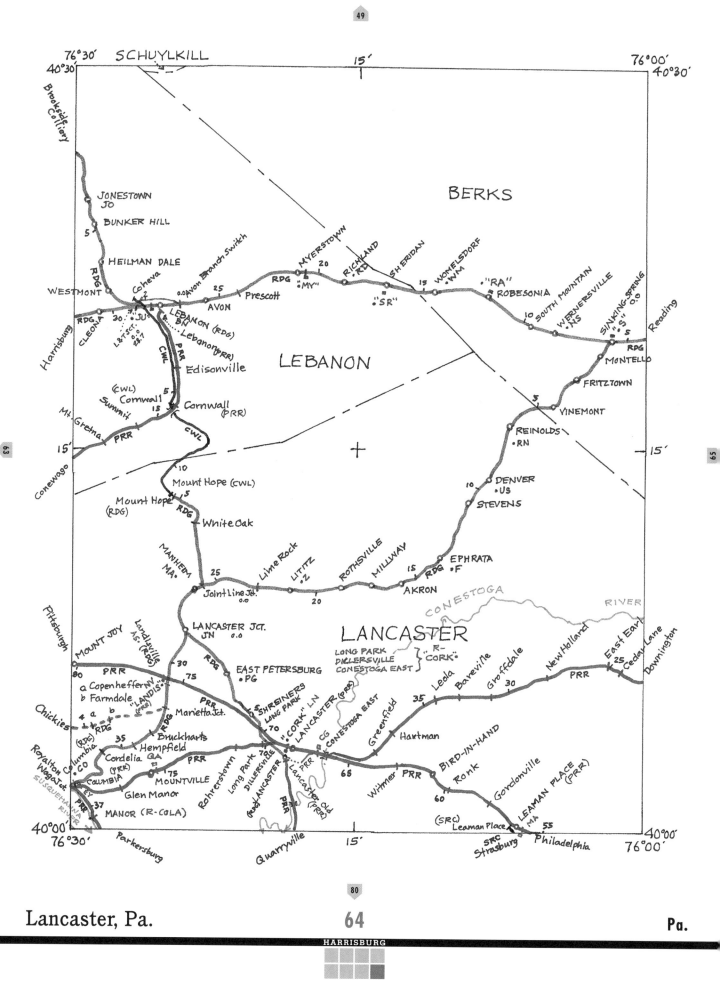

SCHUYLKILL

76°30'
40°30'

15'

76°00'
40°30'

BERKS

Brookside Colliery

JONESTOWN
JO

BUNKER HILL
5

HEILMAN DALE

WESTMONT

RDG

Coleva

Avon Branch Switch

MYERSTOWN
20

RICHLAND
RD

SHERIDAN
15

WOMELSDORF
WM

SOUTH MOUNTAIN

WERNERSVILLE
NS

SINKING SPRING
5

Reading

RDG   30   JU

Harrisburg   RDG   30   JU

CLEONA

L&T.JCT.
0.0

LEBANON (RDG)

Lebanon (PRR)

BH

PRR

CWL

Edisonville

(CWL)
Cornwall
Summit

(CWL)

5

15

Cornwall
(PRR)

Mt. Gretna   PRR   15'

Conewago

CWL

AVON

Prescott

RDG

25

"MY"

"SR"

ROBESONIA

"RA"

10

"S"   0.0

RDG

MONTELLO

FRITZTOWN

VINEMONT

5

LEBANON

+

REINOLDS
RN

Mount Hope (CWL)
10

Mount Hope
(RDG)

RDG

5

White Oak

DENVER
US

STEVENS

10

MANHEIM
MA

25

Lime Rock

LITITZ
Z

ROTHSVILLE

MILLWAY

15

EPHRATA
F

RDG

Joint Line Jct.
0.0

20

AKRON

CONESTOGA   RIVER

LANCASTER

Pittsburgh

LANCASTER JCT.
JN   0.0

LONG PARK
DILLERSVILLE
CONESTOGA EAST

"R-
CORK"

New Holland

East Earl

MOUNT JOY

Landisville
AS. (RDG)

30

RDG

75

EAST PETERSBURG
PG

Leola

Bareville

Groffdale

30

PRR

25   Cedar Lane

Downington

80   PRR

a Copenheffer NV
b Farmdale

"LANDIS"
(PRR)

PRR

SHREINERS
LONG PARK

"CORK" LN

LANCASTER (PRR)

CG   CONESTOGA EAST

Greenfield

35

Hartman

Chickies

4 a b

RDG

(RDG)

Columbia

35

Marietta Jct.

Bruckharts

Hempfield
GA

RDG   70
Long Park 70
DILLERSVILLE

"CORK"

(East) LANCASTER

Lancaster Old
(PRR)

65

Witmer   PRR

BIRD-IN-HAND

Ronk

Gordonville

LEAMAN PLACE
(PRR)

Royalton
Wago Jct.

SUSQUEHANNA RIVER

PRR

Cordelia
(PRR)

COLUMBIA

37

75

MOUNTVILLE

Glen Manor

MANOR (R-COLA)

Rohrerstown

PRR

LANCASTER
PRR

60

(SRC)
Leaman Place

LEAMAN PLACE
MA   55

40°00'
76°30'

Parkersburg

Quarryville

15'

SRC
Strasburg

Philadelphia

40°00'
76°00'

Lancaster, Pa.

Pa.

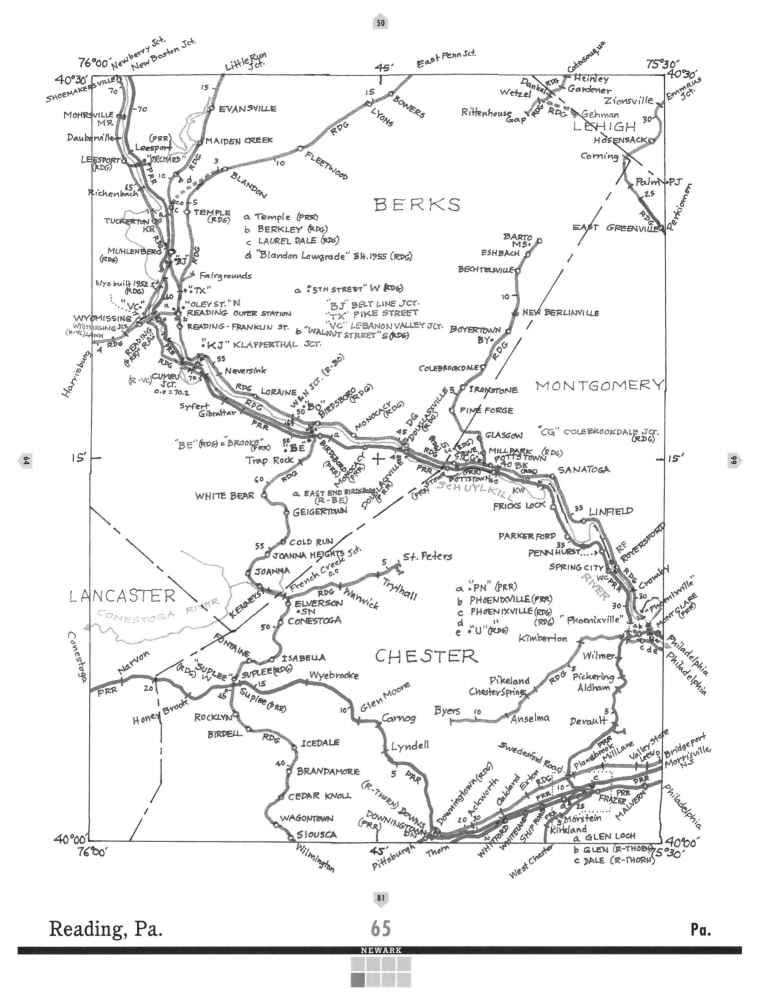

76°00' Newbery Jct. New Boston Jct. Little Run Jct. 45' East Penn Jct. Catosauqua 75°30'

40°30' SHOEMAKERSVILLE 70 15 15 BOWERS Dankel Wetzel Heinley Gardener Zionsville Emmaus Jct. 40°30'

MOHRSVILLE MR ~70 EVANSVILLE LYONS Rittenhouse Gap Gehman 30 LEHIGH

Dauberville MAIDEN CREEK RDG FLEETWOOD HOSENSACK

Leesport (PRR) Corning

LEESPORT "ORCHARD" 10 BLANDON BARTO MS. Palm PJ

(RDG) PRR 3 10 ESHBACH 25 Perkiomen

Richenbach C 300-S BECHTELSVILLE

TUCKERTON 65 TEMPLE a Temple (PRR) NEW BERLINVILLE EAST GREENVILLE

KR 90 (RDG) b BERKLEY (RDG) 10 BOYERTOWN

MUHLENBERG R120 c LAUREL DALE (RDG) BY

(RDG) "BJ" RDG d "Blandon Lowgrade" B4. 1955 (RDG)

Wye built 1952.5 Fairgrounds a "5TH STREET" W (RDG) COLEBROOKDALE RDG MONTGOMERY

(RDG) "TX" "BJ" BELT LINE JCT. BERKS

"VC." 60 "OLEY ST." N "TX" PIKE STREET

WYOMISSING a READING OUTER STATION "VC" LEBANON VALLEY JCT.

WYOMISSING JCT. READING-FRANKLIN ST. b "WALNUT STREET" S (RDG) IRONSTONE

(R-VC) LAWN READING "KJ" KLAPPERTHAL JCT. PINE FORGE "CG" COLEBROOKDALE JCT.

Harrisburg 4 RDG (PRR) RR 55 Neversink GLASGOW (RDG)

(R-VC) CUMRU JCT. 70 RDG LORAINE W&N JCT. (R-30) DOUGLASSVILLE 5 MILL PARK (RDG)

0.0=70.2 RDG Syfert PRR 55 "BO" MONOCACY DG STOWE POTTSTOWN SANATOGA

15' Gibraltar RDG BIRDSBORO (RDG) 45 (RDG) "CG" 40 BK 15'

"BE" (RDG) = "BROOKE" 60 "BE" BIRDSBORO MONOCACY DOUGLASSVILLE PRR POTTSTOWN (RDG)

(PRR) Trap Rock (PRR) (PRR) (PRR) FRICKS LOCK LINFIELD

WHITE BEAR a EAST END BIRDSBORO (PRR) KW 35 RF

(R-BE) GEIGERTOWN SCHUYLKILL PARKER FORD 35 ROYERSFORD

55 COLD RUN PENN HURST RIVER 30 Crombly

JOANNA HEIGHTS Jd. SPRING CITY RDG "Phoenixville"

JOANNA French Creek 0.0 5 St. Peters WC PRR MONTCLARE (PRR)

LANCASTER RDG Warwick Trythall a "PN" (PRR) Philadelphia

CONESTOGA RIVER KENNEYS ELVERSON Kimberton b PHOENIXVILLE (PRR) Philadelphia

FONTAINE SN c PHOENIXVILLE (RDG)

50 CONESTOGA d "U" (RDG) "Phoenixville" Wilmer

"SUPLEE" ISABELLA CHESTER Pikeland RDG Pickering

(RDG) W SUPLEE (RDG) Wyebrooke ChesterSprings Aldham

Narvon 20 15 Suplee (PRR) Glen Moore Byers 10 Anselma Devault

PRR 45 10 Carnog PRR Plankebrook Valley Store Bridgeport

Honey Brook ROCKLYN Lyndell Swedesford Road MillLane Let's Morrisville NJ

BIRDELL RDG PAR Downingtown(RDG) Oakland Exton RDG C PRR

40 ICEDALE 5 Ackworth PRR 10 FRAZER Philadelphia

BRANDAMORE (R-THORN) DOWNS 20 30 25 MALVERN

CEDAR KNOLL DOWNINGTOWN (PRR) Thorn WHITFORD SHIP ROAD Morstein a GLEN LOCH

WAGONTOWN DN WHITELAND Kirkland b GLEN (R-THORN)

40°00' SIOUSCA 30 c DALE (R-THORN) 40°00'

76°00' Wilmington 45' Pittsburgh Thorn West Chester 75°30'

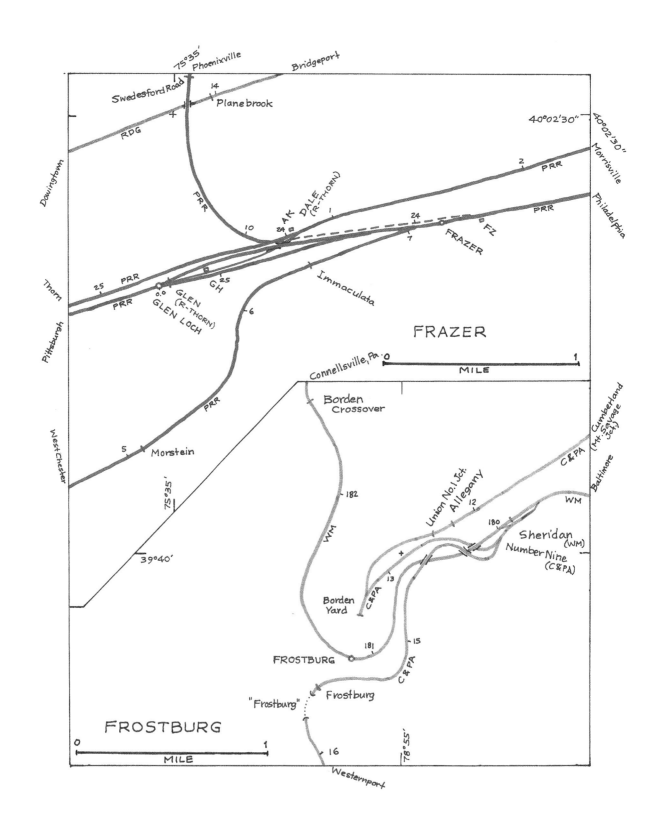

FRAZER

FROSTBURG

65A Frazer, Pa. **75A** Frostburg, Md.

75°30' Emmaus Jct.   Bethlehem   Riegelsville   15'   Manunka Chunk   75°00'
40°30'   40°30'

DILLINGER
RDG 32   45 Hilltop   5   CHESTNUT HILL BRANCHES   N.J.
                                   FORT WASHINGTON BRANCH
LEHIGH   SHELLY   Fort Hill ....   HUNTERDON
                        Pullen   WHITE MARSH   Morrisville
Perkiomen Jct.   Glen Loch 20 SUNNYBROOK
                                              PRR   RDG
                        Quakertown (QB)   c "CHESTNUT HILL" (PRR)   PRR
QUAKERTOWN Q. 40   d CHESTNUT HILL (RDG) LAVEROCK HILL CREST   BYRAM
                        RDG   e GRAVERS   ASBESTOS   RK   RAVEN ROCK
PENNSBURG   f WYNDMOOR   EAST LANE   Trenton
 PS   ROCKHILL   CHESTNUT   Highland 10 GERMANTOWN ROAD
RED HILL   HILL   ST. MARTIN SEDGWICK STENTON
          "Perkasie"   b "ALLEN LANE"   WASHINGTON
20 McLEANS 35 PERKASIE   CW ALLEN LANE   LANE
    Macoby Siding   PK   CARPENTER   GERMANTOWN
    GREEN LANE   SELLERSVILLE   UPSALL   WISTER FISHERS
     GR   TULPEHOCKEN   "WS"
    PERKIOMENVILLE TELFORD   CHELTEN AV.
RDG KRATZ   MS   QUEEN LANE
    HENDRICKS   30   WESTMORELAND
15 SALFORD   SOUDERTON   BUCKS   LAHASKA
    SPRING MOUNT   BYCOT
    ZIEGLERSVILLE HATFIELD   NEW BRITAIN 10 DOYLESTOWN 30 BUCKINGHAM
    SCHWENKSVILLE ORVILLA   D   BU
     SC   5 RDG CHALFONT FARM SCHOOL MONTESSORI SCHOOL
10   COLMAR   CH   WYCOMBE
    GRATERS FORD   CR   RUSHLAND
MONTGOMERY   FORTUNA   25
    RAHNS   25 LANSDALE "MA"   GRENOBLE
    Kneedler   PENNBROOK   TRAYMORE
    West Point   IVYLAND
     WP NORTH WALES   JOHNSVILLE 20 BONAIR   VI
    Acorn   RDG   HB.   CHURCHVILLE Newtown
COLLEDGEVILLE 5 GWYNEDD VALLEY   20   HATBORO   U RDG   SOUTHAMPTON
 CV   Belfry   FULMOR COUNTYLINE Morrisville
5 YERKES   PENLLYN   a FELLWICK (RDG) PSN   30
   ARCOLA   RDG AMBLER MH   Dresher 25 HEATON WOODMONT PRR
New Boston Jct. GR   Fort Hill (PRR) WILLOW GROVE Heaton PAPER MILLS Bound Brook Jct. N.J.
Port Providence PORT WASHINGTON   W PRR ORELAND CRESMONT BRYN ATHYN DA Neshaminy Jct.
OAKS (RDG) FORT INDIAN NQ EARNEST (WA) NORTH HILLS HUNTINGDON NK BETHAYRES Bybery
OAKS PK(PRR) NORRIS NQ PRR PLYMOUTH (RDG) WOTH ROSLYN ARDSLEY FOREST HILLS Somerton BU
OAKS PROTECTORY PRR BETZWOOD Meeting WPH RDG YM GLENSIDE RYDAL VALLEY FALLS Bustleton (PRR)
RDG 25 PK FORT KENNEDY NS Williams Flourtown JO a DB NOBLE WALNUT HILL PHILADELPHIA
VALLEY FORGE PW Corsons (RDG) PRR KE FOX CHASE Holmes
Newberry Jct. RDG NS SN (SEE INSET AT TOP) JENKINTOWN F. RYERS Ashton
Downingtown Chesterbrook New Centreville CHESTNUT HILL (PRR) ELKINS PARK CHELTENHAM Holmesburg Jct.
Glen Loch CHESTER King GR OAK LANE DOBY "MN" Jersey City N.J.
CedarHollow Howellville PRR TABOR FERN ROCK LAWNDALE HG
"PAOLI" Pt Howellville Maple Abrams FRANKFORD SUMMERDALE PA.
DAYLESFORD BERWYN STRAFFORD WAYNE ST. DAVIDS RADNOR WISSINOMING South Amboy
Pittsburgh PAOLI 20 EAGLE DEVON VILLANOVA Gladwyne WH TACONY Palmyra RIVERTON
PRW ROSEMONT "BRYN MAWR" Flat Rock WS N.J. PRR Camden
DELAWARE HAVERFORD ARDMORE SCHUYLKILL RIVER Port Richmond DELAWARE RIVER
40°00'   40°00'
75°30' 65TH ST. WINNEWOOD NARBERTH Philadelphia 15' Philadelphia BRIDESBURG 75°00'
        Philadelphia Philadelphia (PRR)North Phila.(RDG) North Phila.

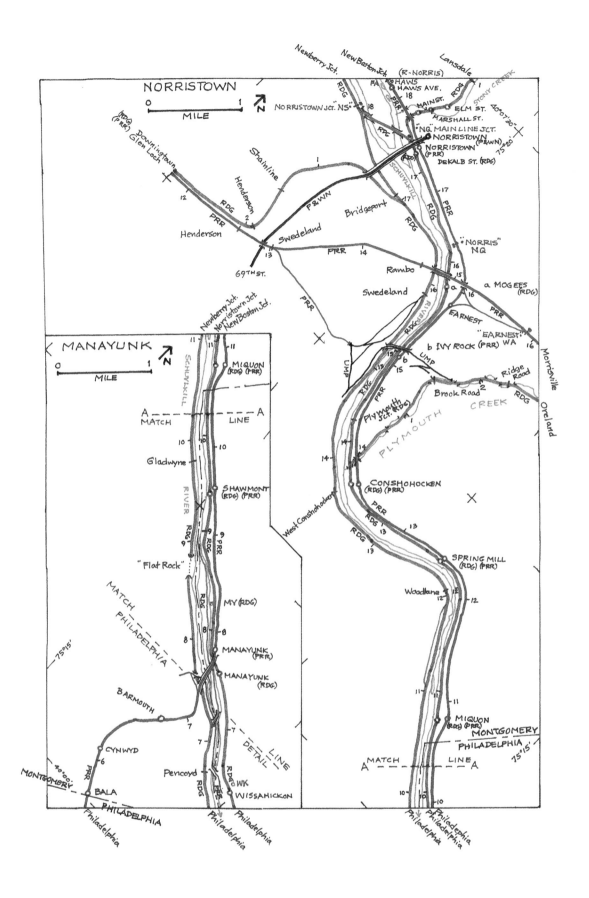

66A Norristown, Pa. **66B** Manayunk, Pa.

Flemington

BoundBrook Jct.

East Millstone

45'

74°30'

75°00'

40°30'

40°30'

Floetown (RV)town

USGOVT

Belle Mead Genl. Depot

Middlebush

PRR

3

Clyde

Millstone Jct.

10 Copper Hill

SOMERSET

50

BELLE MEADE "BM"

HUNTERDON

HARLINGEN

RDG 5080

Jersey City

PRR

Ringoes

5

RDG TOWERS:
"BM" BELLE MEADE
SN SKILLMAN
"SV" HOPEWELL
"GH" GLEN MOORE
CX EWING
"CN" WEST TRENTON

45

SKILLMAN SN

STOUTSBURG

Rocky Hill

DEANS

40

PRR

PRR 20 STOCKTON

"SV"

HOPEWELL

Kingston

PRR "MIDWAY" MK

5

MONMOUTH JCT.

PRR 2

Dayton

Manunka Chunk

FLEMINGTON FJ JCT.

LAMBERTVILLE

40 GLEN MOORE

NH NEWHOPE HOOD

FH 0.0

PRINCETON

CH VO

KS

MIDDLESEX

45

PLAINSBORO

"Plainsboro"

35 REEDER 15

RDG

"GH"

PENNINGTON

Penns Neck

MF

0.0

"NASSAU" CD

PRINCETON JCT.

Hightstown

Philadelphia (Glenside)

MOORE

Titusville

10

CX EWING (R-CN)

35

a PROSPECT ST. (RDG)
b TRENTON UN "
c WARREN ST. (PRR)
d "MG" (PRR)

50

South Amboy

DELAWARE

WASHINGTON CROSSING

GW

Somerset

EWING

WEST TRENTON

Lawrence

Hightstown (PRR) (UTR)

40

K

15'

RIVER

BUCKS

RDG TOWERS:
"DY" DELLA
"RO" ROELOFS

WB

"CN"

Wilburth

"DY"

5

AJAX PARK
AGASOTE
HILLCREST

CADG

RDG

35

Millham (RDG)

MERCER

PRR

"MILLHAM" MO

Windsor

35

Robbinsville

NW

Sharon

20

66

Philadelphia (Cheltenham)

NEWTOWN NS

"RO"

ROELOFS

30

YARDLEY

WS

PRE

A 55

d b

"FAIR" DO

TRENTON

Broad St.

Lalor St.

CS

UN

"MORRIS" SV

15'

68

GEORGE SCHOOL 25

RDG

ST. LEONARD

HOLLAND 35

PRR

Langhorne (PRR)

RDG 5080

Dunlap

MA

40

CF

Nickel

MORRISVILLE

MW MB 45

PRR

PRR

MW MB (R-Morris)

MY

60

MY

PRR

Yardville

30

Imlaystown

MONMOUTH

Davis

15

Cream Ridge

Roxton

JANNEY

WOODBOURNE

GLENLAKE

"PV"

LANGHORNE

PARKLAND

RDG TOWERS:
"PY" LANGHORNE
"JG" NESHAMINY FALLS

TULLYTOWN

BD (R-BO)

5

BORDENTOWN

"BO" BORDENTOWN

CROSSWICKS CREEK

Glen Loch

TREVOSE 20

"JG"

NESHAMINY FALLS

SY

EDGELY

FIELDSBORO

Hornerstown

Philadelphia RDG

Philadelphia (Newtown Jct.)

RDG 5080 18

BD "GREENE"

65

BRISTOL

"Bristol"

RIVER

25

Kinkora

KINKORA

ROEBLING PRR (PA)

Sharp

New Egypt

UTR

PHILADELPHIA

CORNWELLS HEIGHTS

TORRESDALE

ANDALUSIA

CROYDON

EDDINGTON

Co PRR

20

Stevens

CN

East Burlington

FLORENCE

"Florence"

ROEBLING

Columbus

5

Cookstown

UTR

Philadelphia Camden

75

RDG

PRR

15

EDGEWATER PARK

BEVERLY

Wall RopeWorks

RC

PERKINS

RIVERSIDE DRAW

DELANCO

RIVERSIDE

CAMBRIDGE

10

BURLINGTON

MJ BURLINGTON

Fountain Woods

5

Deacon

BURLINGTON PRR

Jobstown 5

Wrightstown

Fort Dix

Juliustown

10

OCEAN

40°00'

Woodlane

Fair Grounds

(PRR) Lewis

Lewis (UTR)

40°00'

75°00'

Mt. Holly

45'

Pemberton

74°30'

Trenton, N.J.

67

N.J., Pa.

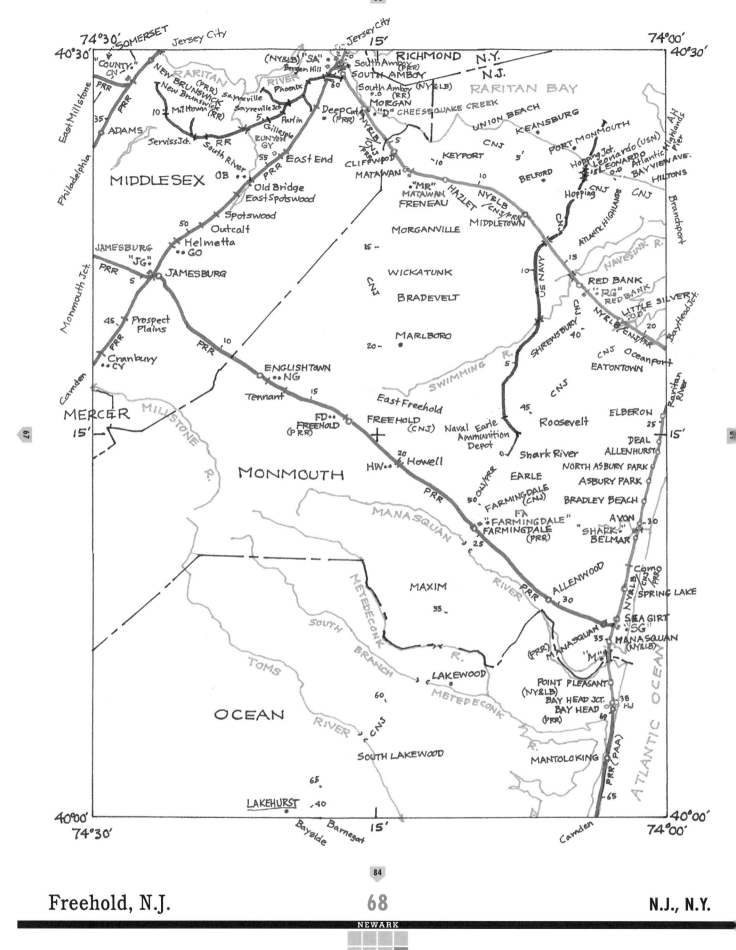

74°30' SOMERSET Jersey City   Jersey City 74°00'
40°30'   15'   40°30'
"COUNTY"   (NY&LB) "SA"   South Amboy RICHMOND N.Y.
CN   Bergen Hill   (PIER)   N.J.
RARITAN RIVER   SOUTH AMBOY   RARITAN BAY
PRR   NEW BRUNSWICK (PRR) Sayreville   South Amboy (NY&LB)
35   New Brunswick   Phoenix   South Amboy (RR)
           Milltown (RR)   MORGAN
10   Sayreville Jct.   "D" CHEESEQUAKE CREEK
           Serviss Jct.   South River Parlin Deep Cut (PRR)   UNION BEACH
                          RR   Gillespie   KEANSBURG
ADAMS   RUNYON GY   PORT MONMOUTH
Philadelphia   MIDDLESEX 55   PRR East End CLIFFWOOD KEYPORT CNJ Hopping Jct. LEONARDO (USN)
                          OB   MATAWAN KEYPORT   5' Belford 15 LEONARDO
                              Old Bridge   "MR"   Hopping Jct. BAYVIEW AVE.
                              East Spotswood   MATAWAN HAZLET NY&LB Hopping CNJ HILTONS
                          Spotswood   FRENEAU /CNJ/PRR CNJ
                    50   Outcalt   MIDDLETOWN
           JAMESBURG   Helmetta   MORGANVILLE ATLANTIC HIGHLANDS NAVESINK R.
           "JG"   GO   US NAVY 15
           PRR   JAMESBURG   WICKATUNK   10 RED BANK
                 5   "RG" RED BANK
Monmouth Jct.   BRADEVELT   CNJ   SHREWSBURY 40 LITTLE SILVER
PRR   45 Prospect   5   20
15'   Plains   MARLBORO   SWIMMING R. CNJ OCEANPORT
PRR   Cranbury   20   EATONTOWN
MERCER CY   MILLSTONE R. ENGLISHTOWN East Freehold 45 Roosevelt ELBERON
15'   NG   FREEHOLD (CNJ) Naval Earle   DEAL 25
              Tennant 10 FD 15 FREEHOLD   Ammunition Depot   ALLENHURST
                          FREEHOLD (PRR)   Shark River   NORTH ASBURY PARK
              MONMOUTH   20 Howell   EARLE ASBURY PARK
                          HW   50 CNJ/PRR FARMINGDALE (CNJ) BRADLEY BEACH
                          PRR   FA AVON 30
                          MANASQUAN   "FARMINGDALE" "SHARK"
                                              FARMINGDALE (PRR) BELMAR
                          METEDECONK MAXIM   RIVER ALLENWOOD Como
                                        55   PRR 30 SPRING LAKE
                          SOUTH BRANCH   SEA GIRT "SG"
                                              (PRR) MANASQUAN 35 MANASQUAN (NY&LB)
                          TOMS R. LAKEWOOD   "M"
                          60   POINT PLEASANT (NY&LB)
                                              BAY HEAD JCT. 38 HJ
                          RIVER CNJ METEDECONK BAY HEAD (PRR)
OCEAN   SOUTH LAKEWOOD   MANTOLOKING PRR (PAA)
                          65   65
40°00'   LAKEHURST 40   ATLANTIC OCEAN 40°00'
74°30'   Bayside Barnegat 15'   Camden 74°00'

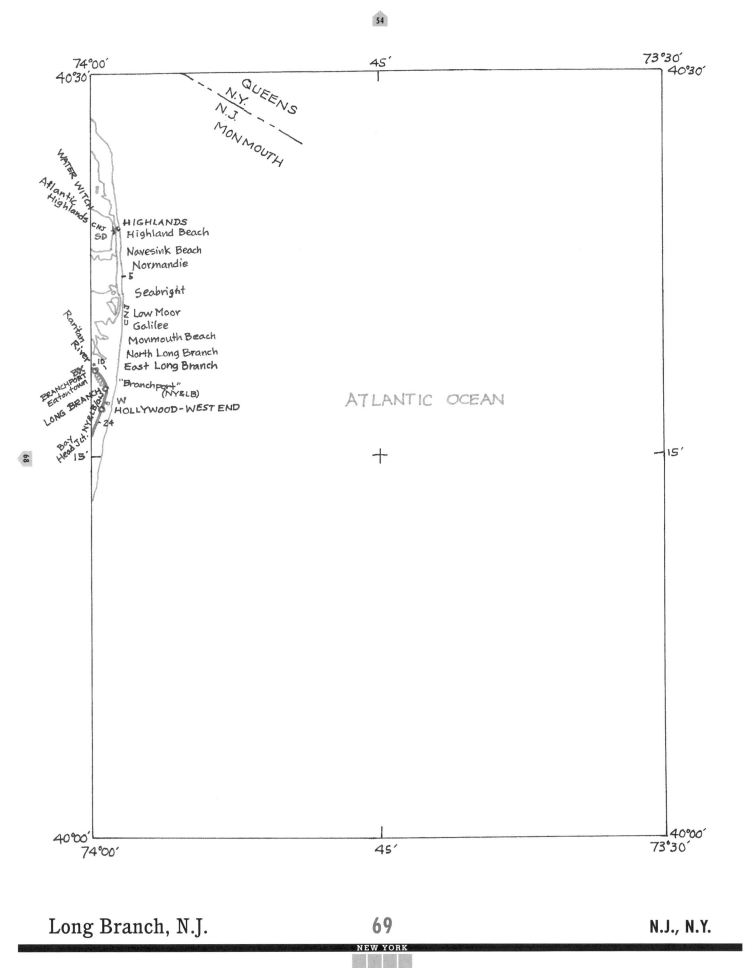

74°00′   40°30′

QUEENS
N.Y.
N.J.
MONMOUTH

45′

73°30′
40°30′

WATER WITCH
Atlantic Highlands

CNJ
SD

HIGHLANDS
Highland Beach
Navesink Beach
Normandie
5
Seabright
Low Moor
Galilee
Monmouth Beach
North Long Branch
East Long Branch
"Branchport"   (NY&LB)
HOLLYWOOD–WEST END

Raritan River

BX
BRANCHPORT
Eatontown
LONG BRANCH

10

W

24

Bay Head Jct. NY&LB&CJ

15′

ATLANTIC   OCEAN

+

15′

89

40°00′
74°00′

45′

40°00′
73°30′

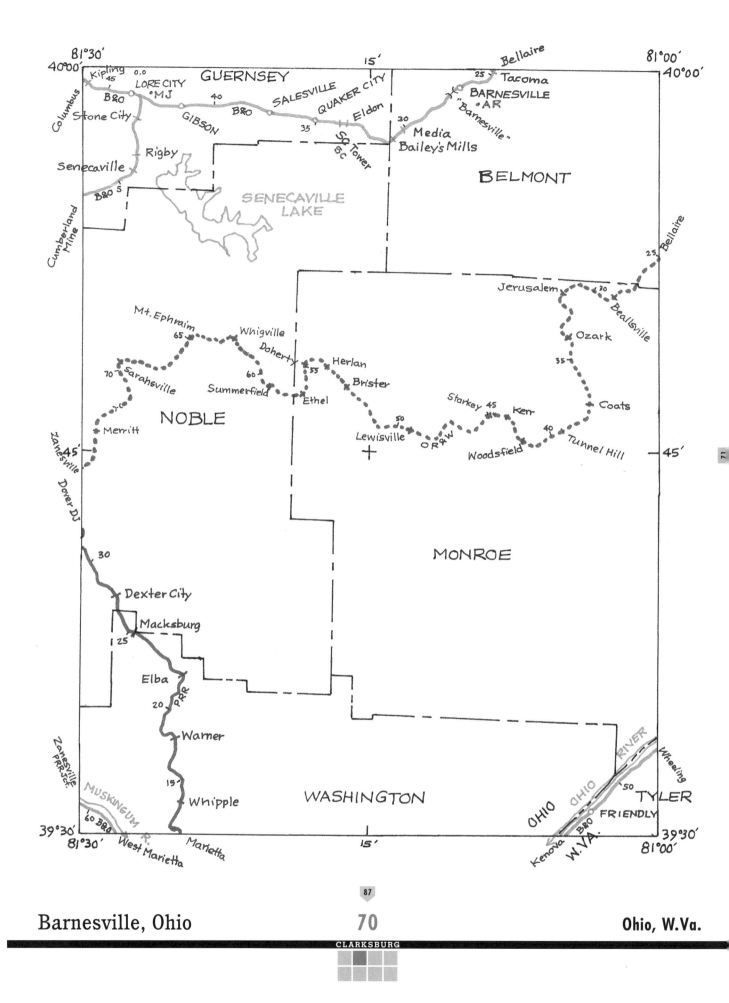

GUERNSEY

81°30'
40°00'

Kipling
45
LORE CITY
MJ
B&O
Columbus
Stone City
GIBSON
B&O
0.0
40
35
Rigby
Senecaville
B&O 5
Cumberland
Mine

SENECAVILLE
LAKE

15'

SALESVILLE
QUAKER CITY
Eldon
SG Tower
SG
30
Media
Bailey's Mills

Bellaire
25
Tacoma
BARNESVILLE
AR
"Barnesville"

81°00'
40°00'

BELMONT

25 Bellaire

Jerusalem
30
Beallsville

Ozark

35

Coats

Mt. Ephraim
65
Whigville
Daherty
70
Sarahsville
Summerfield
60
Herlan
55
Brister
Ethel

Starkey 45
Kerr

Merritt
Zanesville
45'

NOBLE

Lewisville
50
O R & W
Woodsfield
40
Tunnel Hill

45'

71

Dover DJ

MONROE

30

Dexter City

Macksburg

25

Elba
P R R
20

Warner

Zanesville
PRR JCT.

15

Whipple

MUSKINGUM R.
60 B&O
West Marietta
39°30'
81°30'
Marietta

WASHINGTON

15'

OHIO
OHIO RIVER
Wheeling
50
B&O
FRIENDLY
Kenova
W. VA.

TYLER

39°30'
81°00'

87

Barnesville, Ohio

70

Ohio, W.Va.

CLARKSBURG

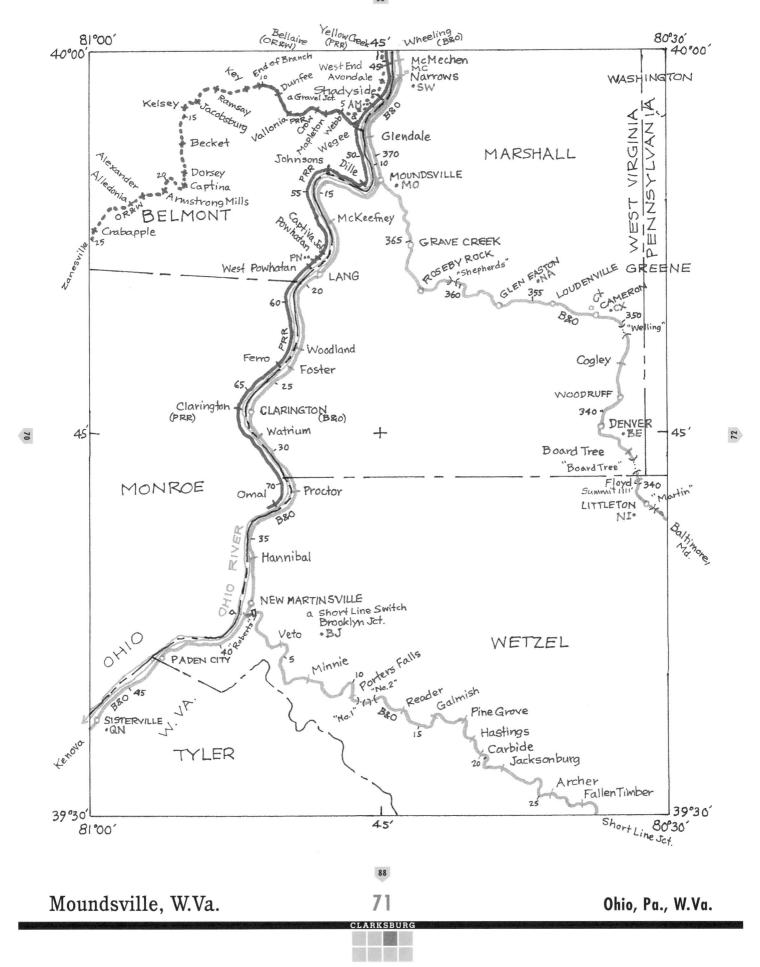

55

81°00'   Bellaire   YellowCreek 45'   Wheeling   80°30'
40°00'   (OR&W)   (PRR)   (B&O)   40°00'
                    McMechen
        Key   End of Branch   West End 45'   MC   WASHINGTON
        10   Dunfee   Avondale   Narrows
                    a Gravel Jct. Shadyside   SW
    Kelsey   Ramsay   5 AM
            Jacobsburg   Vallonia   PRR   Crow   B&O
        15   Mapleton   Webb
        Becket   Wegee   Glendale   MARSHALL
                            50   370
        Dorsey   Johnsons   Dille   10
    Alexander   20   Captina   PRR   MOUNDSVILLE
    Alledonia   Armstrong Mills   55   15   MO
            OR&W   WEST VIRGINIA
    Crabapple   BELMONT   McKeefney   PENNSYLVANIA
    Zanesville 25   Captiva Jct.   GRAVE CREEK
                    Powhatan   365   GREENE
                    PN   ROSEBY ROCK
        West Powhatan   "Shepherds"   GLEN EASTON   LOUDENVILLE
                    LANG   360   355   Ct   CAMERON
                    20   NA   Ct   CX
                60   B&O   350
                                        "Welling"
                PRR
                    Woodland   Cogley
            Ferro   Foster
        65.   25   WOODRUFF
    Clarington   CLARINGTON (B&O)   340
    (PRR)   Watrium   DENVER
45'   30   +   BE   45'
    MONROE   Board Tree
                    70   Proctor   "Board Tree"
            Omal   B&O   Floyd   340
                    35   Summit   "Martin"
            Hannibal   LITTLETON
                                NI   Baltimore, Md.
        OHIO RIVER
                    NEW MARTINSVILLE
                    a Short Line Switch
                    Brooklyn Jct.
                a   Veto   BJ   WETZEL
                    "Roberts"
                40   Minnie   Porters Falls
            PADEN CITY   5   10   "No.2"   Reader   Galmish
        B&O 45   "No.1"   B&O   Pine Grove
    Kenova   SISTERVILLE   15   Hastings
            QN   W. VA.   Carbide   Jacksonburg
        OHIO   TYLER   20
                                Archer   FallenTimber
39°30'   25   39°30'
81°00'   45'   80°30'
                            Short Line Jct.

88

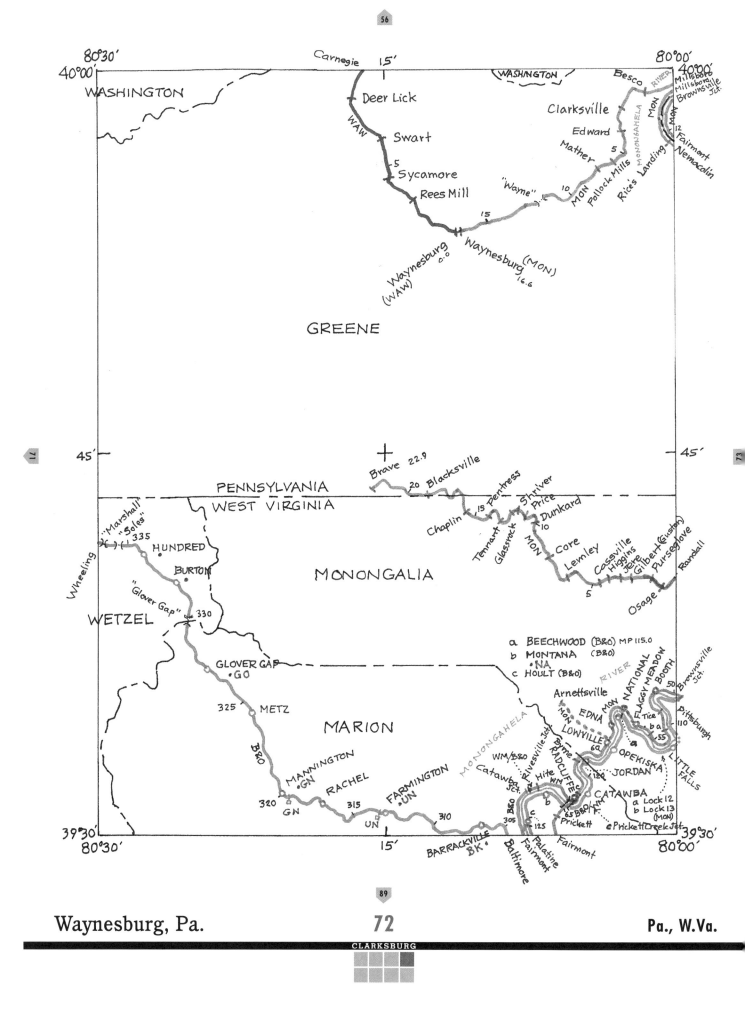

80°30'
Carnegie   15'
80°00'
40°00'   40°00'
WASHINGTON   WASHINGTON   Besco
                                                                   Millsboro
                                    Deer Lick   Clarksville   Millsboro
                                                                     Brownsville Jct.
                              WAW   Edward   12
                                    Swart   Mather   5   Fairmont Nemacolin
                                  5   Rice's Landing
                                    Sycamore   "Wayne"   10   MON Pollock Mills
                                    Rees Mill   )   MON
                                        15
                                          Waynesburg (MON)
                              Waynesburg 0.0   16.6
                              (WAW)

                              GREENE

45'   + Brave 22.9   45'
                              20 Blacksville
              PENNSYLVANIA   15 Pentress   Shriver
              WEST VIRGINIA   Chaplin   Price
                                        Tennant   Dunkard
                    "Marshall" "Soles"   Glassrock 10
)(335   MON Core
                    HUNDRED   Lemley   Cassville Gilbert (Gusty)
Wheeling   BURTON   MONONGALIA   Higgins Jere Purseglove
                                                  5   Osage Randall
              "Glover Gap"   330

WETZEL   a BEECHWOOD (B&O) MP 115.0
              GLOVER GAP   b MONTANA (B&O)   RIVER
                •GO   •NA   NATIONAL 50
                              c HOULT (B&O)   Arnettsville FLAGGY MEADOW Brownsville Jct.
              325 METZ   EDNA MON Tice Pittsburgh
                              MARION   MON   b a   110
                B&O   LOWVILLE 55
                    MANNINGTON RACHEL FARMINGTON WM/B&O Byrne OPEKISKA
                    •GN   •UN   Rivesville Jct. 6a   LITTLE
              320   315   Catawba Hite RADCLIFFE 120 JORDAN FALLS
                GN   UN   Sct. b   CATAWBA a Lock 12
                                    310   305 c 65 B&O b Lock 13 (MON)
39°30'   125 Prickett c Prickett Creek Jct.
80°30'   15'   BARRACKVILLE Palatine Fairmont 80°00'
                              BK Baltimore

Waynesburg, Pa.   72   Pa., W.Va.

CLARKSBURG

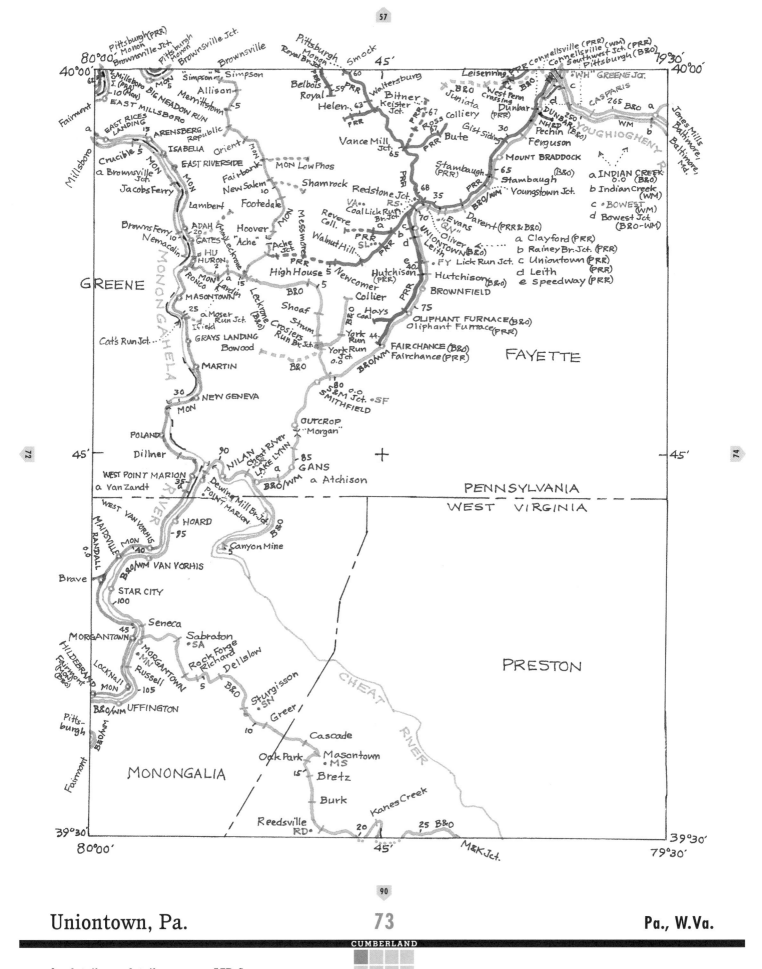

Pittsburgh (PRR)
Monon
Brownsville Jct.
80°00'
40°00'
Pittsburgh Monon
Brownsville Jct.
Brownsville Jct.
Brownsville
"Simpson" Simpson
Fairmont
Millsboro Big MEADOW RUN
I. (Mon)
MON
(Mon)
Allison
Merrittown
Belbois
Royal
Helen
EAST MILLSBORO
Royal Br. Jct.
Pittsburgh Monon
Smock
45'
60
Waltersburg
Bitner
Keister Jct.
59 PRR
63 PRR
67
Juniata
Ross Colliery
67
Connellsville (PRR)
Connellsville (WM) (PRR)
Southwest Jct. (PRR)
Pittsburgh (B&O)
19 30'
40°00'
Leisenring
B&O B&O
West Penn
Crossing
"WH" GREENE Jct.
B&O
CASPARIS
265 B&O

EAST RICES
LANDING
13
ARENSBERG
Republic
ISABELLA
Millsboro
a Brownsville Jct.
Crucible 5 MON
EAST RIVERSIDE
Orient
Fairbank
MON Low Phos
Jacobs Ferry
MON
New Salem 10
Shamrock
Lambert
Footedale
Browns Ferry 10
ADAH
Hoover
GATES "Ache"
Nemacolin
Messmore
Ache Jct.
Walnut Hill
PRR
HU HURON
MON Hardin
RONCO
a 15
MON
MASONTOWN
25
a Moser Run Jct.
I. field
GRAYS LANDING
Bowood
Cat's Run Jct.
MARTIN
30 NEW GENEVA
MON
POLAND
Dillner
90
NILAN
Cheat River
LAKE LYNN
WEST POINT MARION
a Van Zandt
35
Dewing Mill Br. Jct.
POINT MARION
Vance Mill Jct.
65
Redstone Jct.
VA
Coal Lick Run
RS.
Revere Coll.
PRR SL
High House 5
B&O
Shoaf
Strum
Crosiers Run Br. Jct.
B&O
Bute
Gist Siding
Stambaugh (PRR)
68
35
Evans
QN
Oliver
UNIONTOWN (B&O)
Leith
Newcomer
5
Collier
Hays coal
York Run
York Run Jct.
0.0
80 0.0
S&M Jct. SF
SMITHFIELD
OUTCROP
"Morgan"
85
GANS
a Atchison
B&O/WM
Darent (PRR & B&O)
30 Dunbar (PRR)
Dunbar
Pechin
Ferguson
Mount BRADDOCK
65 Stambaugh (B&O)
Youngstown Jct. (WM)
a Clayford (PRR)
b Rainey Br. Jct. (PRR)
c Uniontown (PRR)
d Leith (PRR)
e Speedway (PRR)
70
FY Lick Run Jct.
Hutchison (PRR)
Hutchison (B&O)
BROWNFIELD
75
OLIPHANT FURNACE (B&O)
Oliphant Furnace (PRR)
FAIRCHANCE (B&O)
B&O/WM
Fairchance (PRR)
FAYETTE

GREENE

MONONGAHELA

a INDIAN CREEK
0.0 (B&O)
b Indian Creek (WM)
c BOWEST
d Bowest Jct. (B&O-WM)

YOUGHIOGHENY R.
Jones Mills
Baltimore, Md.
Baltimore

250 B&O
WM
a
b

40'
45'
PENNSYLVANIA
WEST VIRGINIA

72

WEST VAN VORHIS
MAIDSVILLE
RANDALL
0.0
Brave
B&O/WM VAN VORHIS
STAR CITY
100
Seneca
45
MORGANTOWN
MORGANTOWN
MN
LackNall
Russell
MON
105
B&O/WM UFFINGTON
Pittsburgh
Fairmont
HILDEBRAND
Fairmont (Mon) (B&O)
Pittsburgh
Fairmont

HOARD
95
MON
40
Canyon Mine
5

Sabraton
SA
Rock Forge
Richard
Dellslow
B&O
5
Sturgisson
SN
Greer
10
Oak Park
Cascade
Masontown
MS
15
Bretz
Burk
Reedsville
RD
20
Kanes Creek
25 B&O
M&K Jct.

CHEAT RIVER

PRESTON

MONONGALIA

39°30'
80°00'
45'
79°30'
39°30'

74

CUMBERLAND

for detail, see detail map page 57B,C

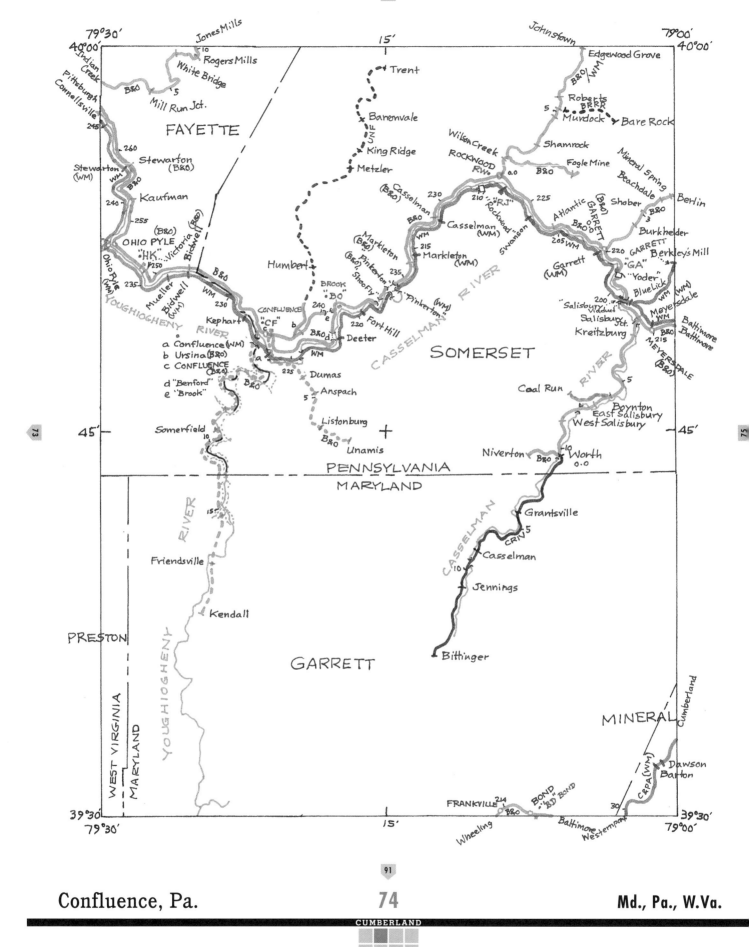

79°30'   Jones Mills
40°00'   15'   Johnstown   79°00' 40°00'
        Rogers Mills   Edgewood Grove
A. Indian   White Bridge   Trent   WM
Creek   B&O
Pittsburgh   Roberts
Connellsville   B&O   Mill Run Jct.   Barronvale   5   BRRR
        245   Murdock   Bare Rock
        260   King Ridge   Shamrock   Mineral Spring
Stewarton   Stewarton   Metzler   Wilson Creek   Beachdale
(WM)   (B&O)   ROCKWOOD   Fogle Mine
        WM   B&O   RW.   0.0   B&O
        240   Kaufman   Casselman   210   225   Atlantic   (B&O)   Shober   Berlin
        255   Markleton   (B&O)   230   "RJ."   Rockwood   GARRETT   3   B&O
        (B&O)   (B&O)   (B&O)   Casselman   WM   Burkhelder
OHIO PYLE   Victoria   Bidwell   Pinterton   B&O   215   Casselman   Swanson   205   220   GARRETT   Berkley's Mill
"HK"   250   Humbert   (B&O)   SheoFly   235   Markleton   (WM)   (WM)   "GA"   "Yoder"
        235   BROOK   240   "BO"   (WM)   Markleton   Garrett   Blue Lick
Mueller   Bidwell   WM   230   "BO"   e   220   Fort Hill   Pinterton   (WM)   200.   "Salisbury"   WM (WN)
YOUGHIOGHENY   (WM)   CONFLUENCE   b   (WM)   Viaduct   Meyersdale
        RIVER   Kephart   "CF"   c   B&O   d   Deeter   Salisbury   WM
        a Confluence (WM)   a   225   Dumas   WM   Jct.   B&O   Baltimore
        b Ursina (B&O)   B&O   5   Anspach   Kreitzburg   Baltimore
        c CONFLUENCE   CASSELMAN RIVER   SOMERSET   215
        (B&O)   Listonburg   MEYERSDALE
        d "Benford"   B&O   Coal Run   RIVER   5   (B&O)
        e "Brook"   Somerfield   Unamis   Boynton
45'   10   East Salisbury
        PENNSYLVANIA   Niverton   10   Worth   West Salisbury   45'
        MARYLAND   B&O   0.0

                                          CASSELMAN
        Friendsville   15'   Grantsville
                                          CRV   5
        RIVER   Casselman
        YOUGHIOGHENY   10
        Kendall   Jennings

PRESTON   GARRETT

WEST VIRGINIA   MARYLAND   Bittinger

                                                      MINERAL   Cumberland

                                                            C&PA (WM)   Dawson
                                                                  Barton
                                    FRANKVILLE   214   BOND   "BD" BOND   30
39°30'   B&O   Baltimore   79°00'   39°30'
79°30'   15'   Wheeling   Westernport

Confluence, Pa.   74   Md., Pa., W.Va.

CUMBERLAND

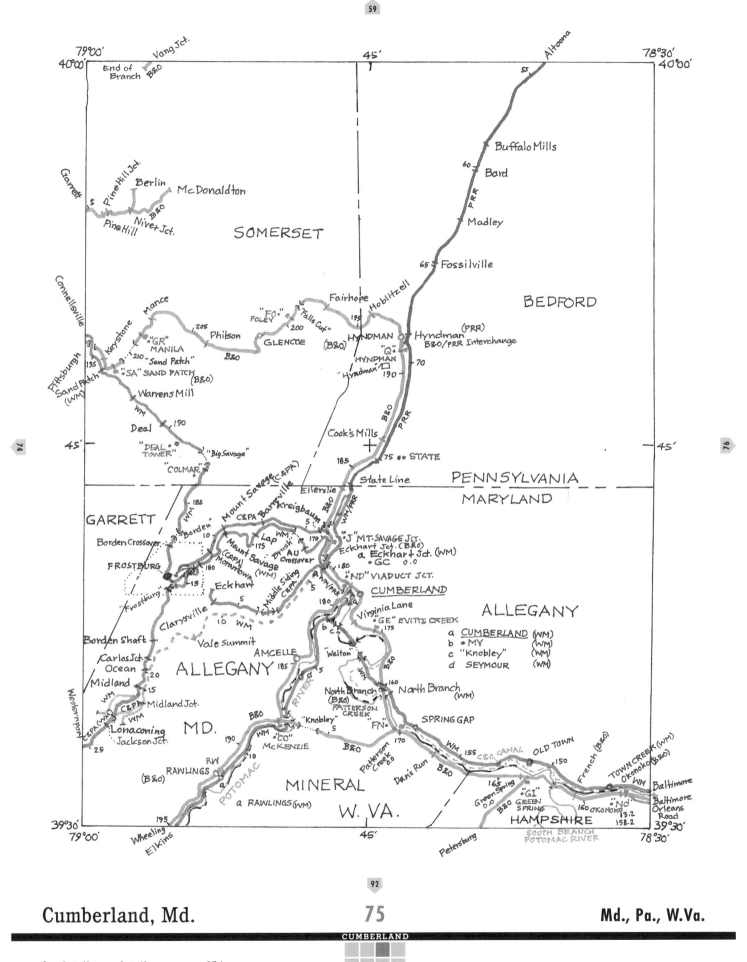

40°00' 79°00' 45' 78°30' 40°00'

Vang Jct. Altoona

End of Branch B&O 55

Buffalo Mills

60 Bard

PRR

Madley

Garrett Pine Hill Jct. Berlin McDonaldton

5 Mance SOMERSET 65 Fossilville

Connellsville Keystone "GR" 205 Philson BEDFORD

MANILA Fairhope Hoblitzell

Pittsburgh "Sand Patch" FOLEY "FO" Falls Cut" 195

195 210 "SA" SAND PATCH (B&O) "Falls Cut" 200 HYNDMAN or Hyndman (PRR)

Sand Patch Warrens Mill GLENCOE (B&O) B&O/PRR Interchange

(WM) WM Deal 190 "Q" 190

B&O HYNDMAN

"Hyndman" 190 70

45' "DEAL "Big Savage" B&O PRR

TOWER" Cook's Mills 45'

"COLMAR" 185 75 •• STATE ALLEGANY

Mount Savage (C&PA) State Line PENNSYLVANIA

185 Ellerslie MARYLAND

GARRETT WM C&PA Barnsville Kreigbaum

Borden 5 B&O "J" MT. SAVAGE JCT.

Borden Crossover 10 175 Lap WM 170 Eckhart Jct. (B&O)

Mount Savage Brush AU a Eckhart + Jct. (WM)

FROSTBURG (C&PA) Morantown Crossover GC 0.0

180 (WM) 180 "ND" VIADUCT JCT.

"Frostburg" 15 Eckhart Middle Siding WM/PRR CUMBERLAND

5 C&PA 5 A WM/PRR 5

Clarysville 10 WM 180 5 "a" Virginia Lane ALLEGANY

Borden Shaft Vale Summit b "GE" EVITTS CREEK a CUMBERLAND (WM)

Carlos Jct. AMCELLE c 175 b •MY (WM)

Ocean 185 Welton c "Knobley" (WM)

Midland 20 5 d SEYMOUR (WM)

Westernport 15 North Branch 160

WM C&PA Midland Jct. (B&O) North Branch

C&PA(WM) WM PATTERSON (WM)

Lonaconing Jackson Jct. "Knobley" CREEK SPRING GAP

25 B&O FN" 170 WM 155 C&O. CANAL OLD TOWN French (B&O)

MD. 190 WM "CO" B&O Patterson 150 TOWN CREEK (WM)

RW McKENZIE Creek Den's Run B&O Okonoko (B&O)

RAWLINGS 10 0.0 165 Baltimore

(B&O) "a" MINERAL Green Spring "NO" WM

a RAWLINGS (WM) 0.0 160 OKONOKO 13.2 Baltimore

W. VA. B&O GREEN 158.2 Orleans

195 SPRING Road

39°30' Wheeling 45' HAMPSHIRE 39°30'

Elkins Petersburg SOUTH BRANCH 78°30'

79°00' POTOMAC RIVER

Cumberland, Md.   75   Md., Pa., W.Va.

CUMBERLAND

for detail, see detail map page 65A

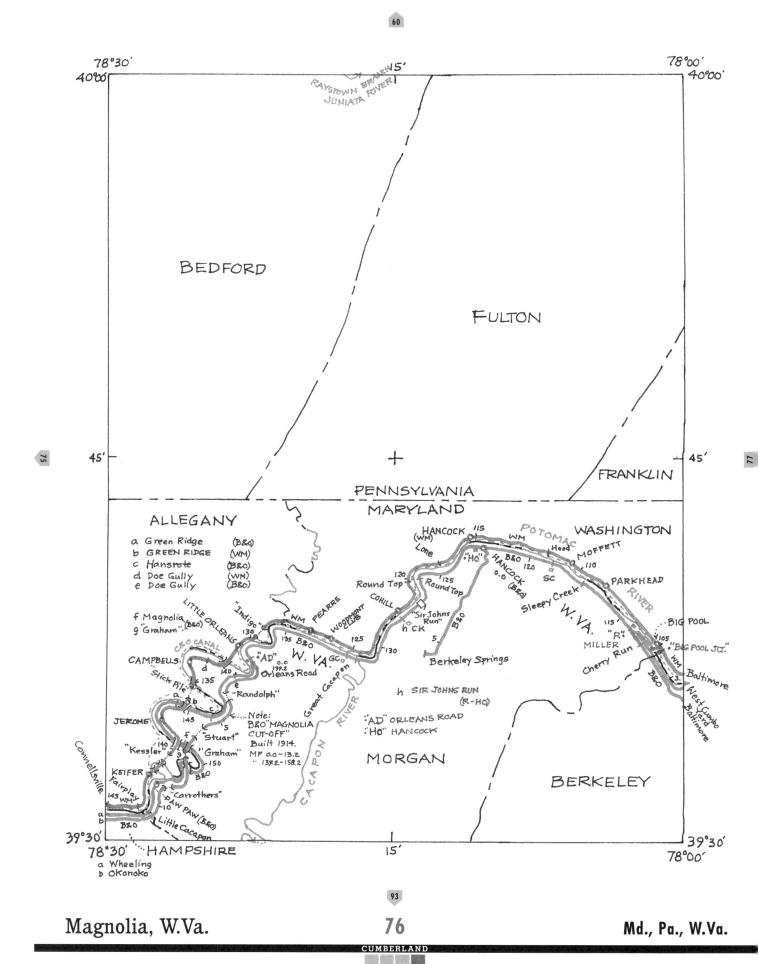

60

78°30'
40°00'

78°00'
40°00'

RAYSTOWN BRANCH
JUNIATA RIVER
15'

BEDFORD

FULTON

75

45'

45'

77

FRANKLIN

PENNSYLVANIA
MARYLAND

ALLEGANY

a Green Ridge   (B&O)
b GREEN RIDGE   (WM)
c Hansrote   (B&O)
d Doe Gully   (WM)
e Doe Gully   (B&O)

HANCOCK
(WM)
Lone

115

POTOMAC

WASHINGTON

WM

Hood

MOFFETT

"HO"

HANCOCK

B&O

120

110

SC

PARKHEAD

Round Top

120
Round Top

125

0.0
(B&O)

Sleepy Creek

W. VA.

RIVER

f Magnolia (B&O)
g "Graham"

LITTLE ORLEANS

"Indigo"

130

WM
135

PEARRE

B&O

WOODMONT
CLUB

COHILL

"Sir Johns
Run"

125

B&O

5.

Berkeley Springs

115

MILLER

"R"

Cherry Run

BIG POOL

105

"BIG POOL JCT."

CAMPBELLS

C&O CANAL

"AD"

0.0
139.2
Orleans Road

W. VA.

GC

130

h CK

WM
Baltimore

B&O

West Cumbo
Yard

Stick Pile

d

140

135

e

"Randolph"

h SIR JOHNS RUN
(R-HO)

Baltimore

JEROME

a

b

c

145

5

Note:
B&O "MAGNOLIA
CUT-OFF"
Built 1914.
MP 0.0-13.2
" 139.2-158.2

"AD" ORLEANS ROAD
"HO" HANCOCK

MORGAN

BERKELEY

Connellsville

f

"Stuart"

"Kessler"

140

9

"Graham"
150

B&O

KEIFER
Fairplay
145 WM

"Carrothers"

"PAW PAW (B&O)
10

B&O

Little Cacapon

39°30'
78°30' HAMPSHIRE

a Wheeling
b Okonoko

Great Cacapon RIVER

CACAPON RIVER

15'

39°30'
78°00'

93

Magnolia, W.Va.   76   Md., Pa., W.Va.

CUMBERLAND

FULTON

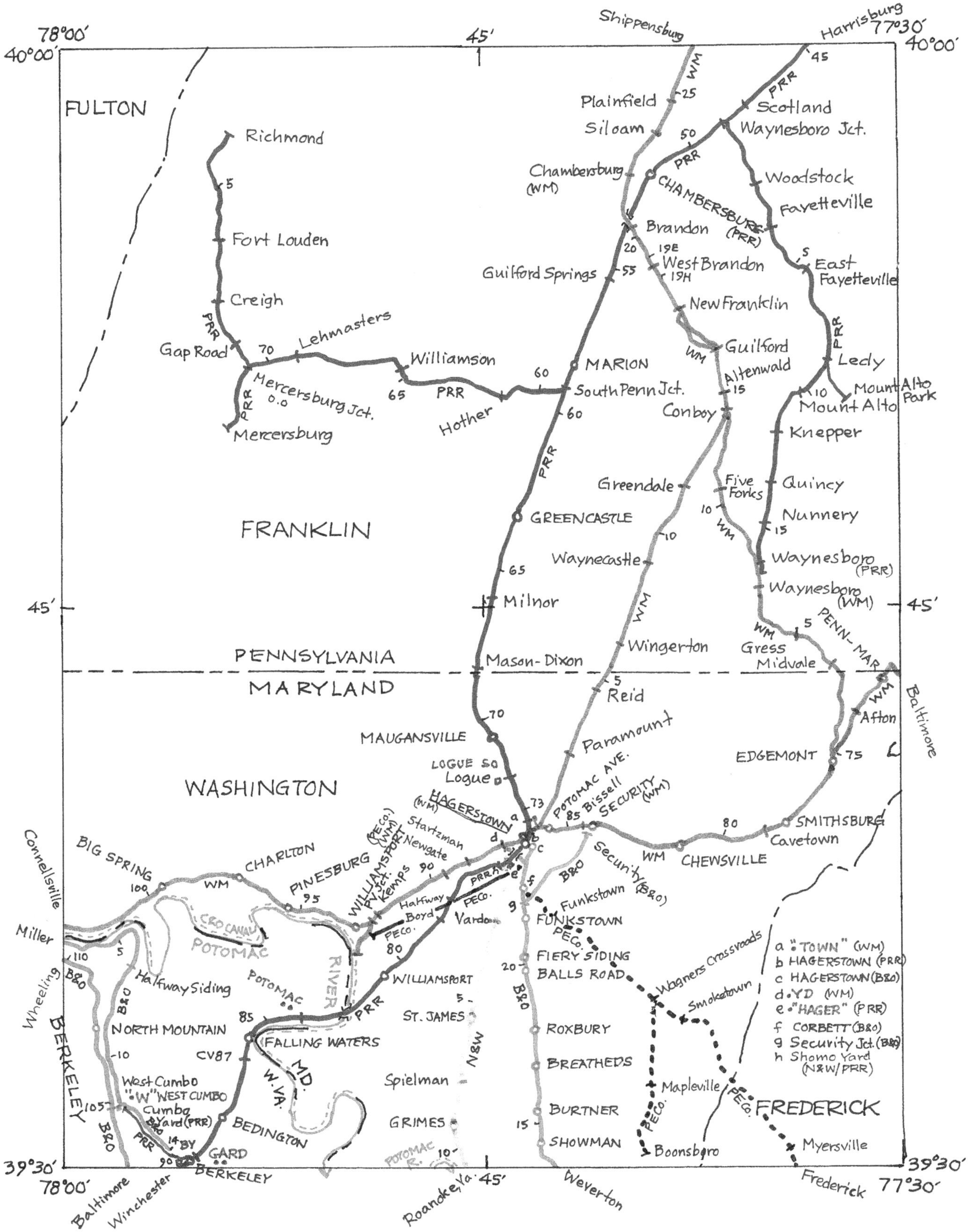

78°00' 45' Shippensburg Harrisburg 77°30'
40°00' 40°00'

Plainfield 25 Scotland
Siloam 50 Waynesboro Jct.
Richmond
5
Chambersburg CHAMBERSBURG Woodstock
(WM) Fayetteville
Fort Louden Brandon 5 East
(PRR) Fayetteville
20 19E
Creigh Guilford Springs 55 West Brandon
19H
Lehmasters New Franklin
Gap Road 70 Williamson WM Guilford Ledy
PRR Mercersburg Jct. 0.0 Altenwald
65 PRR 60 MARION 15 10 Mount Alto
Mercersburg Hother South Penn Jct. Conboy Mount Alto Park
60 Knepper

FRANKLIN PRR Greendale Five Forks Quincy
65 GREENCASTLE 10 Nunnery
Waynecastle 10 15
WM
45' Milnor 10 Waynesboro 45'
65 WM (PRR)
PENNSYLVANIA Mason-Dixon Wingerton Waynesboro PENN-
MARYLAND WM Gress (WM) MAR
5 Reid WM Midvale 5
70 Afton Baltimore
MAUGANSVILLE Paramount 75
LOGUE SQ EDGEMONT
WASHINGTON Logue Potomac Ave. SMITHSBURG
HAGERSTOWN 73 85 Bissell SECURITY 80 Cavetown
(WM) Startzman (WM)
Connellsville CHARLTON Newgate WM CHEWSVILLE
BIG SPRING PINESBURG 90 PRR B&O
WM 95 Kemps Security (B&O)
100 Halfway PECO. Funkstown
Miller Boyd Vardo 9 FUNKSTOWN Wagners Crossroads
110 5 80 PECO. a "TOWN" (WM)
B&O WILLIAMSPORT 20 FIERY SIDING Smoketown b HAGERSTOWN (PRR)
Halfway Siding St. James BALLS ROAD c HAGERSTOWN (B&O)
NORTH MOUNTAIN 85 PRR 5 d YD (WM)
CV87 N&W B&O e "HAGER" (PRR)
10 FALLING WATERS ROXBURY Mapleville f CORBETT (B&O)
West Cumbo Spielman BREATHEDS g Security Jct. (B&O)
"W" WEST CUMBO MD. GRIMES 15 BURTNER FREDERICK h Shomo Yard
Cumbo W. VA. (N&W/PRR)
Yard (PRR) BEDINGTON SHOWMAN
105 14 BY PECO.
B&O PRR 90 GARD POTOMAC R. Boonsboro Myersville
39°30' BERKELEY 39°30'
78°00' Baltimore 45' Weverton Frederick 77°30'
Winchester Roanoke, Va.

BERKELEY

Hagerstown, Md. 77 Md., Pa., W.Va.

BALTIMORE

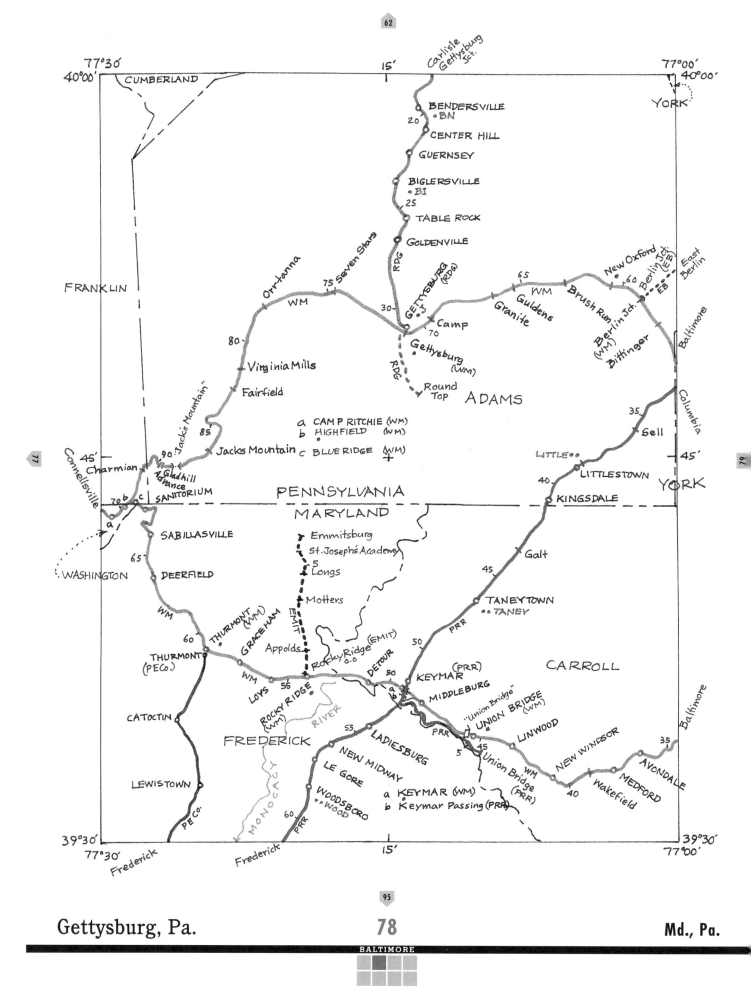

CUMBERLAND

77°30'
40°00'

15'

77°00'
40°00'

Carlisle
Gettysburg Jct.

YORK

BENDERSVILLE
• BN
20

CENTER HILL

GUERNSEY

BIGLERSVILLE
• BI
25

TABLE ROCK

GOLDENVILLE

RDG

FRANKLIN

Orrtanna

Seven Stars
75

WM

30

GETTYSBURG
(RDG)

J

Camp
70

Gettysburg
(WM)

RDG

Round
Top

New Oxford

Berlin Jct.
(EB)

East
Berlin

65

WM

Guldens

Brush Run Jct.

60

EB

Berlin Jct.
(WM)

Bittinger

Granite

Baltimore

ADAMS

Columbia

80

Virginia Mills

Fairfield

85

Jacks Mountain

Jacks Mountain

a CAMP RITCHIE (WM)
b HIGHFIELD (WM)
c BLUE RIDGE (WM)

35

Sell

77

45'

Connellsville

Charmian

90

a
Gladhill
Advance
SANITORIUM

70 b
c

a

PENNSYLVANIA

MARYLAND

LITTLE

40

LITTLESTOWN

KINGSDALE

YORK

45'

79

SABILLASVILLE

Emmitsburg
St. Joseph's Academy

5
Longs

Galt

WASHINGTON

65

DEERFIELD

Motters

45

TANEYTOWN
•• TANEY

CARROLL

WM

THURMONT
(WM)

GRACEHAM

EMIT

Appolds

Rocky Ridge (EMIT)
0.0

DETOUR

50

PRR

60

THURMONT
(PE Co.)

WM

LOYS

55

Rocky Ridge
(WM)

50

KEYMAR (PRR)

MIDDLEBURG

a
b

"Union Bridge"

UNION BRIDGE
(WM)

LINWOOD

CATOCTIN

RIVER

55

PRR

Baltimore

FREDERICK

LADIESBURG

PRR

NEW MIDWAY

5

Union Bridge
(PRR)

NEW WINDSOR

35

AVONDALE

LEWISTOWN

LE GORE

MONOCACY

WOODSBORO
•• WOOD

a KEYMAR (WM)
b Keymar Passing (PRR)

WM

40

Wakefield

MEDFORD

PE Co.

60

PRR

39°30'

77°30'
Frederick

Frederick

15'

39°30'
77°00'

Gettysburg, Pa.

78

Md., Pa.

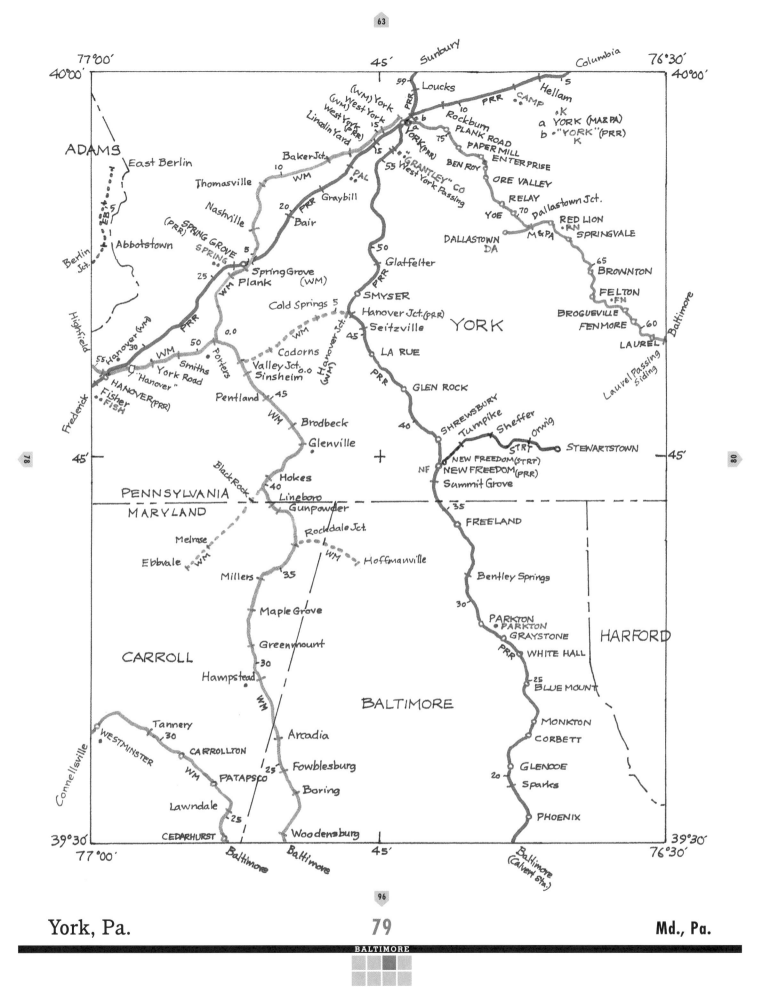

77°00' 45' Sunbury Columbia 76°30'
40°00' 40°00'

ADAMS

East Berlin

(WM) York
West York
(WM)
West York (PRR)
Lincoln Yard

Loucks Hellam 5
59 PRR CAMP
b 10 PRR
a Rockburn a YORK (MA&PA)
15 75 PLANK ROAD b "YORK"(PRR)
YORK(PRR) PAPER MILL K
ENTERPRISE
BEN ROY ORE VALLEY
RELAY

Thomasville Baker Jct.
10 WM PAL
20 PRR Graybill 55 "GRANTLEY" CO
West York Passing
Nashville Bair
50

SPRING GROVE
(PRR)
SPRING Glatfelter YOE 70 Dallastown Jct.
5 RED LION RN
Spring Grove 25 SMYSER DALLASTOWN M&PA SPRINGVALE
Plank (WM) DA 65
Abbotstown Cold Springs 5 BROWNTON
Berlin Hanover Jct.(PRR) FELTON
Jct. WM Seitzville FN
0.0 45 YORK BROGUEVILLE
Highfield PRR LA RUE FENMORE 60
55 Hanover (WM) 30 50 Codorns Baltimore
WM Valley Jct.0.0 GLEN ROCK LAUREL
York Road Sinsheim 40 Laurel Passing
Smiths Pentland 45 Siding
HANOVER(PRR) WM SHREWSBURY Sheffer Orwig
Fisher Brodbeck Turnpike STR
FISH 40 STEWARTSTOWN
Frederick Glenville NEW FREEDOM(STR)
45' NF NEW FREEDOM(PRR) 45'
Black Rock Summit Grove

PENNSYLVANIA Hokes
MARYLAND 40 Lineboro 35
Gunpowder FREELAND

Melrose Rockdale Jct.
Ebbvale WM WM Hoffmanville Bentley Springs
Millers 35 30
HARFORD
Maple Grove PARKTON
CARROLL Greenmount PARKTON
30 GRAYSTONE
Hampstead PRR WHITE HALL
BALTIMORE 25 BLUE MOUNT
Tannery WM
30 Arcadia MONKTON
WESTMINSTER CARROLLTON CORBETT
WM PATAPSCO 25 Fowblesburg
Lawndale Boring 20 GLENOOE
25 Sparks
CEDARHURST Woodensburg PHOENIX
39°30' 39°30'
77°00' Baltimore Baltimore 45' Baltimore 76°30'
(Calvert Sta.)

York, Pa. **79** Md., Pa.

76°30' Niago Jct. Lancaster 15' Pittsburgh Leaman Place 76°00'
40°00' Washington Boro Mellinger PRR SRC Kinzers PRR GAP 40°00'
-35 5 4 STRASBURG 50 Philadelphia
PORT (R-COLA) West Willow Parkesburg
CRES Baumgardner Coopersville 5
(R-COLA) 30 Buzzard Rock PRR 10 Refton
Star Rock LG-30 PRR New Providence 10 PRR Mars Hill
25 PRR SafeHarbor 20 Smithville LG-14 15 Q Summit
WEST HARBOR 25 PRR 15 Quarryville (PRR)
(R-COLA) HARBOR LG-21 SMITH (LO&S)
(R-COLA) Martic Forge Blackburn
Shenk's Ferry LANCASTER Mechanic Grove
-30 Pequea King's Bridge
YORK Tuquan Fairmont White Rock
McCALLS (R-COLA) Fulton House LO&S Spruce Grove Oxford
McCalls Ferry LO&S Goshen Tweedale Philadelphia
York 25 Holtwood Eldora CHESTER
HIGH ROCK MIDWAY (R-COLA) (LO&S) Westbrook
MUDDY CREEK FORKS Fishing Creek Dorsey's Peach Bottom
MC BRUCE 20 Nottingham
55 BRIDGETON PRR 45'
M&PA WOODBINE 50 Haines Sylmar
SOUTHSIDE M&PA -15 55
CASTLE FIN RHo Slate Hill Bald Pilot Rising Sun
BRYANSVILLE DELTA WEST PILOT(R-COLA)
45 Susquehanna PRR
PENNSYLVANIA PILOT Conowingo Colora
MARYLAND CARDIFF 10 Rowlandville 60 GROVE
WHITEFORD Octoraro Liberty Grove
WD 65 WEST ROCK CECIL
PYLESVILLE (R-COLA) Philadelphia
40 ROCK (R-COLA)
STREET Port Deposit QUARRY Philadelphia
TOME (R-COLA) MINNICK Jackson B&O
MINEFIELD SA AIKIN
ROCKS PNo PRR PRINCIPIO
35 SO HARFORD QUARRY 55 Principio
FERN CLIFF MINNICK PERRYVILLE GC
Hornberger's Siding PRINCIPIO 60
M&PA SHARON HAVRE DE GRACE HAVRE DE GRACE
FOREST HILL OAKINGTON (R-PERRYVILLE) HAVRE DE GRACE
30 FR Havre-de-Grace PRR OAKINGTON
BYNUM Swan Creek 60 OAKINGTON
BEL-AIR ABERDEEN B&O Swan Creek CHESAPEAKE BAY
25 KN ABERDEEN "A" 65 ABERDEEN
BALTIMORE 20 Little Gunpowder VALE US ARMY
39°30' FS FALLSTON Washington, D.C. Aberdeen
76°30' LAUREL BROOK Baltimore Proving Ground 76°00'
Baltimore 15'

Perryville, Md. **80** Md., Pa.

76°00' 40°00' 45' 75°30' 40°00'

LANCASTER

Pittsburgh

Cumru Jct.
Reading

Pomeroy
0.0

VALLEY 40

COATESVILLE

Philadelphia Philadelphia

Frazer

Green Hill
2

Fern Hill

CALN "THORN" HD
(R-THORN) "Thondale" THORNDALE

VA 35

CHRISTIANA PARKESBURG
NI "PARK" PG
ATGLEN 45

PRR

CV.

0.0
44.0

"Atglen"

Wago Jct.

3

"WC"
0.0

WEST CHESTER PRR
28.4

25

OAKBOURNE

25

WESTTOWN

LOCKSLEY

CHEYNEY

DELAWARE

MARKHAM

CONCORDVILLE
(PRR)

"CHADD" CF

CHADD's FORD

BRANDYWINE SUMMIT
25
15

Cossart

Granogue
a Guyencourt
b Winterthur
c Montchanin
. DU

Rockland

Greenville

Philadelphia

Bellevue

B&O

25

RDG

PRR

PRR

0.0

Philadelphia

PRR

Modena
South Modena
30

Mortonville

Embreeville
25 . BR

Wawaset
20

Northbrook RDG Lenape

Pocopson
PO

Chadd's Ford Jct.

MENDENHALL

ROSEDALE

SQUARE . . PRR
30

KENNETT SQUARE

PENNSYLVANIA
DELAWARE

Glen Rose

Buck Run

Laurel

Doe Run
5

Springdell

Green Lawn

Clonmel
10

Chatham

Baker

CHESTER

AVONDALE
MV

"TOUGH
TOUGHKENAMON

35

NewGarden
(B&O)

Landenburg

Yorklyn

LowerMill

Ashland

Mt.Cuba
5

10 RDG

R
D
G

B&O

WEST GROVE 40
KEL . .

PRR 45

KELTON

ELKVIEW

LINCOLN UNIVERSITY

Oxford (L&S)

L&S

OXFORD (PRR)
OX

Barnsley

Susquehanna

Rock
45'

50

AVONDALE

15

PRR

PRR

Landenburg
(PRR) Nevin

Yeatman

20

Thompson

Tweed
25

Southwood

MillCreek

Hockessin
Wooddale

Green Bank

"Kiamensi"

Stanton

Landenberg Jct.

RDG
0.0

30

"WJ" 30
LANDENBERG
JCT.

NEWPORT

Farnhurst

5

TASKER

AK

NEW CASTLE
5

B&O

PRR
5

RDG

0.0

PRR PRR
25 5

B&O

PENNSYLVANIA
MARYLAND

Providence
Provident Mills
3

Hillside

SINGERLY

Elk Mills

"SY"
SINGERLY

NEWARK 35

PRR

STANTON

B&O

RUTHBY
"Ruthby"

35

NEW CASTLE

Cooch

NEWARK
DAVIS
NY

IRON HILL

40

1

State Road

Farnhurst

CECIL

40

B&O

45

CHILDS

Eder

Bacon Hill

BIG ELK BQ
(R-DAVIS)

ELKTON

45

PRR

10 PRR

BEAR

Red Lion

Corbit

Reybold

NEW JERSEY

SALEM

DELAWARE

RIVER

Baltimore

Foys Hill

50

LESLIE

B&O

NORTH EAST
NORTH EAST (R-DAVIS)

50

CHARLESTON

55

PRR

45

MARYLAND

DELAWARE

Glasgow

Porter

5

19

PRR

KIRKWOOD

10

Washington
D.C.

NORTHEAST RIVER

ELK RIVER

CHESAPEAKE & DELAWARE CANAL

CANAL

D "CANAL"

PRR

20

MOUNT PLEASANT

Delaware City

Reybold

39°30' 39°30'

76°00' 45' 75°30'
Delmar

Wilmington, Del. 81 Del., Md., N.J., Pa.

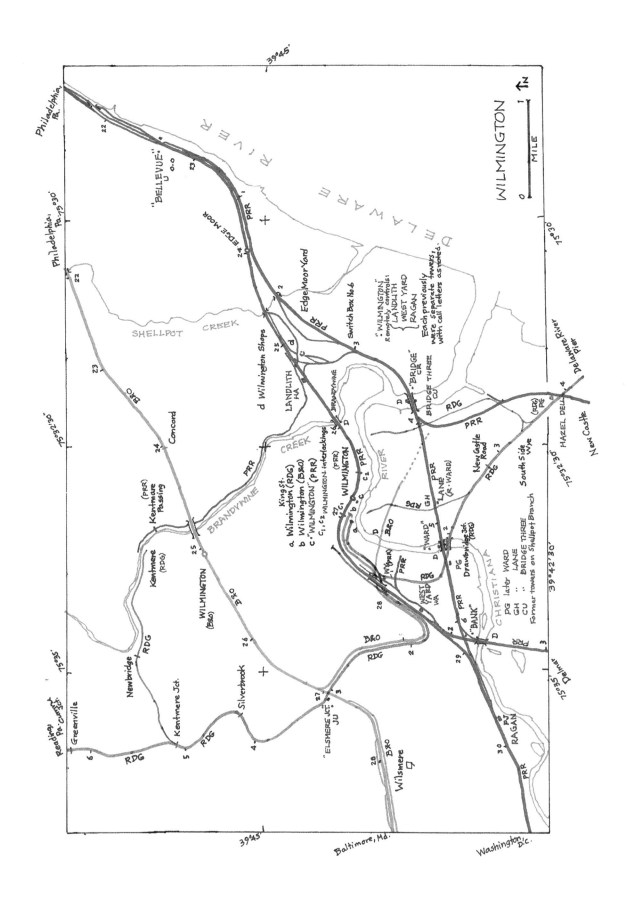

81A Wilmington, Del.

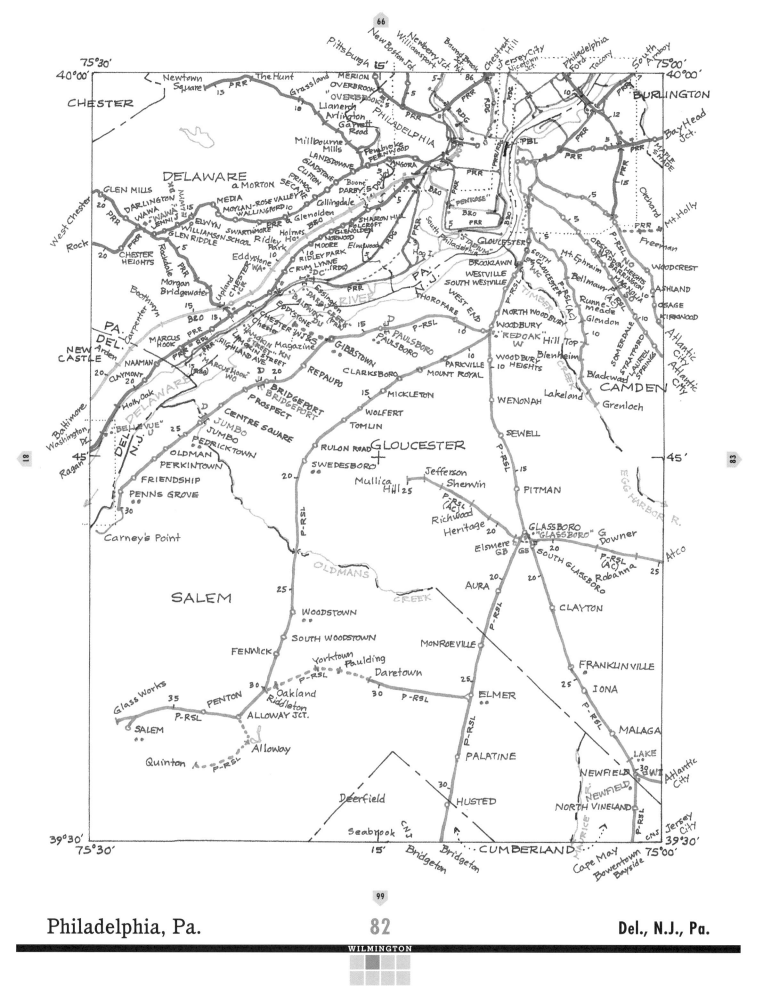

Pittsburgh 15'

Newberry Jct NY
Williamsport
New Boston Jct
Bound Brook
Chestnut Hill
Jersey City
Nicetown Jct
Philadelphia
Ford
Tacony
South Amboy

75°30'
40°00'

CHESTER

Newtown Square
The Hunt
MERION
Grassland
Overbrook
"OVERBROOK"
Llanerch
Arlington
Garrett Road
Millbourne Mills
PHILADELPHIA
Pembroke
FERNWOOD
Landsdowne
Angora
GLADSTONE
CLIFTON
Boone
PRIMOS
Morton
SECANE
DARBY
SHARON HILL
Folcroft
GLENOLDEN
NORWOOD
Elmwood

DELAWARE
GLEN MILLS
Darlington
WAWA
Lenni
ELWYN
MEDIA
MOYLAN-ROSE VALLEY
WALLINGFORD
Collingdale
Glenolden
Holmes
Ridley Park
Moore
RIDLEY PARK

West Chester
Rock
PRR
CHESTER HEIGHTS
Williamson School
Swarthmore
Eddystone

Morgan
Bridgewater
Upland
Rockdale

Boothwyn
Carpenter
Marcus Hook
Essington
CHESTER WJ&S
Chester
Magazine

PA.
DEL.
NEW CASTLE
Arden
NAAMAN
Marcus Hook
HIGHLAND AVE
Repaupo

CLAYMONT
Holly Oak

Baltimore Washington DC
"BELLEVUE"
"JUMBO"
JUMBO
PEDRICKTOWN
OLDMAN
PERKINTOWN
FRIENDSHIP
PENNS GROVE
Carney's Point

CENTRE SQUARE
PROSPECT
BRIDGEPORT
BRIDGEPORT
SWEDESBORO
RULON ROAD

SALEM
WOODSTOWN
SOUTH WOODSTOWN
FENWICK
Yorktown Paulding
Daretown
Oakland Riddleton
PENTON
ALLOWAY JCT.
Glass Works
P-RSL
SALEM
Alloway
Quinton

BURLINGTON
Bay Head Jct.
MAPLE SHADE
Mt. Holly
Freeman
WOODCREST
ASHLAND
OSAGE
KIRKWOOD
Atlantic City

GLOUCESTER
BROOKLAWN
WESTVILLE
SOUTH WESTVILLE
WEST END
THOROFARE
NORTH WOODBURY
WOODBURY
"REDOAK" W
Hill Top
WOODBURY HEIGHTS
Blenheim
Blackwood
Lakeland
Grenloch
WENONAH
SEWELL

Mt. Ephraim
South Gloucester
Barrington
Bellmawr
Runnemede
Glendon
Somerdale
STRATFORD
LAUREL SPRINGS
MAGNOLIA
CAMDEN
Atlantic City

PAULSBORO
GIBBSTOWN
CLARKSBORO
MICKLETON
WOLFERT
TOMLIN
MOUNT ROYAL
GLOUCESTER
Mullica Hill
Jefferson
Sherwin
Richwood
Heritage
GLASSBORO
"GLASSBORO" G
Downer
Elsmere
SOUTH GLASSBORO
Robanna
Atco
AURA
CLAYTON
MONROEVILLE
FRANKLINVILLE
IONA
ELMER
MALAGA
LAKE
PALATINE
NEWFIELD
HUSTED
NORTH VINELAND
Deerfield
Seabrook
Bridgeton
CUMBERLAND
Cape May
Bowentown
Bayside
Jersey City

PITMAN

39°30'
75°30'
15'
Bridgeton
75°00'
39°30'

Philadelphia, Pa.

82

Del., N.J., Pa.

WILMINGTON

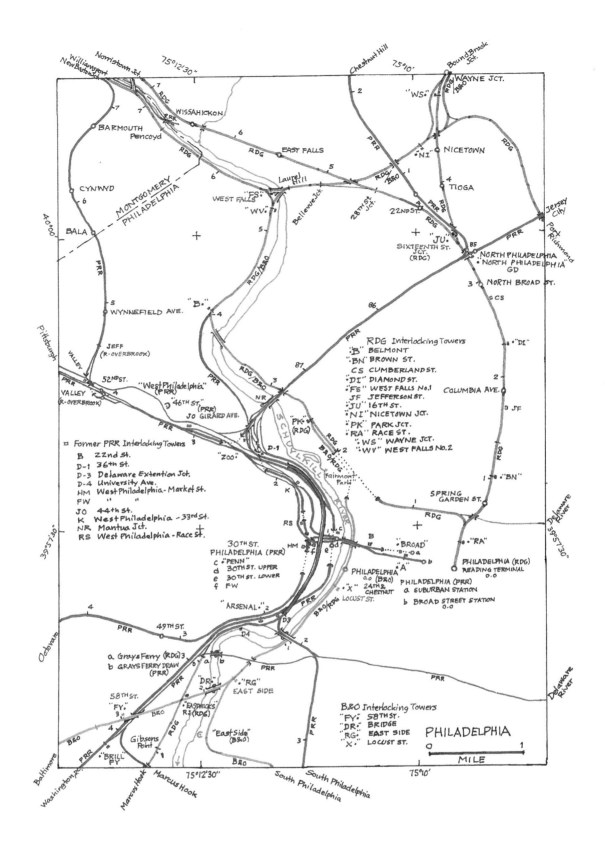

Williamsport
New Baston Jct.
Norristown Jct.
75°12'30"
Chestnut Hill
75°10'
Bound Brook Jct.
RDG WAYNE JCT.
B&O

"WS"
RDG
WISSAHICKON
RDG
"NI"
NICETOWN
RDG
RDG
BARMOUTH
Pencoyd
RDG
EAST FALLS
RDG
PRR
4
TIOGA
PRR
RDG/B&O
CYNWYD
6
MONTGOMERY
PHILADELPHIA
Laurel Hill
West Falls
"FS"
"WV"
Bellevue Jct.
5
28TH ST. JCT.
RDG/B&O
22ND ST.
PRR
RDG
Jersey City
Port Richmond
BALA
PRR
"JU"
SIXTEENTH ST. JCT. (RDG)
B5
NORTH PHILADELPHIA
NORTH PHILADELPHIA GD
5
RDG/B&O
3
NORTH BROAD ST.
"B."
WYNNEFIELD AVE.
4
B6
cs
RDG Interlocking Towers
"B" BELMONT
"BN" BROWN ST.
CS CUMBERLAND ST.
"DI" DIAMOND ST.
"FS" WEST FALLS No.1
JF JEFFERSON ST.
"JU" 16TH ST.
"NI" NICETOWN JCT.
"PK" PARK JCT.
"RA" RACE ST.
"WS" WAYNE JCT.
"WV" WEST FALLS No.2
"DI"
JEFF
(R-OVERBROOK)
VALLEY
52ND ST.
PRR
RDG/B&O
3
NR
87
PRR
2
COLUMBIA AVE.
JF
"West Philadelphia" (PRR)
"46TH ST." (PRR)
JO GIRARD AVE.
SCHUYLKILL RIVER
PK (RDG)
RDG
2
"BN"
VALLEY (R-OVERBROOK)
PRR
3
Former PRR Interlocking Towers
B 22nd St.
D-1 36th St.
D-3 Delaware Extention Jct.
D-4 University Ave.
HM West Philadelphia - Market St.
FW " "
JO 44th St.
K West Philadelphia - 33rd St.
NR Mantua Jct.
RS West Philadelphia - Race St.
"ZOO"
D-1
B&O/RDG
Fairmont Park
SPRING GARDEN ST.
RDG
1
K
"LC"
RS
HM
30TH ST. PHILADELPHIA (PRR)
c "PENN"
d 30TH ST. UPPER
e 30TH ST. LOWER
f FW
d
f
B
"BROAD"
a
b
"RA"
"A"
Philadelphia 0.0 (B&O)
PHILADELPHIA (PRR)
a. SUBURBAN STATION
b BROAD STREET STATION 0.0
PHILADELPHIA (RDG)
READING TERMINAL 0.0
"X" 24th & CHESTNUT
LOCUST ST.
"ARSENAL"
2
B&O/RDG
D3
D4
2
PRR
49TH ST.
3
a Grays Ferry (RDG) 3
b GRAYS FERRY DRAW (PRR)
b a b
PRR
PRR
58TH ST.
"FY"
"DR"
"RG"
EAST SIDE
EASTWICKS RJ (RDG)
RDG
"East Side" (B&O)
B&O Interlocking Towers
"FY" 58TH ST.
"DR" BRIDGE
"RG" EAST SIDE
"X" LOCUST ST.
PHILADELPHIA
Gibsons Point
B&O
0 MILE 1
Baltimore
Washington
PRR
"BRILL"
FY
Marcus Hook
Marcus Hook
75°12'30"
South Philadelphia
South Philadelphia
75°10'

82A Philadelphia, Pa.

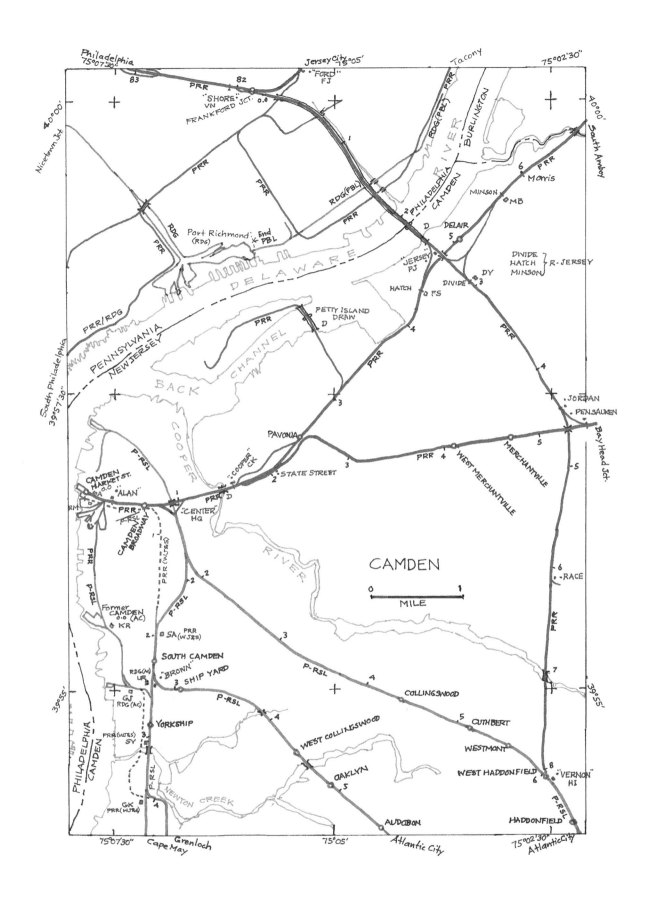

82B Camden, N.J.

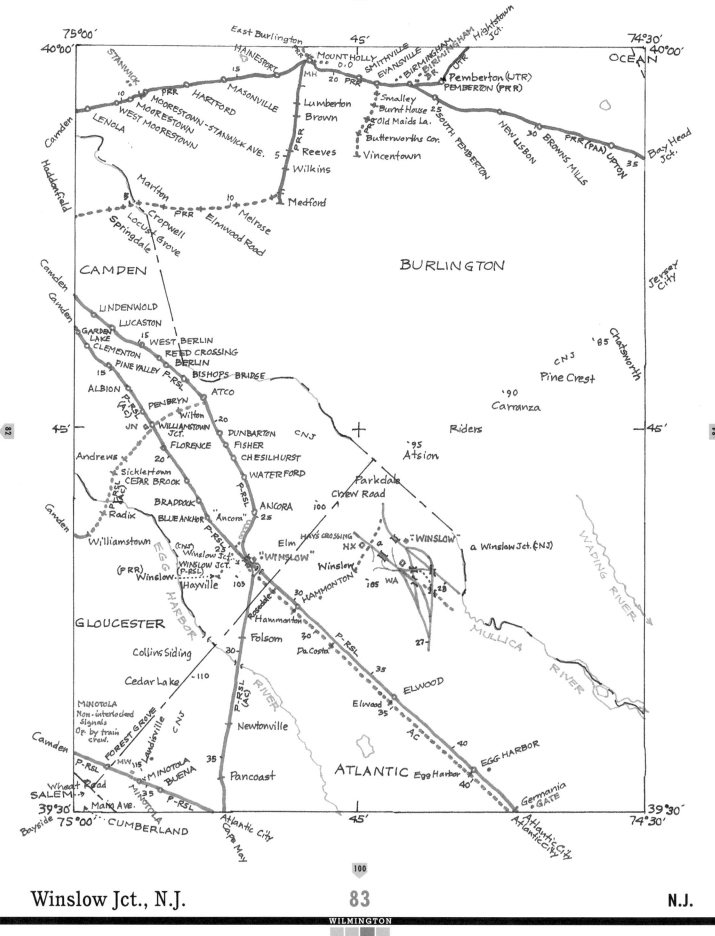

Winslow Jct., N.J.   **83**   N.J.

WILMINGTON

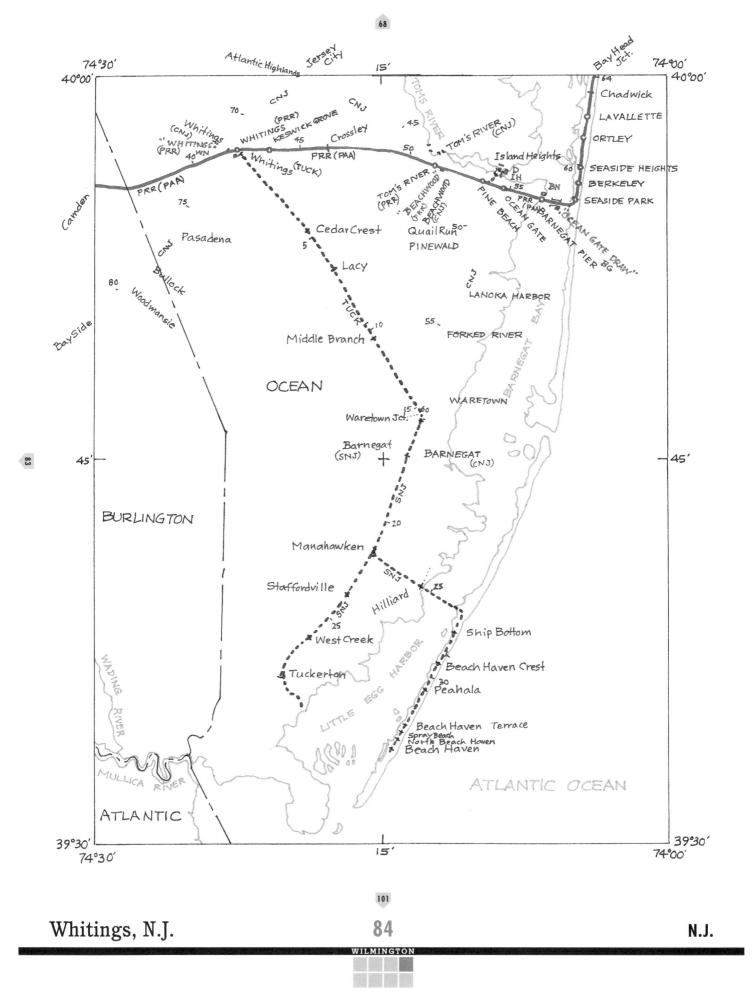

74°30'
40°00'

Atlantic Highlands Jersey City 15' Bay Head Jct. 74°00' 40°00'

64 Chadwick

70 CNJ WHITINGS CNJ .45 LAVALLETTE

Whitings (PRR) KESWICK GROVE TOM'S RIVER (CNJ) ORTLEY

(CNJ) WHITINGS 45 Crossley 50 Island Heights

"WHITINGS" 40 WN Whitings (TUCK) PRR (PAA) D 66 SEASIDE HEIGHTS

(PRR) IH BERKELEY

PRR (PAA) TOM'S RIVER 55 BN SEASIDE PARK

75 (PRR) BEACHWOOD (PRR) PRR "OCEAN GATE DRAW"

(Camden) Pasadena CedarCrest BEACHWOOD (CNJ) PINE BEACH BARNEGAT PIER BG

CNJ 5 QuailRun 50 OCEAN GATE

Bullock Lacy PINEWALD

80 Woodmansie

BaySide TUCK LANOKA HARBOR

10 55 FORKED RIVER

Middle Branch

OCEAN BARNEGAT BAY

WARETOWN

15 60

Waretown Jct.

83 45' Barnegat BARNEGAT (CNJ) 45'

(SNJ)

BURLINGTON CNJ

20

Manahawken

SNJ

Staffordville 23

SNJ Hilliard Ship Bottom

25

West Creek Beach Haven Crest

Tuckerton 30 Peahala

LITTLE EGG HARBOR Beach Haven Terrace

WADING RIVER Spray Beach

North Beach Haven

Beach Haven

MULLICA RIVER ATLANTIC OCEAN

ATLANTIC

39°30' 15' 39°30'

74°30' 74°00'

Whitings, N.J.

N.J.

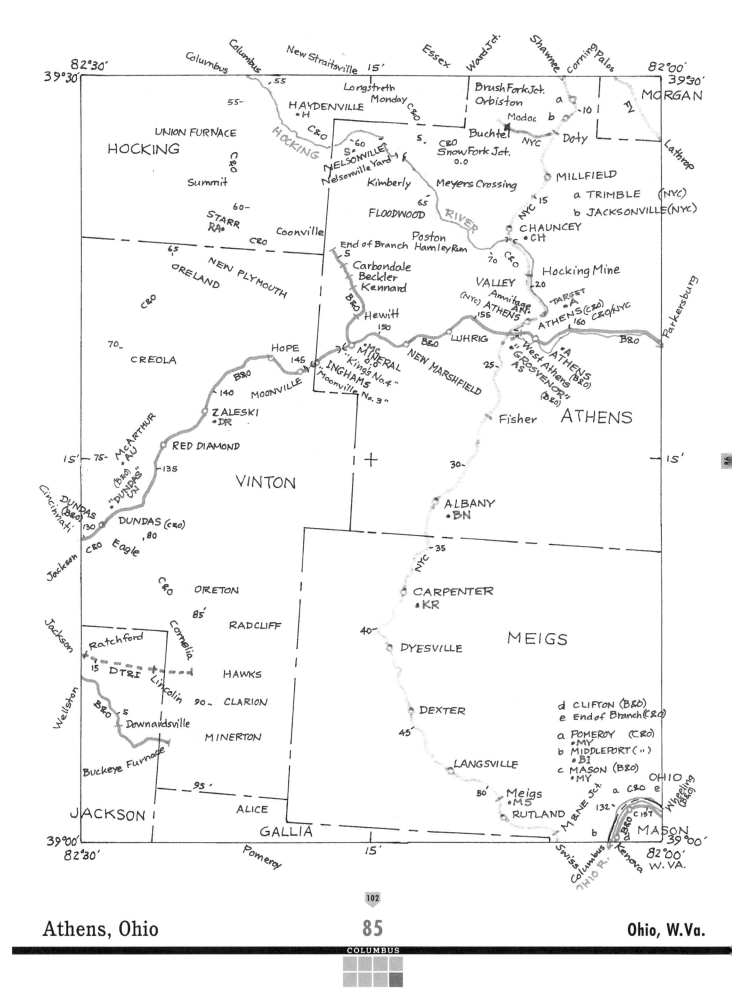

82°30'
39°30'

Columbus

Columbus   New Straitsville   15'

Longstreth
Monday C&O

55
,55

55-

HAYDENVILLE
• H

UNION FURNACE

HOCKING

C&O

Summit

STARR
RA•

60-

C&O

65

ORELAND

NEW PLYMOUTH

Coonville

C&O

70-

CREOLA

HOPE

145

B&O

140
MOONVILLE

ZALESKI
• DR

15'   75-

McARTHUR
• AU

RED DIAMOND

135

"DUNDAS"
UN
(B&O)

DUNDAS
(B&O)

130

DUNDAS (C&O)
,80

Cincinnati

Jackson   C&O   Eagle

Jackson

Wellston

B&O

Ratchford
15   DT&I

5

Lincoln

Cornelia

C&O

ORETON

85'

RADCLIFF

HAWKS

90   CLARION

Downardsville

MINERTON

Buckeye Furnace

95'

JACKSON

ALICE

GALLIA

39°00'
82°30'

Pomeroy

15'

HOCKING

HOCKING

-60
S.
NELSONVILLE
Nelsonville Yard

Kimberly

FLOODWOOD

S

End of Branch

Carbondale
Beckler
Kennard

B&O

Hewitt

150

"Kings No.4"

MS

MINERAL
0.0

INGHAMS

"Moonville No. 3"

VINTON

ALBANY
• BN

35

NYC

CARPENTER
• KR

40-

DYESVILLE

DEXTER

45'

LANGSVILLE

50'

Essex   Ward Jct.   Shawnee   Corning Palos
82°00'
39°30'

Brush Fork Jct.
Orbiston

a

10

MORGAN
Fy

Modac   b

C&O
SnowFork Jct.
0.0

S.

Meyers Crossing

65

RIVER

Buchtel   NYC   Doty

MILLFIELD

NYC   15

a TRIMBLE   (NYC)

b JACKSONVILLE (NYC)

CHAUNCEY
• CH

Lathrop

Poston
HamleyRim

70
C&O

C&O

VALLEY

Hocking Mine

-20

Amitage
AN.
(NYC)
ATHENS
,155

TARGET
• A

ATHENS (C&O)
160   C&O/NYC

B&O   LUHRIG

NEW MARSHFIELD

ATHENS
• A
(B&O)

West Athens
• AS
"GROSVENOR"
(B&O)

25-

Parkersburg

B&O

Fisher

ATHENS

30-

MEIGS

d CLIFTON (B&O)
e End of Branch (C&O)

a POMEROY   (C&O)
• MY
b MIDDLEPORT (")
• BI
c MASON (B&O)
• MY

Meigs
• MS

M&NE Jct.   a C&O

132

RUTLAND

B&O   c 157

b

Swiss Jct.

Columbus

Ohio R.

Kenova

OHIO
e
Wheeling
(B&O)

MASON

39°00'
82°00'
W. VA.

86

15'

15'

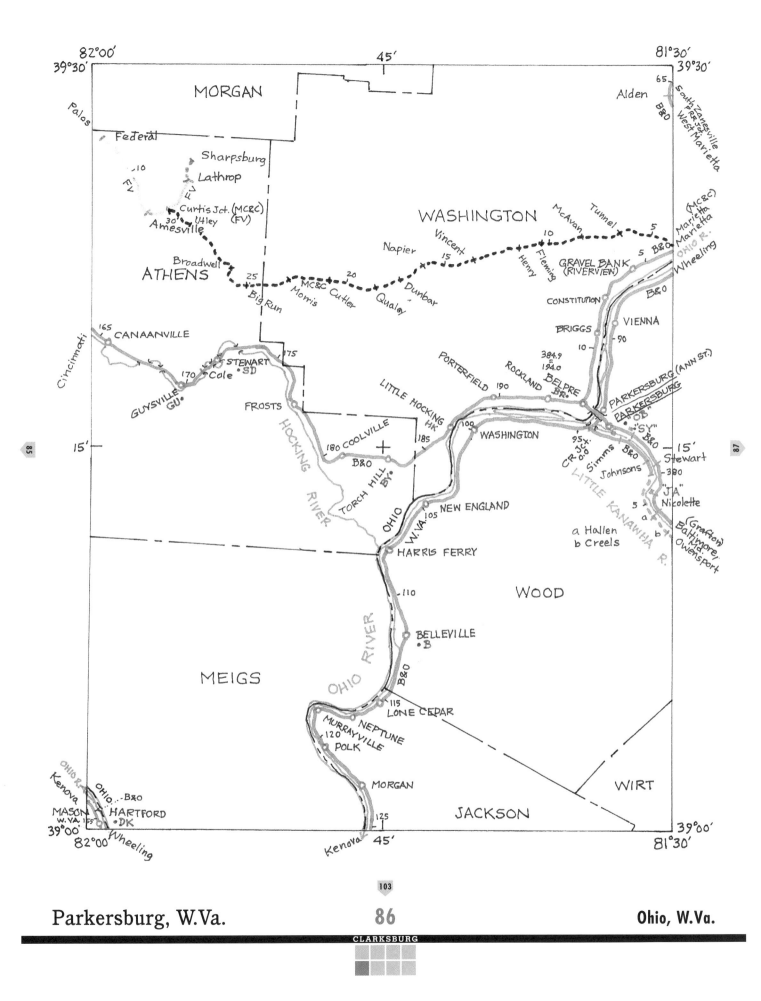

MORGAN

82°00'
39°30'

45'

81°30'
39°30'

Palos

Federal

Sharpsburg

Lathrop

-10

FV

FV

Curtis Jct.(MC&C)
(FV)

30' Utley

Amesville

Broadwell

ATHENS

Big Run

Morris

MC&C Cutter

Qualey

Dunbar

WASHINGTON

Napier

Vincent

15

Henry

Fleming

McAvan

Tunnel

10

5

B&O

Alden

65

B&O

South Zanesville
RR. Jct.
West Marietta

(MC&C)

Marietta

Marietta

OHIO R.

Wheeling

GRAVEL BANK
(RIVERVIEW)

B&O

CONSTITUTION

BRIGGS

10

VIENNA

90

384.9
&
194.0

ROCKLAND

PORTERFIELD

190

BELPRE
BR.

PARKERSBURG (ANN ST.)

165 CANAANVILLE

STEWART
SD

170 Cole

GUYSVILLE
GU

FROSTS

Cincinnati

15'

180 COOLVILLE

B&O

TORCH HILL
BY.

HOCKING RIVER

175

LITTLE HOCKING
HK

185

100

WASHINGTON

95 Jct.
0.0

Simms

B&O

CR. Jct.
0.0

Johnsons

LITTLE KANAWHA R.

PARKERSBURG
"OB"
"SY"

B&O

Stewart

380

"JA"
Nicolette

5

a

b

(Grafton)
Baltimore,
Md.
Owensport

15'

OHIO

W.VA.

105

NEW ENGLAND

HARRIS FERRY

110

a Hallen
b Creels

WOOD

BELLEVILLE
B

OHIO RIVER

B&O

115

LONE CEDAR

NEPTUNE

MURRAYVILLE

120

POLK

MORGAN

125

MEIGS

OHIO R.

Kenova

MASON
W.VA. 155

HARTFORD
DK

OHIO

B&O

39°00'
82°00'

Wheeling

Kenova

45'

JACKSON

WIRT

39°00'
81°30'

85

87

103

Parkersburg, W.Va.   86   Ohio, W.Va.

CLARKSBURG

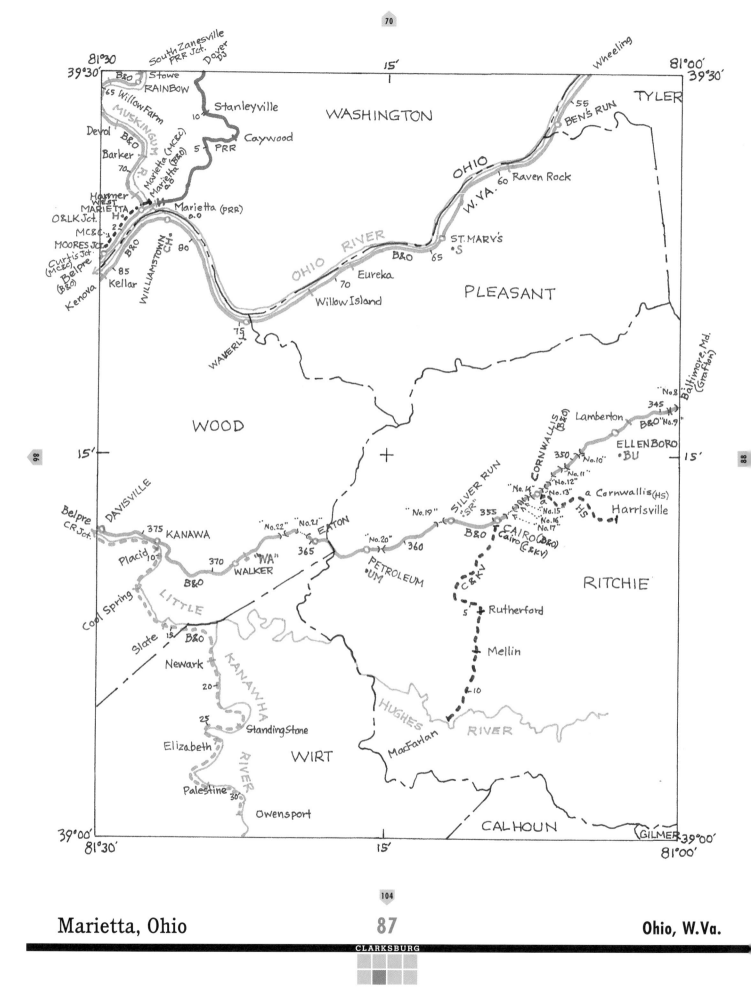

81°30
39°30   Wheeling   81°00'
South Zanesville   15'   39°30'
PRR Jct. Dover
B&O Stowe   D.J.   TYLER
'65 Willow Farm RAINBOW   55
                          Stanleyville   WASHINGTON   BEN'S RUN
MUSKINGUM R. 10   OHIO
Devol   B&O   Caywood   W. VA. 60 Raven Rock
Barker   Marietta (MC&C) 5 PRR
70   Marietta (B&O)
                Marietta (O&LK)
Harmer WEST
MARIETTA MARIETTA   Marietta (PRR)   ST. MARY'S
O&LK Jct. H.   0.0   B&O 65 S
MC&C   PLEASANT
MOORES JCT.   80
Curtis Jct.   WILLIAMSTOWN CH.   Eureka 70 OHIO RIVER
(MC&C) Belpre
(B&O) Kellar 85   Willow Island
Kenova   PLEASANT

                                                              Baltimore, Md.
                                    WOOD   (Grafton)
                                                                          No.8
                                                    Cornwallis (B&O) Lamberton   345   B&O No.9
15'   +   350   No.10   ELLENBORO   15'
98   No.11   BU   88
                                                            No.12
                                              SILVER RUN No.13
                                      No.19   SR   No.14   a Cornwallis (HS)
Belpre   DAVISVILLE   355   No.15   Harrisville
CR Jct.   No.20   B&O CAIRO (B&O) No.16
375 KANAWA   No.22 No.21 EATON   Cairo (C&KV) No.17
Placid 10   365   360
        370   PETROLEUM   RITCHIE
Cool Spring WALKER B&O   UM   C&KV
        LITTLE   Rutherford
Slate 15 B&O   5
Newark   Mellin
        KANAWHA 20   10
                25   Standing Stone
Elizabeth   RIVER   HUGHES   RIVER
Palestine 30   MacFarlan
        Owensport   WIRT
39°00'   CALHOUN   GILMER 39°00'
81°30'   15'   81°00'

Marietta, Ohio   87   Ohio, W.Va.

CLARKSBURG

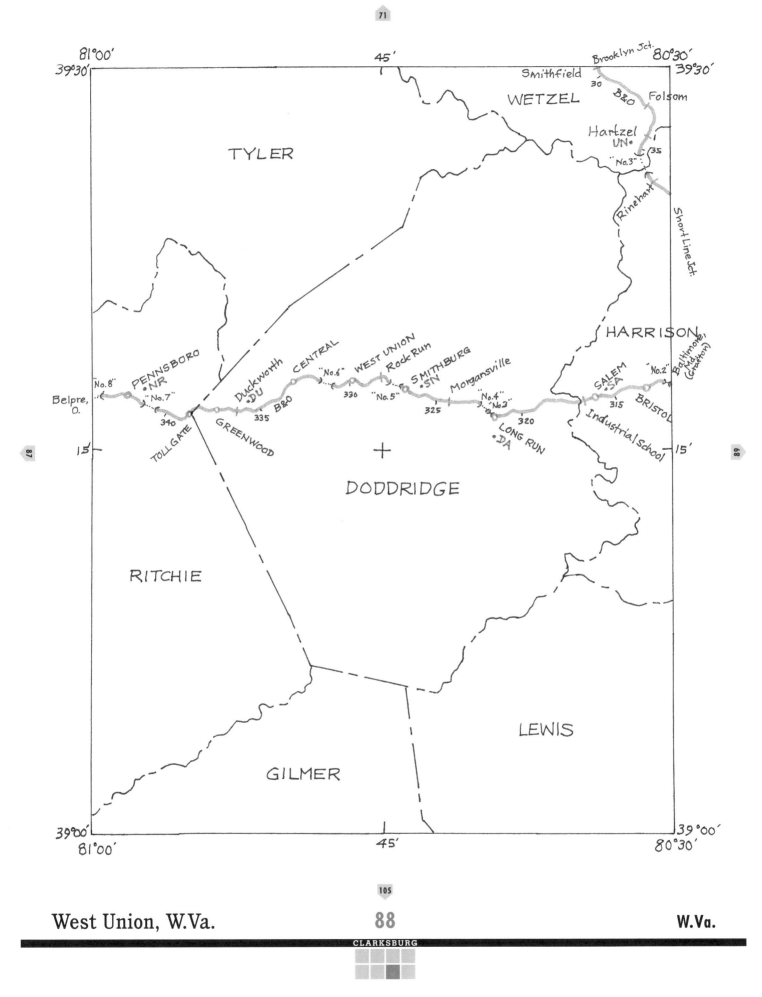

71

81°00'
39°30'

45'

Brooklyn Jct.
Smithfield
WETZEL
30
B&O
Folsom

Hartzel
UN•
"No.3"
35

Rinehart

Short Line Jct.

80°30'
39°30'

TYLER

HARRISON

Baltimore,
(Grafton)

PENNSBORO
•NR

Duckworth
•DU
CENTRAL

"No.6"
WEST UNION
Rock Run

SMITHBURG
•SN
Morgansville

"No.4"
"No.3"

SALEM
•SA
"No.2"

Belpre,
O.

"No.8"

"No.7"

340

335
B&O

330

"No.5"

325

320

315
BRISTOL

Industrial School

TOLLGATE
GREENWOOD

LONG RUN
•DA

15'

15'

+

DODDRIDGE

RITCHIE

LEWIS

GILMER

39°00'
81°00'

45'

80°30'
39°00'

105

87

89

West Union, W.Va.

88

W.Va.

CLARKSBURG

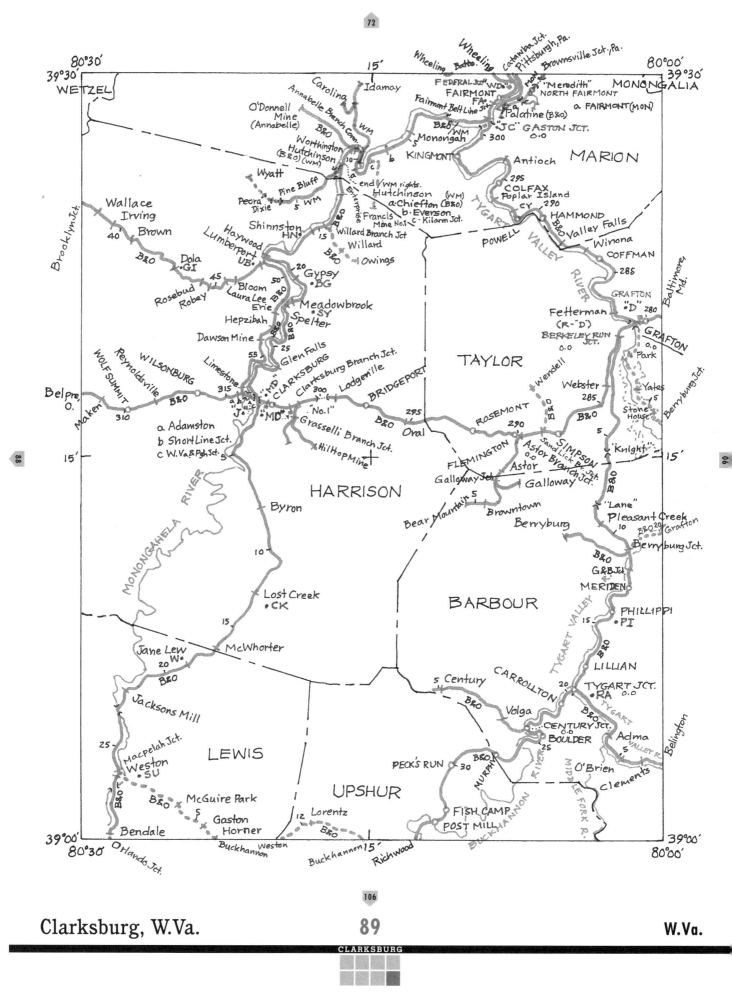

88
90

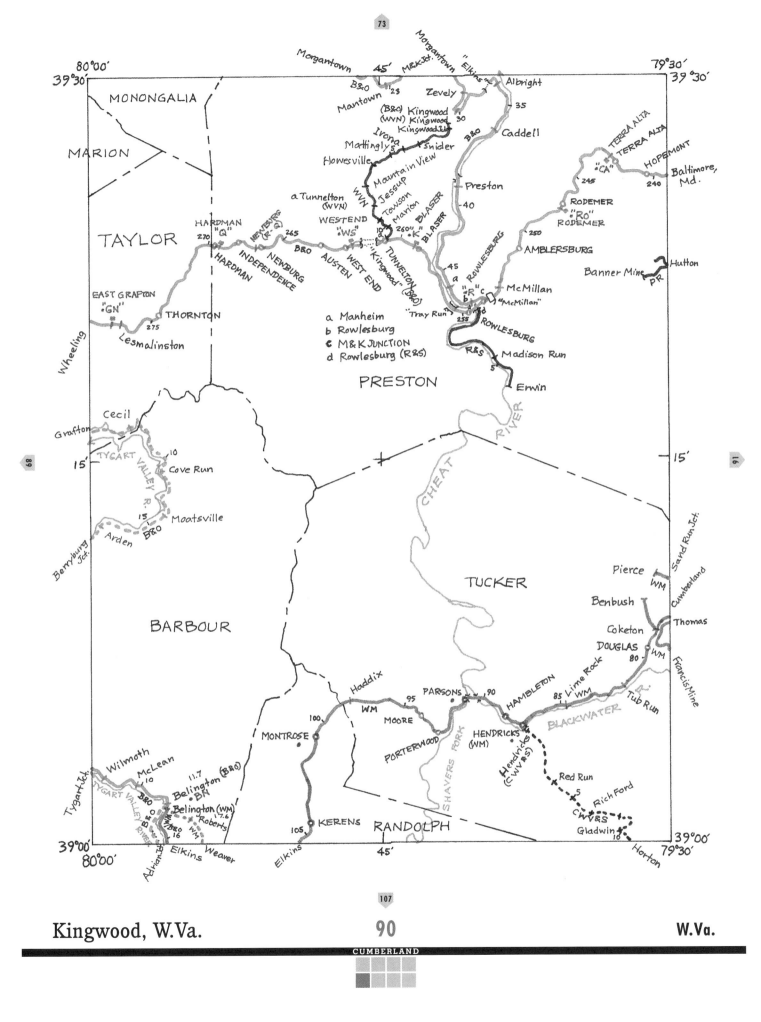

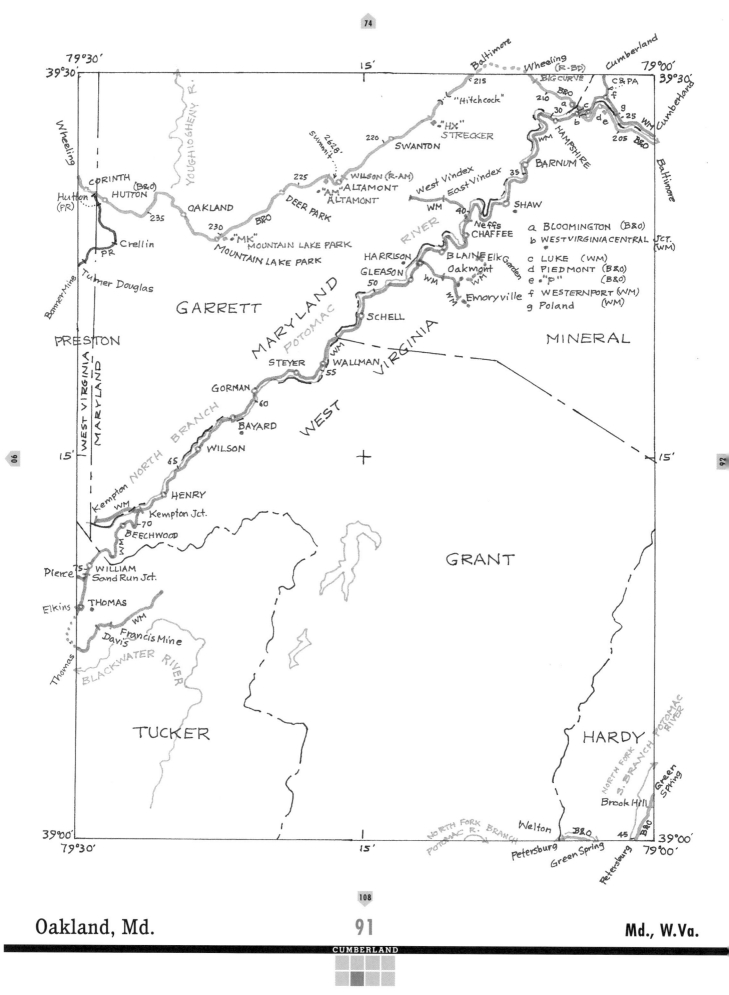

79°30'
39°30'

Baltimore
Wheeling (R-BD)
Cumberland
Cumberland
79°00'
39°30'

15'

215

BIG CURVE
C&PA

210
B&O
f

Wheeling
"Hitchcock"
a
c
g

30
b
d e
25

WM

"HX"
STRECKER

220
SWANTON
WM
HAMPSHIRE
205
B&O

Baltimore

2628'
summit

WILSON (R-AM)
West Vindex
East Vindex
35
BARNUM

225
"AM"
ALTAMONT

CORINTH
(B&O)
HUTTON
OAKLAND
DEER PARK
ALTAMONT
WM
40
SHAW

Hutton
(PR)

235
230
B&O
RIVER
Neffs
CHAFFEE

a BLOOMINGTON (B&O)

Crellin
PR
"MK"
MOUNTAIN LAKE PARK
HARRISON
BLAINE ElkGarden

b WEST VIRGINIA CENTRAL JCT. (WM)
c LUKE (WM)

MOUNTAIN LAKE PARK
GLEASON
50
WM
Oakmont
WM

d PIEDMONT (B&O)
e "P" (B&O)

BannerMine
Turner Douglas
GARRETT
MARYLAND
POTOMAC
WM
Emoryville

f WESTERNPORT (WM)
g Poland (WM)

MINERAL

PRESTON
SCHELL

STEYER
WALLMAN
WM

WEST
VIRGINIA

55

GORMAN
60

15'
WEST
06

BAYARD
VIRGINIA
15'
92

WILSON

NORTH BRANCH
65
+

Kempton
WM
HENRY

Kempton Jct.
-70
GRANT

BEECHWOOD
WM

Pierce
75
WILLIAM
Sand Run Jct.

Elkins
THOMAS

WM
Francis Mine

Thomas
Davis

BLACKWATER RIVER
GRANT

TUCKER
HARDY

NORTH FORK POTOMAC RIVER
Green Spring

Brook Hill

NORTH FORK
POTOMAC R. BRANCH
Welton
B&O
45
B&O

39°00'
79°30'
15'
Petersburg
Green Spring
39°00'
79°00'

Petersburg

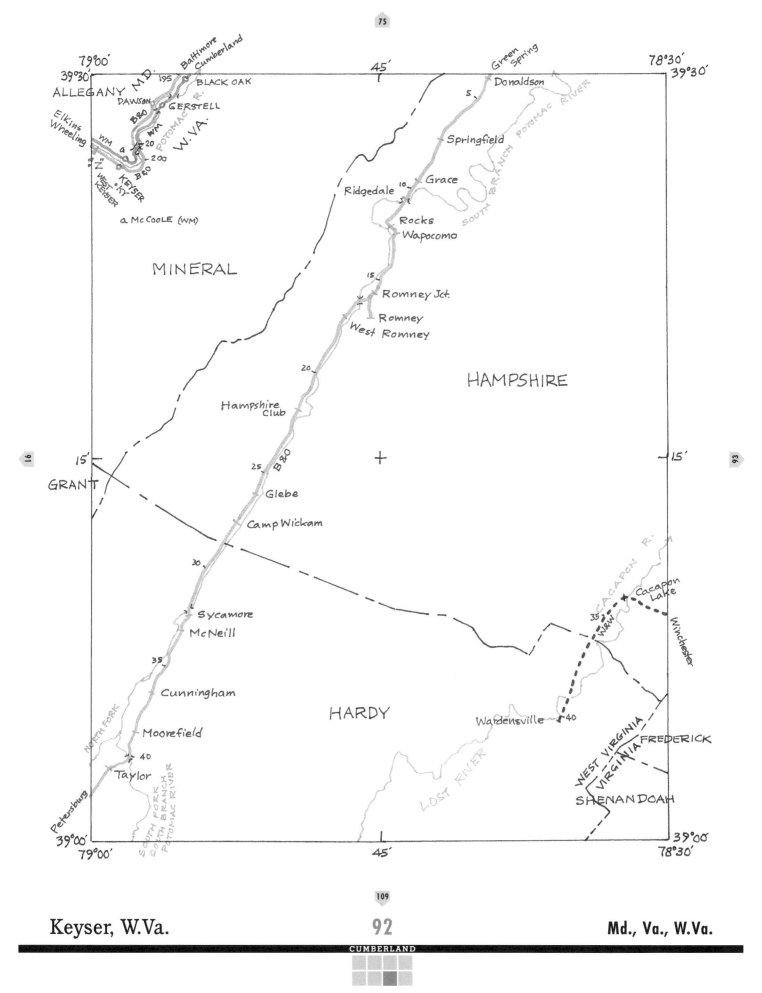

75

79°00'
39°30'
ALLEGANY MD.
Baltimore
Cumberland
45'
Green Spring
78°30'
39°30'

BLACK OAK
DAWSON
GERSTELL
B&O
WM
POTOMAC R.
W. VA.
Donaldson
5
SOUTH BRANCH POTOMAC RIVER
Springfield

Elkins
Wheeling
WM
20
B&O
200
"N"
KEYSER
Grace
10
Ridgedale
Rocks
Wapocomo

WEST KY
KEYSER
a McCOOLE (WM)

MINERAL
15
Romney Jct.
Romney
West Romney
HAMPSHIRE

16
15'
GRANT
20
Hampshire
Club
+
15'
93

25
B&O
Glebe
Camp Wickam

30
CACAPON R.
Cacapon
Lake
35
W&W
Winchester

Sycamore
McNeill
35
Cunningham
NORTH FORK
HARDY
Wardensville
40
WEST VIRGINIA
VIRGINIA
FREDERICK

Moorefield
40
Taylor
LOST RIVER
SHENANDOAH

Petersburg
SOUTH FORK
SOUTH BRANCH
POTOMAC RIVER
45'
39°00'
78°30'

79°00'
39°00'

109

Keyser, W.Va.   92   Md., Va., W.Va.

CUMBERLAND

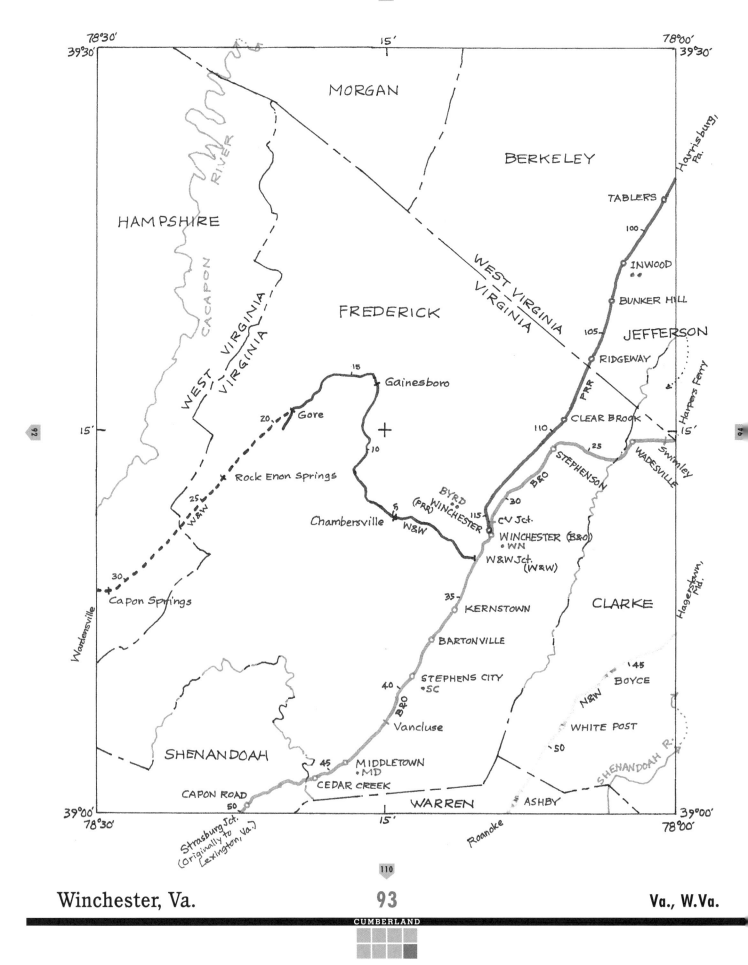

MORGAN

BERKELEY

HAMPSHIRE

CACAPON RIVER

WEST VIRGINIA / VIRGINIA

FREDERICK

WEST VIRGINIA / VIRGINIA

TABLERS

100

INWOOD

BUNKER HILL

105

JEFFERSON

RIDGEWAY

PRR

Harrisburg, Pa.

Harpers Ferry

15'

Gainesboro

15

20 Gore

Rock Enon Springs

25 W&W

10

CLEAR BROOK

110

25

STEPHENSON

B&O

Swimley

WADESVILLE

BYRD WINCHESTER (PRR)

115

CV Jct.

30

Chambersville

5 W&W

WINCHESTER (B&O)
• WN

W&WJct. (W&W)

30

Capon Springs

Wardensville

35

KERNSTOWN

BARTONVILLE

CLARKE

Hagerstown, Md.

STEPHENS CITY
• SC

40

B&O

N&W

45 BOYCE

WHITE POST

50

Vancluse

SHENANDOAH

45

MIDDLETOWN
• MD

CEDAR CREEK

CAPON ROAD

50

WARREN

ASHBY

SHENANDOAH R.

Strasburg Jct.
(Originally to
Lexington, Va.)

Roanoke

78°30' 39°30'   15'   78°00' 39°30'

15'   15'

39°00' 78°30'   15'   39°00' 78°00'

Winchester, Va.   **93**   **Va., W.Va.**

CUMBERLAND

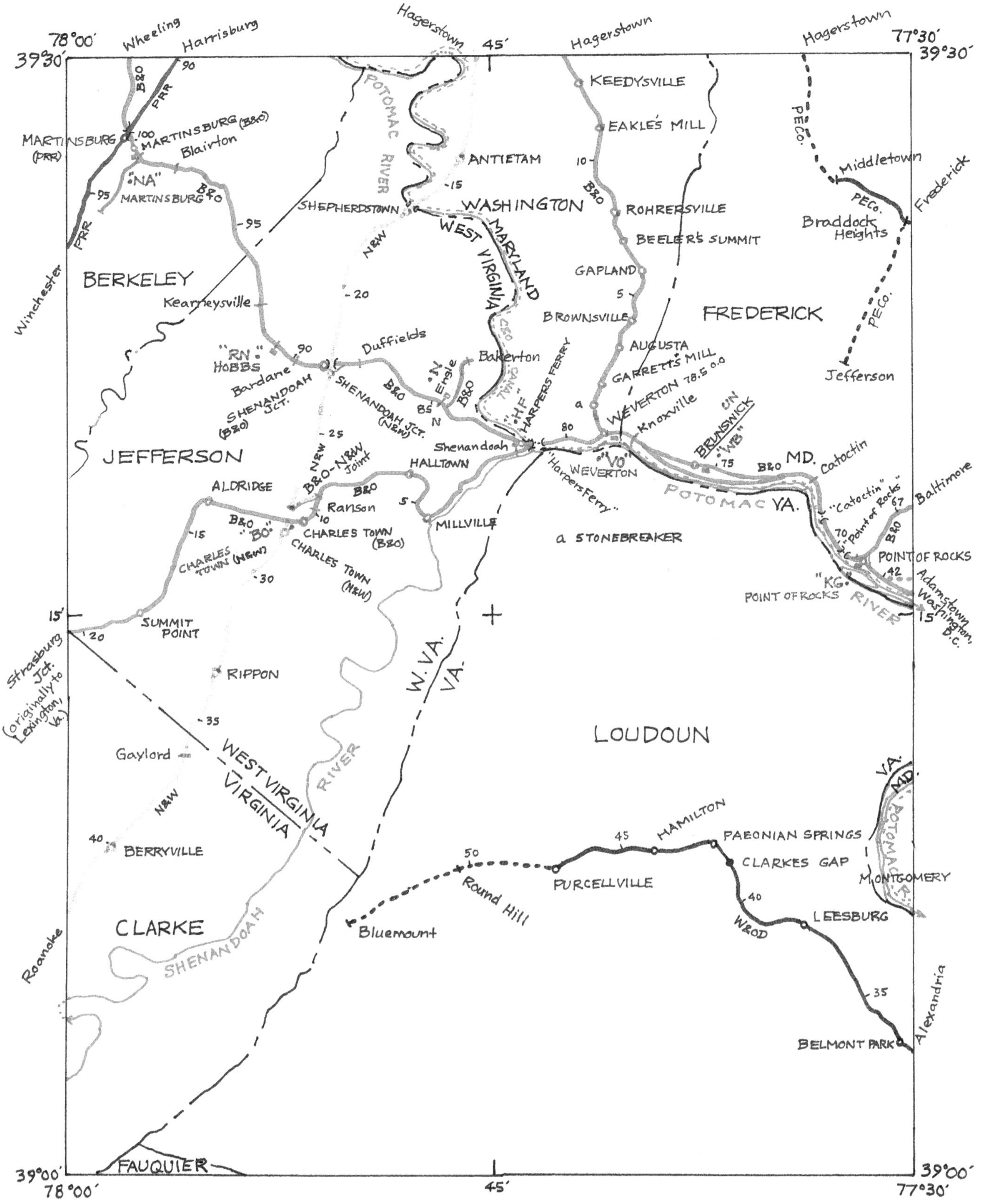

78°00' Wheeling Harrisburg Hagerstown 45' Hagerstown Hagerstown 77°30'
39°30' 39°30'

B&O PRR 90

KEEDYSVILLE

EAKLE'S MILL

MARTINSBURG (B&O) MARTINSBURG 100 Blairton
MARTINSBURG (PRR)

PE Co.

Middletown

Frederick

ANTIETAM 10 B&O
"NA" B&O 15
MARTINSBURG 95

WASHINGTON WEST VIRGINIA MARYLAND ROHRERSVILLE Braddock Heights

SHEPHERDSTOWN 95 BEELER'S SUMMIT

BERKELEY 20 GAPLAND 5 FREDERICK

Kearneysville NEW PE Co.

Duffields BROWNSVILLE Jefferson

"RN" 90 AUGUSTA
HOBBS Bakerton GARRETT'S MILL

Bardane N Engle H F TANG 78.5 0.0

SHENANDOAH SHENANDOAH JCT. 85 B&O WEVERTON
JCT. (B&O) (NEW) N B&O a "N" Knoxville "N" MD. Catoctin

JEFFERSON NEW 80 BRUNSWICK "WB" B&O Baltimore
B&O-N&W Joint Shenandoah WEVERTON 75 67
ALDRIDGE HALLTOWN "Vo" POTOMAC VA. "Catoctin" Point of Rocks
B&O Ranson 5 "HarpersFerry" 70 B&O
"BO" 10 MILLVILLE 42 POINT OF ROCKS
15 CHARLES TOWN "KG" Adamstown,
CHARLES (B&O) a STONEBREAKER RIVER Washington, D.C.
TOWN (NEW) CHARLES TOWN POINT OF ROCKS 15'
30 (NEW)

15' SUMMIT 20 W. VA. LOUDOUN
Strasburg POINT VA.
Jct. RIPPON
(originally to
Lexington,
Va.) 35

Gaylord WEST VIRGINIA VIRGINIA VA. MD.
NEW HAMILTON POTOMAC R.
40 45 PAEONIAN SPRINGS
BERRYVILLE 50 CLARKES GAP
Roanoke RIVER Round Hill PURCELLVILLE 40 MONTGOMERY

CLARKE SHENANDOAH Bluemount W&OD LEESBURG
35 Alexandria
BELMONT PARK

39°00' 39°00'
FAUQUIER
78°00' 45' 77°30'

Martinsburg, W.Va. 94 Md., Va., W.Va.

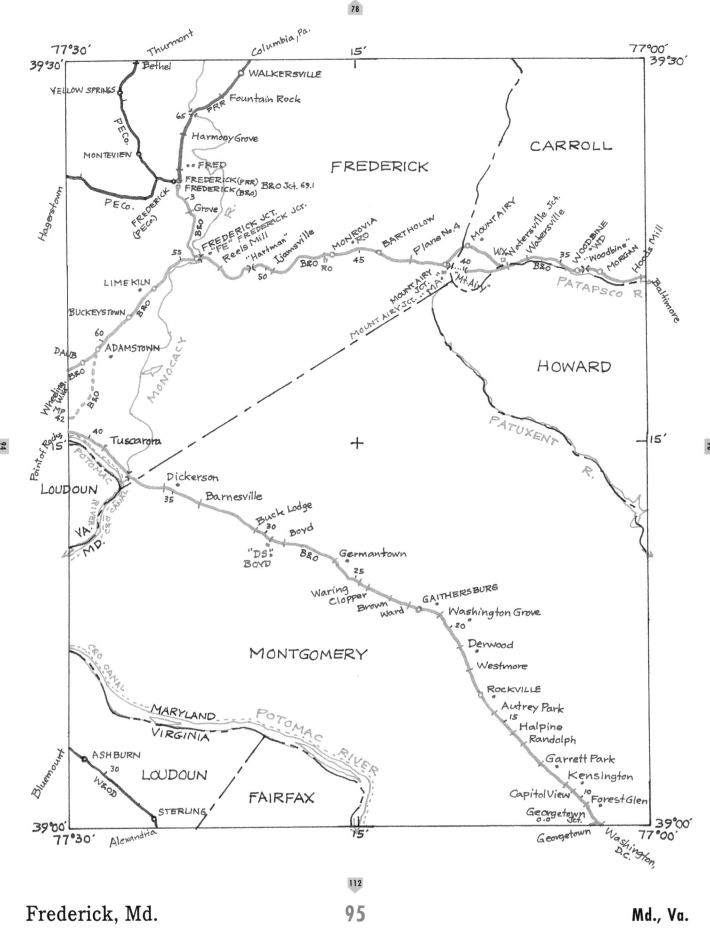

77°30' Thurmont Columbia, Pa. 15' 77°00'
39°30' 39°30'

Bethel
WALKERSVILLE
YELLOW SPRINGS
PRR Fountain Rock
65
CARROLL
P E Co.
Harmony Grove
MONTEVIEW
FRED FREDERICK
P E Co.
FREDERICK (PRR) B&O Jct. 69.1
FREDERICK (B&O)
Hagerstown
3
Grove
FREDERICK (PE Co.)
FREDERICK JCT.
"FE" FREDERICK JCT.
Reels Mill
"Hartman" Ijamsville MONROVIA RD BARTHOLOW Plane No.4 MOUNT AIRY Watersville Jct.
55 71 B&O RO 45 Watersville WOODBINE WD
50 B&O RO 40 WX "Woodbine"
LIME KILN MOUNT AIRY JCT. "MA" B&O MORGAN
B&O MOUNTAIRY JCT. "Mt.Airy" 35 Hoods Mill
BUCKEYSTOWN 60 MOUNT AIRY JCT. "MA" PATAPSCO R. Baltimore

DAUB
ADAMSTOWN HOWARD
Wheeling B&O B&O
MP 42 PATUXENT R.
Point of Rocks 40 Tuscarora
15'
POTOMAC RIVER CANAL
LOUDOUN Dickerson 15'
35 Barnesville
VA. Buck Lodge
MD. 30 Boyd
"DS" B&O Germantown
BOYD 25
Waring GAITHERSBURG
Clopper Washington Grove
Brown Ward 20
MONTGOMERY Derwood
Westmore
ROCKVILLE
Autrey Park
C&O CANAL 15 Halpine
MARYLAND Randolph
VIRGINIA POTOMAC RIVER Garrett Park
Bluemount ASHBURN Kensington
30 LOUDOUN Capitol View 10 Forest Glen
W&OD FAIRFAX Georgetown 0.0 Jct.
STERLING Georgetown Washington, D.C.
39°00' 39°00'
77°30' Alexandria 15' 77°00'

Frederick, Md.   **95**   Md., Va.

BALTIMORE

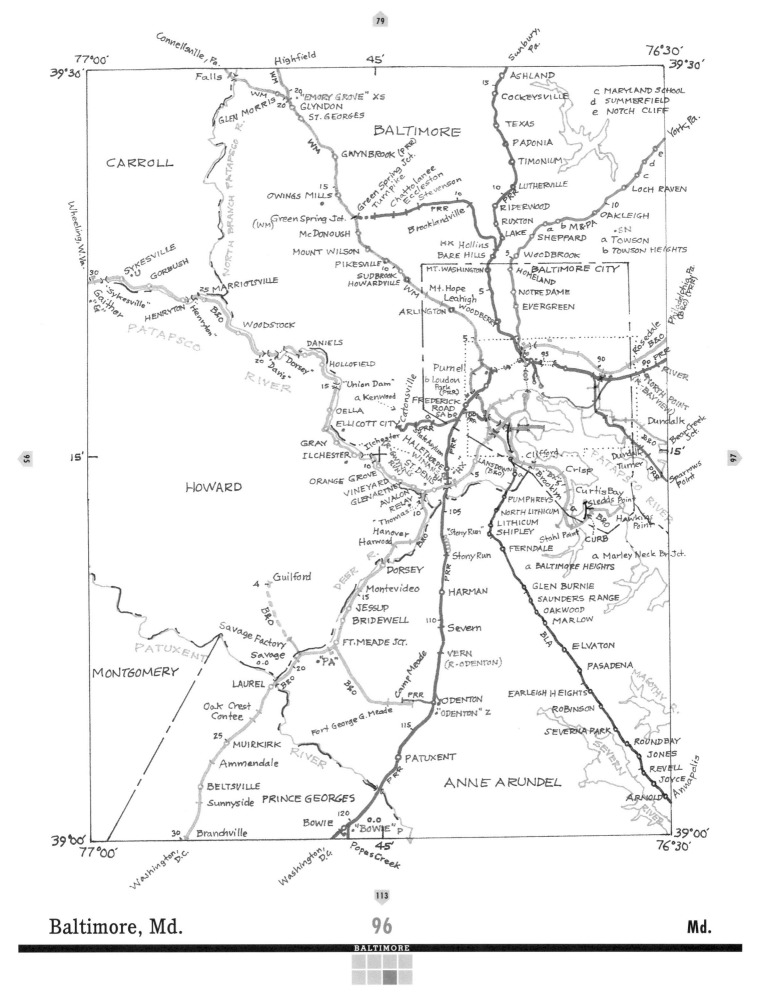

Connellsville, Pa.   Highfield   45'   Sunbury, Pa.   76°30'
77°00'   39°30'
39°30'   Falls   ASHLAND   c MARYLAND SCHOOL
                                                              d SUMMERFIELD
              WM   15 COCKEYSVILLE   e NOTCH CLIFF
        GLEN MORRIS 20 "EMORY GROVE" XS
                      20 GLYNDON   TEXAS
           WM   ST. GEORGES   York, Pa.
                                              PADONIA   e
                    BALTIMORE   d
CARROLL   WM GWYNBROOK (PRR)   TIMONIUM   c
                        Green Spring Jct. 10 LUTHERVILLE   LOCH RAVEN
              15   PRR
        OWINGS MILLS   Chattolanee 10   10 OAKLEIGH
                    Green Spring Turnpike Eccleston   RIDERWOOD
(WM) Green Spring Jct.   Stevenson 10   RUXTON a b M&PA   SN
        McDONOUGH   Brooklandville   LAKE SHEPPARD   a TOWSON
                    PRR   b TOWSON HEIGHTS
        MOUNT WILSON   HK Hollins
                              BARE HILLS   5 WOODBROOK
         PIKESVILLE 10   MT. WASHINGTON   HOMELAND BALTIMORE CITY
30   SYKESVILLE   SUDBROOK   NOTRE DAME
          U GORBUSH   HOWARDVILLE WM Mt. Hope 5 NOTRE DAME
"Sykesville" 25 MARRIOTSVILLE   Leahigh Woodberry EVERGREEN
Gaither "G"   WM   ARLINGTON
        HENRYTON Henryton B&O   Rosedale B&O
PATAPSCO   WOODSTOCK   5   90
                                                    95   90 PRR
        DANIELS   RIVER
RIVER 20 "Dorsey" HOLLOFIELD   Purnel   90 (R.BAYVIEW)
        "Davis"   b Loudon
              15 "Union Dam"   Park   Dundalk
                a Kenwood (PRR)   15
        OELLA Catonsville FREDERICK   B&O Bear Creek Jct.
        ELLICOTT CITY   ROAD 100
HOWARD   State Asylum Sa b c PRR   PRR Sparrows
        GRAY   SPRR HALETHORPE Winans   Dundalk   Point
        ILCHESTER Ilchester   GWYNNS Clifford Turner
                    ST. DENIS RUNS LANSDOWNE (B&O) Crisp
        ORANGE GROVE WINANS HX 5 Brooklyn CurtisBay
        VINEYARD   10   Pumphreys Sledds Point
        GLENARTNEY RELAY   NORTH LITHICUM Hawkins
        AVALON   10 105 LITHICUM   CURB Point
        "Thomas"   SHIPLEY Stahl Point
        Hanover   "Stony Run" FERNDALE a Marley Neck Br.Jct.
        Harwood B&O   Stony Run   a BALTIMORE HEIGHTS
              DEER R. PRR Stony Run   GLEN BURNIE
4 Guilford   DORSEY   SAUNDERS RANGE
        B&O   Montevideo HARMAN   OAKWOOD
                    15   MARLOW
        Savage Factory JESSUP 110 Severn   BLA ELVATON
        Savage 0.0 BRIDEWELL   PASADENA
MONTGOMERY   20 "PA" FT.MEADE JCT. VERN   EARLEIGH HEIGHTS
        LAUREL B&O   Camp Meade (R.ODENTON)   ROBINSON
        Oak Crest   PRR   SEVERNA PARK
        Contee Fort George G. Meade 115 ODENTON   ROUNDBAY
        25 MUIRKIRK   "ODENTON" Z   JONES
        Ammendale RIVER   REVELL
        BELTSVILLE   PRR PATUXENT   JOYCE
        Sunnyside PRINCE GEORGES   ANNE ARUNDEL ARNOLD
                    120 BOWIE 0.0
39°00' 30 Brandhville   "BOWIE" P   39°00'
77°00'   Washington, D.C. 45'   76°30'
        Washington, D.C. Popes Creek

Baltimore, Md.   96   Md.

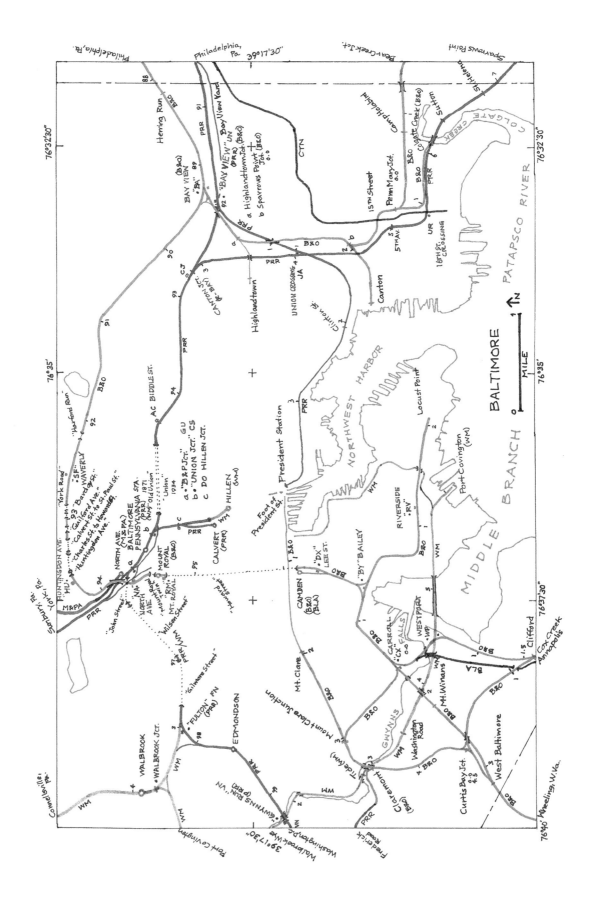

96A Baltimore, Md.

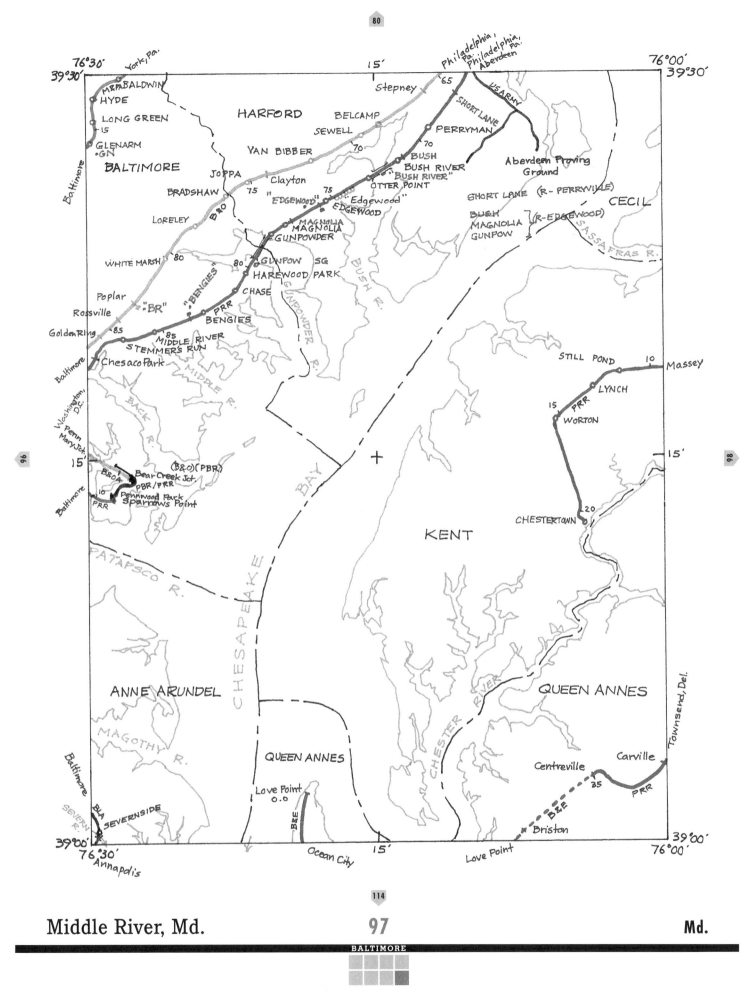

76°30'
39°30'
York, Pa.
MR.A BALDWIN
HYDE
LONG GREEN
15
GLENARM
GN
BALTIMORE

HARFORD

15'
Stepney
65
Philadelphia, Pa.
Philadelphia, Pa.
Philadelphia, Pa.
Aberdeen
76°00'
39°30'

BELCAMP
SEWELL
VAN BIBBER
70
Clayton
75
"Edgewood"
JOPPA
BRADSHAW
Loreley
B&O

SHORT LANE
US ARMY
PERRYMAN

BUSH
BUSH RIVER
"BUSH RIVER"
OTTER POINT
75
"Edgewood"
EDGEWOOD
Aberdeen Proving
Ground

SHORT LANE (R-PERRYVILLE)
BUSH
MAGNOLIA
GUNPOW
(R-EDGEWOOD)
CECIL

MAGNOLIA
MAGNOLIA
GUNPOWDER
WHITE MARSH   80
80   GUNPOW   SG
HAREWOOD PARK
Poplar
Rossville   "BR"   "BENGIES"   CHASE
GoldenRing   85   85   PRR   BENGIES
MIDDLE RIVER
STEMMER'S RUN
Baltimore   Chesaco Park

BUSH
R.

GUNPOWDER
R.

SASSAFRAS R.

STILL POND   10
Massey
15   LYNCH
PRR
WORTON

15'

Washington, D.C.
Penn
Mary Jct.
15'
Baltimore
B&O
PRR
(B&O)(PBR)
Bear Creek Jct.
PBR/PRR
Pennwood Park
Sparrows Point

MIDDLE R.

BACK
R.

BAY

+

KENT

20
CHESTERTOWN

PATAPSCO R.

CHESAPEAKE

ANNE ARUNDEL

MAGOTHY
R.

QUEEN ANNES

QUEEN ANNES

Love Point
0.0

CHESTER RIVER

QUEEN ANNES

Centreville
Carville
35   PRR

Townsend, Del.

Baltimore
BLA
SEVERNSIDE
SEVERN
R.
39°00'
76°30'
Annapolis

B&E
Ocean City   15'   Love Point

B&E
Briston
39°00'
76°00'

Middle River, Md.

97

Md.

Dover, Del.   **98**   Del., Md., N.J.

WILMINGTON

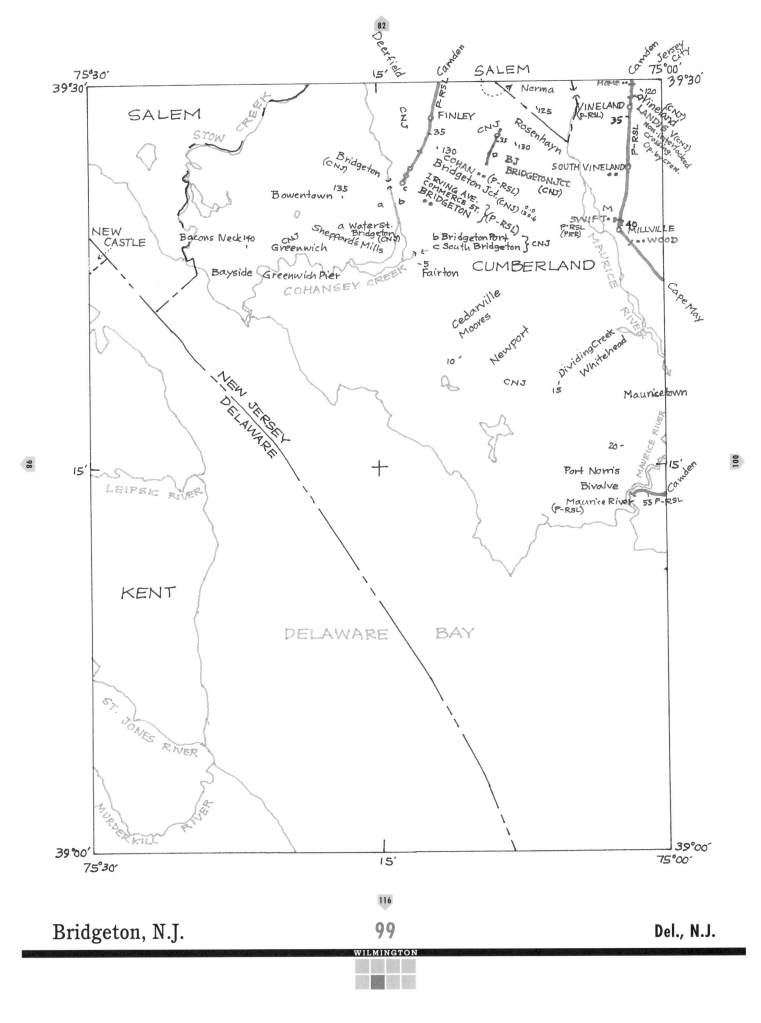

SALEM

SALEM

Camden

CNJ
P-RSL
FINLEY
35

Norma
'125
Rosenhayn
CNJ
35 '130
BJ
COHAN '130
Bridgeton Jct. (P-RSL)
BRIDGETON JCT. (CNJ)

HOME
VINELAND
(P-RSL)
'120 (CNJ)
Vineland V (CNJ)
Vineland LANDIS Non-Interlocked
Crossing
Op. by crew.
35

Camden
Jersey City
75°00'
39°30'

Bridgeton
(CNJ)

Bowentown '135

SOUTH VINELAND

IRVING AVE. (CNJ)
COMMERCE ST. (CNJ) 130.6
BRIDGETON

a
b
c

Bridgeton Jct. (CNJ)
}(P-RSL)

SWIFT
P-RSL
(PRB)
M
40
MILLVILLE
WOOD

NEW
CASTLE

Bacons Neck 140
CNJ
Greenwich
Shepperd's Mills
a Water St.
Bridgeton (CNJ)

b Bridgeton Port
c South Bridgeton
CNJ

CUMBERLAND

MAURICE RIVER

Bayside
Greenwich Pier
COHANSEY CREEK
'5
Fairton

Cape May

Cedarville
Moores

Newport

10

NEW
JERSEY
DELAWARE

CNJ

Dividing Creek
Whitehead

Mauricetown

15

15'
100

88
15'

LEIPSIC RIVER

20

Port Norris
Bivalve
Maurice River
(P-RSL)

15'
Camden
53 P-RSL

KENT

ST. JONES RIVER

DELAWARE   BAY

MURDER KILL RIVER

75°00'
39°30'

Camden   Camden   45'   Camden
                                                    Camden   74°30'
                                                                    39°30'

Richland   RICHLAND   Cologne   45
                                        Brigantine   POMONA
Richland   MIZPAH   Jct.   Ac
        40   45   AC   P-RSL   50
                P-RSL   45   Ac
        40   Brigantine Beach
                MAYS LANDING
    Milmay   50   REEGA   ABSECON   Atlantic
                                                                    City
                                McKEE CITY   55   Atlantic
CUMBERLAND   Dorothy   CARDIFF   50   City
        45   MT. CALVARY   Pleasantville   (AC)
                    Risley   Atlantic
                                            50   60   City
                        ATLANTIC   Pleasantville (P-RSL)
Camden   (AC)
MENANTICO   P-RSL   h g f e d c b a   Atlantic City
    45   (AC)   P-RSL   (P-RSL)
P-RSL   45   i
    MANUMUSKIN   j   P-RSL   Atlantic City
                                                k
    Port   (See   P-RSL   MARGATE
    Elizabeth   50   65   m   below)
    Mauricetown   50   P-RSL   TUCKAHOE RIVER   n o   LONGPORT
        P-RSL   KD TUCKAHOE RIVER   p   W.J&S
    Dorchester   TUCKAHOE   "Tuckahoe"   GARDENS   OCEAN CITY GARDENS
                        BELLEPLAIN   KG   Middletown   4TH ST.
    Leesburg   55   PETERSBURG   65   OCEAN CITY 10TH ST.
    15'   55   55   14TH ST.
Heislerville   P-RSL   (AC)   P-RSL   PALERMO   24TH ST.
        P-RSL   BD   (AC)   P-RSL   60
                WOODBINE   WOODBINE JCT.   (AC)   34TH ST.
    CAPE MAY   PINE   P-RSL   51ST ST.
                    MPP   Mount Pleasant   55TH   CROOK HORN DRAW
                    (WJ&S)   P-RSL   60   ST.
                    JN   Sea Isle Jct.   Strahmere
                DENNISVILLE   P-RSL   a SMITHS LANDING
                                        Central Ave.   b DOLPHIN
                SOUTH DENNIS   c NORTHFIELD
                P-RSL   South Seaville   Prospect St.   d ZION ROAD
                (AC)   W.J&S   e TUDOR TERRACE
                                    Sea Isle City   f BAKERSVILLE
                                                g OAKCREST
        a Cape May Courthouse   Loretto Ave   h LINWOOD COUNTRY CLUB
        b CAPE MAY COURTHOUSE   i LINWOOD
                            65   Townsend Inlet   j BELL HAVEN
                GOSHEN   k SEAVIEW
                    Swain   l OCEAN HEIGHTS
DELAWARE   Avalon   m GLYN NEATH
BAY   n LAUNCH HAVEN
                                    Peermont   W.J&S   o NEW YORK AVENUE
                                                p SOMERS POINT
                    a   ATLANTIC OCEAN
                    70   b
                        P-RSL
cc BURLEIGH BG (AC)   (AC)   Mayville   Stone Harbor
d   WILDWOOD JCT. JU (AC)   DO   70   P-RSL
                        cc   (AC)
        W.J&S   W.J&S   Grassy Sound   a RIO GRANDE
                WHITESBORO   Beach Creek   b WILDWOOD GARDENS
    Rio Grande   75   WILDWOOD JCT.   c GRASSY SOUND DRAW
        75   a   b   Anglesea
                c
39°00'   P-RSL (AC)   39°00'
75°00'   Cape May   45'   74°30'
        Cape May   Wildwood   Cold Spring Harbor

Tuckahoe, N.J.   100   N.J.

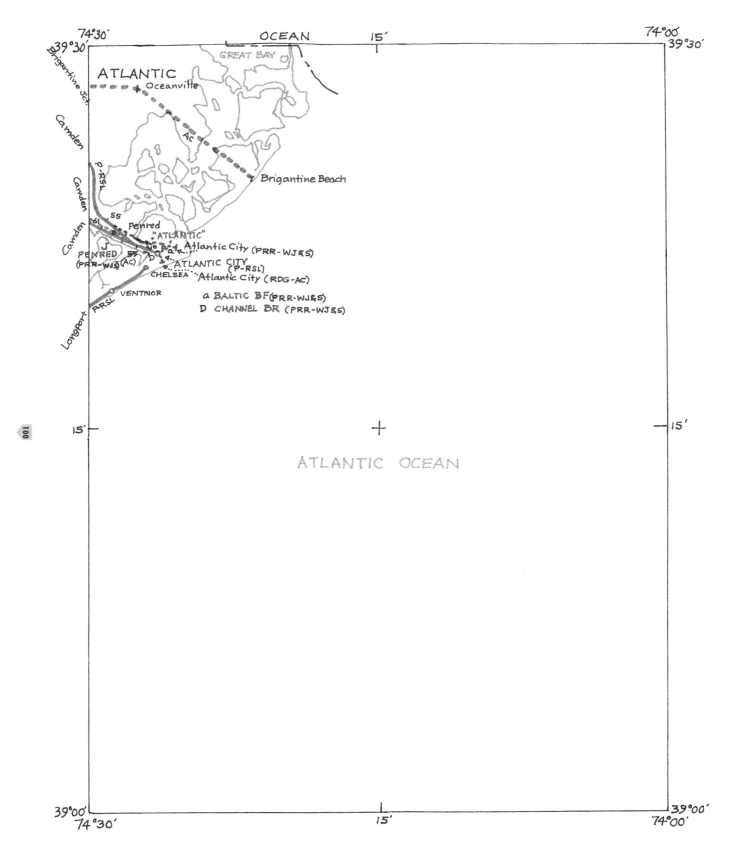

OCEAN

74°30'   15'   74°00'

39°30'   39°30'

GREAT BAY

ATLANTIC

+ Oceanville

Brigantine Jct.

Camden

Ac

Camden

Brigantine Beach

P-RSL

55

Camden

Penred

61

"ATLANTIC"

J

PENRED

Atlantic City (PRR-WJ&S)

(PRR-WJ&S)(Ac)

55

ATLANTIC CITY
(P-RSL)

CHELSEA

Atlantic City (RDG-AC)

VENTNOR

a BALTIC BF (PRR-WJ&S)

D CHANNEL BR (PRR-WJ&S)

Longport

P-RSL

100

15' —

| 15'

ATLANTIC OCEAN

39°00'   39°00'

74°30'   15'   74°00'

WILMINGTON

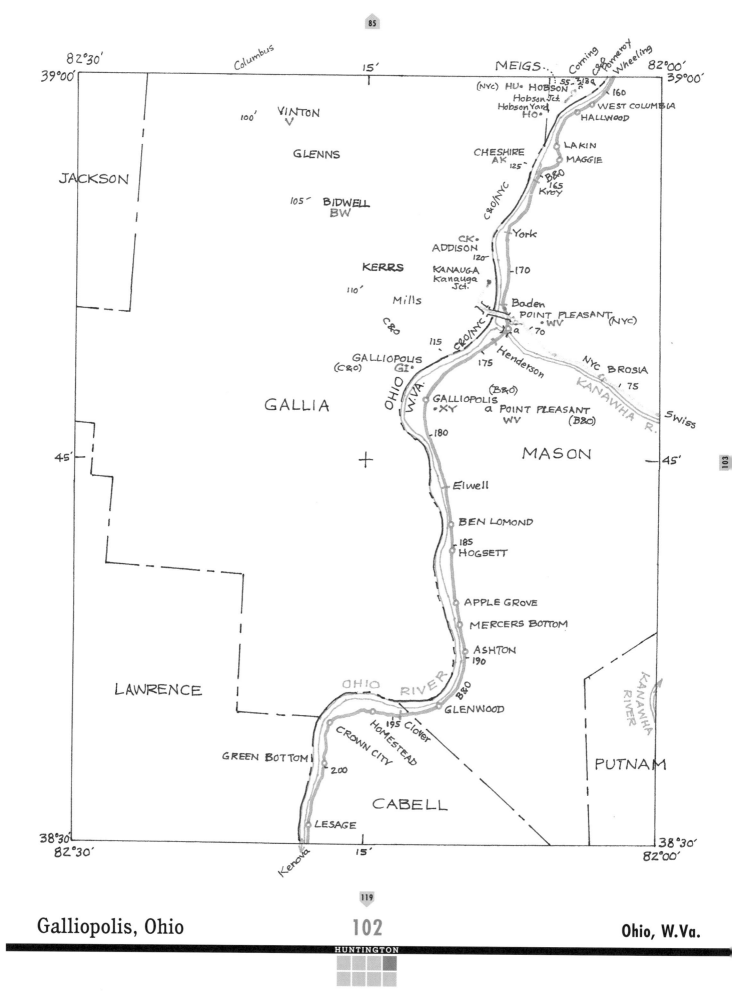

82°30'
39°00'

Columbus

15'

MEIGS

Corning
Pomeroy
Wheeling

82°00'
39°00'

(NYC) HU•HOBSON
Hobson Jct.
Hobson Yard
HO•

SS 2139
2

160

WEST COLUMBIA
HALLWOOD

JACKSON

VINTON
V

100'

GLENNS

CHESHIRE
AK
125'

LAKIN
MAGGIE

105'
BIDWELL
BW

C&O/NYC

B&O
165
Kroy

York

CK•
ADDISON
120'

170

KERRS

KANAUGA
Kanauga
Jct.

110'

Mills

C&O

Baden
POINT PLEASANT (NYC)
•WV
70

GALLIA

115

C&O/NYC

GALLIOPOLIS
(C&O)
GI•

Henderson
175

a

NYC BROSIA
75

KANAWHA R.

Swiss

OHIO

W.VA.

GALLIOPOLIS
•XY

(B&O)

a POINT PLEASANT
WV
(B&O)

45'

180

MASON

45'

Elwell

BEN LOMOND

185
HOGSETT

APPLE GROVE

MERCERS BOTTOM

ASHTON
190

OHIO   RIVER

B&O

LAWRENCE

GLENWOOD

195 Clover

HOMESTEAD

CROWN CITY

GREEN BOTTOM
200

KANAWHA
RIVER

PUTNAM

CABELL

LESAGE

38°30'
82°30'

Kenova

15'

38°30'
82°00'

Galliopolis, Ohio

Ohio, W.Va.

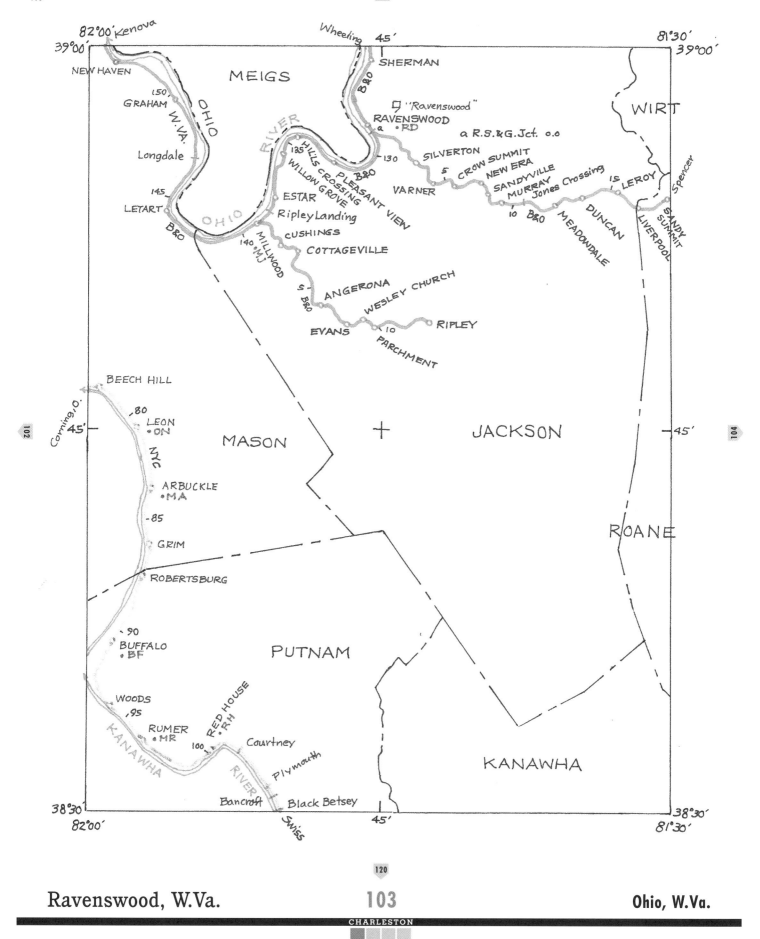

82°00' Kenova

Wheeling   45'

81°30'

39°00'   39°00'

NEW HAVEN   SHERMAN

MEIGS   B&O

GRAHAM   150'

W.VA.   OHIO   "Ravenswood"

RAVENSWOOD

RD   a R.S.&G.Jct. 0.0   WIRT

Longdale   RIVER   SILVERTON   CROW SUMMIT   Spencer

135 HILLS CROSSING   130   5   NEW ERA

WILLOW GROVE   SANDYVILLE

145   PLEASANT VIEW   VARNER   MURRAY   Jones Crossing   15   LEROY

LETART   ESTAR   B&O   10 B&O   DUNCAN   SANDY SUMMIT

OHIO   Ripley Landing   MEADONDALE   LIVERPOOL

B&O   CUSHINGS

140 MILLWOOD   COTTAGEVILLE

MJ

S   ANGERONA   WESLEY CHURCH

B&O

EVANS   10   RIPLEY

PARCHMENT

BEECH HILL

Corning, O.   80   LEON

45'   ON   MASON   JACKSON   45'

NYC

ARBUCKLE

MA

85   ROANE

GRIM

ROBERTSBURG

90

BUFFALO

BF   PUTNAM

WOODS

95   RED HOUSE

RUMER   RH

MR   100   Courtney

KANAWHA   Plymouth   KANAWHA

RIVER

38°30'   Bancroft   Black Betsey   38°30'

82°00'   Swiss   45'   81°30'

Ravenswood, W.Va.   103   Ohio, W.Va.

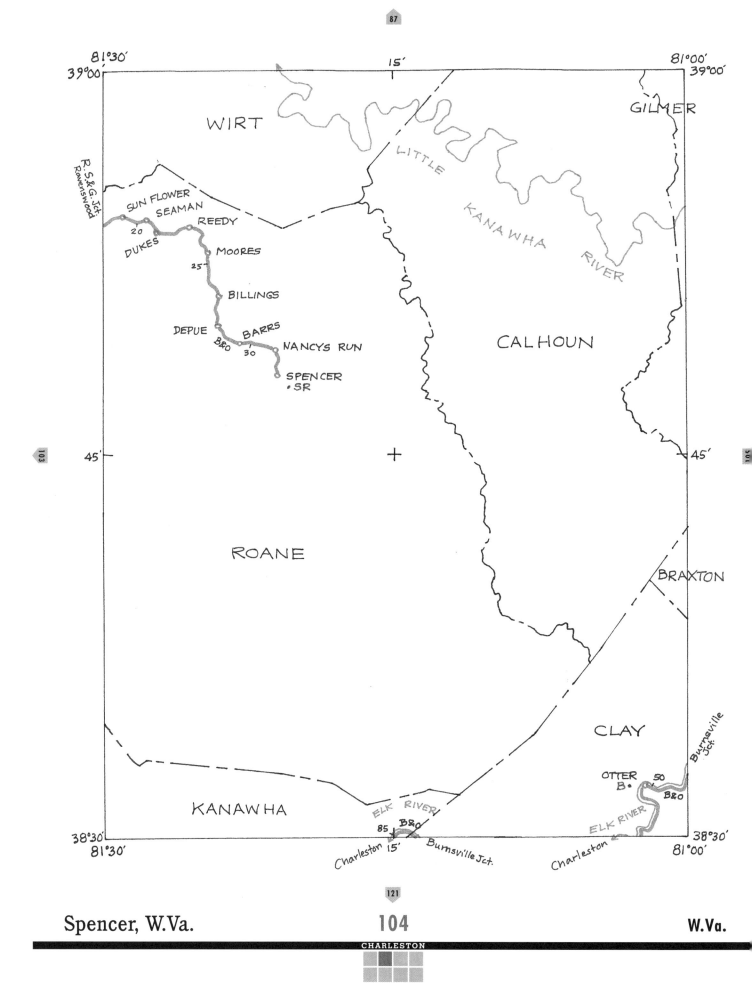

81°30'
39°00'
15'
81°00'
39°00'

WIRT

GILMER

LITTLE

KANAWHA

RIVER

R.S.&G. Jct.
Ravenswood

SUN FLOWER
SEAMAN
REEDY
20
DUKES
MOORES
25
BILLINGS
DEPUE
BARRS
B&O
30
NANCYS RUN
SPENCER
• SR

CALHOUN

103

45'

+

45'

105

ROANE

BRAXTON

CLAY

Burnsville Jct.

OTTER
B•
50
B&O

KANAWHA

ELK RIVER

ELK RIVER

85
B&O

38°30'

Charleston 15'

Burnsville Jct.

Charleston

38°30'
81°00'

81°30'

Spencer, W.Va.

104

W.Va.

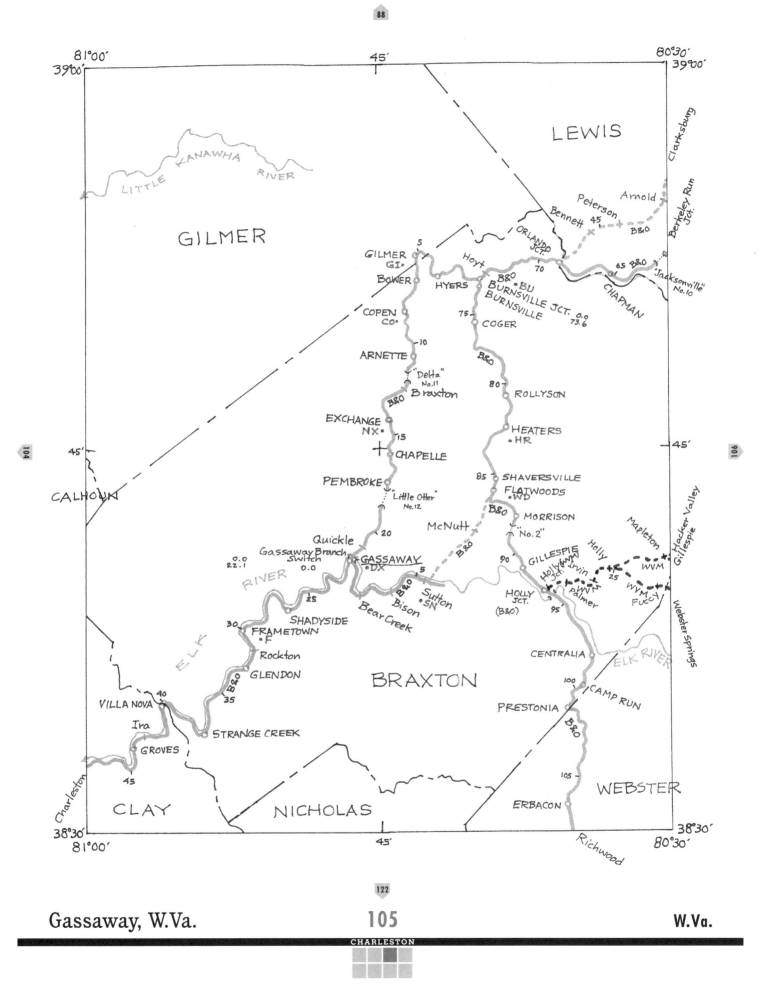

81°00′
39°00′

45′

80°30′
39°00′

LEWIS

Clarksburg

LITTLE KANAWHA RIVER

GILMER

Peterson   Arnold
Bennett   45
ORLANDO JCT.   70   65 B&O
GILMER   5
GI.
BOWER   HYERS   B&O BU
BURNSVILLE JCT. 0.0
BURNSVILLE   73.6
CHAPMAN
"Jacksonville"
No.10

Berkeley Run
B&O   Jct.

COPEN
CO.   75   COGER

10

ARNETTE   B&O

"Delta"   80   ROLLYSON
No.11
Braxton

B&O

EXCHANGE
NX.   15

CHAPELLE   HEATERS
HR.

PEMBROKE   85   SHAVERSVILLE
FLATWOODS
WD.
"Little Otter"
No.12   B&O
McNutt   MORRISON
"No.2"

104   45′   45′   106

CALHOUN

20

Quickle
Gassaway Branch   GASSAWAY   B&O   GILLESPIE   Holly   Mapleton   Hacker Valley
0.0   Switch   DX   90   Holly   WVM   Gillespie
22.1   RIVER   0.0   5   JCT. Irvin   WVM   25   WVM
25   B&O   HOLLY   Palmer   WVM   Fuccy
Sutton   JCT.   95
Bison   SN   (B&O)
Bear Creek
SHADYSIDE
30   FRAMETOWN   BRAXTON   CENTRALIA
F.
Rockton   ELK RIVER
B&O   GLENDON   100   CAMP RUN
VILLA NOVA   40   PRESTONIA
35   B&O   Webster Springs
Ira
STRANGE CREEK
GROVES
105
45   WEBSTER

Charleston   CLAY   NICHOLAS   ERBACON   38°30′
38°30′   80°30′
81°00′   45′   Richwood

Gassaway, W.Va.   **105**   W.Va.

CHARLESTON

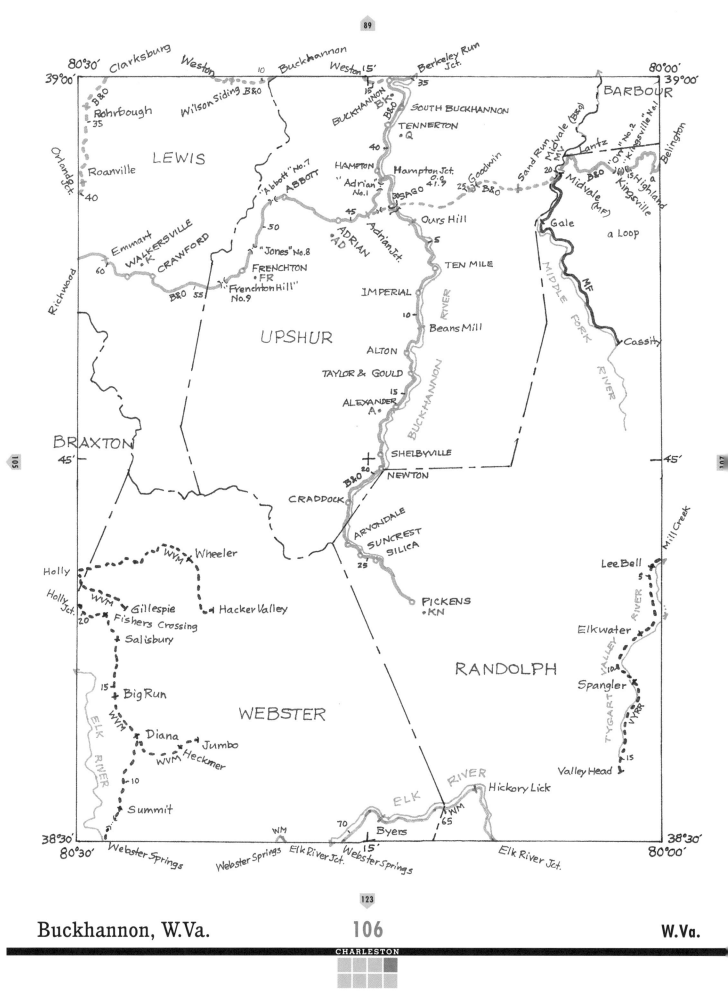

80°30' Clarksburg Weston 10 Buckhannon Weston 15' Berkeley Run Jct. 80°00'
39°00'   39°00'
B&O   BARBOUR
Rohrbough Wilson Siding B&O   35
-35   15   BUCKHANNON SOUTH BUCKHANNON
                                                    BK B&O
Orlando Jct.   TENNERTON Sand Run Midvale (B&O) Lantz "O" No.2 "Kingsville No.1"
LEWIS   40   •Q M.V.   Belington
Roanville   HAMPTON Hampton Jct. 20 Midvale "S" Highland
-40   "Abbott" No.7 "Adrian" 0.0 25 Goodwin B&O (MF) Kingsville
                            ABBOTT No.1 41.9 B&O   a Loop
Emmart WALKERSVILLE   50   SAGO Ours Hill Gale
Richwood •K CRAWFORD   ADRIAN Adrian Jct. 5
60   "Jones" No.8 AD   MF
                        FRENCHTON TEN MILE   Cassity
B&O 55   •FR   MIDDLE FORK RIVER
                "Frenchton Hill" IMPERIAL
                No.9   10   BUCKHANNON
                        UPSHUR   Beans Mill RIVER
                                    ALTON
                                TAYLOR & GOULD
                                        15
                            ALEXANDER
                            A.
BRAXTON   SHELBYVILLE
45'   B&O 20 NEWTON 45'
                            CRADDOCK
                                ARVONDALE
                                SUNCREST
                                SILICA
                        Holly WVM Wheeler   25   Lee Bell Mill Creek
                        Holly   5
                        Jct. WVM Gillespie   PICKENS RANDOLPH Elkwater VALLEY RIVER
                        20 Fishers Crossing Hacker Valley •KN   Spangler 10A
                            Salisbury   WVRR
                        15 Big Run   WEBSTER   Valley Head 15
                        WVM Diana Jumbo
                        WVM Heckmer ELK RIVER Hickory Lick
                        10   WVM 65
                        Summit   70 Byers
38°30'   WM   WM   38°30'
80°30' Webster Springs Webster Springs Elk River Jct. Webster Springs 15' Elk River Jct. 80°00'

ELK RIVER

Buckhannon, W.Va.   **106**   **W.Va.**

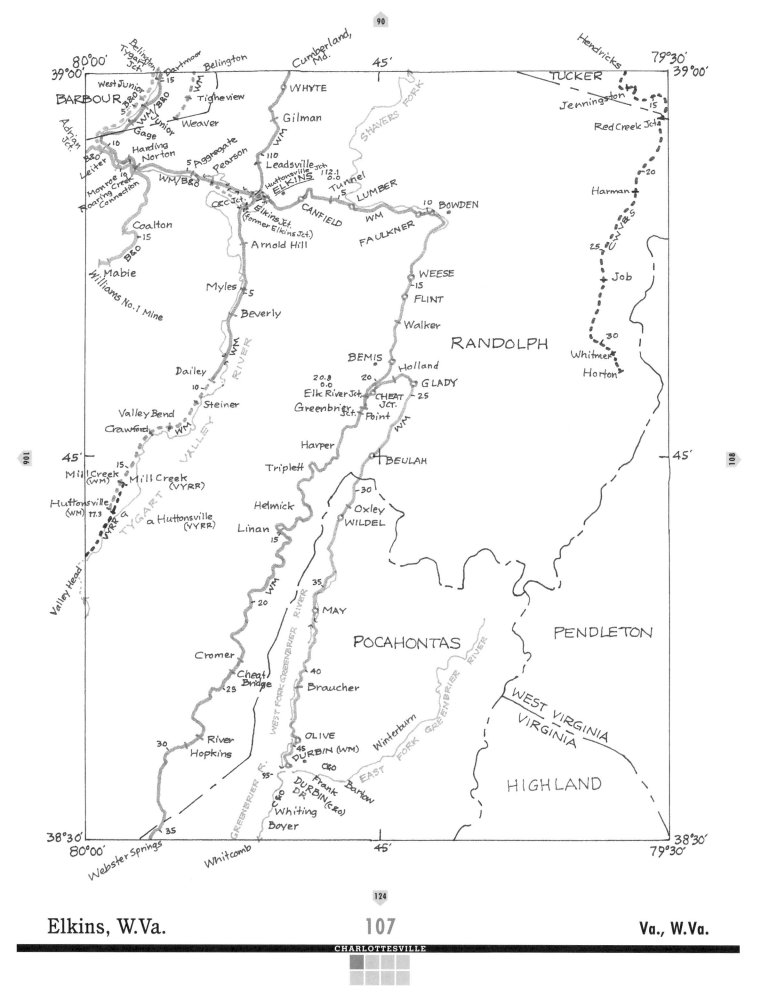

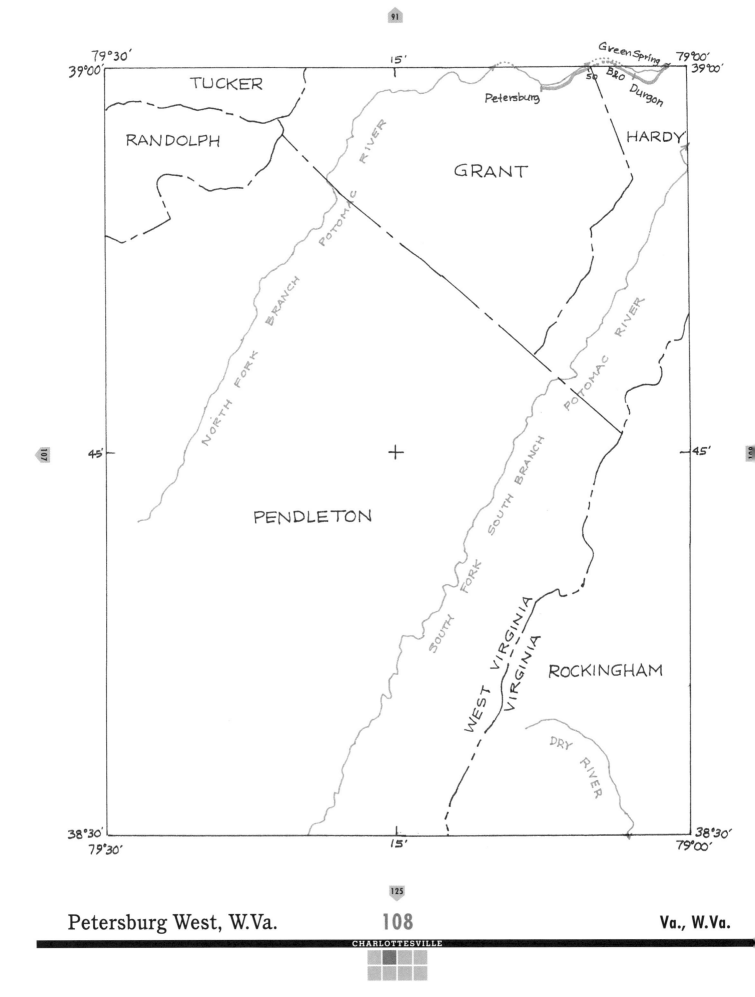

Petersburg West, W.Va.   **108**   Va., W.Va.

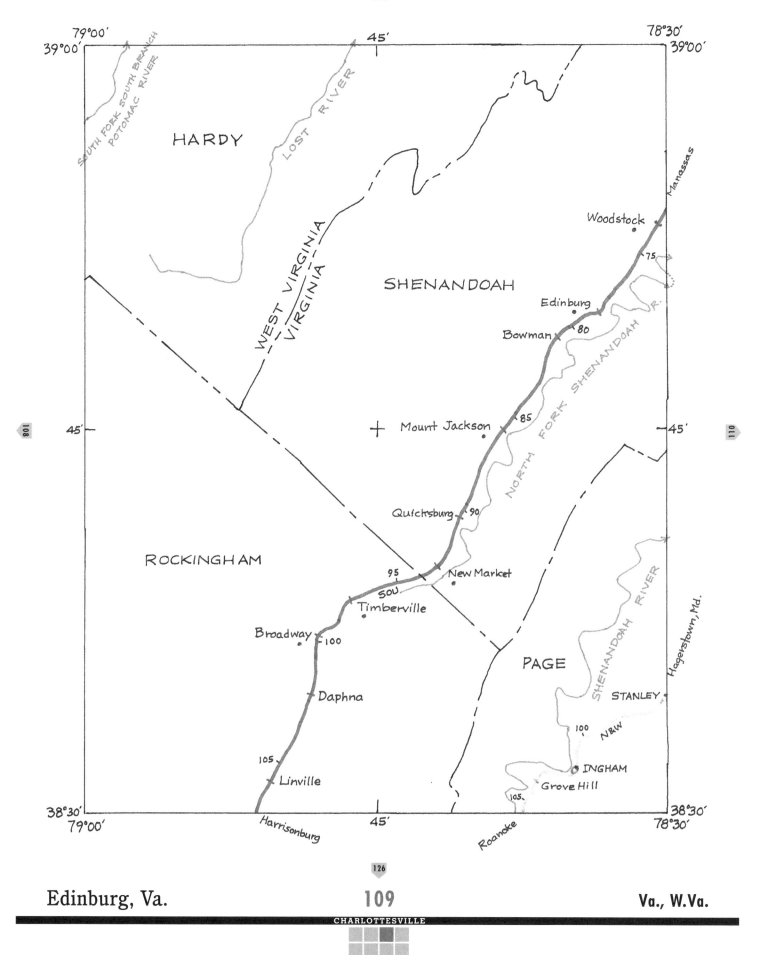

Edinburg, Va.   **109**   Va., W.Va.

CHARLOTTESVILLE

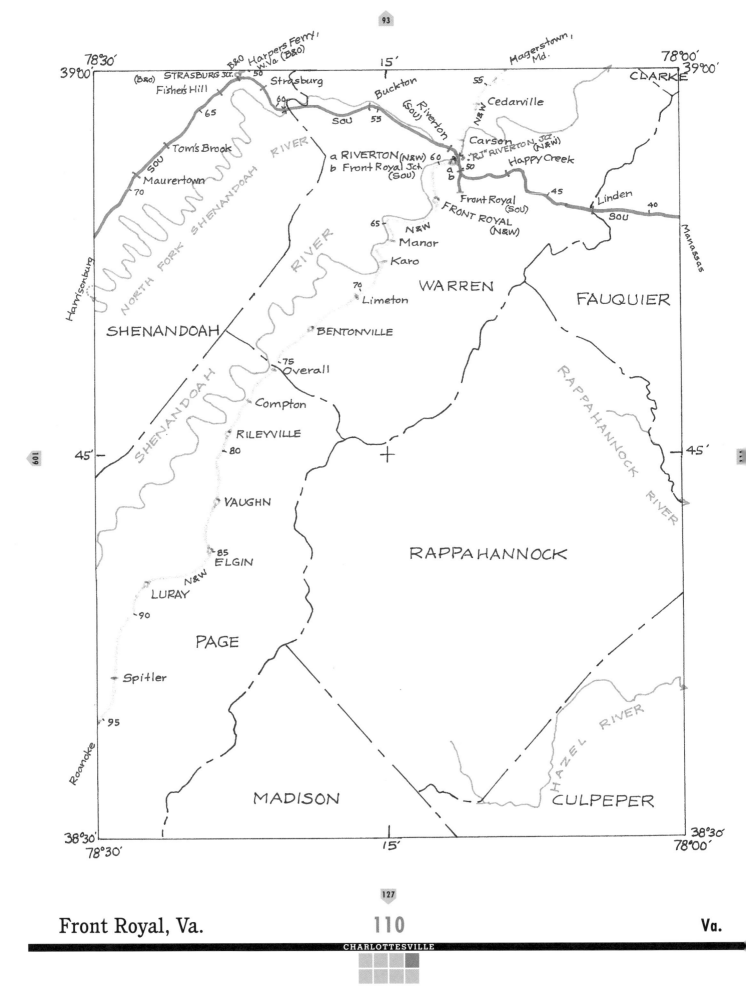

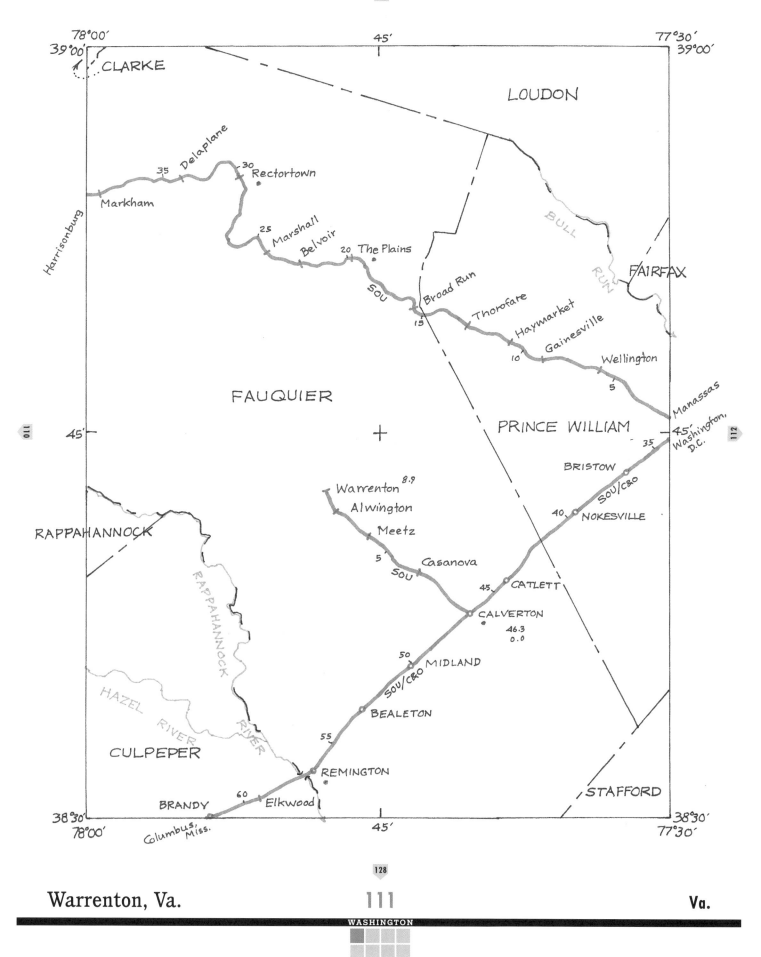

CLARKE

LOUDON

78°00'
39°00'
45'
77°30'
39°00'

35 Delaplane
30 Rectortown
Markham
Harrisonburg
25 Marshall
Belvoir
20 The Plains
SOU
Broad Run
15
Thorofare
Haymarket
Gainesville
10
Wellington
5

BULL RUN

FAIRFAX

FAUQUIER

+

45'

PRINCE WILLIAM

Manassas
45'
Washington, D.C.
35

BRISTOW
SOU/C&O
40
NOKESVILLE

Warrenton 8.9
Alwington
Meetz
5
SOU
Casanova
45
CATLETT
CALVERTON
46.3
0.0
50
SOU/C&O MIDLAND
BEALETON
55
RAPPAHANNOCK

RAPPAHANNOCK
RIVER

HAZEL RIVER

CULPEPER

REMINGTON
60
BRANDY
Elkwood
STAFFORD

38°30'
78°00'
Columbus, Miss.
45'
77°30'
38°30'

Warrenton, Va.   111   Va.

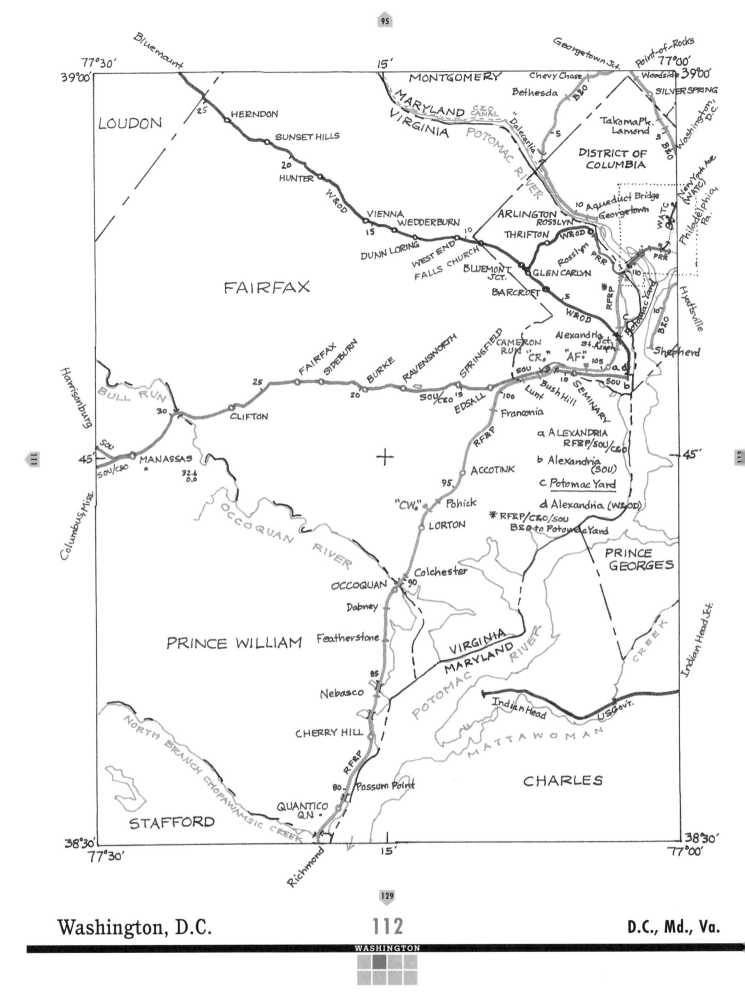

95

Bluemount

77°30'   39°00'   15'

MONTGOMERY

MARYLAND
VIRGINIA

Georgetown Jct.   Point-of-Rocks
Chevy Chase   77°00'
Bethesda   B&O   Woodside 39°00'
C&O CANAL   SILVER SPRING
Dalecarlia   TakomaPk.
Lamond   Washington, D.C.
5   5   B&O

LOUDON

HERNDON   25

SUNSET HILLS

HUNTER   20

W&OD

VIENNA   WEDDERBURN
15

DUNN LORING   10
WEST END
FALLS CHURCH

FAIRFAX

DISTRICT OF
COLUMBIA

10 Aqueduct Bridge   New York Ave.
Georgetown   (WATC)
ARLINGTON   Philadelphia,
ROSSLYN   WATC   Pa.
THRIFTON   W&OD   PRR
Rosslyn   PRR
BLUEMONT   GLEN CARLYN   110
JCT.   Potomac Yard
BARCROFT   5   C   10
W&OD   * RF&P   B&O
Hyattsville
Alexandria Jct.   Shepherd
St. Asaph

Harrisonburg

BULL RUN

FAIRFAX
SIDEBURN
BURKE   RAVENSWORTH   SPRINGFIELD
25   20   SOU/C&O   15
EDSALL   100
CLIFTON   SOU   Lunt   Bush Hill
30   Franconia
SOU

CAMERON
RUN "CR."   "AF."
105   SOU   a d
10   SOU b
SEMINARY

a ALEXANDRIA
RF&P/SOU/C&O

b Alexandria
(SOU)

SOU   SOU/C&O
45'   MANASSAS
32.6
0.0

Columbus, Miss.

RF&P

ACCOTINK
95

"CW."   Pohick

LORTON

OCCOQUAN RIVER

c Potomac Yard

d Alexandria (W&OD)

* RF&P/C&O/SOU
B&O to Potomac Yard

45'

PRINCE
GEORGES

Colchester
OCCOQUAN   90

Dabney

PRINCE WILLIAM   Featherstone

VIRGINIA
MARYLAND
POTOMAC RIVER

85

Nebasco

NORTH BRANCH CHOPAWAMSIC CREEK

CHERRY HILL

RF&P

80   Possum Point

QUANTICO
QN.

STAFFORD

38°30'
77°30'   15'

Indian Head   US Govt.

MATTAWOMAN

CHARLES

Indian Head Jct.

CREEK

38°30'
77°00'

Richmond

129

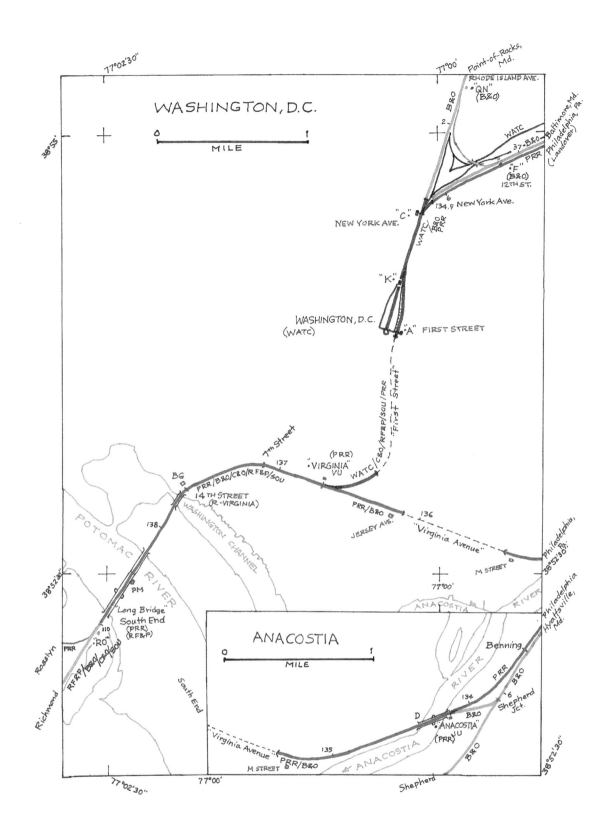

112A Washington, D.C. **113A** Anacostia, D.C.

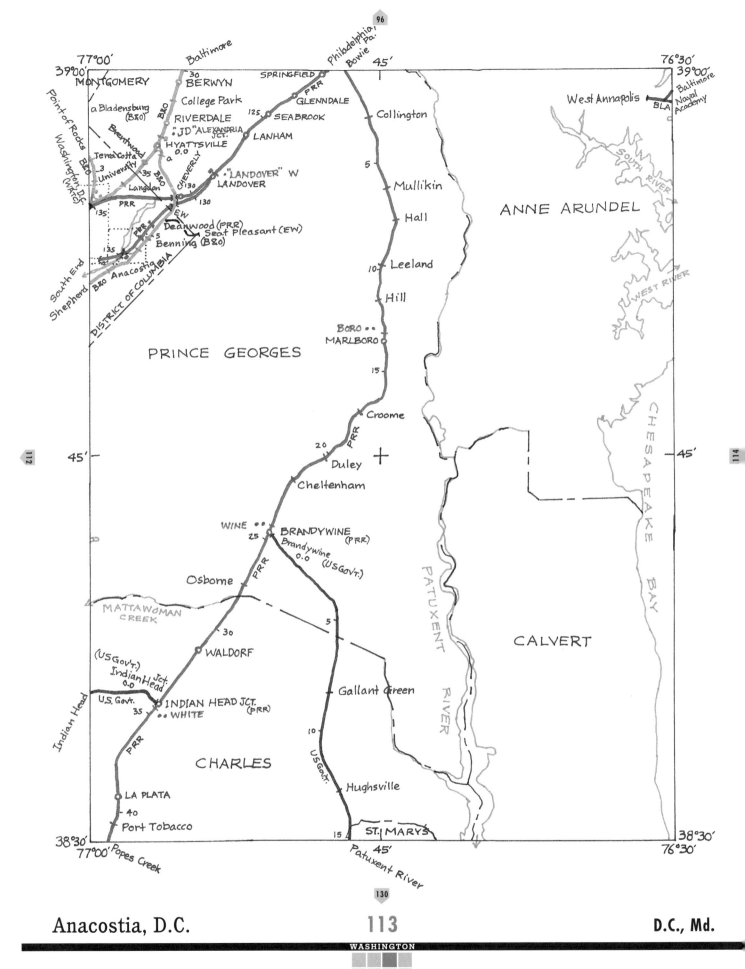

96

77°00'
39°00'
MONTGOMERY
Point of Rocks
Washington B&O
Baltimore
a Bladensburg (B&O)
BERWYN
College Park
RIVERDALE
30
B&O
SPRINGFIELD
Philadelphia, Pa.
Bowie
45'

West Annapolis
76°30'
39°00'
Baltimore
Naval
Academy
BLA

Brentwood
"JD" ALEXANDRIA
JCT.
GLENNDALE
SEABROOK
125
LANHAM
Collington
Washington D.C. (WRC)
Terra Cotta
University
3
B&O
35
HYATTSVILLE
0.0
a
CHEVERLY
130
"LANDOVER" W
LANDOVER
5
Mullikin
Langdon
PRR
135
130
Hall
South End
PRR
135
EW
Deanwood (PRR)
Seat Pleasant (EW)
Benning (B&O)
5
Leeland
10
Hill

Shepherd B&O Anacostia
DISTRICT OF COLUMBIA
PRINCE GEORGES

ANNE ARUNDEL

SOUTH RIVER

WEST RIVER

BORO
MARLBORO
15
Croome
20
PRR
Duley
Cheltenham
45'
45'

CHESAPEAKE BAY

WINE
25
BRANDYWINE (PRR)
Brandywine 0.0 (US GOV'T.)
Osborne
PRR
MATTAWOMAN CREEK
5
30
WALDORF
(US GOV'T.)
Indian Head 0.0
JCT.
U.S. Govt.
35
INDIAN HEAD JCT. (PRR)
WHITE
Indian Head
PRR
CHARLES
Gallant Green
10
US Govt.
PATUXENT RIVER
CALVERT

LA PLATA
40
Port Tobacco
Hughsville
15
ST. MARYS
38°30'
76°30'
38°30'
77°00' Popes Creek
45'
Patuxent River

Anacostia, D.C.   113   D.C., Md.

WASHINGTON

for detail, see detail map page 112A

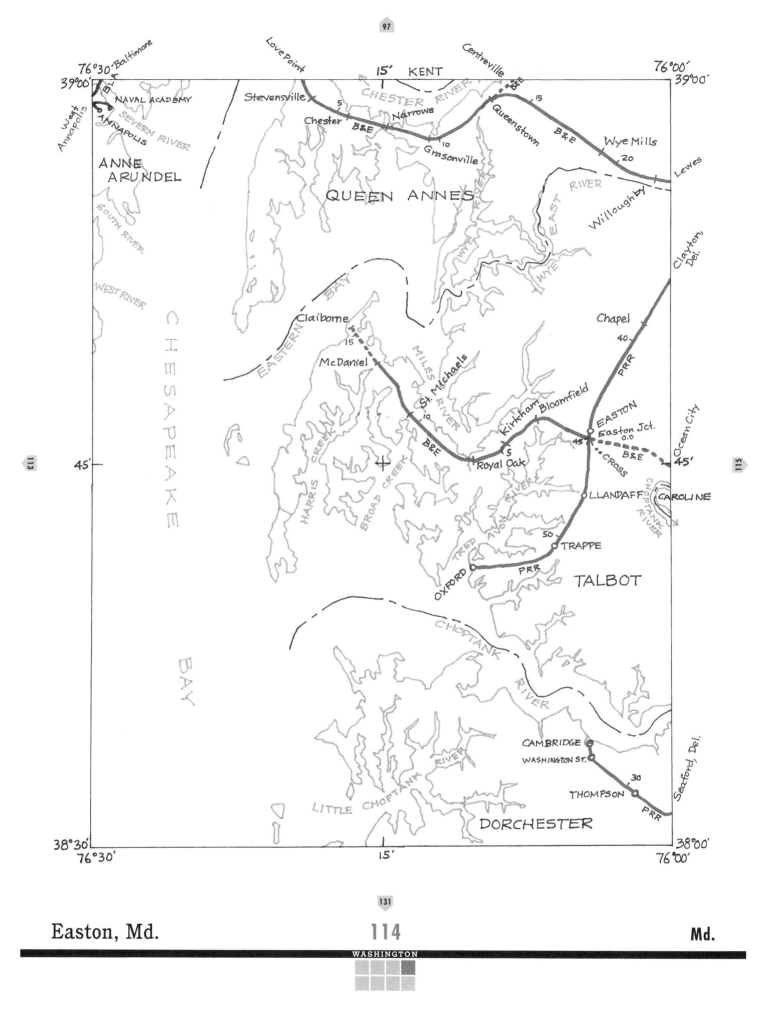

76°30' Baltimore
39°00'

LovePoint

15' KENT

Centreville

76°00'
39°00'

NAVAL ACADEMY

Stevensville

CHESTER RIVER

B&E

15

West Annapolis

SEVERN RIVER

Chester

B&E

Narrows

Queenstown

B&E

Wye Mills

ANNAPOLIS

5

10

Grasonville

20

Lewes

ANNE
ARUNDEL

SOUTH RIVER

QUEEN ANNES

WYE

EAST RIVER

Willoughby

Clayton, Del.

WEST RIVER

EASTERN BAY

Claiborne

Chapel

40

15

McDaniel

MILES RIVER

St. Michaels

PRR

CHESAPEAKE

Kirkham

Bloomfield

EASTON

10

Easton Jct.
0.0

Ocean City

45'

HARRIS CREEK

B&E

5

45

B&E

45'

Royal Oak

CROSS

LLANDAFF

CAROLINE

BROAD CREEK

TRED AVON RIVER

50

CHOPTANK RIVER

OXFORD

Trappe

TALBOT

PRR

BAY

CHOPTANK RIVER

CAMBRIDGE

WASHINGTON ST.

LITTLE CHOPTANK RIVER

30

THOMPSON

Seaford, Del.

PRR

DORCHESTER

38°30'
76°30'

15'

38°00'
76°00'

Easton, Md.

114

Md.

113

115

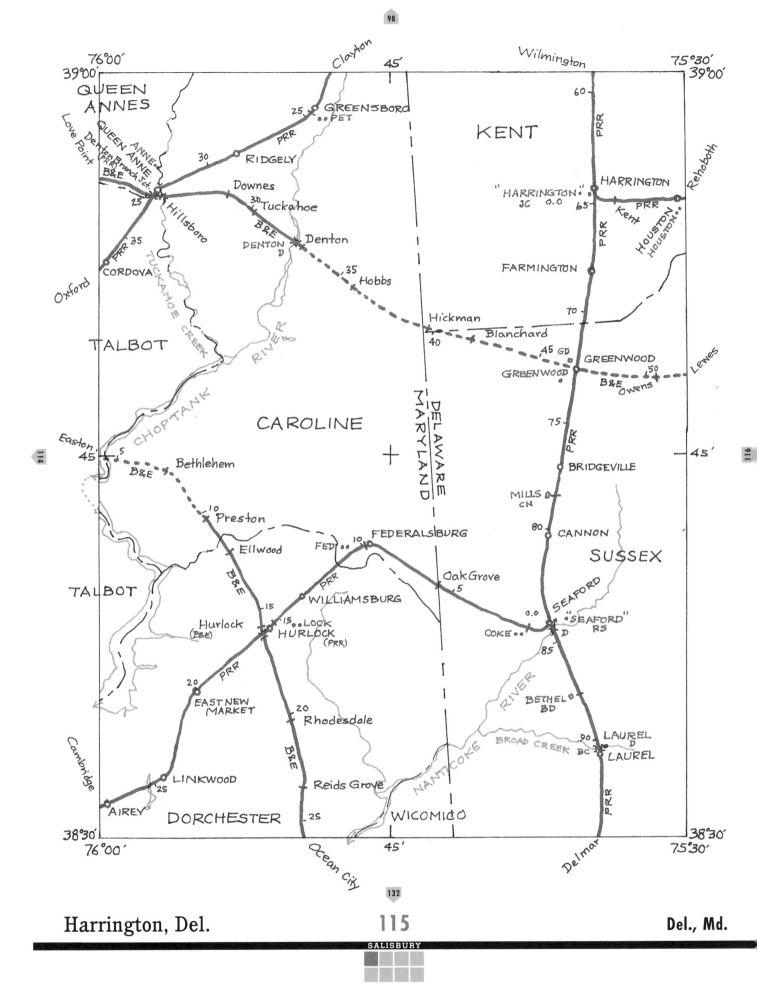

Clayton
45'
Wilmington
75°30'
39°00'

QUEEN
ANNES

76°00'
39°00'

KENT

Love Point

25 GREENSBORO
PET

PRR

60

PRR

Denton Branch Jct.
QUEEN ANNE
B&E

30 RIDGELY

Downes

HARRINGTON

"HARRINGTON"
JC 0.0

25 Hillsboro

30 Tuckahoe

B&E

65 Kent

PRR

Houston
HOUSTON

Rehoboth

PRR 35

DENTON
D

Denton

FARMINGTON

Oxford

CORDOVA

35 Hobbs

TALBOT

TUCKAHOE CREEK

RIVER

Hickman
40

Blanchard

45 GD

70

GREENWOOD

Lewes

GREENWOOD
B&E Owens 50

Easton
45'

5

CHOPTANK

CAROLINE

+

DELAWARE
MARYLAND

75

PRR

45'

116

B&E Bethlehem

BRIDGEVILLE

10 Preston

MILLS D
CN

Ellwood

FED 10 FEDERALSBURG

Oak Grove
5

80 CANNON

SUSSEX

B&E

PRR

WILLIAMSBURG

SEAFORD

TALBOT

15

0.0

15 LOCK
HURLOCK
(PRR)

"SEAFORD"
RS

COKE

D

Hurlock
(B&E)

85

PRR

20 EAST NEW
MARKET

20 Rhodesdale

RIVER

BETHEL
BD

90 LAUREL
D

BROAD CREEK BC

LAUREL

Cambridge

B&E

LINKWOOD
25

Reids Grove

NANTICOKE

PRR

AIREY

DORCHESTER

25

WICOMICO

38°30'

76°00'

45'

Ocean City

Delmar

75°30'
38°30'

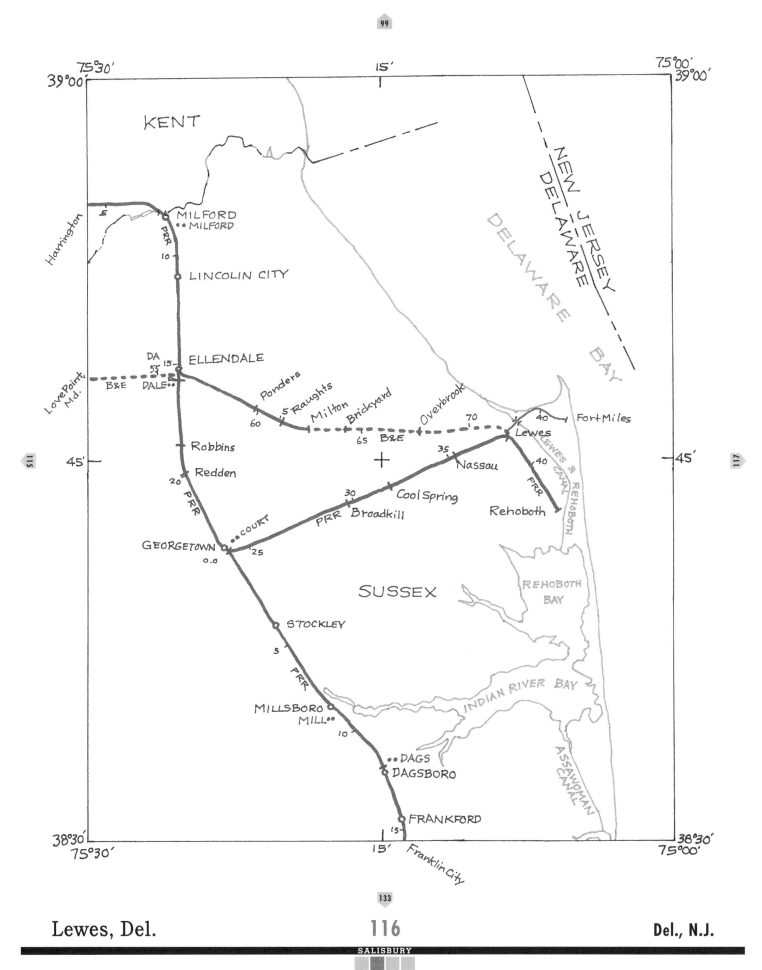

SALISBURY

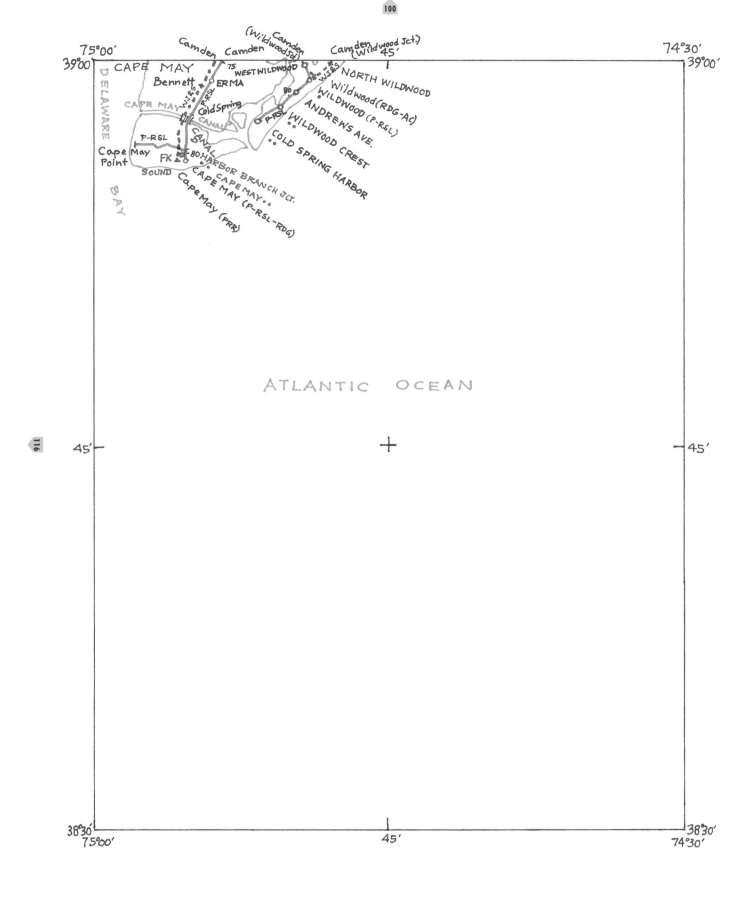

75°00′
39°00′

(Wildwood Jct.)

Camden
Camden Camden

Camden Wildwood Jct.)

74°30′
39°00′
45′

CAPE MAY

Bennett

75

WEST WILDWOOD

W.P.R.S.

P-RSL

ERMA

W.P.R.S.

NORTH WILDWOOD

Cold Spring

80

Wildwood (RDG-Ac)

WILDWOOD (P-RSL)

CAPE MAY

CANAL

P-RSL

ANDREWS AVE.

WILDWOOD CREST

COLD SPRING HARBOR

Cape
Point

May

P-RSL

CANAL

P-RSL

80

HARBOR BRANCH JCT.

FK

SOUND

Cape May

CAPE MAY
CAPE MAY (P-RSL-RDG)

DELAWARE

BAY

ATLANTIC OCEAN

45′

+

45′

38°30′
75°00′

45′

74°30′
38°30′

SALISBURY

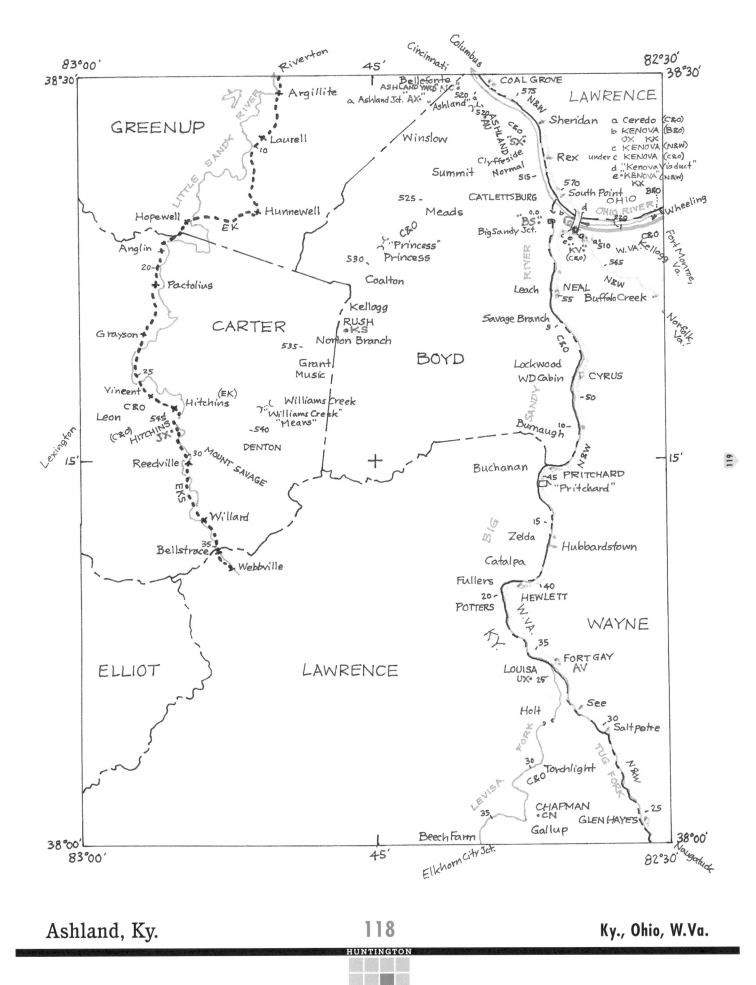

GREENUP

Riverton

Argillite

LITTLE SANDY RIVER

Laurell

10

Hopewell

Hunnewell

EK

Anglin

20

Pactolius

Grayson

CARTER

25

Vincent

Hitchins (EK)

Leon
(C&O)

C&O

545

HITCHINS
JX

Reedville

30 MOUNT SAVAGE

EKS

Willard

35

Bellstrace

Webbville

ELLIOT

LAWRENCE

Cincinnati

Columbus

45

Bellefonte

ASHLAND YARD "C"

a Ashland Jct. "AX"

520

"Ashland" "J"

520
AU

520

Winslow

Clyffeside
Normal

Summit

515

525

Meads

C&O

"Princess"

530

Princess

Coalton

Kellogg

RUSH
KS

Norton Branch

535

Grant
Music

Williams Creek

"Williams Creek"

540 "Means"

DENTON

BOYD

COAL GROVE

575

N&W

LAWRENCE

Sheridan

a Ceredo (C&O)
b KENOVA (B&O)
OX   KX
c KENOVA (N&W)
under c KENOVA (C&O)
d "Kenova Viaduct"
e "KENOVA" (N&W)
KX

Rex

ASHLAND SX

C&O

Summit

CATLETTSBURG

570

South Point
OHIO

BS

Big Sandy Jct.

a
KV
(C&O)

510

OHIO RIVER

530

Wheeling

B&O

b

e c a d

W.VA. Kellogg
Va.

565

Fort Monroe

Leach

NEAL
55

N&W

Buffalo Creek

Norfolk,
Va.

Savage Branch

5

C&O

Lockwood
WD Cabin

CYRUS

50

SANDY

Burnaugh

10

N&W

Buchanan

45 PRITCHARD

"Pritchard"

15

BIG

Zelda

Hubbardstown

Catalpa

Fullers

40

20

HEWLETT

POTTERS

W.VA.

WAYNE

35

KY.

FORT GAY
AV

LOUISA
UX 25

See

Holt

30 Saltpetre

TUG FORK

30

Torchlight

C&O

N&W

LEVISA FORK

35

CHAPMAN
CN

GLEN HAYES

25

Gallup

Beech Farm

Elkhorn City Jct.

Naugatuck

119

15'

15'

45

Ashland, Ky.

Ky., Ohio, W.Va.

82°30'
38°30'
15'
82°00'
38°30'

LAWRENCE

MASON

Wheeling

205
COX LANDING

CABELL

RIVER

Hurricane
KX

Ft. Monroe, Va.

WEST HUNTINGTON

HUNTINGTON (B&O)
"HU"

GUYANDOTTE (B&O)

OHIO RIVER OHIO

B&O
W. VA.

C&O

490
Ona

MUD

RIVER

Yates Crossing

MILTON
MI

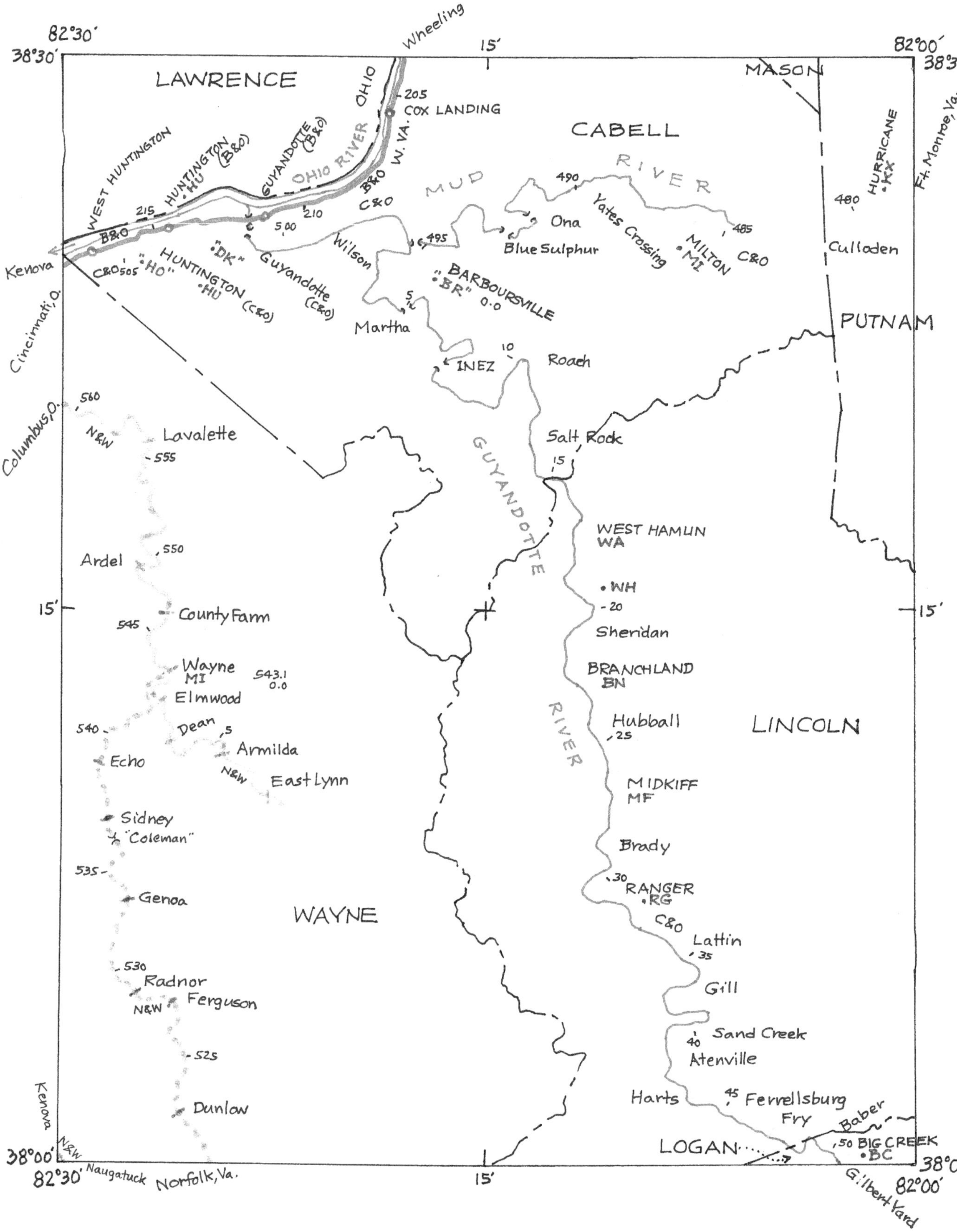

480

Culloden

485
C&O

B&O

215

210

5.00

Blue Sulphur

495

PUTNAM

Kenova

Wilson

C&O 505 "HO"

"DK"

Guyandotte
(C&O)

HUNTINGTON (C&O)
"HU"

BARBOURSVILLE
"BR" 0.0

Cincinnati, O.

Martha

5

INEZ

10

Roach

Salt Rock

560

Columbus, O.

N&W

Lavalette

555

15

GUYANDOTTE

WEST HAMLIN
WA

WH
20

15'

15'

Ardel

550

Sheridan

County Farm

545

BRANCHLAND
BN

LINCOLN

Wayne
MI
543.1
0.0

Hubball
25

Elmwood

540

Dean

5

Armilda

MIDKIFF
MF

Echo

N&W

East Lynn

RIVER

Brady

Sidney

"Coleman"

30
RANGER
RG

535

Genoa

WAYNE

C&O

Lattin
35

530

Radnor

Gill

Ferguson

N&W

525

Sand Creek
40

Atenville

Dunlow

Harts

45
Ferrellsburg

Fry

Baber

50 BIG CREEK
BC

38°00'

82°30'
Naugatuck

Norfolk, Va.

15'

82°00'

LOGAN

Gilbert Yard

38°00'

Kenova

N&W

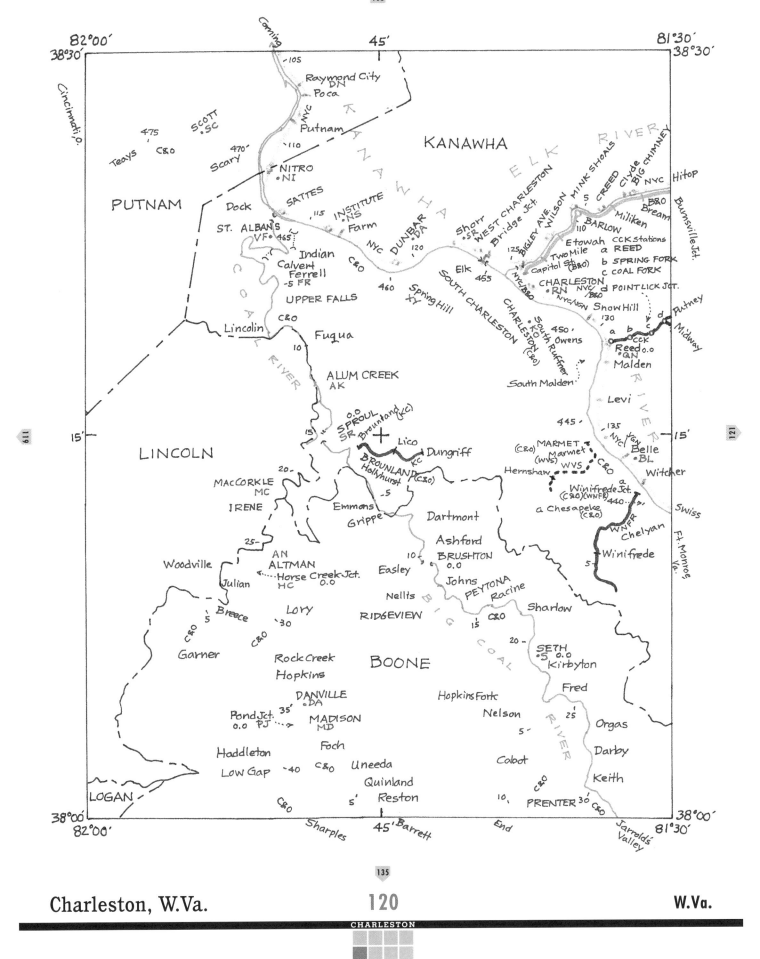

82°00' 45' 81°30'
38°30' 38°30'

Corning
-105

Raymond City
DN
Poca

SCOTT
·SC
NYC
Putnam

Teays 475
C&O

Scary 470-
-110

KANAWHA

ELK RIVER

MINK SHOALS
CREED Clyde BIG CHIMNEY
NYC Hitop

PUTNAM

NITRO
·NI

SATTES

Dock

115

INSTITUTE
·NS
Farm

ST. ALBANS
VF· 465'

Indian
Calvert
Ferrell
-5 FR

UPPER FALLS

NYC

DUNBAR
DA

120

WEST CHARLESTON
Shorr Bridge Jct.
·SR
BIGLEY AVE.
WILSON
Capitol St.
(B&O)
125

BARLOW Miliken
Bream
110 B&O

Etowah CCK stations
Two Mile a REED
b SPRING FORK
c COAL FORK

Burnsville Jct.

Lincoln
C&O

Fuqua
10

Elk 455
C&O
460
Spring Hill
XY

SOUTH CHARLESTON

CHARLESTON
·RN NYC
/B&O
d POINTLICK JCT.

NYC/NGN Snow Hill
130

CHARLESTON
KOH
South Charleston
(C&O)

450'
Owens

a b c d Putney
CCK Midway
Reed 0.0
·QN
Malden

South Ruffner

ALUM CREEK
AK

South Malden

Levi

RIVER

0.0
SPROUL Brounland (KC)
SR
Lico
15' BROUNLAND
Holyhurst
(C&O)
Dungriff
KC

445
135
NYC NGN
C&O
15'
Belle
·BL
Witcher

LINCOLN

MacCORKLE
MC
20-
-5

MARMET
(C&O) Marmet
(WVS) WVS
Hernshaw C&O

IRENE

Emmons
Grippe

Dartmont

Winifrede Jct.
(C&O)(WNFR) 440
a Chesapeke
(C&O)

Swiss

Woodville

25-

AN
ALTMAN
Horse Creek Jct.
HC 0.0

Easley
10
Ashford
BRUSHTON
0.0

Johns PEYTONA
Racine

WNFR
Chelyan

Winifrede
-5

Ft. Monroe,
Va.

Julian

Breece
5'

LORY
·30

Nellis

RIDGEVIEW

Sharlow
C&O
15'

BIG COAL

20-

SETH
·5 0.0
Kirbyton

Garner

C&O
C&O

Rock Creek
Hopkins

BOONE

Hopkins Fork

Nelson

Fred
25'
5-

Orgas

DANVILLE
·DA

Pond Jct. 35'
0.0 PJ

MADISON
MD

Foch

Cabot

Darby

Keith

Haddleton
Low Gap -40
C&O

Uneeda
Quinland
Reston
5'

C&O
10, PRENTER 30
C&O

LOGAN

C&O

Sharples

RIVER

38°00' 45' Barrett End Jarrolds 38°00'
82°00' 81°30' Valley

611 15' 15' 121

CHARLESTON

81°30′ 38°30′

Burnsville Jct.
15 · Charleston
Burnsville Jct. 81°00′ 38°30′
55

Barren Creek
CAMP
Spread

ELK RIVER
CLENDENIN
REAMER
95 · B&O
WALGROVE
100
a BLUE CREEK
· BF TARGET
(B&O)

PORTERS
80
QUEEN SHOALS
90
Birch Run
MARNE
Rouzer
70 ELKHURST
DUNDON DN (B&O)
DUNDON QO (BC&G)
SAND FORK
60
· Widen
5

ELKVIEW K.
NYC
NORTH PINCH
SANDY
a
105
B&O
Jarretts Ford
Copenhaver
PINCH
Q
FALLING ROCK
· FA
BLUE CREEK
· BC
OILSIDING
THREE MILE
· VICTOR
15

KANAWHA

75
SHELTON
DORFEE
B&O
65
CLAY
· SQ
AVOCA
BC&G

HARTLAND (B&O)
Hartland (McK)

Charleston (B&O)
a Charleston
Bigley Ave. Jct. (NYC)
SCHRADER
COCO
PENTACRE
PN
NYC
20
QUICK
COALRIDGE
SANDERSON
ACUP
25
MIDDLE FORK
KENDALIA
30
CLAY

· Bickmore
McK

Reed
TAD 5
CINCO
RENSFORD CCK
MIDWAY 10
EIGHT MILE
MILL HOLLOW
PUTNEY
AMELIA
POND FORK
WILLS HOLLOW
BLAKELEY BY
HITOP
BENTREE
GREENDALE
NICHOLAS

15′
Dixie
C&O
C&O 10 Belva (NYC)
BeechGlen
NYC/C&O 170
Swiss (172.8 NYC)
Swiss Jct. (NF&G)
"Koontz"
15′

Mammoth
Ward
Dickinson (NYC)
Shrewsbury
NYC/VGN
Monarch
CedarGrove (NYC & KCR&W)
UG TARGET
KC&NW
KEC Glasgow
Midwest
RS
PRATT (C&O)
Hugheston
London
150
Lock 2
Cannelton
NYC
VGN
Smithers
OPEN FORK JCT.
(C&O)
BELVA
Rick Creek Jct.
Wyndal
165
NYC
C&O
-S
Gamoca
Vanetta
GAULEY
NF&G/C&O
NYC
Koontz
GAULEY RIVER
Meadow Creek

Corningo
Cincinnati, O.
NYC/VGN
140 -
435 c
a (see below)
d C&O e
b
CABIN CREEK JCT.
· CA
-145
Chelyan
Dry Branch
Ronda
Sharon
Miami
Dawes
Giles
HANDLEY RD
Holly Grove
430
Morris Jct.
Livingston
Standard
GALLAGHER
MONTGOMERY
Eagle
425
Longacre
Harewood
155 Boomer
Alloy
DB Tower
Falls View
160
Gauley Bridge
Glen Ferris
GauleyBridge
GB
GAULEY
"GU" GAULEY
NEW RIVER
KANAWHA FALLS
C&O 415
McDougal
C&O
Anstead
"MA"
HAWK'S NEST

Main Line Stations
a Coalburg
b East Bank
c Black cat
d Crown Hill
e Hansford
Coal
5 -
C&O
Ohley
CANE FORK
· CJ
Eskdale
10 -
C&O
10
Nuckolls
WHITAKER
GlenHuddy
Ridenour
Columbia
MOUNT CARBON
Elkridge Jct. 5
5 · b
C&O
a
WEST DEEPWATER
"Deepwater"
DEEPWATER
· VN
Robson
420
BEARDS JCT.
COTTON HILL
410
405
Ames
-405
FAYETTE

CHEROKEE
RED WARRIOR JCT.
LEEWOOD
Holly 0.0
C&O
Quarrier
Acme
Laing
DECOTA
South Carbon
C&O
15
BURNWELL a Elkridge
b Powellton
Collinsdale
MAHAN
Coalfield
MILBURN
15
430 ·
VGN
3
Beard's Fork
PAGE
PAGE (R-MULLENS)
425
Ingram Branch
"Cow Hollow"
WRISTON
VGN
South Fayette
405
Fayette
Kaymoor
Elverton
Bridge Jct.
C&O
Nuttall
Keeneys Creek
C&O Lookout
400
-400
Caperton

BOONE
KAYFORD
Tunnel Siding
15
Kayford
End
-15 5- -5
Wevaco
C&O
HAMILTON (R-MULLENS)
Summerlee
5
Lochgelly
VGN/C&O
OakHill Jct.
Fort Monroe, Va.

38°00′
81°30′
Whitesville
W.VA. No. 2
Republic
Kingston
15′
Norfolk, Va.
38°00′ 81°00′

Gauley Bridge, W.Va.   121   W.Va.

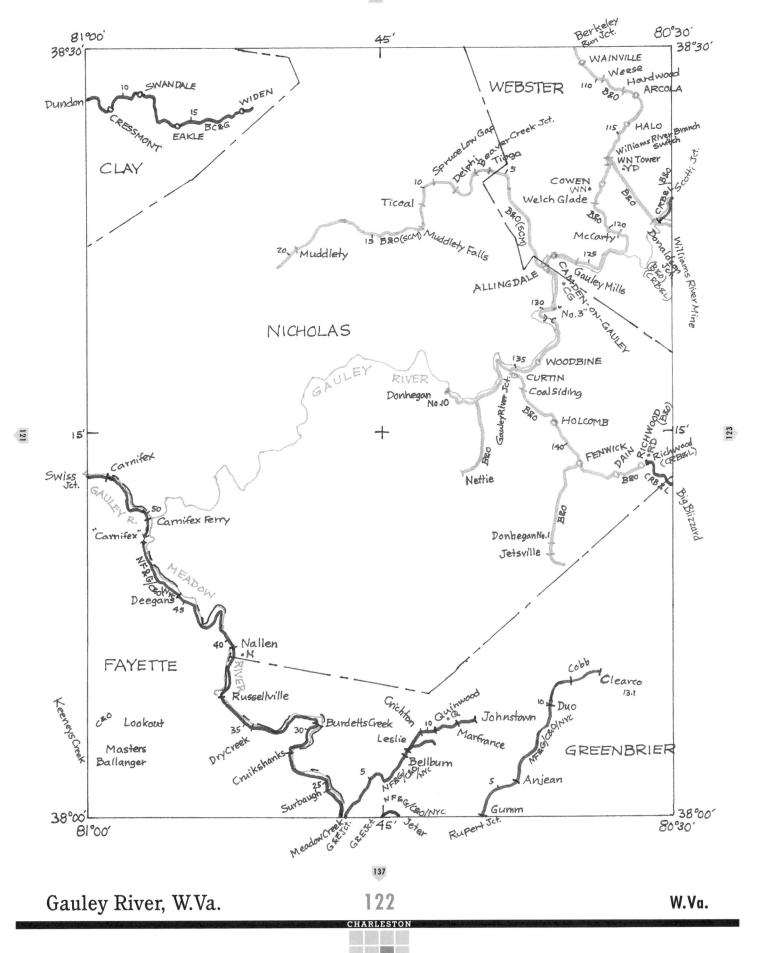

81°00'
38°30'

SWANDALE
10
Dundon
CRESSMONT
EAKLE
15
BC&G
WIDEN

CLAY

Berkeley
Run Jct.
80°30'
38°30'

WAINVILLE
Weese
Hardwood
ARCOLA
110
B&O

WEBSTER

115
HALO
Williams River Branch
Switch
Scotti Jct.
WN Tower
·YD
Spruce Low Gap
Delphi Beaver Creek Jct.
Tioga
10
5
B&O
COWEN
WN·
Welch Glade
CR B&L
B&O
B&O

Ticoal
15 B&O(SCM) Muddlety Falls
B&O(SCM)
120
McCarty
Donaldson
Jct.
(B&O)
(CRB&L)

20 Muddlety

125
Williams River Mine

NICHOLAS
ALLINGDALE
Gauley Mills
CAMDEN-ON-GAULEY
·CG
130
·C No.3"

GAULEY RIVER
WOODBINE
135
CURTIN
Coal Siding
Gauley River Jct.
B&O
HOLCOMB
B&O
140
B&O
Donhegan
No.10

15'
+
Carnifex
Swiss
Jct.
GAULEY R.
50
Carnifex Ferry
"Carnifex"
NF&G/C&O/NYC
Deegans
45
FENWICK
DAIN
RICHWOOD B&O
·RD Richwood
B&O (CRB&L)
CR B&L
Big Blizzard
15'

Nettie

MEADOW
Donhegan No.1
Jetsville

40
Nallen
·N

FAYETTE
RIVER
Russellville
C&O Lookout
Masters
Ballanger
35
DryCreek
30
BurdettsCreek
Cruikshanks
25
Surbaugh
Crichton
10 Quinwood
·Q
Leslie
5
Bellburn
NF&G/C&O/NYC
NF&G/C&O/NYC
Jeter
MeadowCreek
G&EJct.
G&EJct. 45'
Johnstown
Marfrance
Cobb
Clearco
13.1
10 Duo
NF&G/C&O/NYC
GREENBRIER
5 Anjean
Gumm
Rupert Jct.

Keeneys Creek

38°00'
81°00'
38°00'
80°30'

121   123

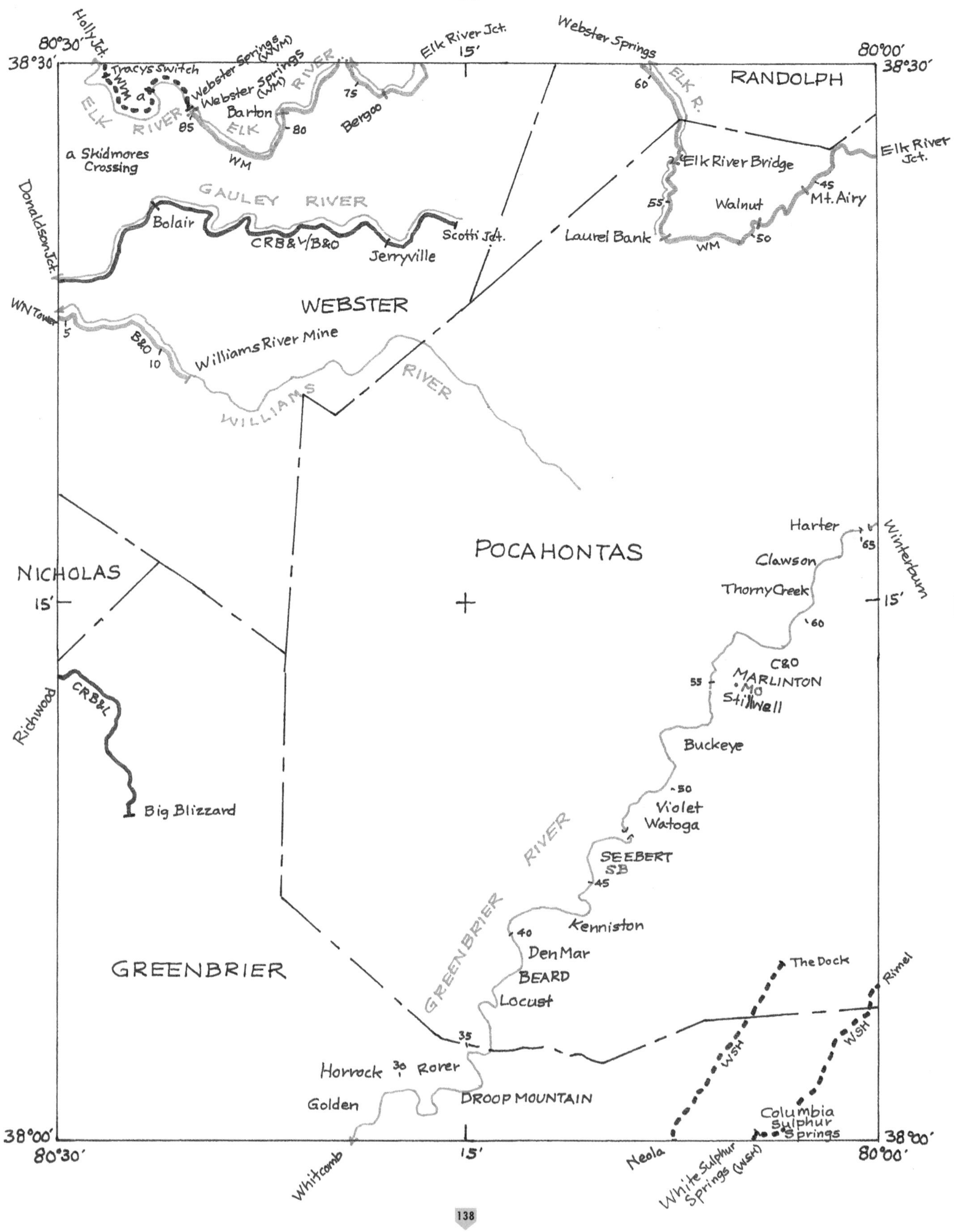

80°30'

38°30'

Holly Jct.

Tracys Switch

Webster Springs (WM)

Webster Springs (WM)

ELK RIVER

ELK

Barton

RIVER

Elk River Jct.
15'

Webster Springs

ELK R.

RANDOLPH

80°00'

38°30'

WM a

85

80

75

Bergoo

60

a Skidmores
Crossing

WM

Elk River Bridge

Walnut

Mt. Airy

45

Elk River
Jct.

Donaldson Jct.

GAULEY RIVER

55

Laurel Bank

WM

50

Bolair

CRB&Y/B&O

Jerryville

Scotti Jct.

WEBSTER

WN Tower

5

B&O

10

Williams River Mine

WILLIAMS

RIVER

POCAHONTAS

Harter

Winterburn

Clawson

Thorny Creek

'63

NICHOLAS

15'

+

'60

'15'

C&O

55

MARLINTON

MO

Stillwell

Richwood

CRB&L

Big Blizzard

Buckeye

'50

Violet

Watoga

GREENBRIER RIVER

SEEBERT
SB

45

Kenniston

40

Den Mar

GREENBRIER

BEARD

Locust

The Dock

Rimel

35

30

WSH

WSH

Horrock

Rover

Golden

DROOP MOUNTAIN

Columbia
Sulphur
Springs

38°00'

80°30'

Whitcomb

15'

Neola

White Sulphur
Springs (WSH)

38°00'

80°00'

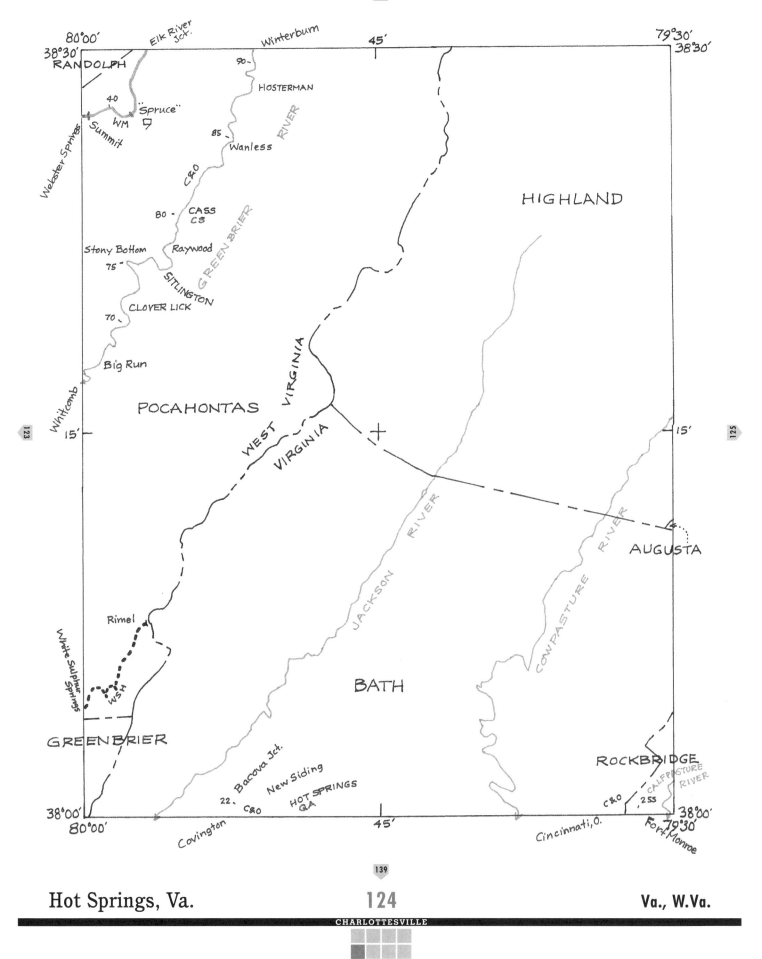

Hot Springs, Va.

124

Va., W.Va.

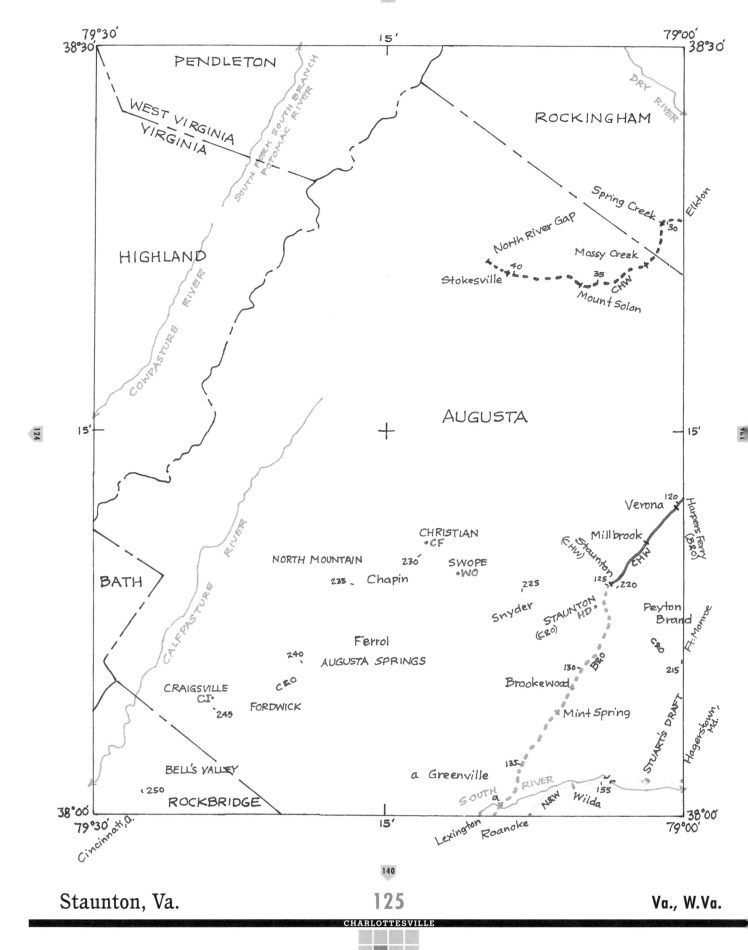

124

126

79°30′
38°30′

15′

79°00′
38°30′

PENDLETON

WEST VIRGINIA
VIRGINIA

SOUTH FORK SOUTH BRANCH POTOMAC RIVER

ROCKINGHAM

DRY RIVER

Spring Creek

Elkton

North River Gap

30

HIGHLAND

COWPASTURE RIVER

Massy Creek

Stokesville

40

35

CHW

Mount Solon

15′

AUGUSTA

15′

Verona   120

Harpers Ferry (B&O)

CHRISTIAN
•CF

Millbrook

Staunton (CHW)

NORTH MOUNTAIN

230′

SWOPE
•WO

CHW

235′ Chapin

225

Staunton (C&O)

125 220

Snyder

STAUNTON
HD.

Peyton Brand

Ft. Monroe

CALFPASTURE RIVER

Ferrol

C&O

215

BATH

AUGUSTA SPRINGS

240

130

B&O

Brookewood

CRAIGSVILLE
CI•

C&O

Mint Spring

FORDWICK

245

STUART'S DRAFT

Hagerstown, Md.'

BELL'S VALLEY

a Greenville

135

155

250

SOUTH   RIVER

N&W Wilda

ROCKBRIDGE

a

38°00′
79°30′

Cincinnati, O.

15′

Lexington Roanoke

38°00′
79°00′

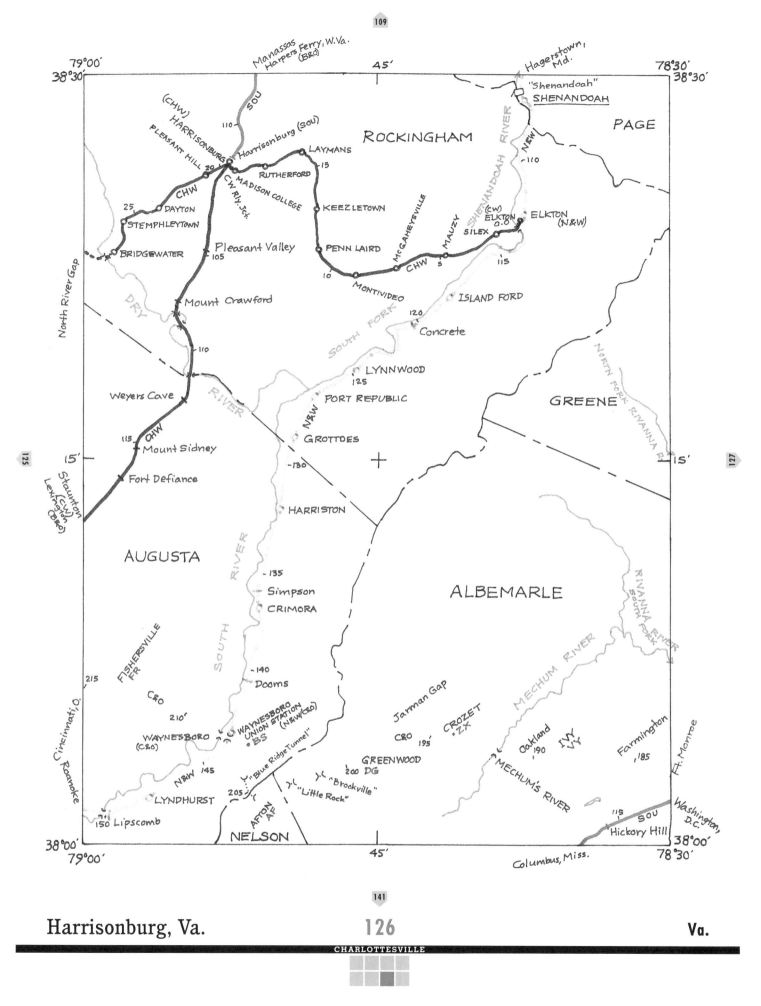

Manassas Ferry, W.Va.
Harpers Ferry, W.Va. (B&O)

Hagerstown, Md.

SOU

"Shenandoah"
SHENANDOAH

79°00'
38°30'

45'

78°30'
38°30'

110

(CHW)
HARRISONBURG
PLEASANT HILL
Harrisonburg (SOU)

ROCKINGHAM

PAGE

SHENANDOAH RIVER

N&W
110

LAYMANS
20
RUTHERFORD
15
CW Rly. Jct.
MADISON COLLEGE

CHW
25 DAYTON
STEMPHLEYTOWN

KEEZLETOWN

McGAHEYSVILLE

MAUZY

(CW)
ELKTON
0.0
SILEX

ELKTON
(N&W)

Pleasant Valley
105

PENN LAIRD

BRIDGEWATER

North River Gap

DRY

Mount Crawford

110

CHW
10

MONTIVIDEO

CHW
5

115

SOUTH FORK

ISLAND FORD

120

Concrete

LYNNWOOD
125

RIVER

PORT REPUBLIC

GREENE

NORTH FORK RIVANNA R.

Weyers Cave

N&W

15'

15'

125

115 CHW
Mount Sidney

Fort Defiance

Staunton
(CW)
Lexington
(B&O)

GROTTOES

-130

HARRISTON

AUGUSTA

SOUTH RIVER

-135

Simpson
CRIMORA

ALBEMARLE

RIVANNA RIVER SOUTH FORK

127

MECHUM RIVER

-140
Dooms

FISHERSVILLE
FR
215
C&O
210'

WAYNESBORO
(C&O)

WAYNESBORO
UNION STATION
BS (N&W/C&O)

"Blue Ridge Tunnel"

"Brookville"
"Little Rock"

Jarman Gap
C&O
195'

CROZET
ZX

Oakland
190

IVY
VY

Farmington
185

MECHUM'S RIVER

Cincinnati, O.
Roanoke

N&W 145

LYNDHURST

150 Lipscomb

AFTON
AF

205

GREENWOOD
200 DG

NELSON

115
Hickory Hill

SOU

Ft. Monroe

Washington, D.C.

79°00'
38°00'

45'

Columbus, Miss.

78°30'
38°00'

Harrisonburg, Va.

126

Va.

CHARLOTTESVILLE

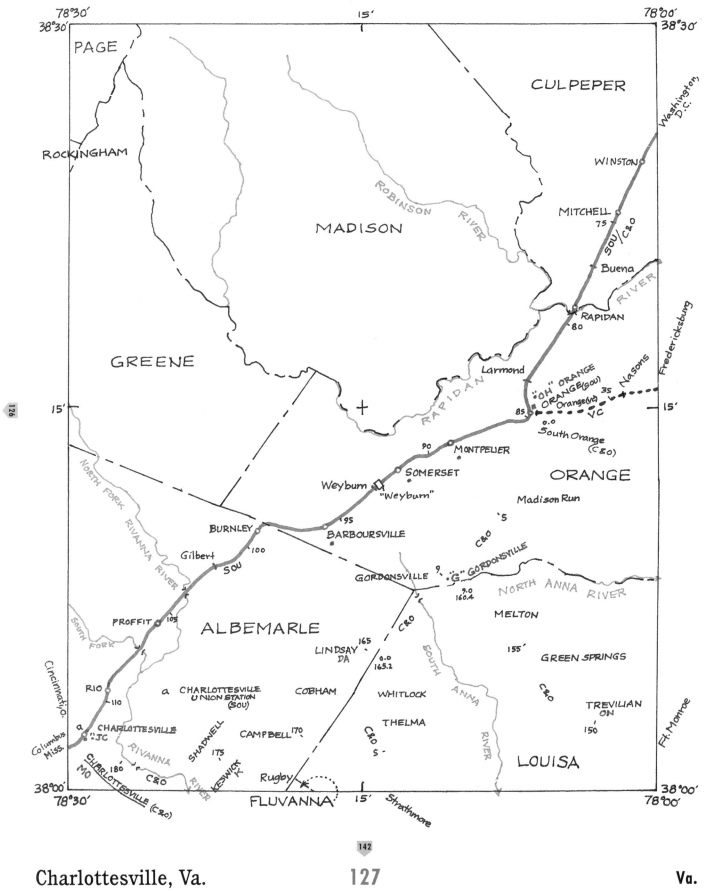

78°30'
38°30'

PAGE

ROCKINGHAM

GREENE

15'

CULPEPER

MADISON

ROBINSON RIVER

WINSTON

MITCHELL
75

SOU/C&O

Buena

RAPIDAN
80

RIVER

RAPIDAN

Larmond

"OH" ORANGE
ORANGE(SOU)
Orange(vа)   35
85   VC

0.0
South Orange
(C&O)

90
MONTPELIER

Weyburn
"Weyburn"

SOMERSET

ORANGE

Madison Run

'5

C&O

BURNLEY

95
BARBOURSVILLE

Gilbert
SOU

100

'9
GORDONSVILLE   "G" GORDONSVILLE

9.0
160.4

NORTH ANNA RIVER

NORTH FORK RIVANNA RIVER

PROFFIT   105

ALBEMARLE

LINDSAY
DA   165

C&O

0.0
165.2

MELTON

155'

GREEN SPRINGS

C&O

SOUTH FORK

RIO
110

a CHARLOTTESVILLE
UNION STATION
(SOU)

a CHARLOTTESVILLE
"JC

COBHAM

WHITLOCK

SOUTH ANNA RIVER

THELMA

C&O

TREVILIAN
ON

150

Cincinnati,O.

Columbus
Miss.

CHARLOTTESVILLE
MO   (C&O)

SHADWELL
175

KESWICK
K

180

RIVANNA RIVER

C&O

CAMPBELL   170

Rugby

LOUISA

FLUVANNA

15'
Strathmore

38°00'
78°00'

38°00'
78°30'

Washington,
D.C.

Fredericksburg

Nasons

15'

Ft. Monroe

Charlottesville, Va.

127

Va.

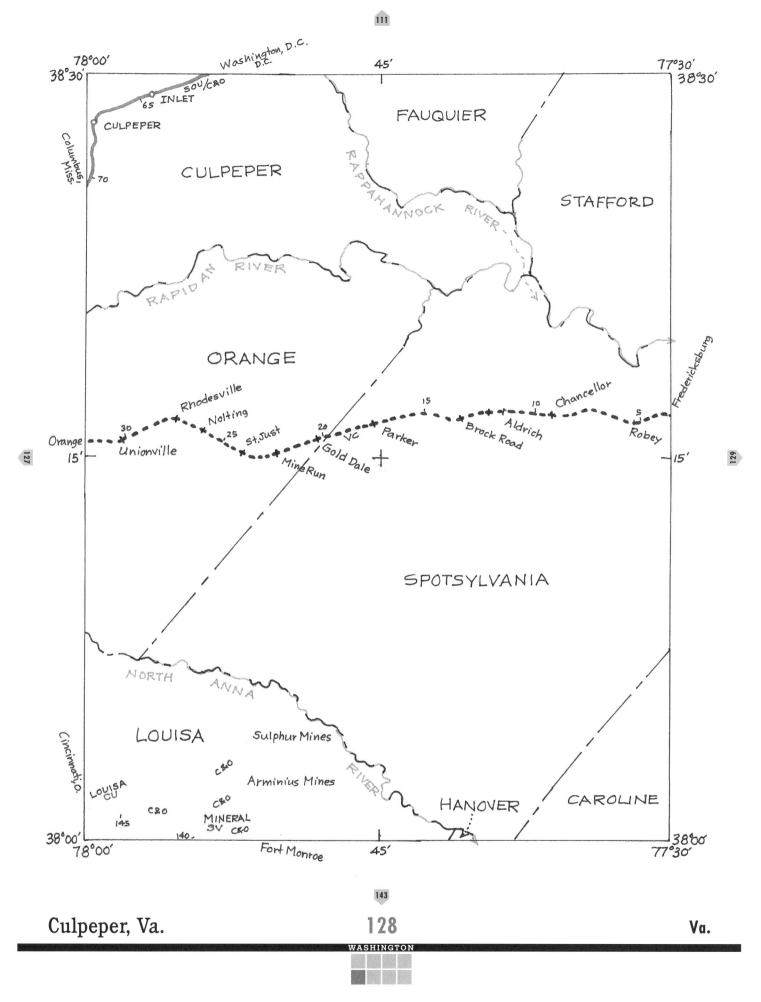

Culpeper, Va.   **128**   Va.

WASHINGTON

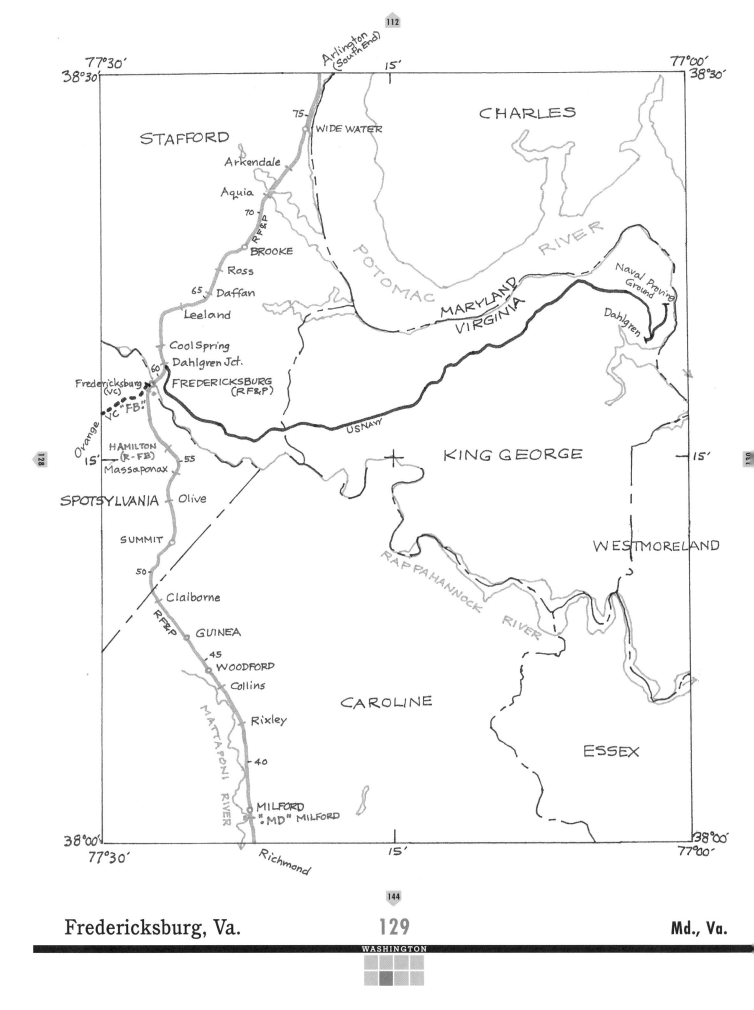

Arlington
(South End)

77°30'
38°30'

15'

77°00'
38°30'

75

CHARLES

STAFFORD

WIDE WATER

Arkendale

Aquia

70

RF&P

BROOKE

Ross

65   Daffan

Leeland

POTOMAC

RIVER

MARYLAND

VIRGINIA

Naval Proving
Ground

Dahlgren

Cool Spring

Dahlgren Jct.

Fredericksburg
(Vc)

60

VC "FB."

FREDERICKSBURG
(RF&P)

USNavy

Orange

15'

HAMILTON
(R-FB)

Massaponax

55

KING GEORGE

15'

SPOTSYLVANIA   Olive

SUMMIT

50

WESTMORELAND

Claiborne

RF&P

GUINEA

45

WOODFORD

Collins

MATTAPONI   RIVER

RAPPAHANNOCK   RIVER

Rixley

40

CAROLINE

ESSEX

MILFORD
"MD" MILFORD

38°00'

77°30'

Richmond

15'

38°00'

77°00'

Fredericksburg, Va.   **129**   Md., Va.

WASHINGTON

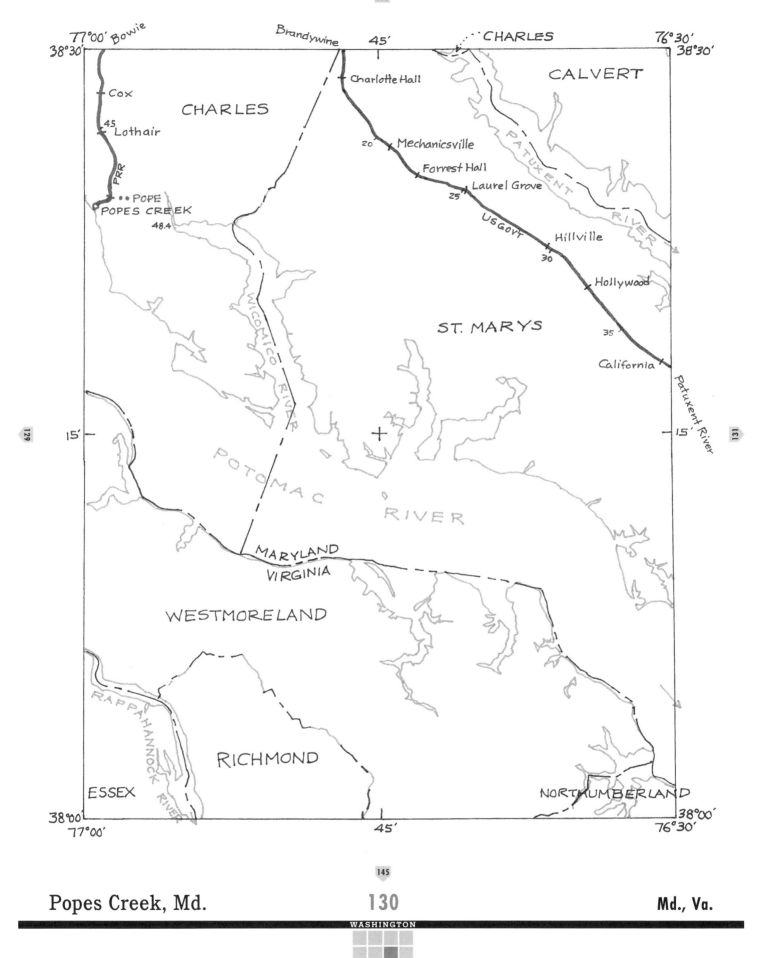

Popes Creek, Md.

77°00' Bowie   Brandywine   45'   CHARLES   76°30'
38°30'   38°30'

Cox   CHARLES   Charlotte Hall   CALVERT

45 Lothair

PRR   20   Mechanicsville

Forrest Hall

25   Laurel Grove

POPE   UsGovt   Hillville
POPES CREEK   30
48.4   Hollywood

35

ST. MARYS   California   Patuxent River

WICOMICO RIVER   PATUXENT RIVER

15'   +   15'

POTOMAC   RIVER

MARYLAND
VIRGINIA

WESTMORELAND

1

RAPPAHANNOCK RIVER   RICHMOND

ESSEX   NORTHUMBERLAND
38°00'   38°00'
77°00'   45'   76°30'

129   131

Popes Creek, Md.   **130**   Md., Va.

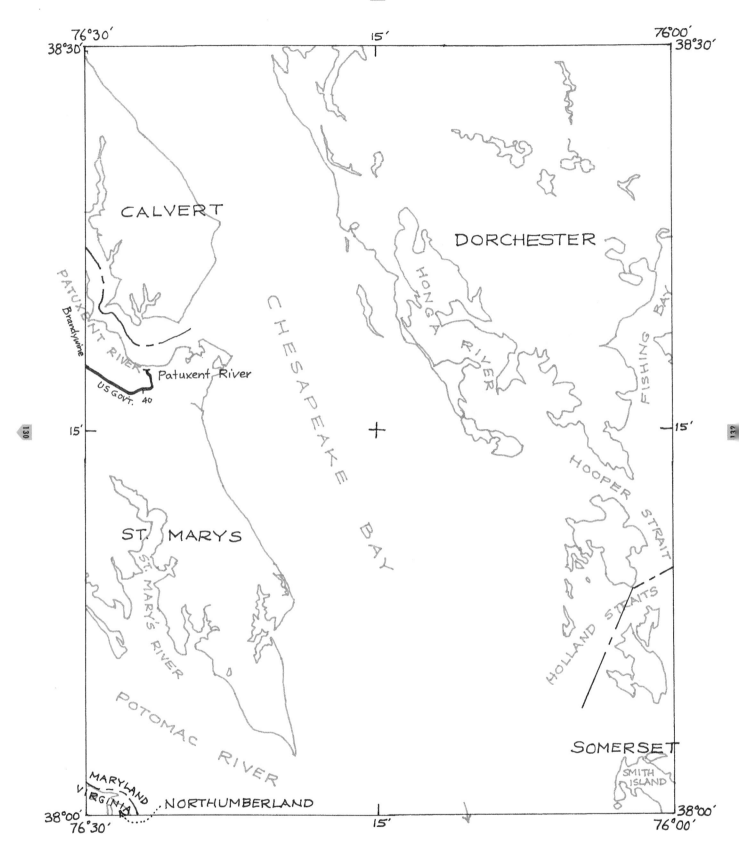

WASHINGTON

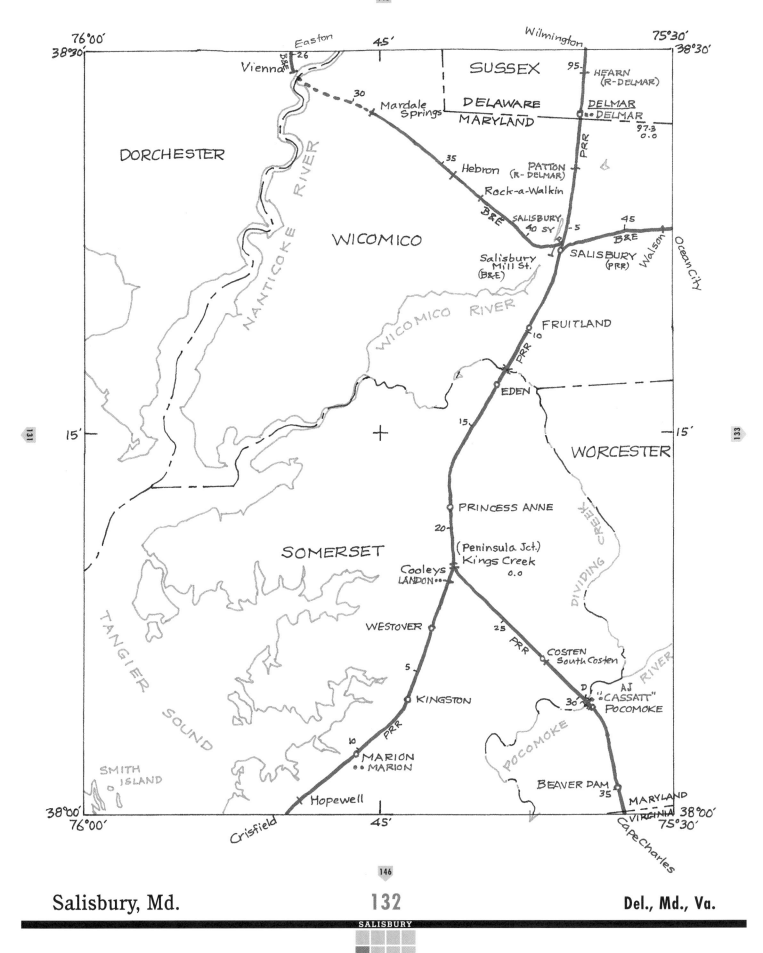

76°00'
38°30'
Easton
45'
Wilmington
75°30'
38°30'

Vienna
26
B&E
SUSSEX
95
HEARN
(R-DELMAR)

DELAWARE
MARYLAND
DELMAR
DELMAR

30
Mardale
Springs
97.3
0.0

DORCHESTER
35
Hebron
PATTON
(R-DELMAR)
PRR

Rock-a-Walkin

B&E
SALISBURY
40 SY
-5
45
B&E
Walson
Ocean City

WICOMICO
Salisbury
Mill St.
(B&E)
SALISBURY
(PRR)

NANTICOKE RIVER

WICOMICO RIVER
FRUITLAND
10

PRR

131
15'
EDEN

15
DIVIDING CREEK
15'
133

WORCESTER

PRINCESS ANNE
20

SOMERSET
(Peninsula Jct.)
Kings Creek
0.0

Cooleys
LANDON

WESTOVER
25
PRR

5
COSTEN
South Costen
RIVER

TANGIER SOUND
KINGSTON
D
AJ
"CASSATT"
POCOMOKE
30

PRR
10
POCOMOKE

SMITH
ISLAND
MARION
MARION

BEAVER DAM
35
MARYLAND

38°00'
76°00'
Hopewell
Crisfield
45'
VIRGINIA
Cape Charles
38°00'
75°30'

Salisbury, Md.   **132**   Del., Md., Va.

SALISBURY

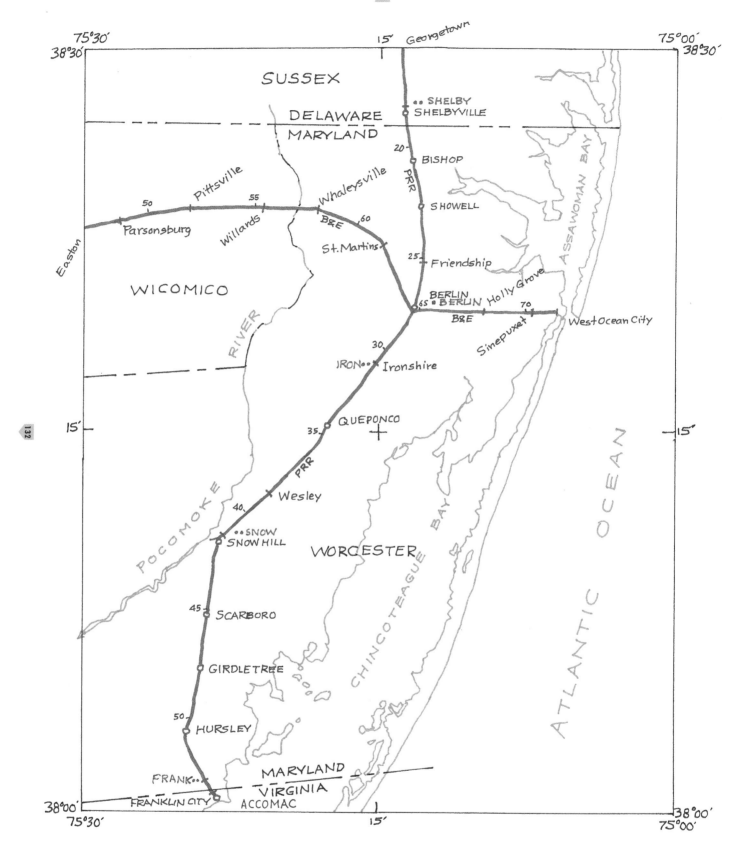

116

SUSSEX

DELAWARE

MARYLAND

SHELBY
SHELBYVILLE

Georgetown

15'

75°30'
38°30'

75°00'
38°30'

20
BISHOP
PRR

Pittsville
50
55
Whaleysville

SHOWELL

Parsonsburg
Willards
B&E
60

Easton
St. Martins

25
Friendship

WICOMICO

BERLIN
65 ● BERLIN Holly Grove

B&E
70

Sinepuxet
West Ocean City

RIVER

30
IRON ● ● Ironshire

ASSAWOMAN BAY

15'
35
QUEPONCO

15'

PRR

Wesley

40

SNOW
SNOW HILL

WORCESTER

CHINCOTEAGUE BAY

POCOMOKE

ATLANTIC OCEAN

45
SCARBORO

GIRDLETREE

50
HURSLEY

FRANK ● ●
FRANKLIN CITY

MARYLAND
VIRGINIA
ACCOMAC

38°00'
75°30'

15'

38°00'
75°00'

132

Berlin, Md.   **133**   **Del., Md., Va.**

SALISBURY

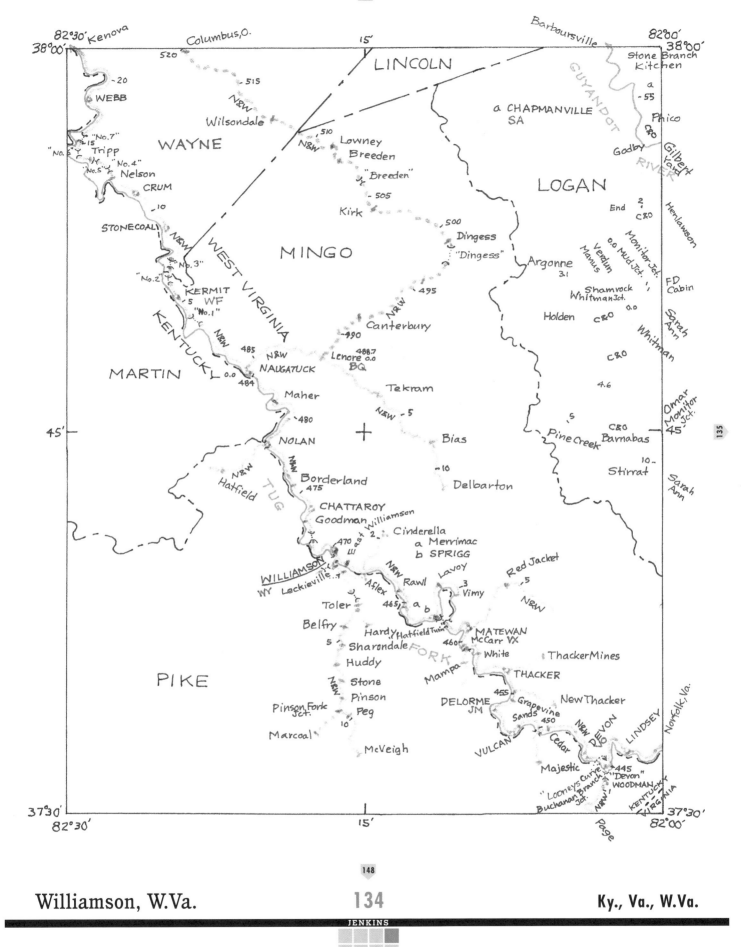

82°30' Kenova
38°00'
Columbus,O.
15'
Barboursville
82°00'
38°00'
520
LINCOLN
Stone Branch
Kitchen
515
~20
a
~55
WEBB
N&W
a CHAPMANVILLE
SA
Phico
Wilsondale
510
Godby C&O
Gilbert
Jct.
"No.7"
WAYNE
N&W
Lowney
Breeden
"No.6" Tripp
LOGAN
"No.4"
"Breeden"
Henlawson
No.5" Nelson
~505
End
2
CRUM
Kirk
C&O
FD
Monitor Jct.
~10
500
Verdun 0.0 Mud Jct.
Cabin
STONECOAL
Dingess
Manus
N&W
Argonne
Shamrock
"No.3"
"Dingess"
3.1
Whitman Jct.
Sarah
C&O
Ann
"No.2"
MINGO
495
Holden C&O
WEST VIRGINIA
5 KERMIT
WF
Canterbury
N&W
C&O
Whitman
"No.1"
490
4.6
MARTIN KENTUCKY
N&W
485
BQ 488.7 Lenore 0.0
Omar
Monitor Jct.
C&O Barnabas 45'
0.0 NAUGATUCK
Tekram
.5
Pine Creek
484
Maher
N&W ~5
10
Stirrat
45'
480
Bias
Sarah
Ann
NOLAN
N&W Delbarton
N&W
Borderland
~10
475
N&W Hatfield
TUG
CHATTAROY
Goodman Williamson
Cinderella
a Merrimac
East Williamson
b SPRIGG
2
470
Red Jacket
WILLIAMSON
N&W Rawl Lovoy
.5
WY Leckieville
Aflex
Vimy
N&W
Toler
465 a b
MATEWAN
Belfry
Hardy Hatfield Twin
McCarr VX
5 Sharondale
460
White
ThackerMines
FORK
Huddy
Mampa
THACKER
PIKE
Stone
455
New Thacker
N&W
Pinson
DELORME
JM
Grapevine
Sands
Pinson Fork
Peg
450 N&W DEVON
Jct.
10
VULCAN Cedar
LINDSEY
Marcoal
Majestic
445
McVeigh
"Devon"
Looneys Curve
WOODMAN
Buchanan Branch
N&W KENTUCKY
Jct.
VIRGINIA
37°30'
15'
82°00'
37°30'
82°30'
Page

Williamson, W.Va.
134
Ky., Va., W.Va.

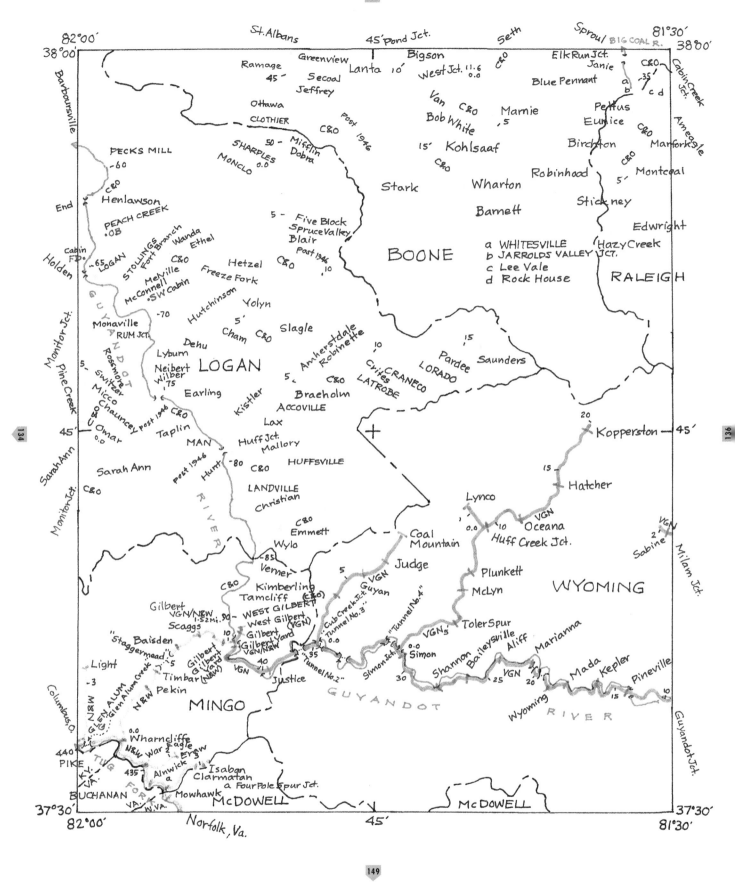

St. Albans
45' Pond Jct.
Seth
Sproul BIG COAL R.
81°30'
38°00'

82°00'
38°00'

Ramage
Greenview
Lanta 10'
Bigson
Elk Run Jct.
Janie
C&O

45'
Secoal
West Jct. 11.6
C&O
a
C&O
-35
Cabin Creek Jct.

Jeffrey
0.0
Blue Pennant
b
c d

Ottawa
Van
C&O
Marnie
Pettus
Ameagle

CLOTHIER
Bob White
.5
Eunice
C&O
Manfork

Barboursville
Post 1946
Birchton
C&O

PECKS MILL
SHARPLES
50 - Mifflin
C&O
15'
Kohlsaaf
Robinhood
5'
Montcoal

-60
MONCLO
Dobra
Stark
Wharton
Stickney

0.0
Barnett
Edwright

End
C&O
Henlawson
Five Block
HazyCreek

Cabin
PEACH CREEK
5 - Spruce Valley
a WHITESVILLE
JARROLDS VALLEY JCT.

Holden
Fk .08
Fort Branch Wanda
Blair
b JARROLDS VALLEY
RALEIGH

-65 LOGAN
STOLLINGS
Ethel
Post 1946
BOONE
c Lee Vale

Melville
C&O
Hetzel
.10
d Rock House

Monitor Jct.
McConnell
Freeze Fork
C&O

RUM JCT.
SW Cabin
Hutchinson
Yolyn
5'

Monaville
-70
Cham
C&O
Slagle
Amherstdale
10
Pardee
15

5 - Rossmore
Dehu
5'
Robinette
Crites CRANECO
LORADO
Saunders

Pine Creek
Switzer
Lybum
Kistler
LATROBE

Micco
Neibert
LOGAN
Braeholm
20

Chauncey Post 1946 C&O
Wilber
Earling
ACCOVILLE
Kopperston
45'

Omar
Taplin
.75
Lax
15

0.0
MAN
Huff Jct.
Hatcher

Sarah Ann
Post 1946
Mallory
Lynco

Monitor Jct.
C&O
Hurt -80 C&O
HUFFSVILLE
0.0
10
Oceana
VGN

C&O
LANDVILLE
Huff Creek Jct.
VGN
2

RIVER
Christian
Sabine
Milam Jct.

C&O
Coal
Plunkett

Emmett
Mountain
McLyn
WYOMING

Wylo
5
Judge

-85
VGN
Toler Spur

Verner
Guyan
Marianna

Gilbert
C&O
Kimberling
Cub Creek Jct.
VGN 5
Aliff
Mada Kepler

VGN/N&W
Tamcliff (C&O)
0.0
Shannon Baileysville
Pineville

1.52Mi. 90-
WEST GILBERT
Simon
VGN

Scaggs
West Gilbert (VGN)
"Tunnel No.4"
30
25 VGN 20
15
6

Baisden
10 Gilbert
Simon Jct.
0.0

Light
"Staggermead"
Gilbert Yard
0.0
Wyoming RIVER

-3
Gilbert (VGN/N&W)
40
"Tunnel No.2"
35
Guyandot Jct.

Timbar Gilbert Yard (N&W)
VGN
Justice
"Tunnel No.3"

Columbus,O.
N&W Pekin
MINGO
GUYANDOT

GLEN ALUM
Glen Alum Creek

440
0.0 Wharncliff
PIKE
N&W War Eagle EY&W

KY. 435
Alnwick Isaban

VA.
Clarmatan

BUCHANAN Mowhawk
a FourPole Spur Jct.

VA. W.VA.
McDOWELL
McDOWELL

37°30'
37°30'

82°00'
Norfolk, Va.
45'
81°30'

134
136

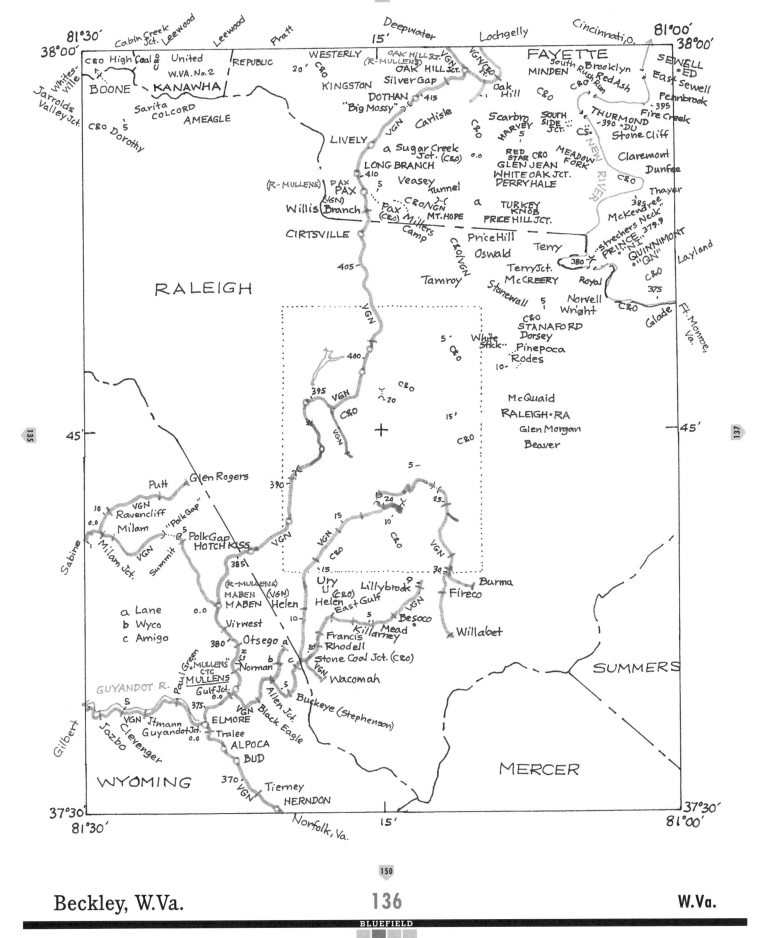

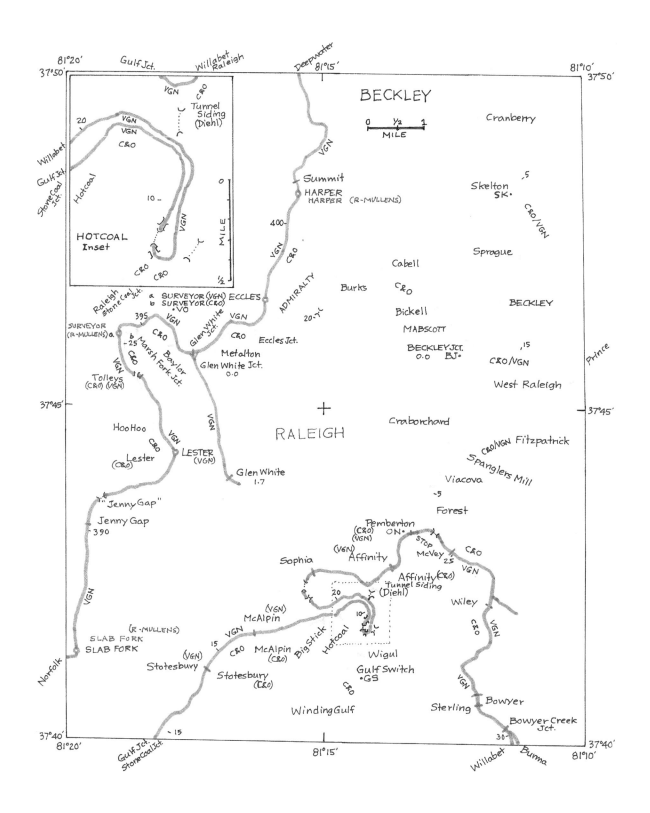

136A Beckley, W. Va.

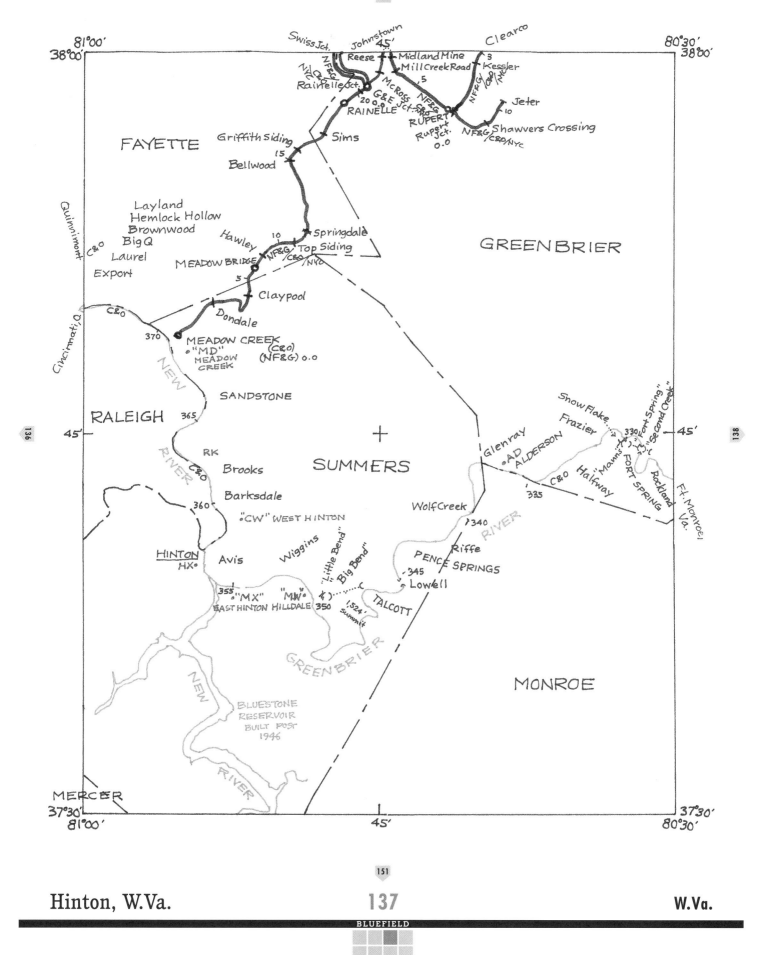

81°00'
38°00'

Swiss Jct.   Johnstown   Clearco
45'

Reese   Midland Mine
NF&G   Mill Creek Road
NF&G   McRoss
Rainelle Jct.   G&E Jct.
200.0
RAINELLE

Kessler
3
NF&G / C&O / NYC

Jeter
10
RUPERT   Shawvers Crossing
Rupert Jct.   NF&G / C&O/NYC
0.0

80°30'
38°00'

FAYETTE   Griffith Siding   Sims

GREENBRIER

Bellwood
15

Layland
Hemlock Hollow
Brownwood
Big Q
Laurel
Export

Hawley   10   Springdale
Top Siding
NF&G / C&O / NYC
MEADOW BRIDGE
5
Claypool
Dondale
370
MEADOW CREEK
"MD"   (C&O)
MEADOW   (NF&G) 0.0
CREEK

Quinnimont
C&O

Cincinnati, O

C&O

RALEIGH
365
SANDSTONE

45'

RK
C&O
Brooks
Barksdale
360

SUMMERS

Show Flake
Frazier   "Fort Spring"
330   Second Creek
45'

Glenray   Manns
"AD"   ALDERSON
C&O   Halfway
335
FORT SPRING
Rockland
Ft. Monroe!

NEW RIVER

"CW" WEST HINTON

HINTON   Avis   Wiggins
HX
355   "MX"   "MW"
EAST HINTON   HILLDALE   350
"Little Bend"
Big Bend
524
summit
TALCOTT

Wolf Creek
340
Riffe
PENCE SPRINGS
345
Lowell

GREENBRIER RIVER

MONROE

NEW RIVER

GREENBRIER

BLUESTONE
RESERVOIR
BUILT POST
1946

MERCER
37°30'
81°00'

45'

37°30'
80°30'

80°30'
38°00'

Winterburn

15'

The Dock

Rimel

80°00'
38°00'

RENICK
RN

25

WSH

WSH

BATH

SPRING CREEK

20

RIVER

Neola

Shryock

WEST VIRGINIA

VIRGINIA

GREENBRIER

C&O
15'

Woodman

ANTHONY

Alvon

WSH

ALLEGHANY

KEISTER

Loopemount

10

Boy's Home °BS

Ft. Monroe

GREENBRIER

WhiteSulphur
Springs
(WSH)

Callaghan

295 "Mud"

Covington

Hopper

5

WHITE SULPHUR SPRINGS

Moss Run

JACKSON RIVER

Camp Alleghany

"WS."

"White Sulphur"

(C&O)

300

C&O

Backbone

NORTH CALDWELL

Caldwell

"Moore's"
"Lake"

WHITCOMB

315

310'

"Kellys"

RONCEVERTE
•"RV"

320

WHITCOMB
(R-RV)

C&O

"Lewis"

Jerry's Run

Blue Spring
Run

45'

Cincinati, O

C&O

325

RONCEVERTE

Hart's Run

Tuckahoe

305

ALLEGHANY
°"A" ALLEGHANY

"Alleghany" Summit 2072'

Arritt

C&O

45'

Jordan Mines

Bess

MONROE

POTTS CR.

BOTETOURT

WEST VIRGINIA

VIRGINIA

EagleRock

Paint Bank

CRAIG

Charlton

Given 20

BARBOUR'S
CREEK

C&O

35

N&W

Marshalltown

Virginia Mineral
Springs

25

Laurel Branch

Pine Top

30

NEW CASTLE
CY

BOTETOURT

37°30'
80°30'

Ripplemead

15'

37°30'
80°00'

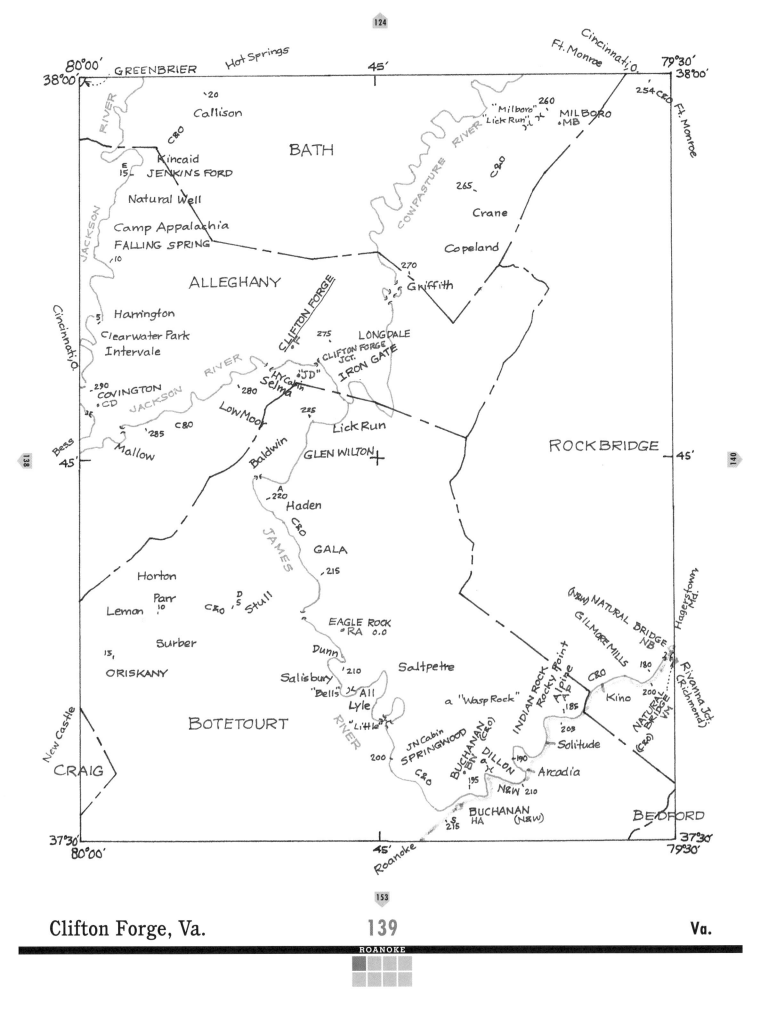

80°00' GREENBRIER Hot Springs 45' Cincinnati,O. 79°30' 38°00'
38°00' Ft. Monroe Ft. Monroe

'20
Callison
C&O BATH "Milboro" 260 MILBORO
"Lick Run" •MB 254 C&O
Kincaid JENKINS FORD C&O
E C&O 265 Crane
15
Natural Well Copeland
Camp Appalachia FALLING SPRING 270
'10 Griffith
ALLEGHANY
5 Harrington CLIFTON FORGE LONGDALE
Clearwater Park 275 CLIFTON FORGE IRON GATE
Intervale RIVER JCT.
'290 COVINGTON HYCabin "JD"
•CD '280 Selma
25 JACKSON ROCKBRIDGE
'285 C&O 225 45'
Bess LowMoor Lick Run
45' Mallow Baldwin GLEN WILTON

A Haden
'220
C&O
JAMES GALA
Horton '215
Parr D
Lemon '10 C&O Stull EAGLE ROCK
Surber •RA 0.0
15 Dunn Saltpetre
ORISKANY '210 a "Wasp Rock"
Salisbury Alpine (New) NATURAL BRIDGE
"Bells" All GILMORE MILLS NB
Lyle INDIAN ROCK C&O 180
BOTETOURT "Little" Rocky Point Kino 200
JN Cabin '185 NATURAL
200 SPRINGWOOD '205 BRIDGE VM
CRAIG C&O BUCHANAN Solitude (C&O)
BN DILLON Arcadia
BUCHANAN (C&O) '195 190
N&W 210
'S BUCHANAN (N&W) BEDFORD
37°30' '215 HA 37°30'
80°00' 45' Roanoke 79°30'

New Castle
Cincinnati,O.

138 140

153

Clifton Forge, Va.   139   Va.

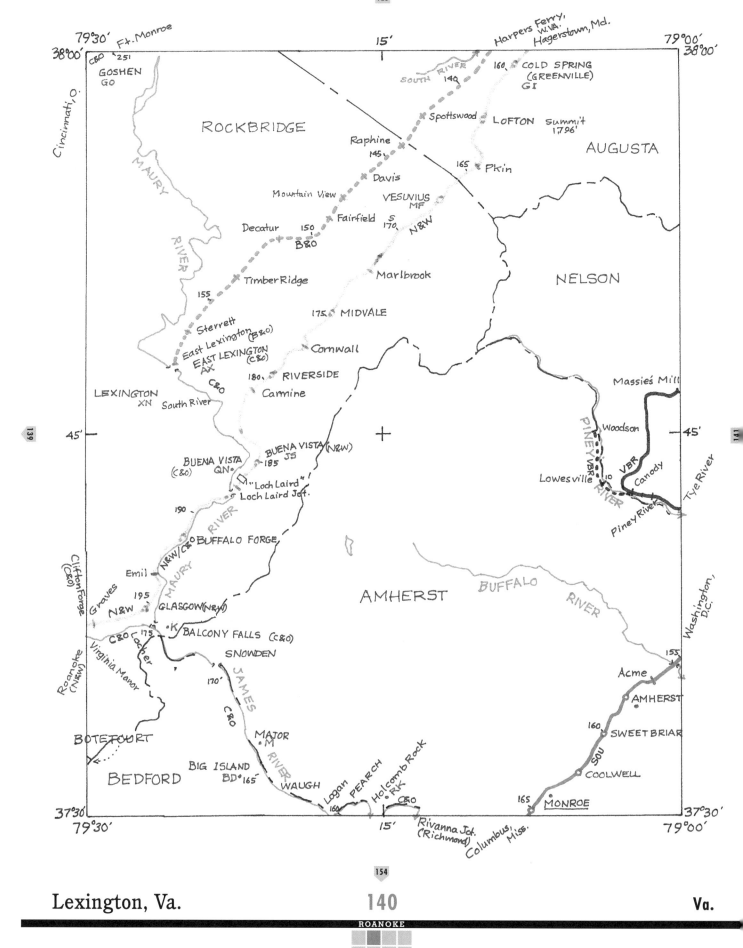

79°30' Ft. Monroe
38°00'

C&O 251
GOSHEN
GO

Cincinnati, O.

Harpers Ferry, W.VA.
Hagerstown, Md.
79°00'
38°00'

SOUTH RIVER 140 160 COLD SPRING
(GREENVILLE)
GI

ROCKBRIDGE

Spottswood LOFTON Summit
1796'

Raphine
145

165 Pkin

AUGUSTA

Davis

Mountain View

VESUVIUS
MF

Fairfield S
170 N&W

Decatur 150
B&O

Marlbrook

NELSON

Timber Ridge

155

175 MIDVALE

Sterrett
East Lexington (B&O)
EAST LEXINGTON
(C&O)
AX

Cornwall

Massie's Mill

C&O

180 RIVERSIDE

Carmine

Woodson

LEXINGTON
XN South River

15'

+

45'

BUENA VISTA (N&W)
JS

BUENA VISTA 185
(C&O) QN

PINEY VBR RIVER

VBR

Lowesville 10 VBR Canody

Tye River

"Loch Laird"
Loch Laird Jct.

190

MAURY RIVER

BUFFALO RIVER

N&W/C&O Buffalo Forge

Emil

AMHERST

Piney River

Clifton Forge
(C&O)

Graves

195
N&W

GLASGOW (N&W)

Washington, D.C.

Roanoke (N&W)

C&O 175 Locher
Virginia Manor

·K BALCONY FALLS (C&O)

SNOWDEN

155
Acme

AMHERST

170'

JAMES RIVER

C&O

160 SWEET BRIAR

BOTETOURT

MAJOR
·M

SOU COOLWELL

BEDFORD

BIG ISLAND
BD·165' WAUGH

Logan PEARCH Holcomb Rock
RK

165 MONROE

37°30'
79°30'

160

15'

C&O

Rivanna Jct.
(Richmond)
Columbus, Miss.

79°00'
37°30'

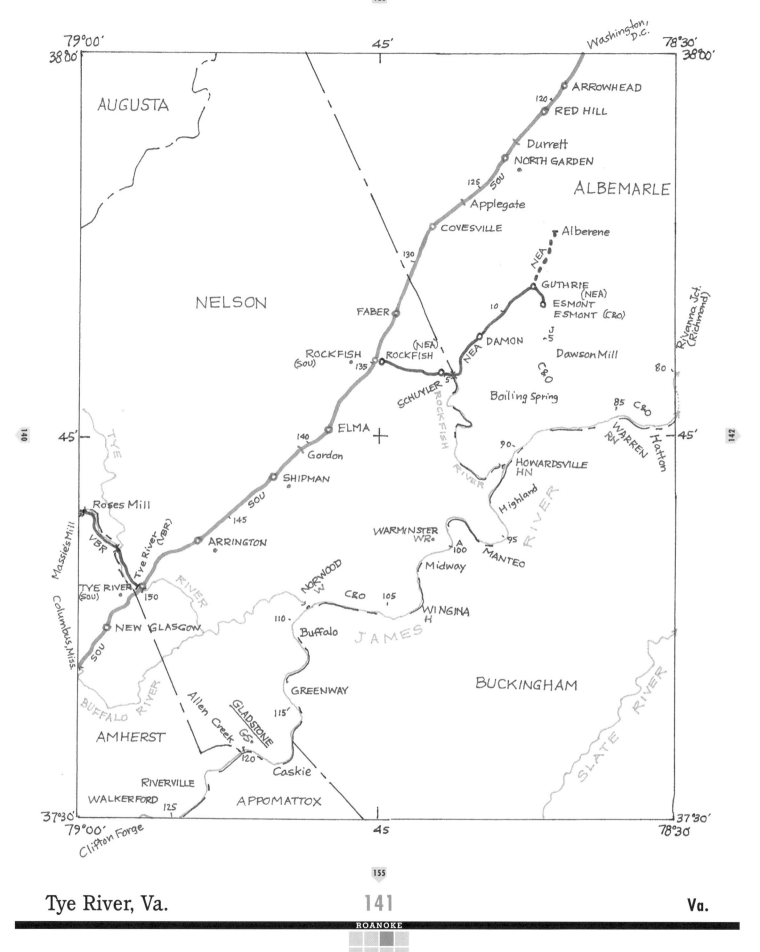

79°00'
38°00'
45'
Washington, D.C.
78°30'
38°00'

AUGUSTA

ARROWHEAD
120
RED HILL

Durrett
NORTH GARDEN

ALBEMARLE

125
SOU
Applegate

COVESVILLE

Alberene

130

NELSON

NEA

FABER

GUTHRIE
(NEA)
ESMONT (NEA)
ESMONT (C&O)

10

Dawson Mill

ROCKFISH
(SOU)

ROCKFISH

ROCKFISH
(NEA)

NEA DAMON

J
·5

135

SCHUYLER

C&O

80

Boiling Spring

85 C&O

45'

ROCKFISH

140

ELMA

Gordon

RIVER

90

WARREN
RN

Hatton

45'

SOU

SHIPMAN

HOWARDSVILLE
HN

Rivanna Jct.
(Richmond)

Roses Mill

145

Highland

TYE

Massies Mill

VBR

Tye River (VBR)

ARRINGTON

WARMINSTER
WR

A
100

RIVER

95

MANTEO

Midway

TYE RIVER
(SOU)

150

NORWOOD
W

C&O

105

WINGINA
H

Columbus, Miss.

SOU

NEW GLASGOW

110

Buffalo

JAMES

BUCKINGHAM

BUFFALO

RIVER

GREENWAY

SLATE RIVER

AMHERST

Allen Creek

GLADSTONE
GS·

115'

120

Caskie

RIVERVILLE

WALKERFORD

125

APPOMATTOX

37°30'
79°00'
Clifton Forge

45

37°30'
78°30'

Tye River, Va.

141

Va.

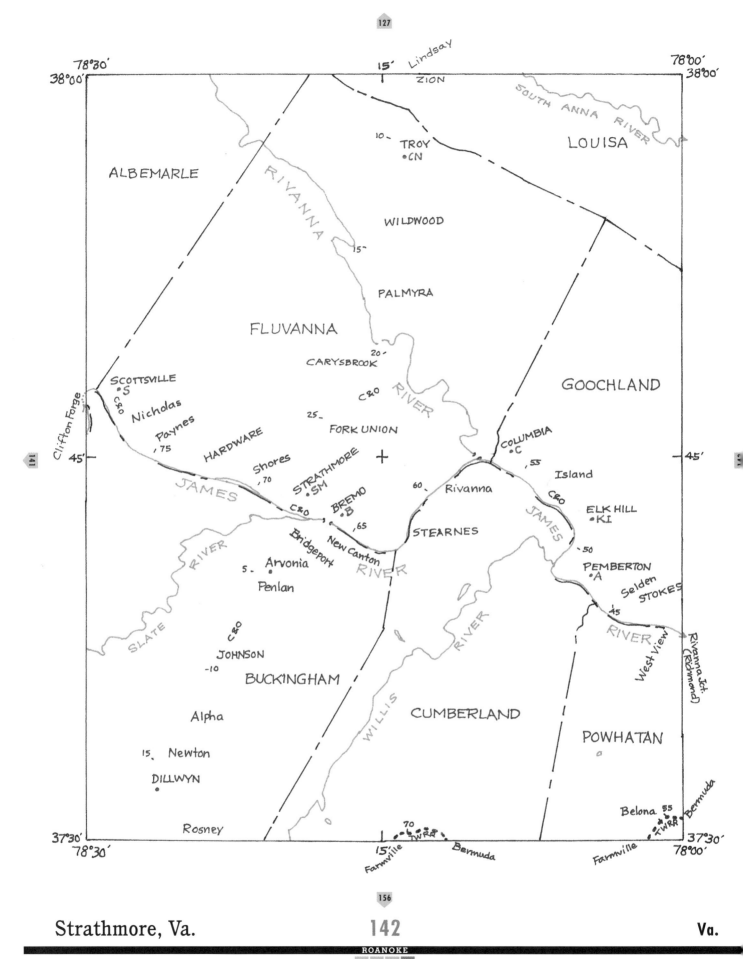

78°30'
38°00'

15'
Lindsay
ZION

78°00'
38°00'

SOUTH ANNA RIVER

LOUISA

ALBEMARLE

10 -
TROY
•CN

RIVANNA

WILDWOOD

15"

PALMYRA

FLUVANNA

20 -
CARYSBROOK

RIVER

GOOCHLAND

SCOTTSVILLE
•S

C&O

C&O

25 -
FORK UNION

+

COLUMBIA
•C

55

Nicholas

Paynes

, 75

HARDWARE

Shores

, 70

STRATHMORE
•SM

Island

141

Clifton Forge

45'

JAMES

C&O

BREMO
•B

, 65

60 -

Rivanna

C&O

45'

ELK HILL
•KI

- 50

Bridgeport

New Canton

STEARNES

45'

PEMBERTON
•A

Selden
STOKES

SLATE

RIVER

Arvonia

5 -

Penlan

RIVER

45

JAMES

West View

RIVER

Rivanna Jct.
(Richmond)

C&O

JOHNSON

-10

BUCKINGHAM

WILLIS

RIVER

CUMBERLAND

POWHATAN

Alpha

15 - Newton

DILLWYN

Rosney

Belona 55
TWRR
Bermuda

37°30'
78°30'

15'
Farmville

70
TWRR
Bermuda

Farmville

37°30'
78°00'

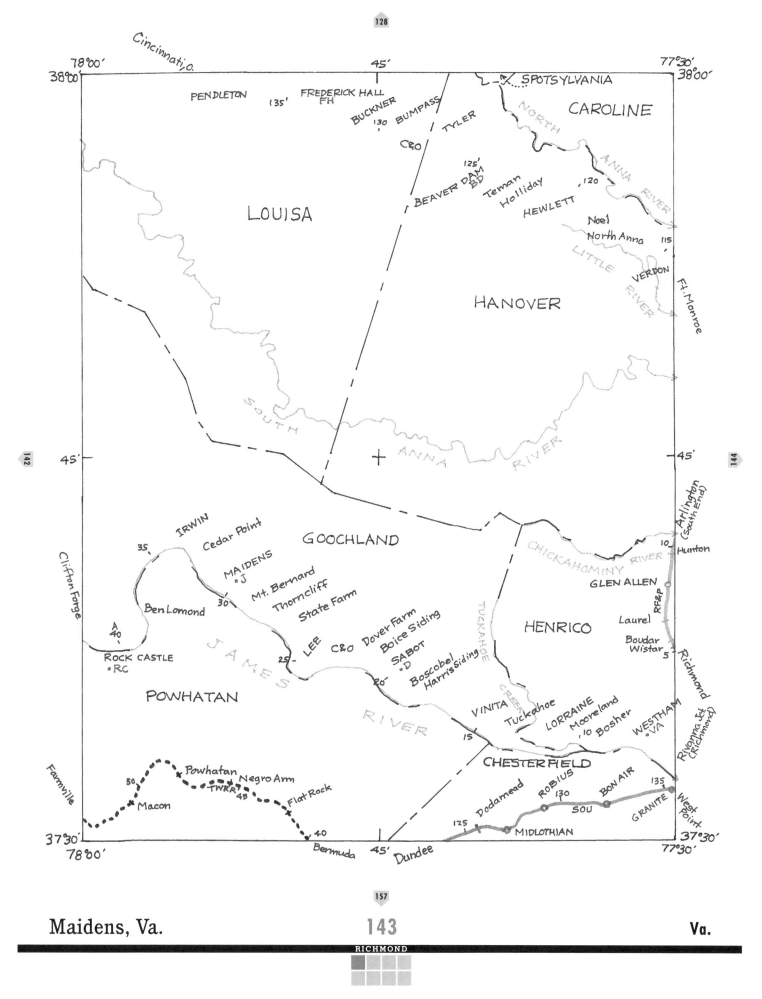

Cincinnati, O.

78°00'
38°00'

PENDLETON   135'   FREDERICK HALL
FH

45'

BUCKNER
'130   BUMPASS

TYLER

SPOTSYLVANIA

CAROLINE

77°30'
38°00'

C&O

LOUISA

125'
BEAVER DAM
BD

Teman

Holliday

HEWLETT

'120

NORTH

Noel

North Anna

VERDON

115

ANNA RIVER

Ft. Monroe

HANOVER

LITTLE RIVER

142

45'

SOUTH

ANNA   RIVER

45'

144

Clifton Forge

IRWIN

35'

Cedar Point

GOOCHLAND

MAIDENS
• J

Mt. Bernard

Thorncliff

State Farm

Ben Lomond

30'

A
40

ROCK CASTLE
• RC

JAMES

LEE

25'

C&O

Dover Farm

Boice Siding

SABOT
• D

Boscobel

Harrissiding

20'

Arlington
(South End)

CHICKAHOMINY   RIVER

Hunton

10

GLEN ALLEN

Laurel

RF&P

Boudar
Wistar

5

HENRICO

Richmond

TUCKAHOE CREEK

POWHATAN

VINITA

Tuckahoe

LORRAINE

Mooreland

'10 Bosher

WESTHAM
• VA

Richmond

Rivanna Jct.
(Richmond)

Farmville

50

Powhatan

TWRR

Negro Arm

48

Macon

Flat Rock

CHESTERFIELD

Dodamead

ROBIUS
'130

BON AIR

135

GRANITE

West
Point

40

Bermuda

45'

Dundee

'125

MIDLOTHIAN

SOU

37°30'

78°00'

37°30'

77°30'

Maidens, Va.   **143**   **Va.**

RICHMOND

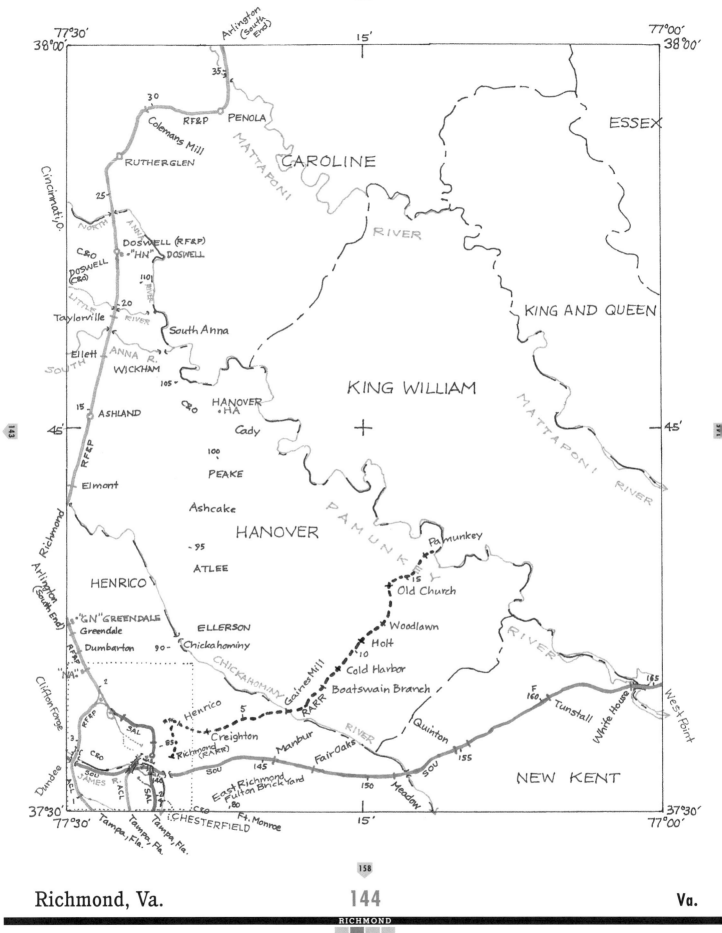

Arlington (South End)

77°30'
38°00'

15'

77°00'
38°00'

35

30

Colemans Mill

RF&P

PENOLA

RUTHERGLEN

MATTAPONI

CAROLINE

ESSEX

25

Cincinnati, O.

NORTH ANNA

C&O

DOSWELL (RF&P)
"HN"
DOSWELL

DOSWELL
(C&O)

RIVER

110

KING AND QUEEN

20

Taylorville

LITTLE RIVER

Ellett

South Anna

SOUTH ANNA R.

WICKHAM

105

KING WILLIAM

ASHLAND

C&O

HANOVER
·HA

Cady

45'

15

100

PEAKE

Elmont

RF&P

Ashcake

MATTAPONI RIVER

45'

143

HANOVER

PAMUNKEY

95

ATLEE

Pamunkey

HENRICO

15

Old Church

Richmond

"GN" GREENDALE
Greendale

ELLERSON

Woodlawn

RIVER

Dumbarton

90

Chickahominy

Holt
10

Arlington (South End)

RF&P

"NA"

2

Henrico

CHICKAHOMINY

Cold Harbor

Boatswain Branch

F
160

Clifton Forge

5

Tunstall

RARR

165

RF&P

SAL

Creighton

Quinton

White House

3

C&O

85

Richmond (RARR)

Manbur

Fair Oaks

RIVER

West Point

Dundee

JAMES R.

SOU

ACL

SAL

ACL

2

East Richmond
Fulton Brick Yard
80

SOU

145

SOU

150

Meadow

155

NEW KENT

37°30'

77°30'

Tampa, Fla.

Tampa, Fla.

Tampa, Fla.

C&O

CHESTERFIELD

Ft. Monroe

15'

37°30'

77°00'

Richmond, Va.

144

Va.

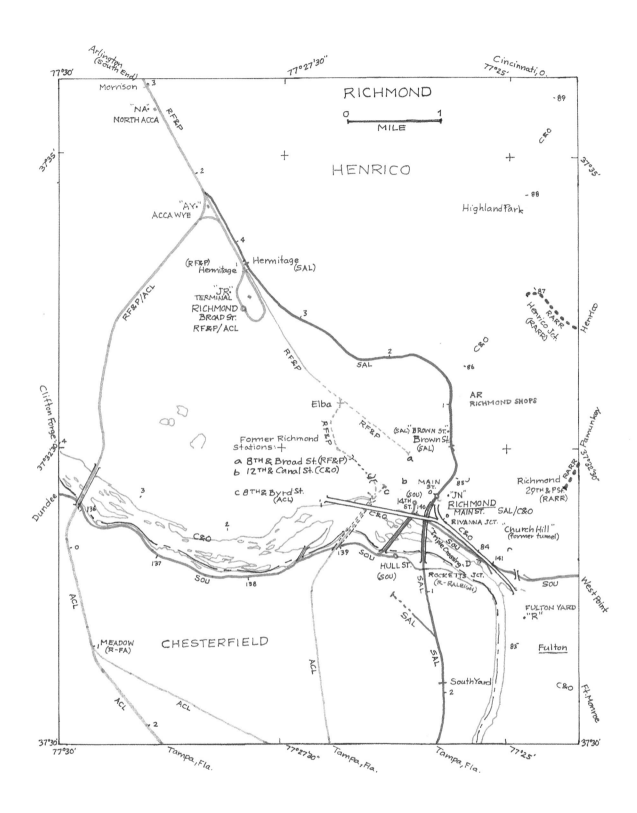

RICHMOND

0 |_____| 1
MILE

HENRICO

Arlington
(South End)
Morrison
"NA"
NORTH ACCA

"AY"
ACCA WYE

RF&P

(RF&P)
Hemitage

Hemitage
(SAL)

"JR"
TERMINAL
RICHMOND
BROAD ST.
RF&P/ACL

RF&P/ACL

RF&P

SAL

Cincinnati, O.

Highland Park

C&O

RARR
Henrico Jct.
(RARR)

C&O

AR
RICHMOND SHOPS

Clifton Forge

Elba

Former Richmond
Stations:
a 8TH & Broad St. (RF&P)
b 12TH & Canal St. (C&O)
c 8TH & Byrd St.
(ACL)

(SAL) BROWN ST.
Brown St.
(SAL)

RF&P

RF&P

Dundee
136

C&O

137

SOU

138

ACL

C&O

139

MEADOW
(R-FA)

ACL

ACL

CHESTERFIELD

ACL

Tampa, Fla.

Tampa, Fla.

SOU

HULL ST.
(SOU)

Triple Crossing, D

SAL

SAL

SAL

Tampa, Fla.

b
c
MAIN
ST.
(SOU)
14TH
ST. 140

"JN"
RICHMOND
MAIN ST.   SAL/C&O
Rivanna Jct.

C&O

SOU

84

Richmond
29TH & P ST.
(RARR)

RARR

Church Hill
(former tunnel)

141

ROCKETTS JCT.
(R-RALEIGH)

SOU

West Point

FULTON YARD
"R"

Fulton

C&O

SouthYard

Ft. Monroe

144A Richmond, Va.

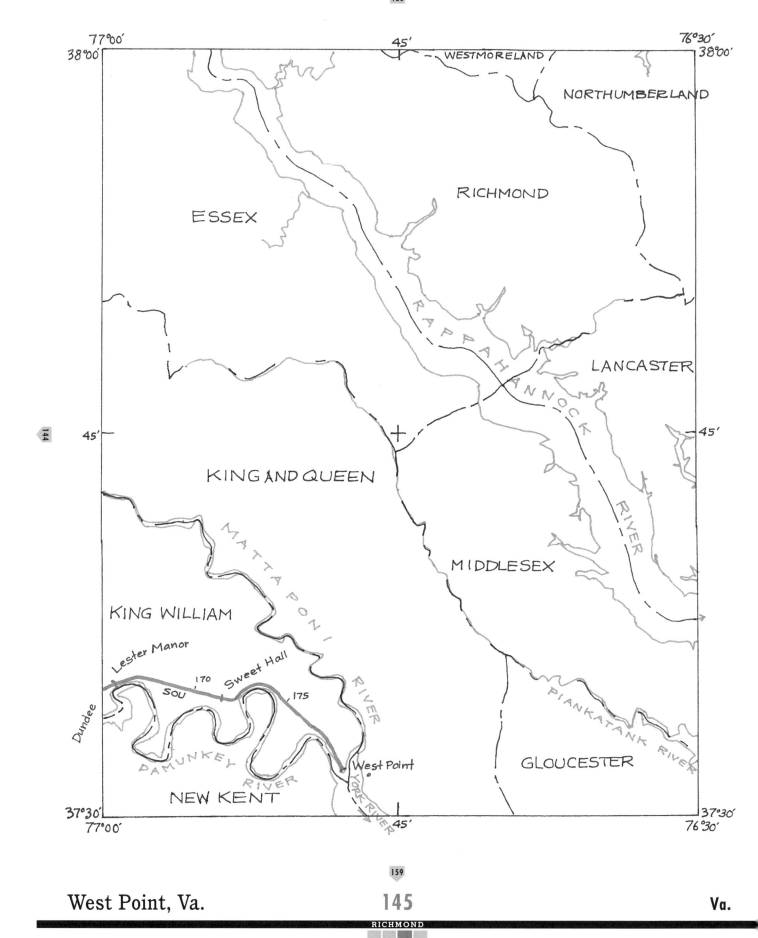

77°00'
38°00'

45'

WESTMORELAND

76°30'
38°00'

NORTHUMBERLAND

RICHMOND

ESSEX

LANCASTER

RAPPAHANNOCK RIVER

45'

KING AND QUEEN

45'

MATTAPONI RIVER

MIDDLESEX

KING WILLIAM

Lester Manor

Dundee

SOU

170

Sweet Hall

175

PAMUNKEY RIVER

PIANKATANK RIVER

West Point

GLOUCESTER

NEW KENT

YORK RIVER

37°30'
77°00'

45'

37°30'
76°30'

West Point, Va.

145

Va.

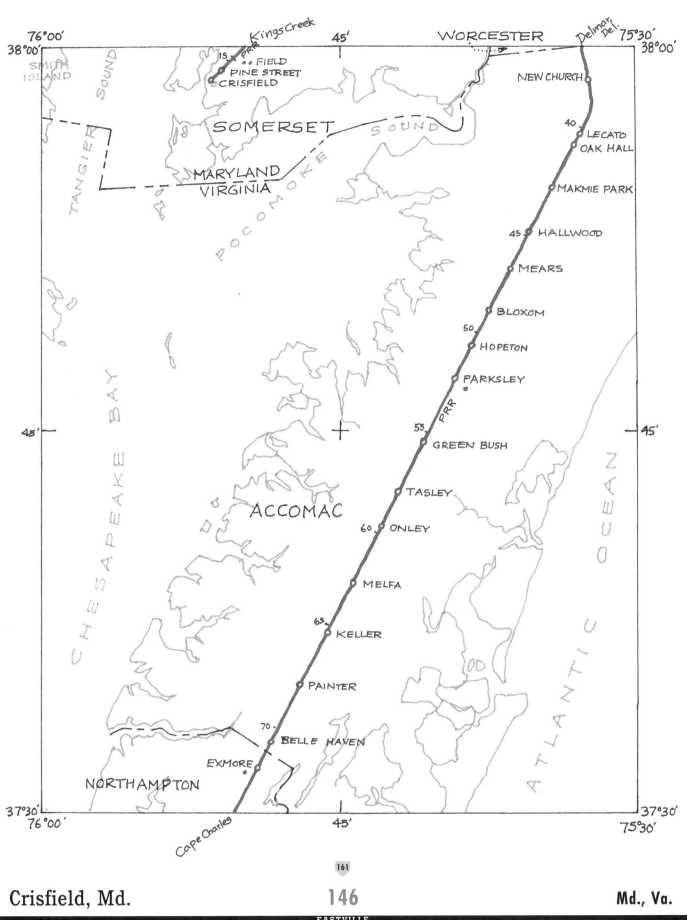

76°00'
38°00'
Kings Creek
45'
WORCESTER
Delmar, Del.
75°30'
38°00'

SMITH
ISLAND

TANGIER SOUND

15
FIELD
PINE STREET
CRISFIELD

PRR

SOMERSET

POCOMOKE SOUND

NEW CHURCH

40
LECATO
OAK HALL

MAKMIE PARK

MARYLAND
VIRGINIA

45
HALLWOOD

MEARS

BLOXOM

50
HOPETON

PARKSLEY

45'
CHESAPEAKE BAY

ACCOMAC

PRR

55
GREEN BUSH

TASLEY

60
ONLEY

45'

ATLANTIC OCEAN

MELFA

65
KELLER

PAINTER

70
BELLE HAVEN

EXMORE

NORTHAMPTON

37°30'
76°00'
Cape Charles
45'
37°30'
75°30'

Crisfield, Md.

146

Md., Va.

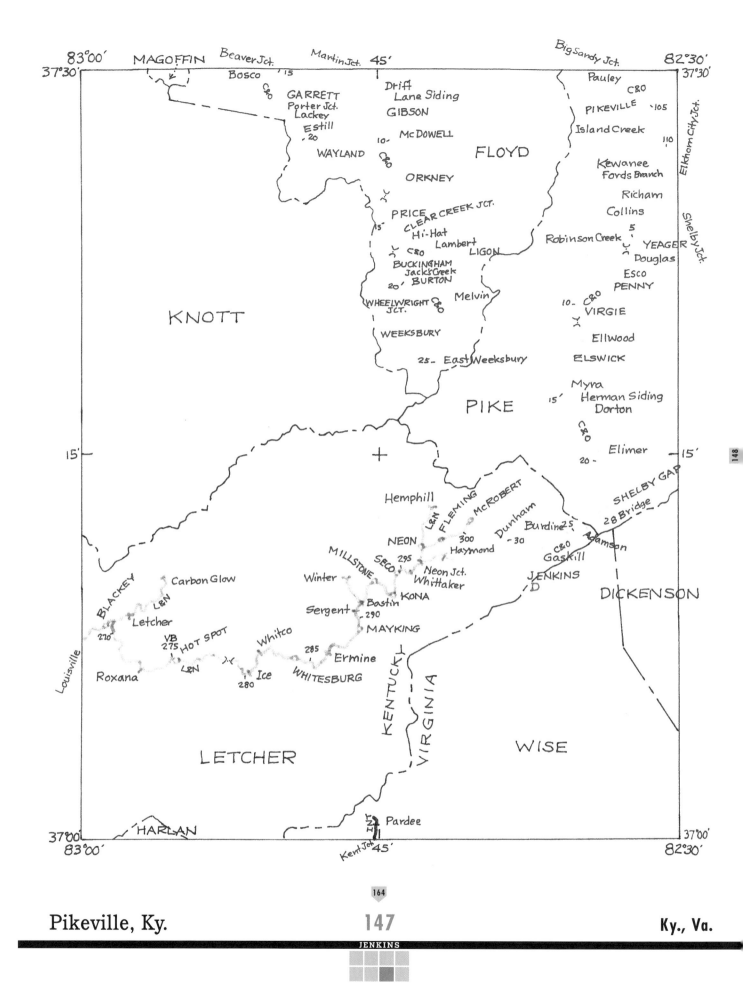

Bosco 15
Pauley
C&O
Drift
Lane Siding
GIBSON
PIKEVILLE 105
Island Creek
GARRETT
Porter Jct.
Lackey
Estill
20
McDOWELL
FLOYD
110
WAYLAND 10
C&O
Kewanee
Fords Branch
Richam
ORKNEY
Collins
PRICE CLEAR CREEK JCT.
15
Hi-Hat
Lambert
C&O
BUCKINGHAM
Jacks Creek
20 BURTON
LIGON
Robinson Creek 5
YEAGER
Douglas
Esco
PENNY
Melvin
10 C&O
VIRGIE
KNOTT
WHEELWRIGHT
JCT. C&O
WEEKSBURY
25 East Weeksbury
Ellwood
ELSWICK
PIKE
Myra
15' Herman Siding
Dorton
C&O
20 Elimer

148

15' 15'
SHELBY GAP

Hemphill
L&N
FLEMING
McROBERT
Dunham
Burdine 25
28 Bridge
NEON 295
300
30
C&O
Adamson
SECO
Haymond
Gaskill
MILLSTONE
Neon Jct.
Whittaker
JENKINS
D
Winter
KONA
DICKENSON
Sergent
Bastin
290
Carbon Glow
BLACKEY
L&N
Letcher
MAYKING
270
VB
275 HOT SPOT
Whitco 285
Ice
Ermine
WHITESBURG
L&N
280
Roxana

KENTUCKY
VIRGINIA
WISE

LETCHER

HARLAN
Pardee
37°00' 37°00'

Pikeville, Ky. **147** Ky., Va.

JENKINS

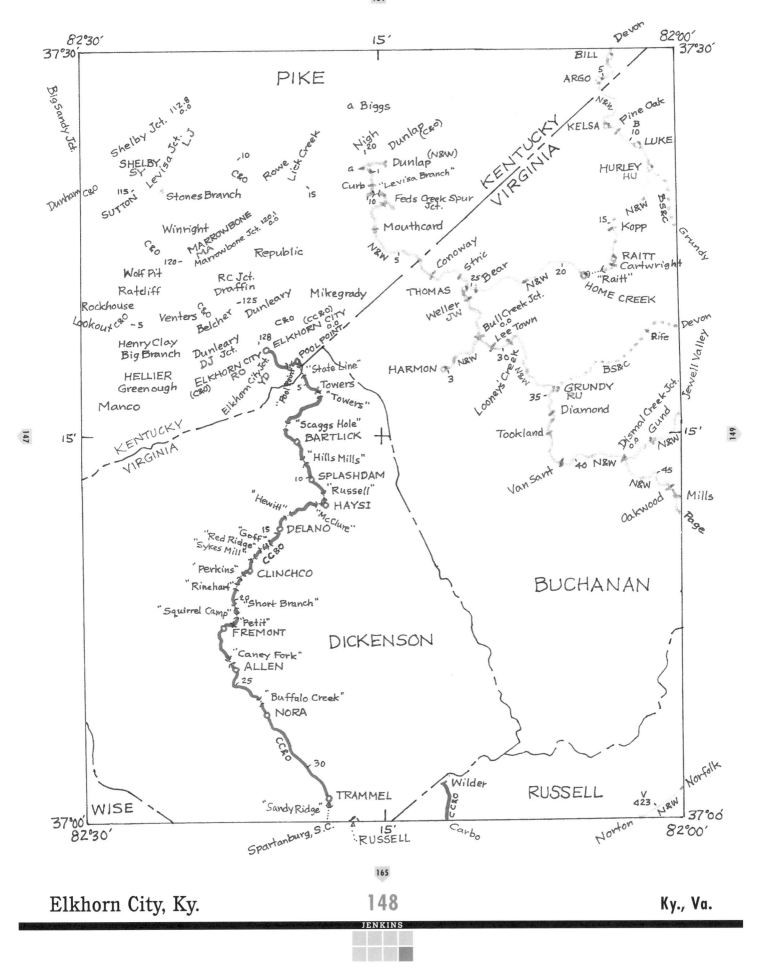

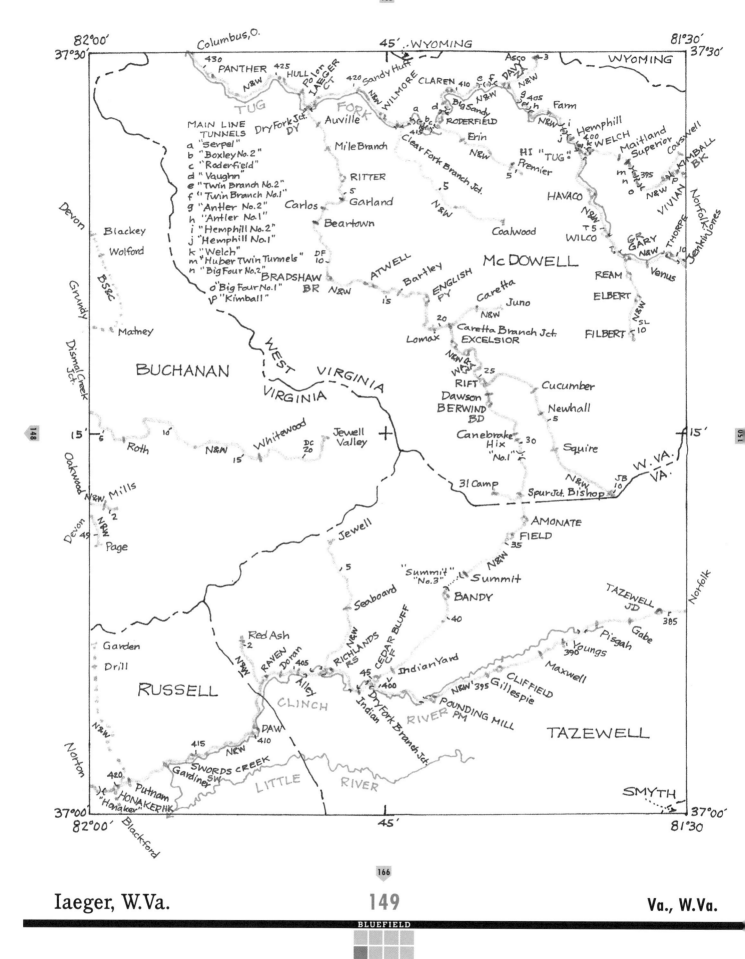

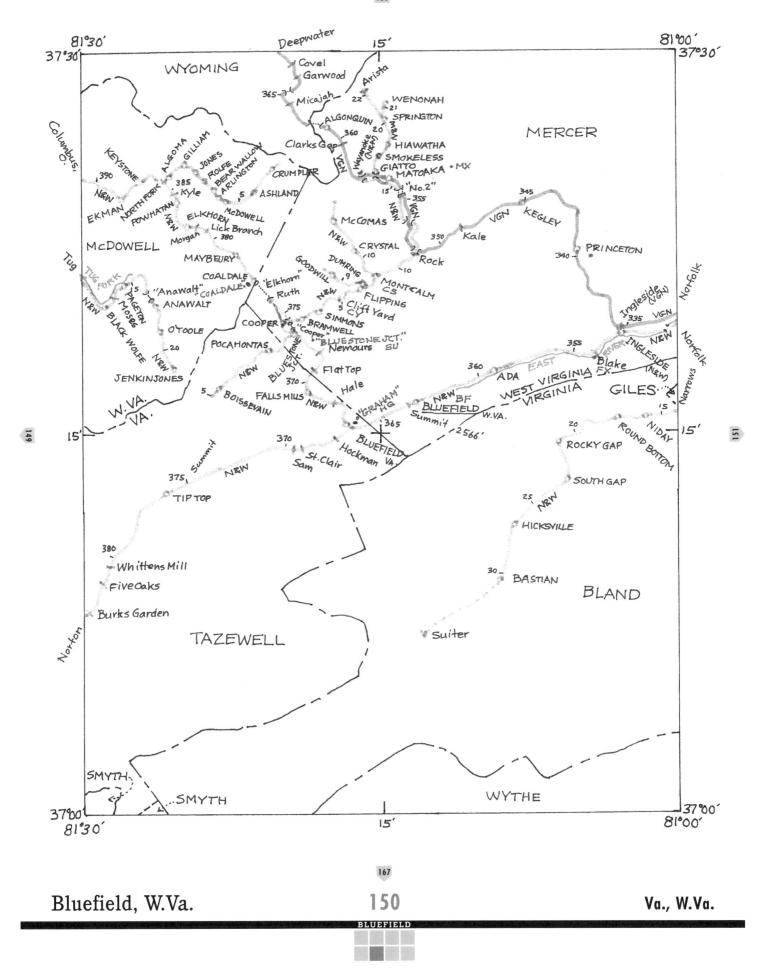

81°30'
37°30'
15'
81°00'
37°30'

Deepwater

WYOMING

Covel
Garwood
Micajah
365
2.6
Arista
22
ALGONQUIN
21
WENONAH
SPRINGTON
360
20
Clarks Gap
HIAWATHA
SMOKELESS
GIATTO
MATOAKA • MX
"No.2"
355
15'

MERCER

KEYSTONE
ALGOMA
GILLIAM
JONES
ROLFE
BEAR WALLOW
ARLINGTON
390
NORTH FORK
385
Kyle
CRUMPLER
ASHLAND
5
345
KEGLEY
EKMAN
POWHATAN
ELKHORN
McDOWELL
Lick Branch
McCOMAS
350
Kale
VGN
PRINCETON
340

McDOWELL
Morgan
380
CRYSTAL
10
Rock
N&W

MAYBEURY
GOODWILL
DUMRING
9
MONTCALM
CS
Ingleside
(VGN)
335
VGN

COALDALE
"Elkhorn"
FLIPPING
Norfolk

Tug
TUG FORK
"Anawalt"
COALDALE
Ruth
375
N&W
Cliff Yard
5
INGLESIDE
5
PAGETON
ANAWALT
SIMMONS
355
Blake
Narrows
MOSES
BRAMWELL
FX
River
BLACK WOLFE
COOPER
"Cooper"
"BLUESTONE JCT."
EAST
ADA
15'
15'
O'TOOLE
20
POCAHONTAS
Nemours
SU
360
WEST VIRGINIA
GILES
NIDAY
JENKINJONES
N&W
Flat Top
N&W
BF
VIRGINIA
ROUND BOTTOM
W. VA.
BLUESTONE JCT.
Hale
"GRAHAM"
BLUEFIELD
20
VA.
5
370
HQ
Summit
W.VA.
ROCKY GAP
BOISSEVAIN
FALLS MILLS N&W
365
2566'
SOUTH GAP
15'
15'
Summit
370
BLUEFIELD
25
375
St. Clair
Hockman Va.
N&W
HICKSVILLE
TIP TOP
Sam

380
30
BASTIAN
BLAND
Whittens Mill
Five Oaks
Burks Garden
Suiter

Norton
TAZEWELL

SMYTH
SMYTH
WYTHE

37°00'
81°30'
15'
81°00'
37°00'

Bluefield, W.Va.

Va., W.Va.

BLUEFIELD

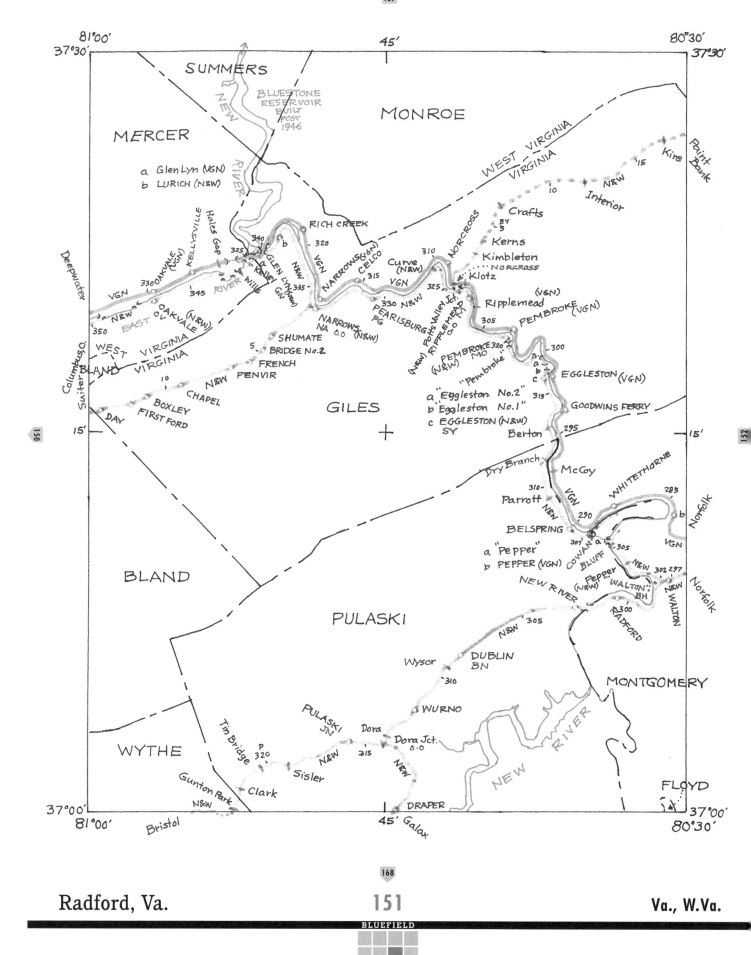

SUMMERS

MONROE

MERCER

BLUESTONE RESERVOIR BUILT POST 1946

NEW RIVER

WEST VIRGINIA

VIRGINIA

a Glen Lyn (VGN)
b LURICH (N&W)

Point Bank

Kire

N&W   15

10   Interior

RICH CREEK

Crafts

BY 5

NORCROSS

Kerns

Kimbleton

NORCROSS

Deepwater

Hales Gap

340   b   320

325   GLEN LYN

Curve (N&W)

310

VGN

CELCO

NARROWS (VGN)

315

VGN

Klotz

Ripplemead   (VGN)

PEMBROKE (VGN)

3300 OAKVALE (VGN)

KELLYSVILLE

VGN

345

NEW RIVER MILLS

335

Keller GN

Glen Lyn (N&W)

325

Potts Valley Jct.

305

300

N&W

350

EAST   OAKVALE (N&W)

OV

330   N&W

PEARISBURG

PG

RIPPLEMEAD (N&W)

PEMBROKE (N&W)   320

MO

"Pembroke"

EGGLESTON (VGN)

Columbus O.

WEST   VIRGINIA

BLAND

NARROWS NA 0.0 (N&W)

SHUMATE BRIDGE No.2

a "Eggleston No.2"

315   GOODWINS FERRY

Suiter O.

DAY

5

FRENCH

10   CHAPEL

N&W   PENVIR

GILES

b "Eggleston No.1"
c EGGLESTON (N&W)
      SY

BOXLEY   FIRST FORD

295

Berton

Dry Branch

McCoy

WHITETHORNE

285

310

VGN

Parrott

290

NEW

BELSPRING

307

301

COWAN

305

BLUFF

302 297

N&W

Norfolk

b

VGN

Norfolk

a "Pepper"
b PEPPER (VGN)

BLAND

PULASKI

NEW RIVER

Pepper (N&W)

"WALTON" BH

N&W   WALTON

RADFORD

300

NEW

N&W   305

MONTGOMERY

Wysor

DUBLIN BN

310

NEW   RIVER

WURNO

Tim Bridge

PULASKI JN

Dora

Dora Jct. 0.0

WYTHE

P   320

N&W   215

NEW

Sisler

Gunton Park   Clark

N&W

FLOYD

DRAPER

Radford, Va.   **151**   Va., W.Va.

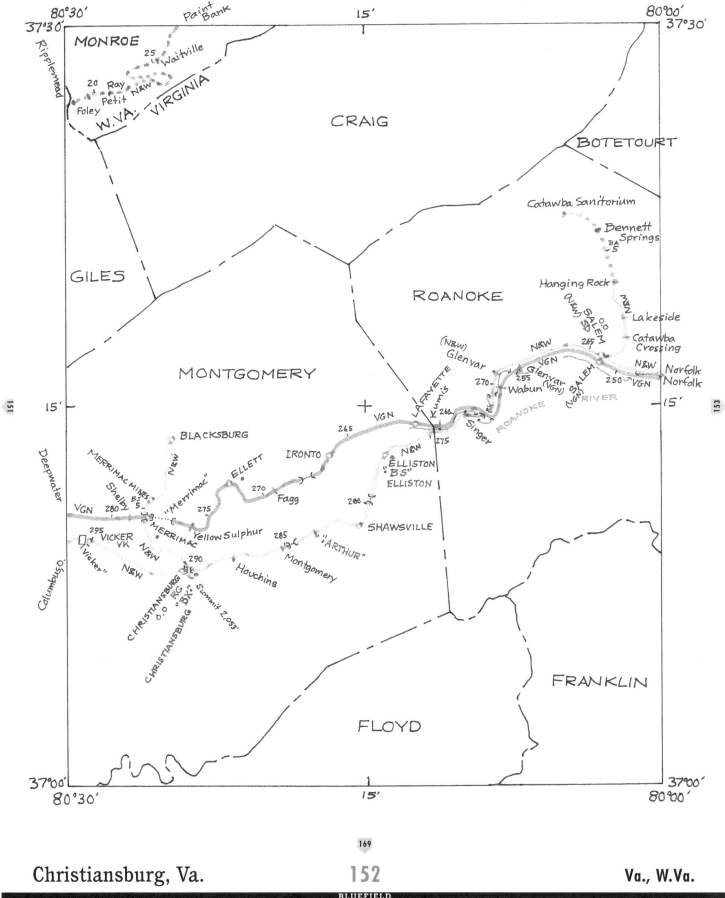

Christiansburg, Va.

152

Va., W.Va.

BLUEFIELD

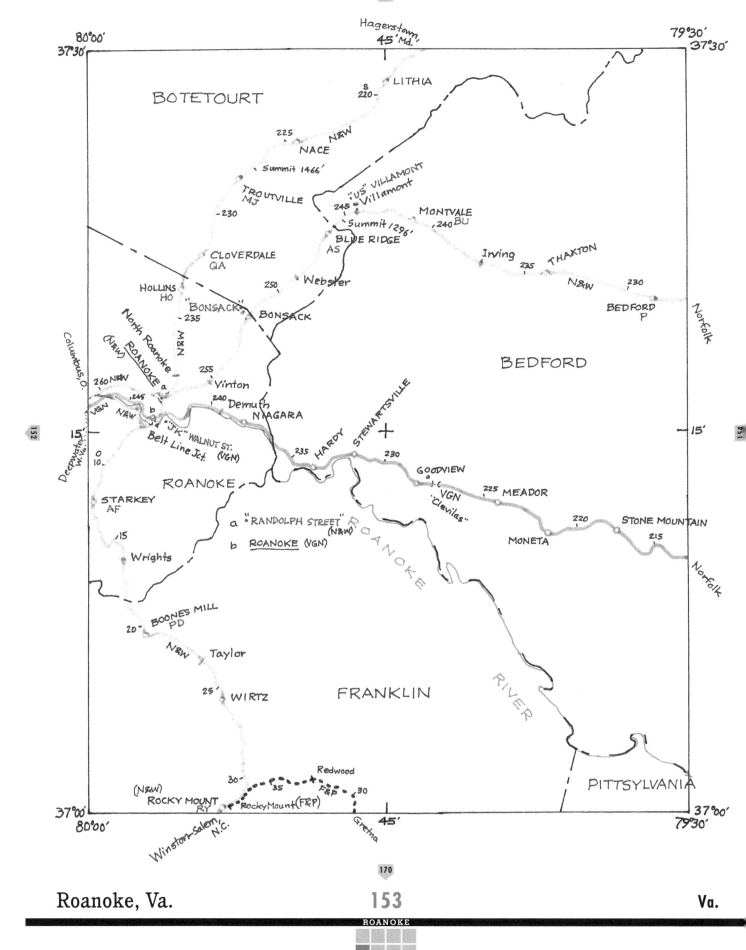

BOTETOURT

Hagerstown,
45' Md.

S
220- LITHIA

225
N&W
NACE

Summit 1466'

"US" VILLAMONT
TROUTVILLE   Villamont
MJ   245 -
-230   Summit 1296'   MONTVALE
                                    240 BU

CLOVERDALE   BLUE RIDGE   Irving   THAXTON
QA   AS   235   N&W
                                                            230
HOLLINS   250   Webster   BEDFORD
HO   P

"BONSACK"   BONSACK   BEDFORD
- 235

Columbus, O.
North Roanoke
(N&W)
ROANOKE   255
260 N&W   Vinton
245   .240 Demuth
VGN   N&W   NIAGARA
O   "JK" WALNUT ST.
10   Belt Line Jct. (VGN)   235 HARDY STEWARTSVILLE
Deepwater, W.Va.   230

ROANOKE   GOODVIEW
                                              C
STARKEY   VGN
AF   a "RANDOLPH STREET"   "Clevilas"   225 MEADOR
                      (N&W)
.15   b ROANOKE (VGN)   220   STONE MOUNTAIN
Wrights   MONETA   215

BOONES MILL
PD
20-
N&W   Taylor

25' WIRTZ   FRANKLIN

                                                        PITTSYLVANIA

(N&W)   30-   Redwood
ROCKY MOUNT   35   F&P
RY   RockyMount(F&P) 30
37°00'   Gretna 45'
Winston-Salem,
N.C.

15'

ROANOKE
RIVER
Norfolk
Norfolk

Roanoke, Va.   **153**   Va.

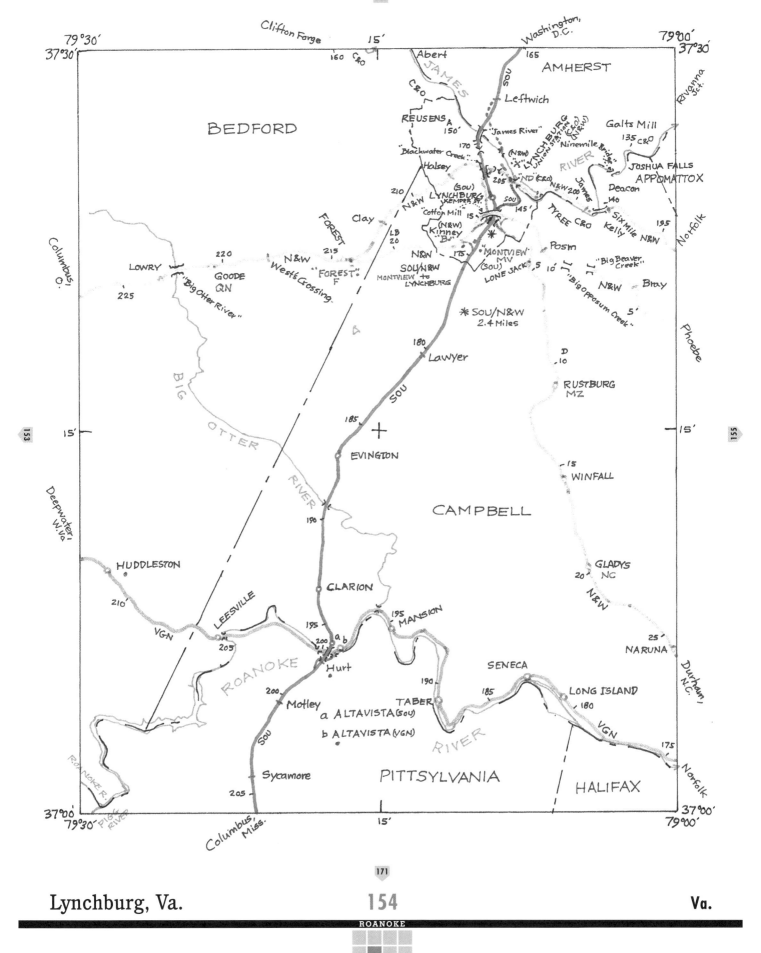

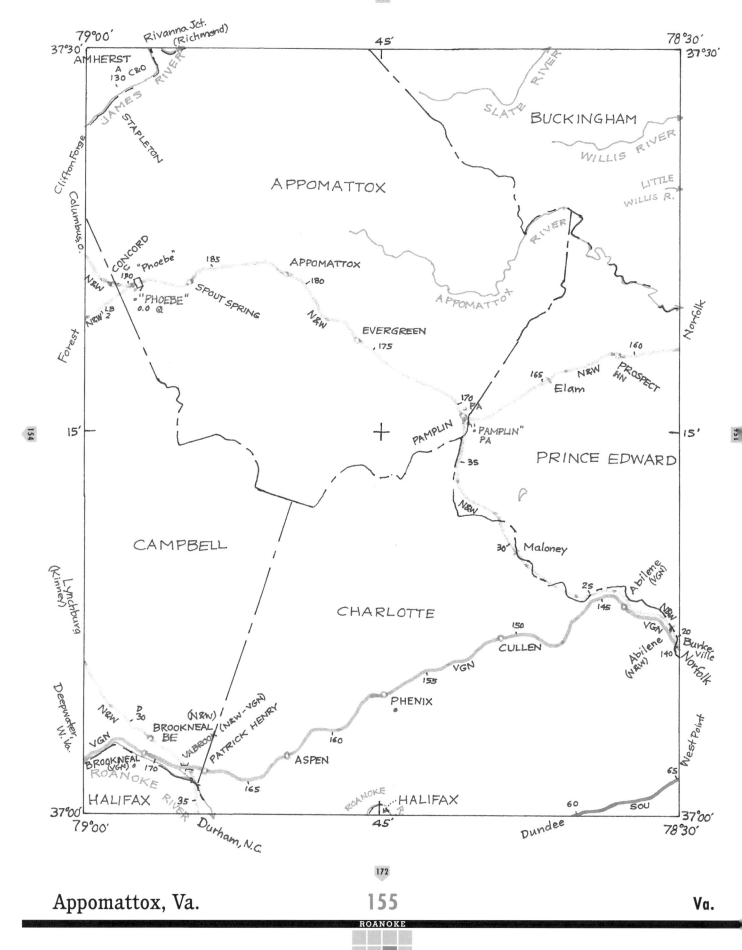

Appomattox, Va.

155

Va.

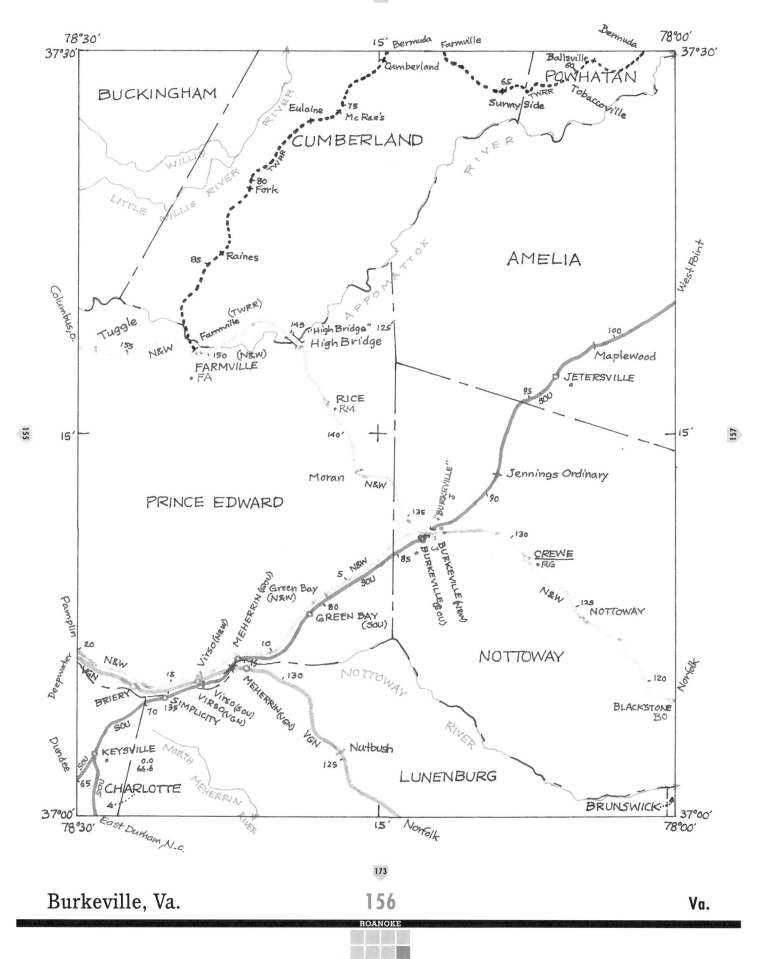

78°30'
37°30'

15' Bermuda   Farmville   Bermuda   78°00'
37°30'

BUCKINGHAM

Cumberland   Ballsville 60
POWHATAN

Eulaine   75   65   TWRR   Tobaccoville
McRae's   Sunny Side

CUMBERLAND

WILLIS RIVER

80
Fork

TWRR

LITTLE WILLIS RIVER

85   Raines

AMELIA

West Point

APPOMATTOX RIVER

100

Columbus, O.

(TWRR)   145   "High Bridge" 125'   Maplewood
Tuggle   Farmville   High Bridge   JETERSVILLE

155   N&W   95   SOU

150 (N&W)

FARMVILLE
FA   RICE
RM

15'   140'   15'

157

Moran   Jennings Ordinary

N&W   90

PRINCE EDWARD   135   "BURKEVILLE"   130

CREWE
RG

5 N&W   85 BURKEVILLE (N&W)
MEHERRIN (SOU)   SOU   BURKEVILLE (SOU)   N&W   125
Green Bay   NOTTOWAY
(N&W)

80   NOTTOWAY
Virso (N&W)   GREEN BAY
(SOU)   120

Pamplin   10

20   BLACKSTONE
N&W   130   BO

Deepwater   Virso (SOU)   MEHERRIN (VGN)   NOTTOWAY RIVER   Norfolk
VGN   15   VIRSO (VGN)   VGN   Nutbush
BRIERY   SIMPLICITY
70 135   125
SOU   Dundee   KEYSVILLE   NORTH MEHERRIN RIVER   LUNENBURG
65   SOU   0.0
66.6
CHARLOTTE
4   37°00'
37°00'   BRUNSWICK
78°30'   East Durham, N.C.   15' Norfolk   78°00'

POWHATAN

Farmville 45' West Point

78°00' 37°30' 37°30' 77°30'

MOSELEY (SOU) TWRR 120 90U HALLSBORO

Dorset Moseley (TWRR) 35

Clayville CHESTERFIELD

115 Pilkinton

Skinquarter

CHULA 110 Coalboro 25 Summit 20 Nash Bermuda

SOU Winterham 105 Fendley 15

Dundee AMELIA Winterpock TWRR Perdue Beach

APPOMATTOX

AMELIA RIVER

15' + 15'

NAMOZINE CREEK Sutherland Norfolk

90 Poe

95 "JACK" 88.3 84

NOTTOWAY CHURCH ROAD

100 Poole NORTH BURGESS Burgess Richmond

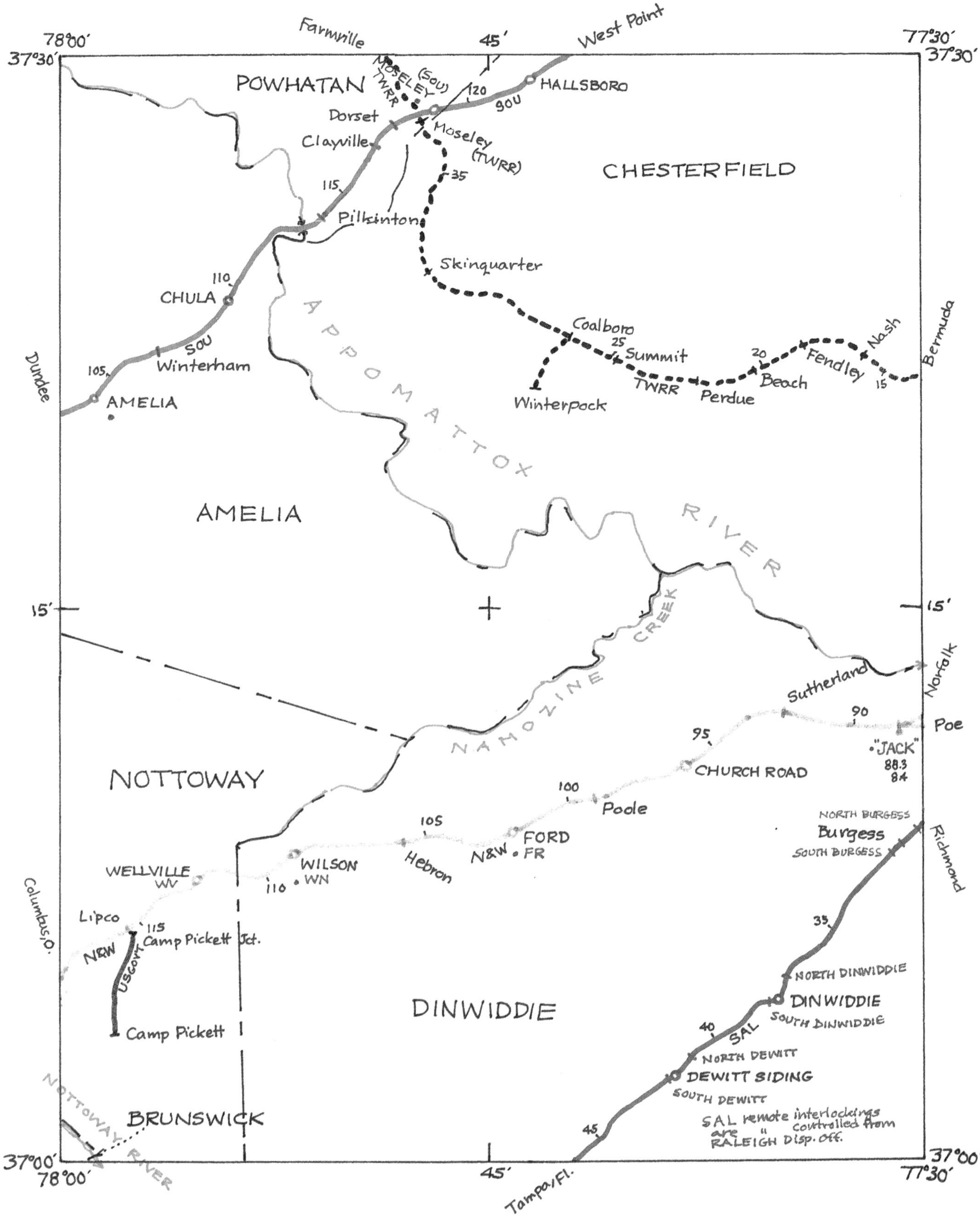

105 SOUTH BURGESS

WELLVILLE WV WILSON WN Hebron N&W FR FORD 35

110

Lipco 115

NEW Camp Pickett Jct. US Govt NORTH DINWIDDIE

DINWIDDIE

Camp Pickett DINWIDDIE SOUTH DINWIDDIE

Columbus, O. 40 SAL

NORTH DEWITT

DEWITT SIDING

SOUTH DEWITT

NOTTOWAY RIVER BRUNSWICK 45 SAL remote interlockings are " controlled from RALEIGH Disp. Off.

78°00' 37°00' 45' 37°00' 77°30'

Tampa, Fl.

Camp Pickett, Va. **157** Va.

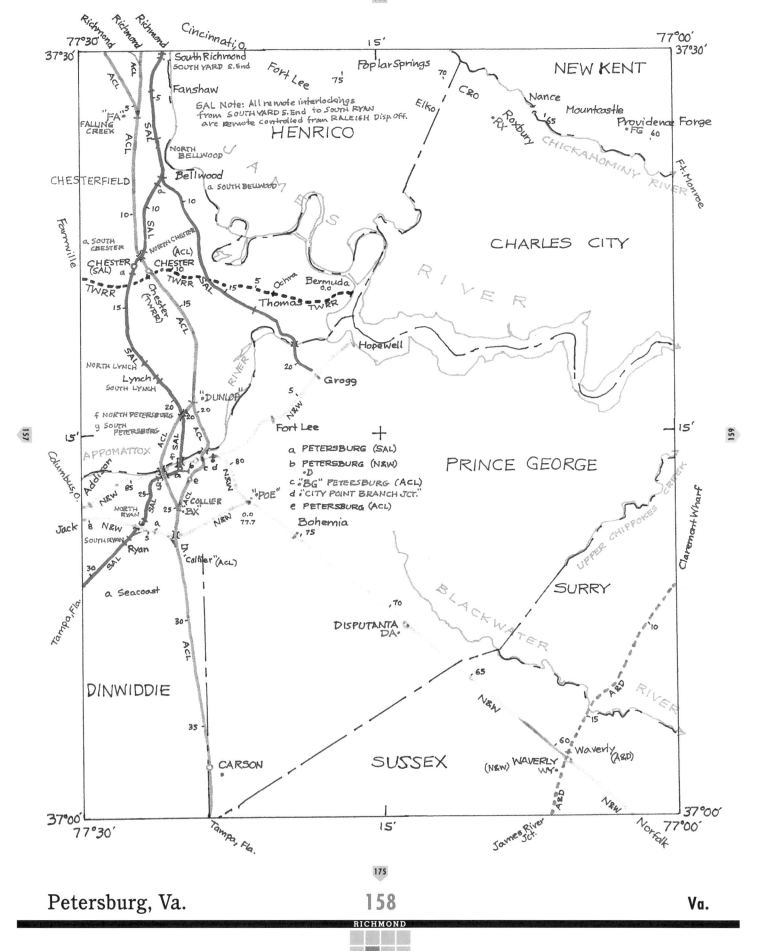

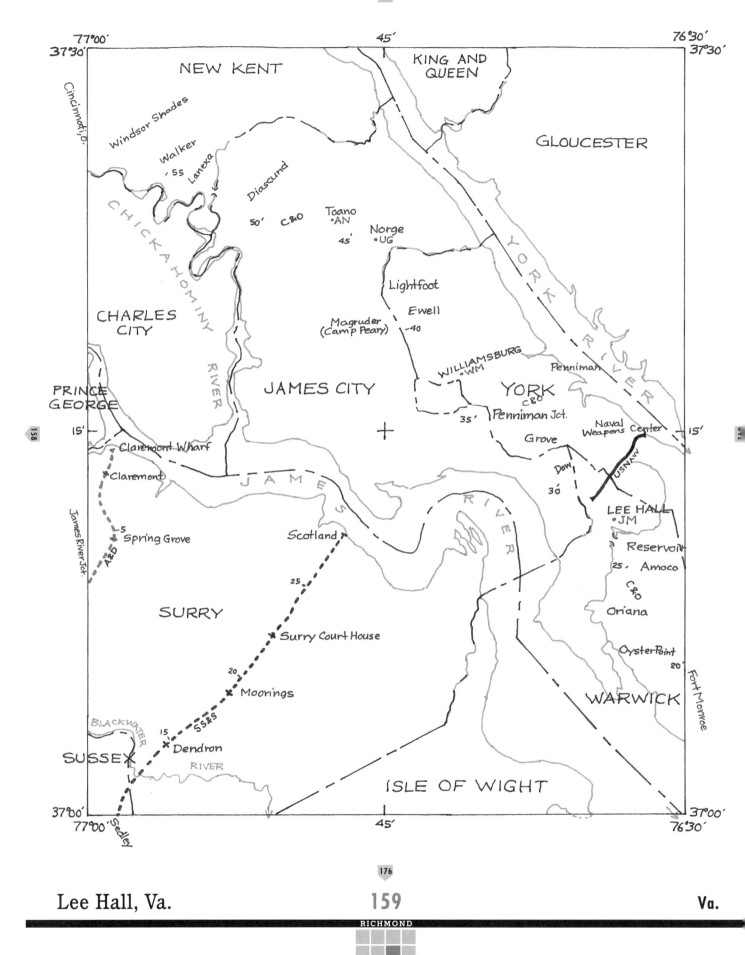

Lee Hall, Va.   **159**   Va.

RICHMOND

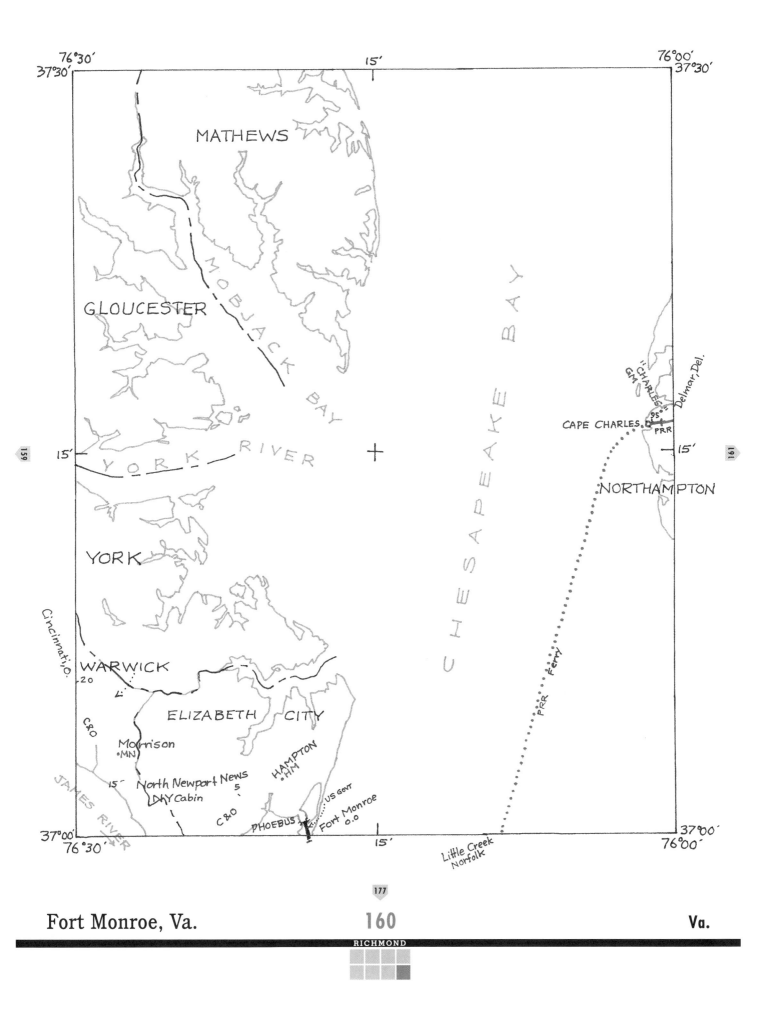

MATHEWS

GLOUCESTER

MOBJACK BAY

CHESAPEAKE BAY

YORK RIVER

YORK

WARWICK

Cincinnati,O.

.20

C&O

ELIZABETH CITY

JAMES RIVER

Morrison
•MN

15´

North Newport News
DKY Cabin

C&O

PHOEBUS

HAMPTON
•HM

US GOVT

Fort Monroe
0.0

CAPE CHARLES

"CHARLES
GM

Delmar, Del.

95
PRR

NORTHAMPTON

PRR Ferry

Little Creek
Norfolk

76°30´   37°30´   15´   76°00´   37°30´

159   161

15´   15´

37°00´   76°30´   15´   76°00´   37°00´

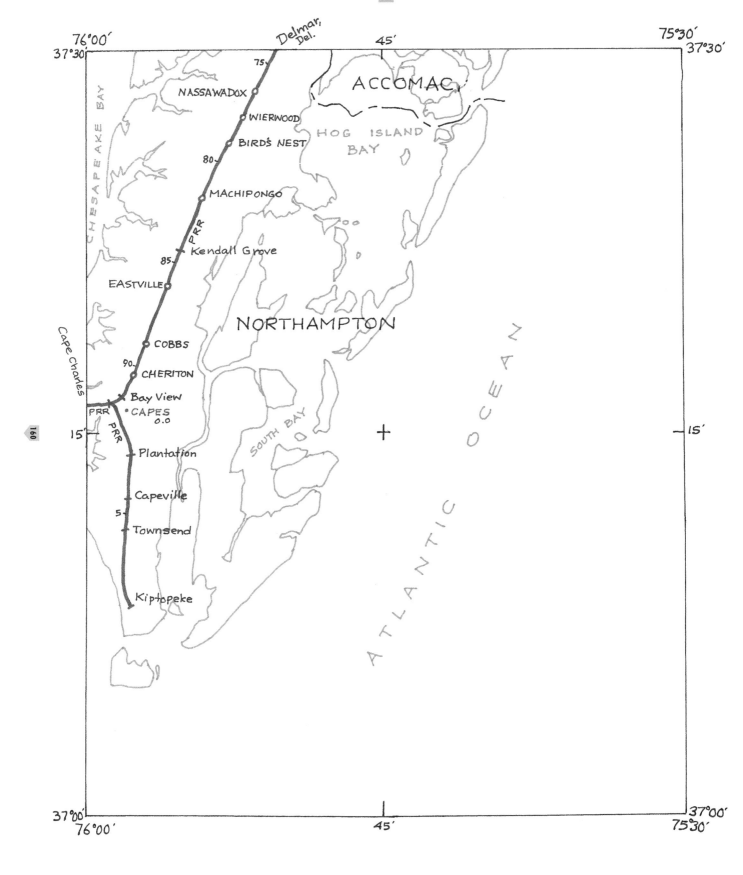

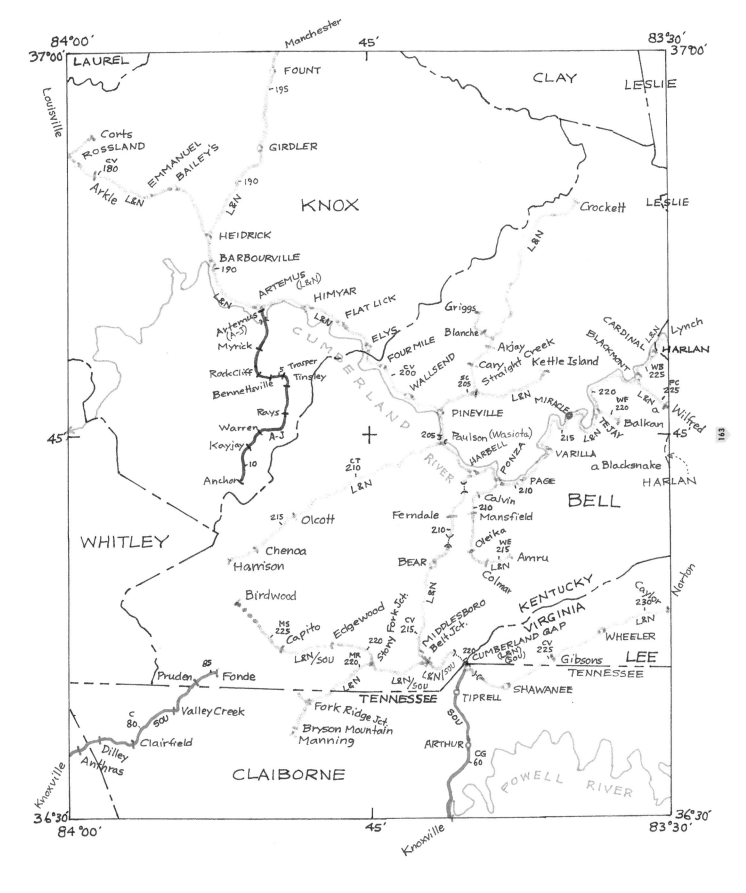

LAUREL

84°00'   Manchester   45'   83°30'
37°00'   37°00'

FOUNT
-195

CLAY

LESLIE

Corts
ROSSLAND
CV
180

EMMANUEL
BAILEY'S

GIRDLER

Crockett

LESLIE

Louisville

Arkle L&N

-190

KNOX

L&N

L&N

HEIDRICK

BARBOURVILLE
-190

ARTEMUS
(L&N)

HIMYAR

FLAT LICK

Griggs

CARDINAL

Lynch

L&N

Artemus
(A-J)

Myrick

L&N

ELYS

FOUR MILE

Blanche

Arjay

Straight Creek

BLACKMONT

HARLAN

WB
225

CUMBERLAND

WALLSEND

CV
200

Cary

Kettle Island

L&N a

PC
225

RockCliff

5 Trosper

Tinsley

SC
205

L&N

MIRACLE

220

WF
220

Wilfred

Bennettsville

PINEVILLE

Balkan

TEJAY

45'   Rays

RIVER

205

Paulson (Wasiota)

215

L&N

45'

Warren

A-J

HARBELL

PONZA

VARILLA

a Blacksnake

163

Kayjay

CT
210

PAGE
210

HARLAN

Anchor

10

L&N

Calvin

BELL

215   Olcott

Ferndale

-210
Mansfield

WHITLEY

Chenoa

Harrison

BEAR

210

L&N

Oleika
WE
215

Colmar

Amru

Birdwood

MS
225

Capito

Edgewood

CV
215

MIDDLESBORO
Belt Jct.

KENTUCKY

VIRGINIA

Ca w/o a
230 ∧

L&N

Norton

WHEELER

L&N/SOU

MR
220

220

Stony Fork Jct.

220

CUMBERLAND GAP

L&N
(L&N)
(SOU)

CV
225

Gibsons

LEE

85

Pruden   Fonde

L&N

L&N/SOU

L&N/SOU

TENNESSEE

C
80

SOU

Valley Creek

Fork Ridge Jct.

Bryson Mountain
Manning

TIPRELL

SHAWANEE

SOU

Clairfield

Dilley
Anthras

CLAIBORNE

ARTHUR

CG
60

POWELL RIVER

Knoxville

36°30'   36°30'
84°00'   45'   83°30'

Knoxville

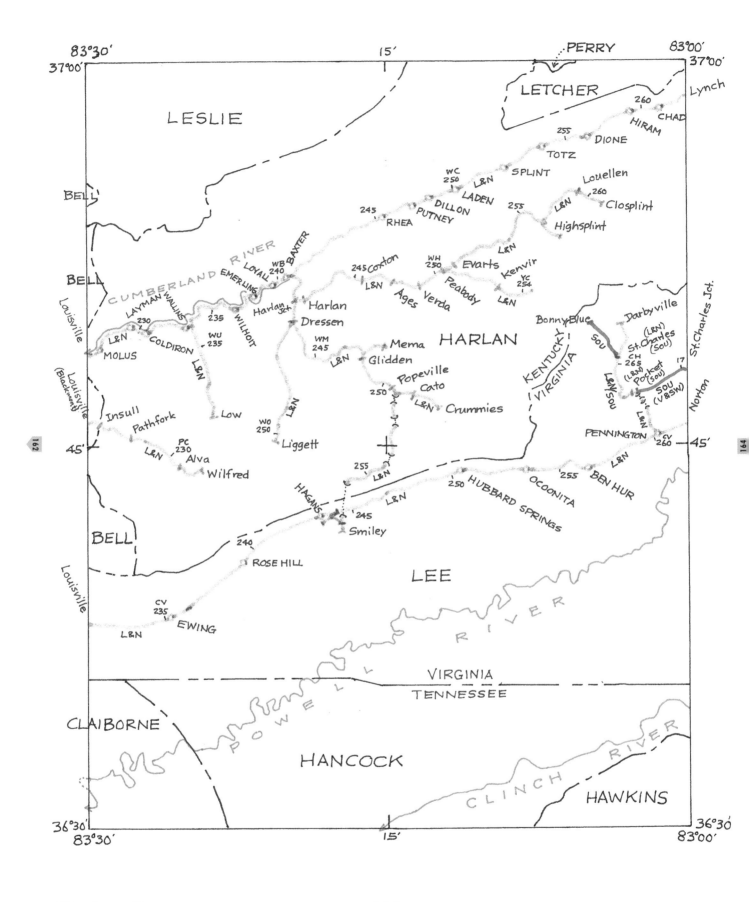

LETCHER

LESLIE

Lynch

260

HIRAM CHAD

255 DIONE

TOTZ

WC L&N SPLINT Louellen

250 LADEN 260

DILLON 255 L&N Closplint

245 PUTNEY Highsplint

RHEA

BELL

CUMBERLAND RIVER BAXTER 245 Coxton WH L&N Evarts Kenvir

EMERLING LOYALL WB L&N 250 YC

240 254

BELL Harlan Harlan Ages Verda Peabody L&N

LAYMAN WALLINS Jct. Dressen

230 235 Bonny Blue Darbyville

L&N WILHOIT Mema HARLAN SOU (L&N)

MOLUS COLDIRON WU WM Glidden St.Charles (SOU)

235 245 L&N CH 265 Pocket

Insull Low Popeville KENTUCKY (L&N) (SOU) 17

Pathfork WO L&N 250 Cato VIRGINIA L&N/SOU SOU

PC 250 Liggett L&N Crummies (V&SW)

230 L&N PENNINGTON CV

Alva 255 260

Wilfred L&N L&N

255 250 OCOONITA 255 BEN HUR

HAGANS L&N HUBBARD SPRINGS

240 245 Smiley

Louisville ROSE HILL LEE

CV

235 VIRGINIA

L&N EWING TENNESSEE

CLAIBORNE

HANCOCK

POWELL RIVER

CLINCH RIVER HAWKINS

162 164

Louisville

Louisville (Blackmont)

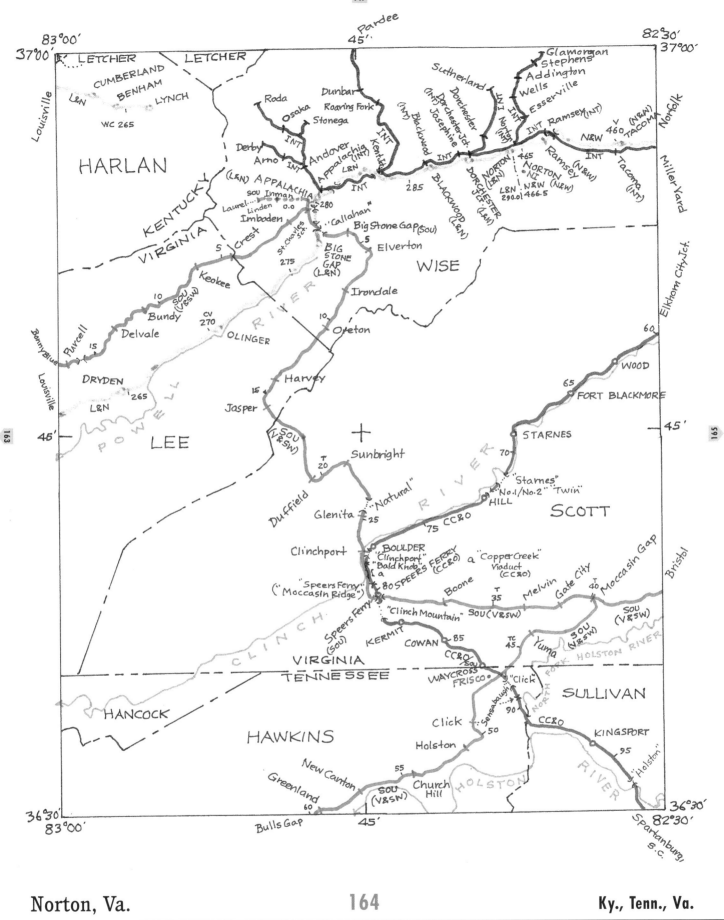

83°00'
37°00'

45'
Pardee

82°30'
37°00'

LETCHER   LETCHER

CUMBERLAND
BENHAM
LYNCH

Louisville

L&N

WC 265

Glamorgan
Stephens
Addington
Sutherland
Wells
Esserville

Roda   Dunbar
Osaka   Roaring Fork
Stonega

INT N&I
INT
INT
Ramsey (INT)
INT
460
N&W
TACOMA

N&W

Norfolk

HARLAN

Derby
Arno   INT
Andover
Appalachia (INT)
INT

L&N

INT

Dorchester
(INT)
Josephine
(INT)
Blackwood

NI North
(N&I)

Ramsey
(N&W)
Tacoma

Miller Yard

Tacoma
(INT)

KENTUCKY

(L&N)   APPALACHIA
SOU Inman
Laurel
Linden   0.0
Imboden   280

INT
L&N
INT
285

INT
INT
DORCHESTER JCT
(L&N)
NORTON
(L&N)
BLACKWOOD

465
NORTON
(N&W)
L&N N&W
466.5

290.01

VIRGINIA

Crest   5
St Charles
Jct
275

Callahan
Big Stone Gap (SOU)
5

BIG
STONE
GAP
(L&N)   Elverton

WISE

Keokee
10   SOU
(V&SW)
Bundy
Delvale
CV
270

Purcell
15
Barnyblue

OLINGER

Irondale
10
Oreton

60
WOOD

Elkhorn City Jct.

Louisville
DRYDEN
L&N   265

Harvey

65
FORT BLACKMORE

LEE
POWELL

15
Jasper

SOU
(V&SW)

Sunbright

STARNES

45'

45'

163   165

36°30'

Duffield
T
20

Glenita
Natural
25

70

Starnes
No.1/No.2 Twin
HILL

SCOTT

CC&O
75

Clinchport

BOULDER
Clinchport
Bald Knob
a

Copper Creek
Viaduct
(CC&O)

Moccasin Gap

Bristol

Speers Ferry
Moccasin Ridge

80 SPEERS FERRY
(CC&O)

Boone
Melvin

Gate City
T
40   Moccasin Gap

SOU
(V&SW)

Speers Ferry
(SOU)
KERMIT

Clinch Mountain
SOU (V&SW)
T
35

SOU
(V&SW)

COWAN   85
CC&O
SOU

TC
45   Yuma

SOU
(V&SW)
HOLSTON RIVER

CLINCH

VIRGINIA
TENNESSEE

WAYCROSS
FRISCO   Click

NORTH FORK

SULLIVAN

HANCOCK

Click
Holston
50

90

CC&O
KINGSPORT

95

HAWKINS

New Canton
55

Holston

Greenland
Church
Hill
60
Bulls Gap   45'

SOU
(V&SW)   HOLSTON

RIVER

36°30'
82°30'
Spartanburg,
S.C.

83°00'
36°30'

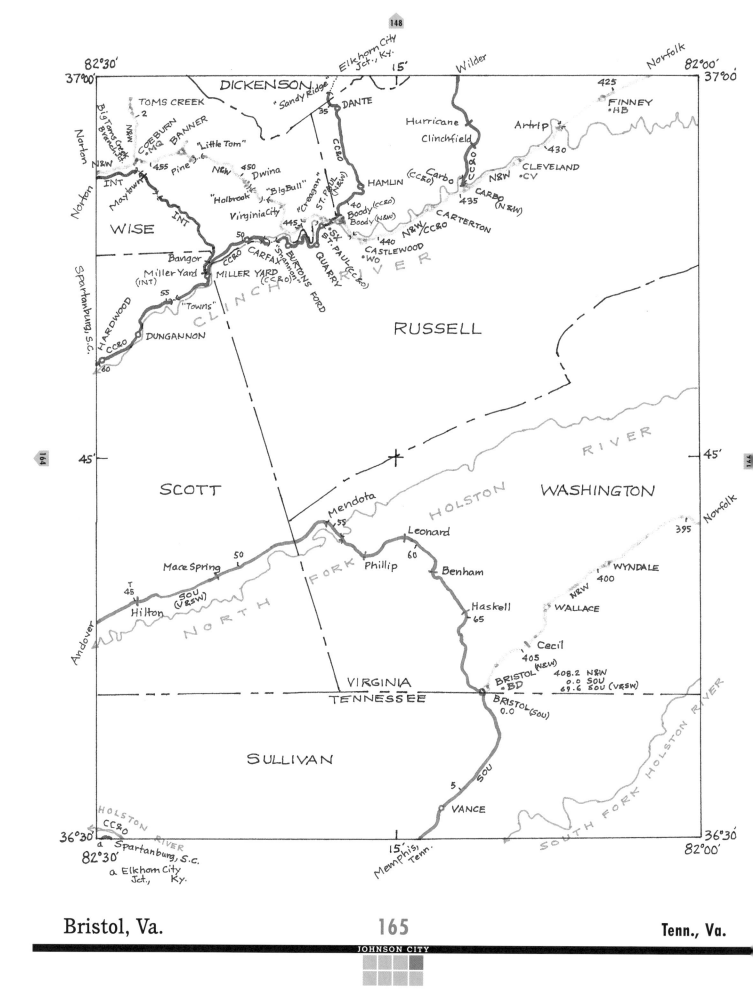

82°30'   37°00'

DICKENSON.   Elkhorn City   Wilder   Norfolk   82°00'   37°00'
                Jct., Ky.   15'

TOMS CREEK   "Sandy Ridge"   35   DANTE   Hurricane   Artrip   425   FINNEY
                                                                    •HB

Norton   2   COEBURN   BANNER   "Little Tom"   Clinchfield   '430   CLEVELAND
          MQ   Carbo   •CV

N&W   '455   Pine J.6   N&W   450   Dwina   HAMLIN   (CC&O)   N&W   CARBO
INT   '435   (N&W)

Maytown   "Holbrook"   "BigBull"   "Creganat"   40   Boody (CC&O)
WISE   INT   VirginiaCity   ST. PAUL   Boody (N&W)   CARTERTON   N&W/CC&O
                          445   (N&W)   '440   SX.
                    50   ST.PAUL (CC&O)   CASTLEWOOD
Bangor   CC&O   CARFAX   "Shannon"   •WO
Miller Yard   MILLER YARD   BURTONS FORD   CLINCH   RIVER
(INT)   (CC&O)   QUARRY   RUSSELL
Spartanburg, S.C.   55   "Towns"
HARDWOOD   CC&O   DUNGANNON
'60

SCOTT   45'   WASHINGTON   45'
                Mendota   HOLSTON   RIVER
                '55   Leonard   395   Norfolk
Mace Spring   50   Phillip   60   WYNDALE
T   SOU   NORTH   FORK   Benham   N&W   400
45   (V&SW)   WALLACE
Hilton   Haskell
Andover   '65   Cecil
                              405
                              (N&W)   408.2 N&W
VIRGINIA   BRISTOL   0.0 SOU
TENNESSEE   •BD   69.6 SOU (V&SW)
                        BRISTOL (SOU)
                        0.0
SULLIVAN
                        5   SOU
HOLSTON RIVER   VANCE
CC&O   36°30'
a Spartanburg, S.C.   15'   SOUTH FORK HOLSTON RIVER   36°30'
82°30'   Memphis, Tenn.   82°00'
a Elkhorn City
  Jct., Ky.

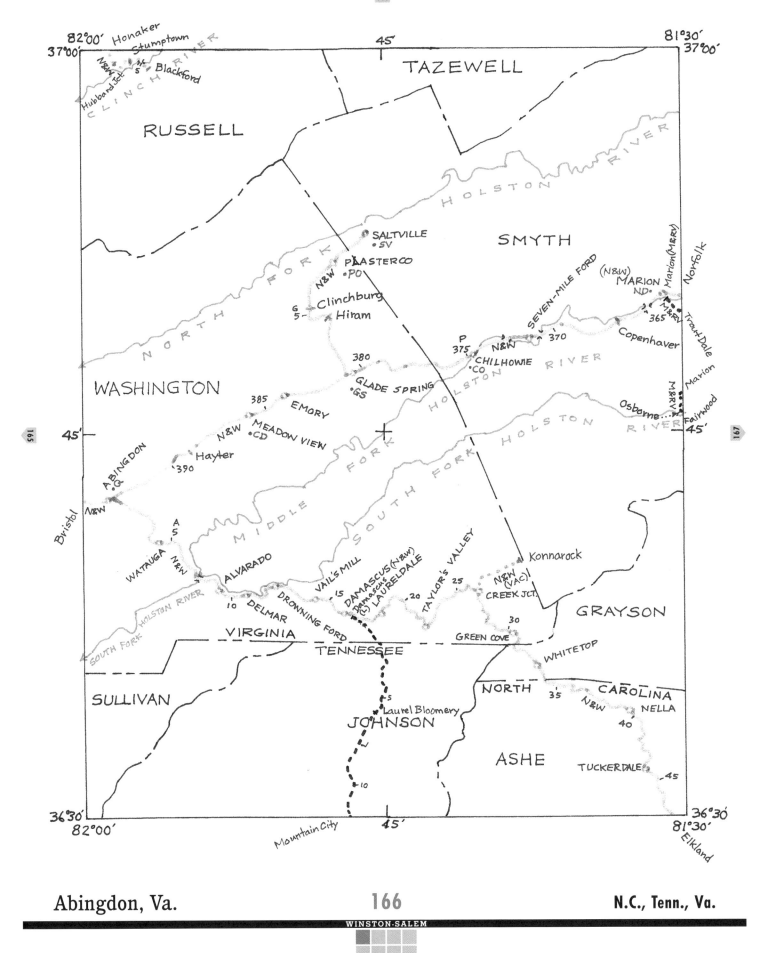

Abingdon, Va.   **166**   **N.C., Tenn., Va.**

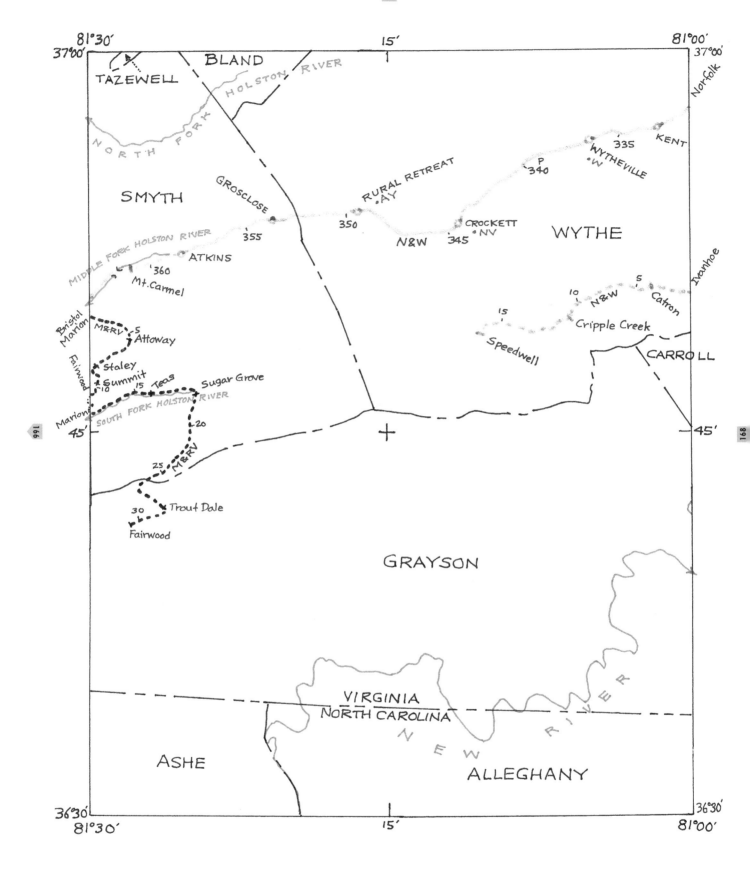

BLAND
TAZEWELL
HOLSTON RIVER
81°30' 37°00'   15'   81°00' 37°00'
Norfolk
NORTH FORK
335 KENT
WYTHEVILLE
P ·W
340
SMYTH
GRosclose
RURAL RETREAT
·AY
350
355
CROCKETT
·NV
N&W   345
WYTHE
MIDDLE FORK HOLSTON RIVER
ATKINS
·360
Mt.Carmel
5
10 N&W Catron Ivanhoe
Cripple Creek
15
Speedwell
Bristol
Marion M&RV 5 Attoway
Fairwood
Staley
Summit
10 15 Teas
Sugar Grove
CARROLL
SOUTH FORK HOLSTON RIVER
45' 20
Marion
45'
168
25 M&RV
30 Trout Dale
Fairwood
GRAYSON
VIRGINIA
NORTH CAROLINA
NEW RIVER
ASHE
ALLEGHANY
36°30'   15'   81°00' 36°30'
81°30'

Rural Retreat, Va.   167   N.C., Va.

WINSTON-SALEM

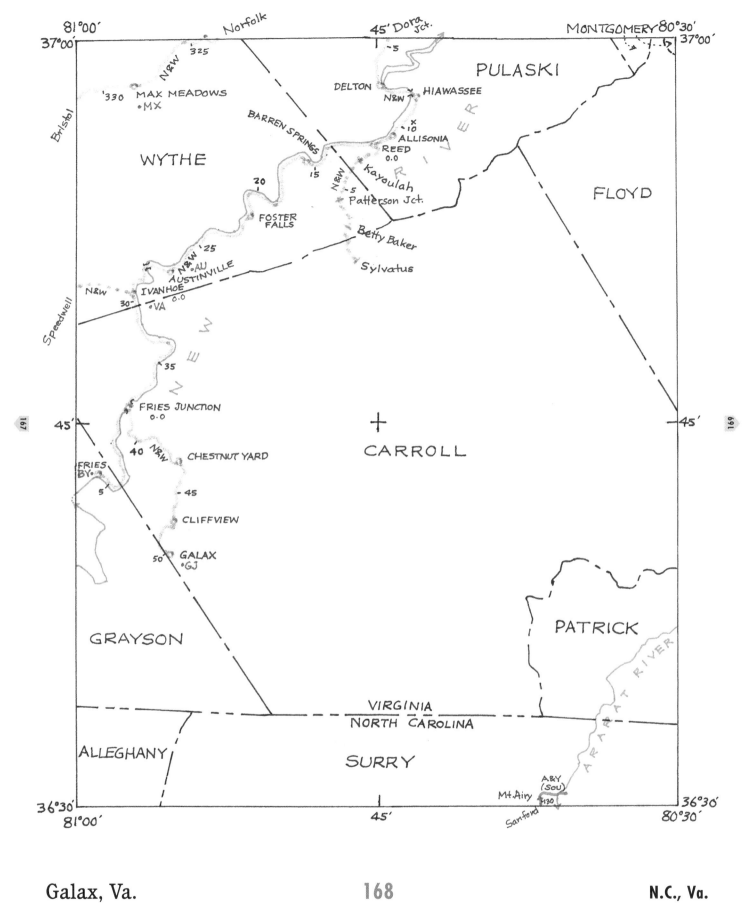

Galax, Va. 168 N.C., Va.

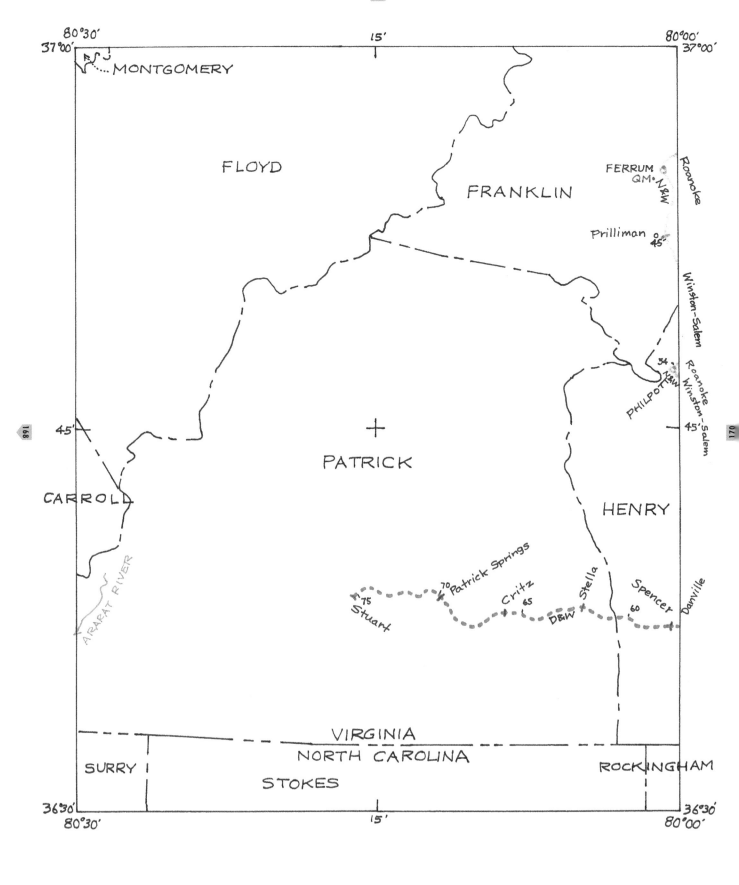

80°30'
37°00'

15'

80°00'
37°00'

MONTGOMERY

FLOYD

FRANKLIN

FERRUM
QM·N&W

Roanoke

Prilliman 0
45

Winston-Salem

54
N&W

Roanoke
Winston-Salem

PHILPOTT

45'

168

45'

PATRICK

HENRY

170

CARROLL

ARARAT RIVER

Patrick Springs

70

75
Stuart

Critz
65

Stella

D&W

Spencer

60

Danville

VIRGINIA

NORTH CAROLINA

ROCKINGHAM

SURRY

STOKES

36°30'
80°30'

15'

36°30'
80°00'

Rocky Knob, Va.

169

N.C., Va.

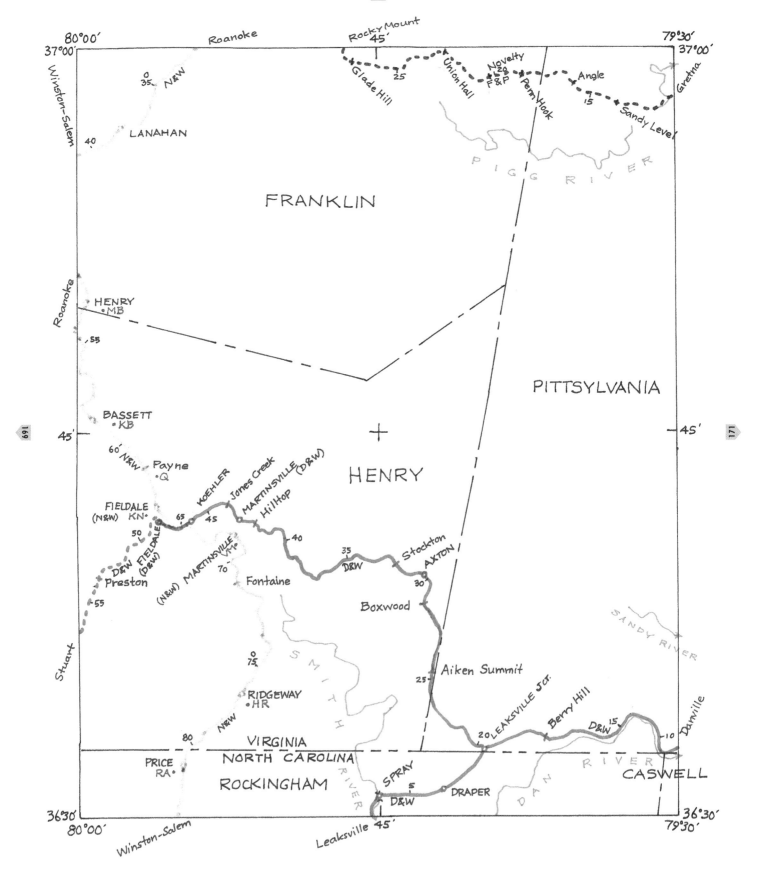

Roanoke

Rocky Mount
45'

80°00'
37°00'

79°30'
37°00'

Winston-Salem

0
35   N&W

Union Hall

Novelty
20
F&P

Angle

Gretna

Glade Hill   25

Penn Hook

15

Sandy Level

40

LANAHAN

PIGG RIVER

FRANKLIN

Roanoke

HENRY
•MB

55

PITTSYLVANIA

199

45'

45'

171

BASSETT
• KB

+

60' N&W   Payne
•Q

HENRY

KOEHLER   Jones Creek MARTINSVILLE (D&W)

SANDY RIVER

FIELDALE
(N&W) KN•

65   45   MARTINSVILLE   Hilltop

50

40

Stockton

D&W FIELDALE
(D&W)

70

D&W

AXTON

Preston

(N&W) MARTINSVILLE
(N&W)

Fontaine

30

55

Boxwood

0
75

Aiken Summit

25

RIDGEWAY
• HR

LEAKSVILLE JCT.

Berry Hill

Danville

N&W

20

D&W 15

10

80

VIRGINIA

Dan River

NORTH CAROLINA

CASWELL

PRICE
RA. •

ROCKINGHAM

SPRAY
5

DRAPER

D&W

36°30'
80°00'

Leaksville 45'

79°30'
36°30'

Winston-Salem

GREENSBORO

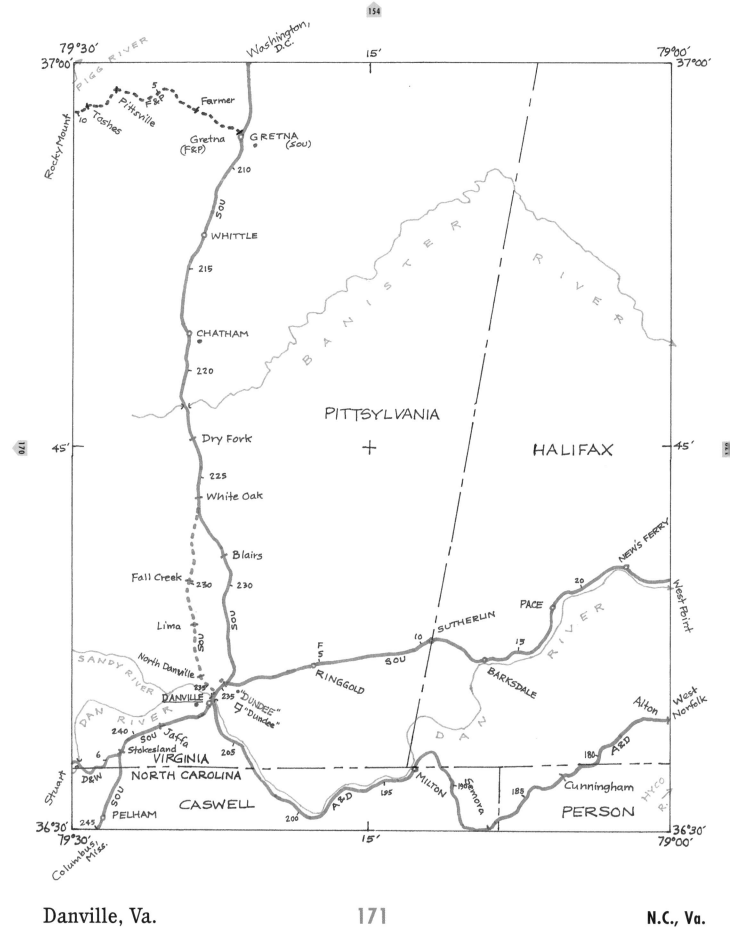

79°30'
37°00'

PIGG RIVER

Washington, D.C.

79°00'
37°00'

Rocky Mount

5
F&P
Farmer

Pittsville

Tashes

10

Gretna (F&P)

GRETNA (SOU)

210

SOU

WHITTLE

215

CHATHAM

220

BANISTER RIVER

PITTSYLVANIA
+

HALIFAX

170

45'

Dry Fork

225

White Oak

Blairs

Fall Creek   230   230

Lima

SOU   SOU

North Danville

SANDY RIVER

235

DANVILLE   235

DAN RIVER

240   SOU Jaffa

6   Stokesland   205

Stuart   D&W   VIRGINIA
NORTH CAROLINA

CASWELL

SOU   PELHAM

245

Columbus, Miss.

79°30'
36°30'

"DUNDEE"
"Dundee"

F
S

RINGGOLD

SOU

SUTHERLIN

10

PACE   20

15   RIVER

BARKSDALE

DAN

A&D   195

200   MILTON   190   Semora

185   Cunningham

PERSON

New's Ferry

West Point

Alton   West Norfolk

180   A&D

HYCO R.

45'   170

15'
36°30'
79°00'

Danville, Va.   171   N.C., Va.

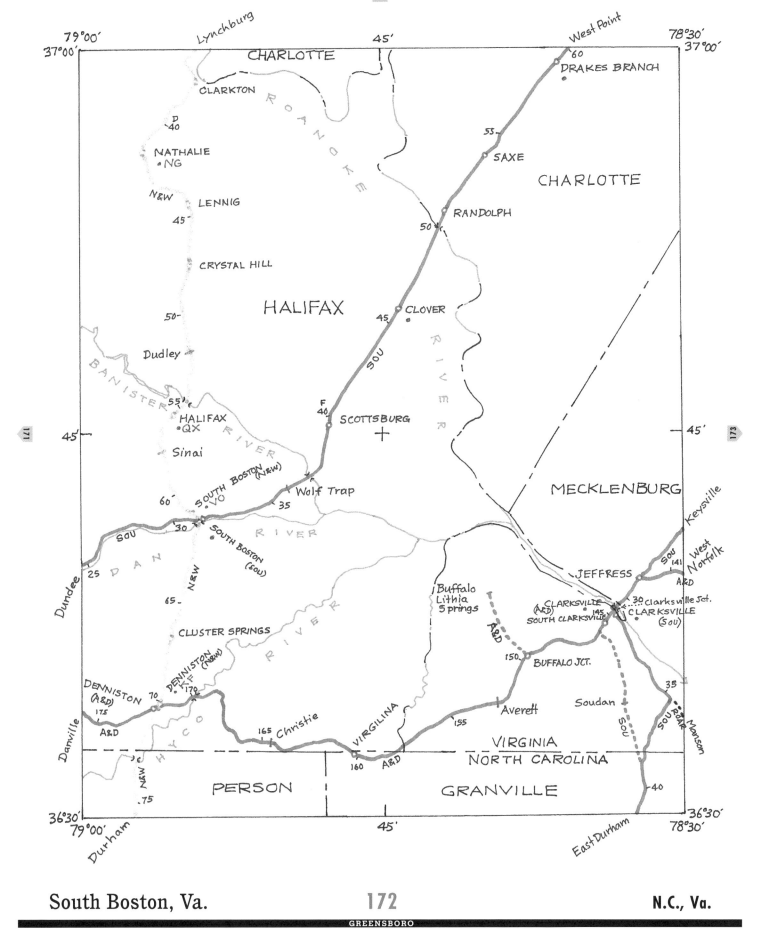

155

79°00'
37°00'

LYNCHBURG

CHARLOTTE

CLARKTON

D
40

NATHALIE
NG

N&W   LENNIG

45'

CRYSTAL HILL

50'

HALIFAX

Dudley

BANISTER RIVER

55'

HALIFAX
QX

Sinai

45'

60'

SOUTH BOSTON (N&W)
VO

30'

SOU

25

Dundee

DAN

NEW

65'

CLUSTER SPRINGS

DENNISTON
(N&W)
F
170

DENNISTON
(A&D)
175

70

A&D

Danville

NEW

75

36°30'
79°00'

Durham

PERSON

HYCO RIVER

165   Christie

160   A&D

45'

ROANOKE

45'

60

DRAKES BRANCH

55

SAXE

RANDOLPH

50

CLOVER
45'

SOU

RIVER

F
40   SCOTTSBURG

Wolf Trap

35

RIVER

Buffalo
Lithia
Springs

A&D

150   BUFFALO JCT.

Averett

155

VIRGILINA

VIRGINIA
NORTH CAROLINA

GRANVILLE

West Point

CHARLOTTE

MECKLENBURG

JEFFRESS

CLARKSVILLE
(A&D)
145
SOUTH CLARKSVILLE

Soudan

SOU

30   Clarksville Jct.
CLARKSVILLE
(SOU)

Keysville

SOU   141   West
Norfolk
A&D

35

SOU   R&R

Manson

40

East Durham

78°30'
37°00'

45'

MECKLENBURG

78°30'
36°30'

171

173

South Boston, Va.   172   N.C., Va.

GREENSBORO

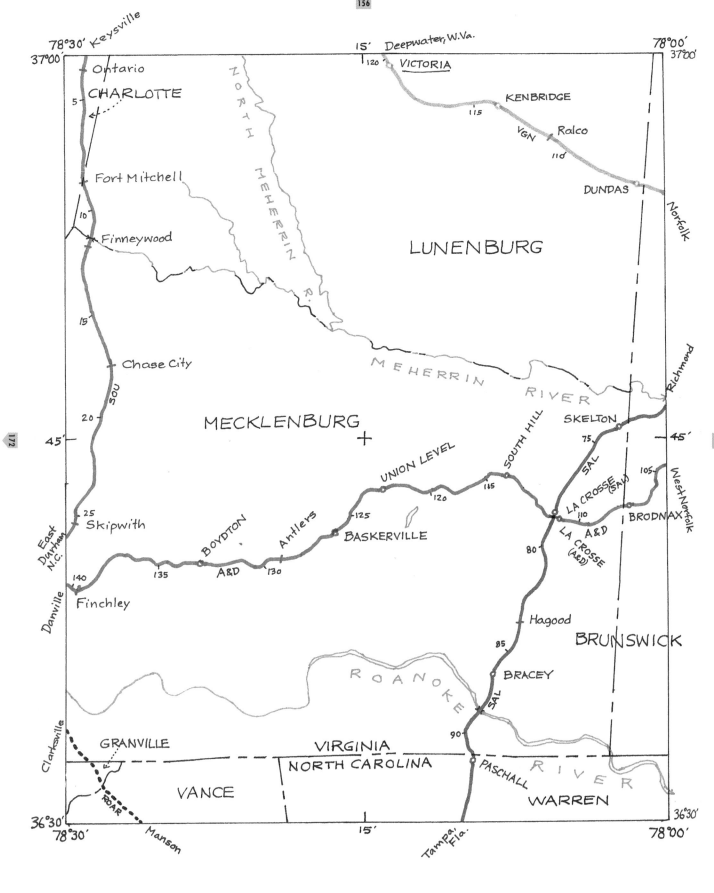

78°30' Keysville

37°00'

Ontario

CHARLOTTE

5

Fort Mitchell

10

Finneywood

15

SOU

20

45'

25 Skipwith

East Durham N.C.

140 Finchley

Danville

Clarksville

GRANVILLE

ROAR

36°30'
78°30'

Manson

15' Deepwater, W.Va.

120 VICTORIA

KENBRIDGE

115

VGN Ralco

110

DUNDAS

LUNENBURG

MEHERRIN R.

MEHERRIN RIVER

NORTH MEHERRIN R.

Chase City

MECKLENBURG +

UNION LEVEL

SOUTH HILL

SKELTON

75 SAL

45'

LA CROSSE (SAL)

105

West Norfolk

Richmond

Norfolk

78°00'
37°00'

120

115

125

BOYDTON Antlers

A&D 130

135

BASKERVILLE

LA CROSSE 110 A&D
(A&D)

80 BRODNAX

Hagood

85

BRUNSWICK

BRACEY

SAL

ROANOKE

90

VIRGINIA
NORTH CAROLINA

PASCHALL

RIVER

VANCE

WARREN

36°30'
78°00'

15'

Tampa, Fla.

Victoria, Va.   **173**   **N.C., Va.**

GREENSBORO

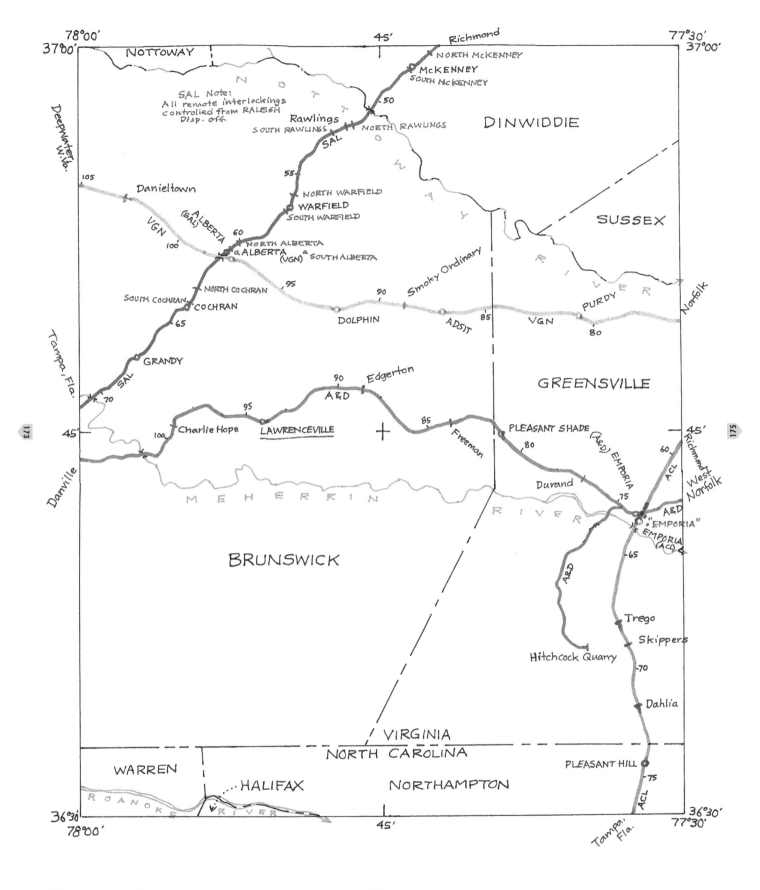

78°00′   37°00′

NOTTOWAY

45′   Richmond

NORTH McKENNEY
McKENNEY
SOUTH McKENNEY

77°30′   37°00′

SAL Note:
All remote interlockings
controlled from RALEIGH
Disp. off.

50

Rawlings
SOUTH RAWLINGS   SAL   NORTH RAWLINGS

DINWIDDIE

Deepwater, W. Va.

105

Danieltown

55

NORTH WARFIELD
WARFIELD
SOUTH WARFIELD

SUSSEX

VGN   ALBERTA (SAL)

100   60

NORTH ALBERTA
ALBERTA   SOUTH ALBERTA
(VGN)

Norfolk

95

SOUTH COCHRAN   NORTH COCHRAN

90   Smoky Ordinary

PURDY

COCHRAN

65

DOLPHIN   ADSIT   85   VGN   80

GRANDY

SAL   70

90   Edgerton
A & D

GREENSVILLE

Tampa, Fla.

95

45′

100   Charlie Hope   LAWRENCEVILLE

85   Freeman

PLEASANT SHADE (A&D) EMPORIA

60   ACL   Richmond
West Norfolk

45′

80

Durand

75

A&D
"EMPORIA"   A&D

Danville

MEHERRIN   RIVER

EMPORIA (ACL)

65

BRUNSWICK

A&D

Trego

Skippers

Hitchcock Quarry

70

Dahlia

VIRGINIA
NORTH CAROLINA

WARREN   HALIFAX   NORTHAMPTON

PLEASANT HILL

75

ROANOKE   RIVER

ACL

36°30′
78°00′   45′   Tampa, Fla.   36°30′
77°30′

Emporia, Va.   174   N.C., Va.

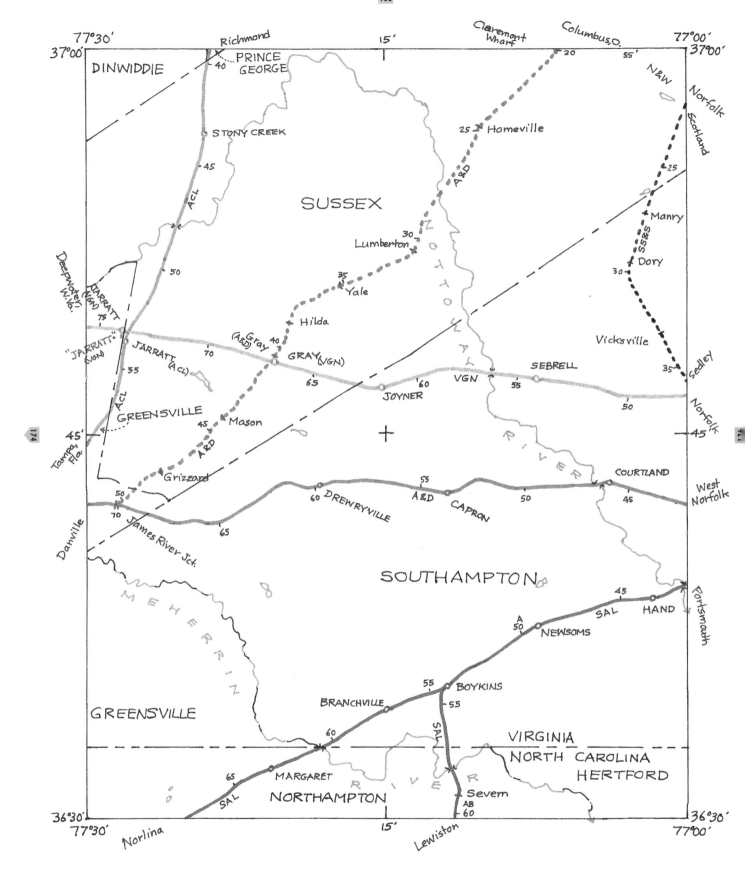

77°30' Richmond 15' Claremont Wharf Columbus, O. 77°00'
37°00' 37°00'

DINWIDDIE PRINCE GEORGE
40
20
N&W Norfolk
25 Homeville 55'
STONY CREEK
Scotland
45
SUSSEX
ACL
25
Manry
50
Lumberton 30
SS&S Dory
Deepwater, W.Va. Yale 35 30
JARRATT (VGN) 75 Hilda Vicksville
"JARRATT" (VGN) Gray (A&D) 40 GRAY (VGN) SEBRELL 35 Sedley
JARRATT (ACL) 70
55 65 JOYNER 60 VGN 55 Norfolk
GREENSVILLE 45 Mason 50
45' ACL 45'
Tampa, Fla. A&D
Grizzard COURTLAND West Norfolk
50 55
70 DREWRYVILLE A&D CAPRON 50 45
Danville James River Jct. 60 65
SOUTHAMPTON Portsmouth
45
SAL HAND
A 50 NEWSOMS
MEHERRIN
55 BOYKINS
GREENSVILLE BRANCHVILLE 55
SAL
VIRGINIA
60 NORTH CAROLINA
HERTFORD
65 MARGARET
SAL NORTHAMPTON Severn
36°30' AB 36°30'
77°30' Norlina 15' Lewiston 60 77°00'

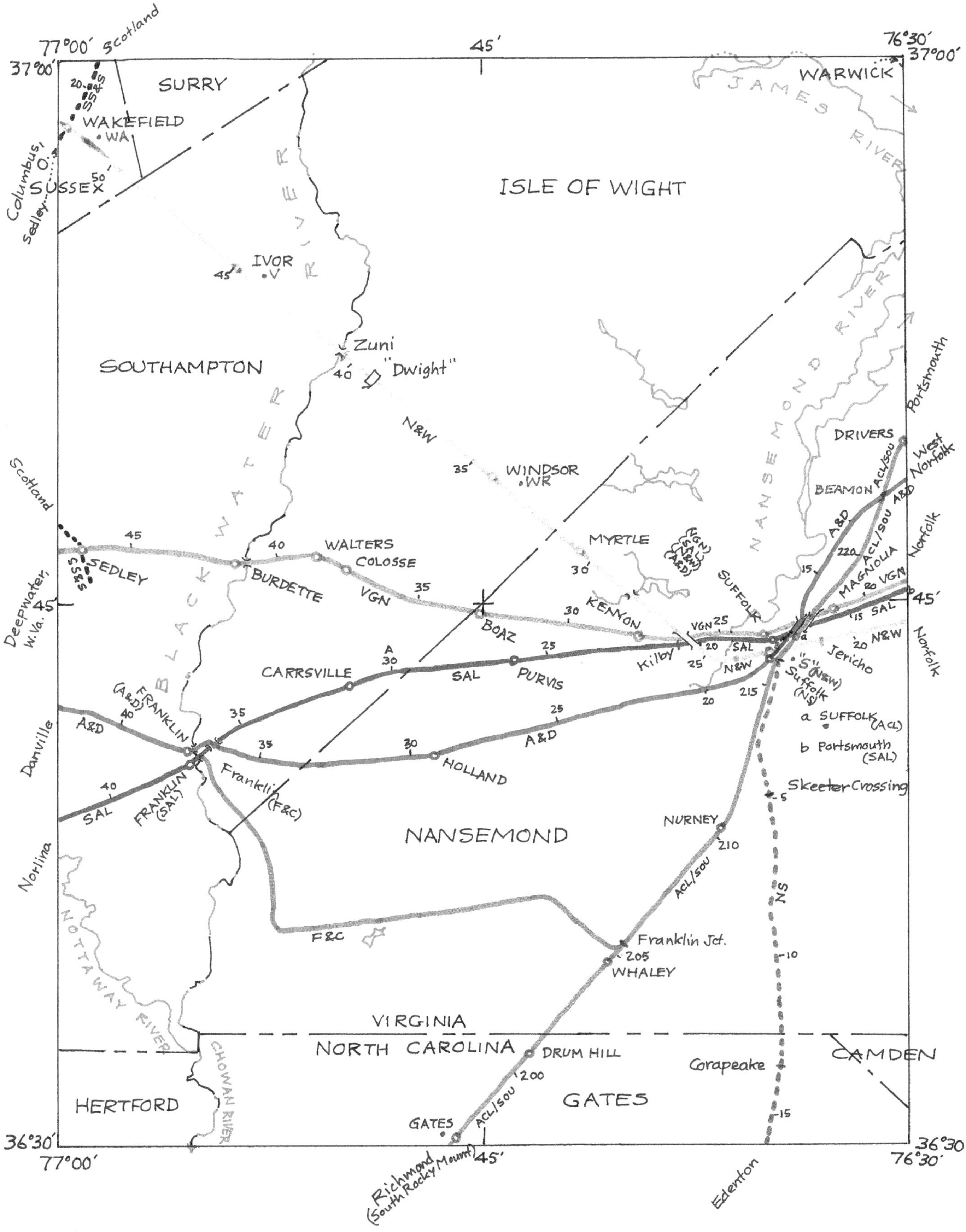

77°00' Scotland

45'

76°30'
37°00'

SURRY

WARWICK

JAMES RIVER

20 S&S
SS
WAKEFIELD
• WA

Columbus, O.
Sedley

SUSSEX 50'

ISLE OF WIGHT

45' IVOR
• V

SOUTHAMPTON

BLACK WATER RIVER

NANSEMOND RIVER

Zuni
40 "Dwight"

N&W

DRIVERS
Portsmouth

West
Norfolk

BEAMON
ACL/Sou
A&D

WINDSOR
35' WR

A&D

Norfolk

MYRTLE
(VGN)
(N&W)
(A&D)

220
ACL/Sou
MAGNOLIA
20 VGN
15

Scotland

SEDLEY
S&S
45

45 WALTERS
40 COLOSSE
VGN 35

30
KENYON

SUFFOLK

VGN 25
BOAZ 25
20 SAL
N&W
25'
20

N&W
45'
15 SAL
20 N&W

Jericho

Deepwater, W.Va.

BURDETTE

30
Kilby

"S"(N&W)
Suffolk
(NS)

215

a SUFFOLK (ACL)

CARRSVILLE
A 30
SAL
PURVIS 25

b Portsmouth
(SAL)

Danville
A&D
FRANKLIN
(A&D) 40

35
35

A&D

Skeeter Crossing
5

Norlina

FRANKLIN
(SAL)
40
SAL

Franklin
(F&C)

30
HOLLAND 25

NURNEY
ACL/Sou
210

NS

NANSEMOND

NOTTAWAY RIVER

F&C

10

CHOWAN RIVER

Franklin Jct.
205
WHALEY

VIRGINIA
NORTH CAROLINA

CAMDEN

HERTFORD

DRUM HILL
200

Corapeake

GATES

15

36°30'
77°00'

GATES

ACL/Sou
45'

Richmond
(South Rocky Mount)

Ederton

76°30'
36°30'

Suffolk, Va.   176   N.C., Va.

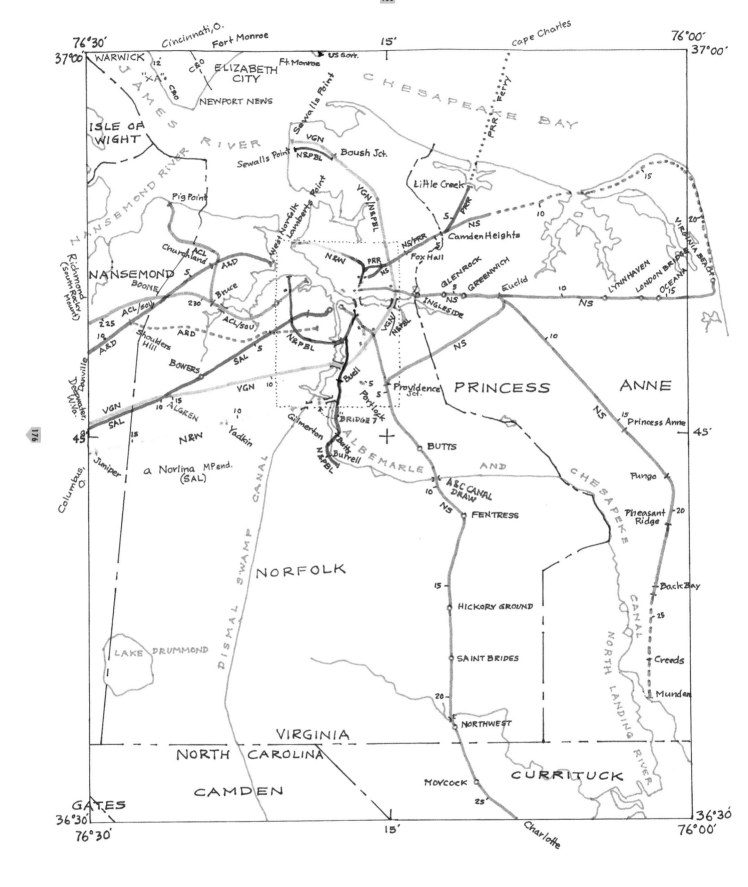

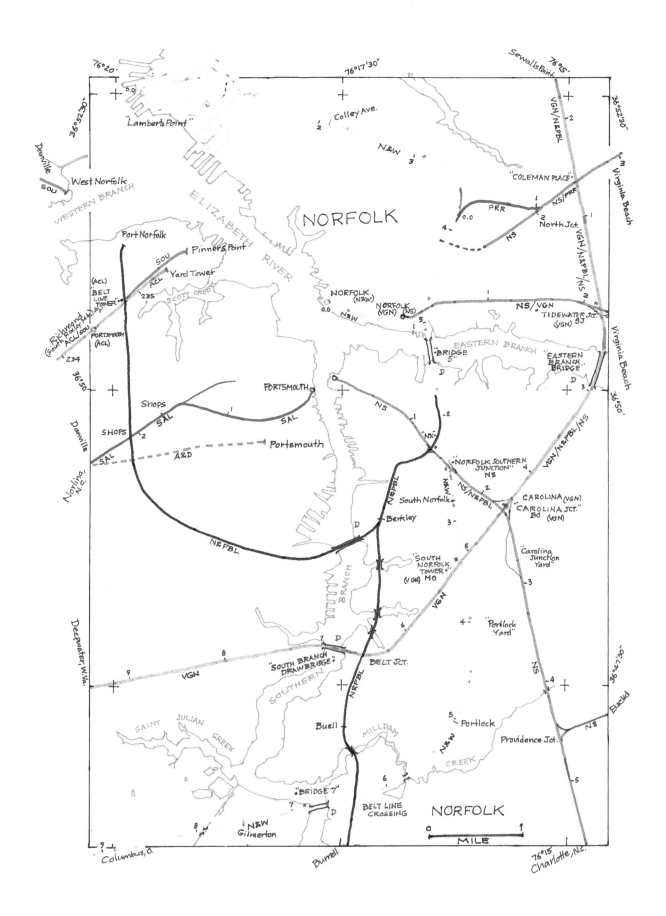

177A Norfolk, Va.

Appendix

RAILROADS IN THE ATLAS

The abbreviations "Rly." and "RR" are used for "Railway" and "Railroad," respectively.

A&D: Atlantic & Danville Rly.

A&S: Addison & Susquehanna RR

A&V: Atlantic & Yadkin Rly.

AC: Atlantic City RR

ACL: Atlantic Coast Line RR

A-J: Artemus-Jellico RR

ALQS: Aliquippa & Southern RR

AT: Allentown Terminal RR

B&E: Baltimore & Eastern RR

B&H: Bath & Hammondsport RR

B&LE: Bessemer & Lake Erie RR

B&M: Brownstone & Middletown RR

B&O: Baltimore & Ohio RR

B&S: Buffalo & Susquehanna RR

BC&G: Buffalo Creek & Gauley RR

BCICE Co.: Bear Creek Ice Company

BEDT: Brooklyn Eastern District Terminal RR

BFC: Bellefonte Central RR

BLA: Baltimore & Annapolis RR

BRRR: Bare Rock RR

BS&C: Big Sandy & Cumberland RR

BSRR: Bloomsburg & Sullivan RR

BWC: Benwood & Wheeling Connecting Rly.

C&KV: Cairo & Kanawha Valley Rly.

C&I: Cambria & Indiana RR

C&BL: Conemaugh & Black Lick RR

C&O: Chesapeake & Ohio Rly.

C&PA: Cumberland & Pennsylvania RR

CC&O: Clinchfield RR

CCK: Campbells Creek RR

CHH: Cheswick & Harmar RR

CHR: Chestnut Ridge Rly.

CHW: Chesapeake Western Rly.

CLAR: Clarion River Rly.

CNJ: Central RR Co. of New Jersey

CNYW: Central New York & Western RR

CPA: Coudersport & Port Allegheny RR

CRB&L: Cherry River Boom & Lumber Co's. RR

CRIV: Castleman River RR

CRPA: Central RR Co. of Pennsylvania

CT&D: Cherry Tree & Dixonville RR

CTN: Canton RR

CURB: Curtis Bay RR

CWL: Cornwall RR

CWV&S: Central West Virginia & Southern RR

D&H: Delaware & Hudson RR

D&N: Delaware & Northern RR

D&W: Danville & Western Rly.

DL&W: Delaware, Lackawanna & Western RR

DT&I: Detroit, Toledo & Ironton RR

DV: Delaware Valley Rly.

E&M: Etna & Montrose RR

EB: East Berlin RR

EBT: East Broad Top RR & Coal Co.

EEC: East Erie Commercial RR

EK: Eastern Kentucky Rly.

EKS: Eastern Kentucky Southern Rly.

EM: Eagles Mere RR

EMIT: Emmitsburg RR

ERIE: Erie RR

EW: East Washington Rly.

F&C: Franklin & Carolina RR

F&P: Franklin & Pittsylvania Rly.

FV: Federal Valley RR

H&BT: Huntingdon & Broad Top Mountain RR & Coal Co.

H&M: Hudson & Manhattan RR

HMR: Hoboken Manufacturers RR

HS: Harrisville Southern RR

HV: Hickory Valley RR

INT: Interstate RR

IRN: Ironton RR

J&SC: Johnstown & Stony Creek RR

JW&NW: Jamestown, Westfield & Northwestern RR

K&E: Kane & Elk RR

KC: Kanawha Central Rly.

KC&NW: Kelly's Creek & Northwestern RR

209

KEC: Kelly's Creek RR
KV: Kishacoquillas Valley RR

L: Laurel Rly.
L&HR: Lehigh & Hudson River RR
L&N: Louisville & Nashville RR
L&WV: Lackawanna & Wyoming Valley RR
LE&E: Lake Erie & Eastern RR
LEF&C: Lake Erie, Franklin & Clarion RR
LGV: Ligonier Valley RR
LI: Long Island RR
LNE: Lehigh & New England RR
LO&S: Lancaster, Oxford & Southern RR
LV: Lehigh Valley RR

M&PA: Maryland & Pennsylvania RR
M&RV: Marion & Rye Valley Rly.
M&U: Middletown & Unionville RR
MC&C: Marietta, Columbus & Cleveland RR
MCK: Middle Creek RR
MF: Middle Fork RR
MGAC: Monongahela Connecting RR
MHM: Mount Hope Mineral RR
MKC: McKeesport Connecting RR
MON: Monongahela Rly.
MT&E: Morristown & Erie RR
MTJK&R: Mount Jewett, Kinzua & Riterville RR
MTR: Montour RR

N&PBL: Norfolk & Portsmouth Belt Line RR
N&W: Norfolk & Western Rly.
NB: Northampton & Bath RR
NEA: Nelson & Albermarle Rly.
NF&G: Nicholas, Fayette & Greenbrier RR
NH: New York, New Haven & Hartford RR
NH&D: New Haven & Dunbar RR
NJ&NY: New Jersey & New York RR
NKP: New York, Chicago & St. Louis RR
NNJ: Northern RR of New Jersey

NS: Norfolk Southern Rly.
NY&GL: New York & Greenwood Lake Rly.
NY&LB: New York & Long Branch RR
NY&P: New York & Pennsylvania Rly.
NYC: New York Central RR
NYCN: New York Connecting RR
NYD: New York Dock Rly.
NYO&W: New York, Ontario & Western Rly.
NYS&W: New York, Susquehanna & Western RR
NYW&B: New York, Westchester & Boston Rly.

OR&W: Ohio River & Western Rly.

P&LE: Pittsburgh & Lake Erie RR
P&NJ: Pennsylvania & New Jersey RR
P&OV: Pittsburgh & Ohio Valley Rly.
P&S: Pittsburgh & Susquehanna RR
P&WN: Philadelphia & Western Rly.
P&WV: Pittsburgh & West Virginia Rly.
PAA: Pennsylvania & Atlantic RR
PAM: Pittsburgh, Allegheny & McKees Rocks RR
PBL: Philadelphia Belt Line RR
PBNE: Philadelphia, Bethlehem & New England RR
PBR: Patapsco & Back Rivers RR
PCY: Pittsburgh, Chartiers & Youghiogheny Rly.
PE Co.: Potomac Edison Co.
PL&W: Pittsburgh, Lisbon & Western RR
PR: Preston RR
PRAT: Prattsburgh Rly.
PRR: Pennsylvania RR
P-RSL: Pennsylvania-Reading Seashore Lines
PS: Pittsburgh & Shawmut RR
PS&N: Pittsburgh, Shawmut & Northern RR

QB: Quakertown & Bethlehem RR

R&FC: Reynoldsville & Falls Creek RR
R&S: Rowlesburg & Southern RR
RARR: Richmond & Rappahannock River RR
RDG: Reading Co.
RF&P: Richmond, Fredericksburg & Potomac RR
ROAR: Roanoke River Rly.
RR: Raritan River RR
RV: Rahway Valley RR

S&NY: Susquehanna & New York RR
S&T: Sheffield & Tionesta Rly.
SAL: Seaboard Air Line Rly.
SBK: South Brooklyn Rly.
SCM: Stroud's Creek & Muddlety RR
SCO: Scootac Rly.
SE&P: South Easton & Phillipsburg RR
SH: Sharpsville RR
SIRT: Staten Island Rapid Transit Rly.
SMR: Sterling Mountain Rly.
SNJ: Southern New Jersey RR
SOU: Southern Rly.
SR&W: Susquehanna River & Western RR
SRC: Strasburg RR
SS&S: Surry, Sussex & Southampton Rly.
STRT: Stewartstown RR
SV&SL: South Vandalia & State Line RR

TIV: Tionesta Valley Rly.
TUCK: Tuckerton RR
TV: Tuscarora Valley RR
TWRR: Tidewater & Western RR

UMP: Upper Merion & Plymouth RR
UNF: Ursina & North Fork Rly.
UNI: Unity Rlys.
URR: Union RR
US ARMY: United States Army

US GOV'T: United States Government
US NAVY: United States Navy
UTR: Union Transportation Co.
UV: Unadilla Valley Rly.

V&SW: Virginia & Southwestern Rly.
VAC: Virginia-Carolina Rly.
VAL: Valley RR
VBR: Virginia Blue Ridge Rly.
VC: Virginia Central Rly.
VGN: Virginian Rly.
VYRR: Valley River RR

W&B: Wellsville & Buffalo RR
W&LE: Wheeling & Lake Erie Rly.
W&NB: Williamsport & North Branch Rly.
W&NO: Wharton & Northern RR
W&OD: Washington & Old Dominion Rly.
W&W: Winchester & Western RR
WAL: Western Allegheny RR
WATC: Washington Terminal Co.
WAW: Waynesburg & Washington RR
WBC: Wilkes-Barre Connecting RR
WBE: Wilkes-Barre & Eastern RR
WC&PC: Wellsville, Coudersport & Pine Creek RR
WJ&S: West Jersey & Seashore RR
WM: Western Maryland Rly.
WNF: Winfield RR
WNFR: Winifrede RR
WNY&P: Western New York & Pennsylvania Rly.
WSH: White Sulphur & Huntersville RR
WT: Wheeling Terminal Rly.
WVM: West Virginia Midland Rly.
WVN: West Virginia Northern RR
WVS: West Virginia & Southern RR

Y&S: Youngstown & Southern Rly.
YOR: Youngstown & Ohio River RR

Notes on the Maps

1. Erie, Pa.: Location of the General Electric Company's Erie locomotive works. In 1946, diesel and electric locomotives were built here for both foreign and domestic railroads. The East Erie Commercial Railroad, created in 1910, provided a test track for these new locomotives.

2. Westfield, N.Y.: Junction between the New York Central Railroad Buffalo-to-Chicago main line at Westfield and the Jamestown, Westfield and Northwestern electric interurban railroad. This electric interurban line ran south along Chautauqua Lake, some 30 miles to the city of Jamestown, N.Y., where a connection could be made with the Erie Railroad main line.

3. Dunkirk, N.Y.: Dunkirk, on Lake Erie, was the western terminus of the original Erie Railroad, built to a 6-foot track gauge in 1851, westward from Piermont Dock on the Hudson River, just north of New York City. This 6-foot gauge, later converted to standard gauge, meant that bridge clearances on the Erie were higher and wider than those of other railroads.

4. Salamanca, N.Y.: Salamanca was the location of the zero milepost of one of the group of original railroad companies which were later merged to form the greater Erie. This railroad, the Atlantic and Great Western, was completed in 1864 and ran from Salamanca, N.Y., to Dayton, Ohio, some 390 miles to the west.

5. Olean, N.Y.: Shown on this map is the "River Line" of the Erie Railroad. Built in 1910, it ran between River Junction and Cuba Junction (SW). Situated on this line was the 3,119-foot Belfast Viaduct. The River Line permitted freight trains to bypass two severe 1 percent grades, at Tip Top and at Summit, on the main line between Hornell and Cuba Junction.

6. Hornell, N.Y.: Called the "Heart of the Erie," Hornell was a crew change point and also the location of a large freight yard and the steam locomotive shops of the Erie Railroad.

7. Corning, N.Y.: Both the Erie Railroad main line and New York Central branch to Williamsport ran through the streets of downtown Corning (SE).

8. Elmira, N.Y.: Four Class I railroads ran through Elmira in 1946. Today, only the Norfolk Southern runs through Elmira, using the tracks of the former Erie.

9. Owego, N.Y.: The Lackawanna Railroad used a switchback (NW) to reach Elmira. Route 17 now runs in what was once the Lackawanna main line roadbed.

10. Binghamton, N.Y.: The junction (SW) of two east-west major railroads—the Lackawanna and the Erie—and the north-south "bridge" railroad,—the Delaware & Hudson Railroad. (See also detail map 10A, **Binghamton**.)

11. Sidney, N.Y.: Oneonta (NE) was a yard and crew change point and a locomotive and car shop on the Delaware & Hudson Railroad.

12. Ashtabula, Ohio: Ashtabula (NW) and Conneaut (NE) were both lake ports for iron ore bound for Pittsburgh and Mahoning Valley steel mills.

13. Meadville, Pa.: Northeast of Meadville, at Cambridge Springs (NE), an Erie Railroad monument marks the halfway point between New York and Chicago.

14. Corry, Pa.: Titusville (SE), the site of the first U.S. oil wells, was served by both the New York Central and the Pennsylvania Railroads.

15. Warren, Pa.: During 1906–1908, the Erie Railroad built a low-grade line through Lottsville (NW) to avoid the severe westbound grade via the original Brady Lake route (Crist, "Erie Memories," p. 39).

16. **Kinzua, Pa.:** In 1900 the Erie Railroad built a steel viaduct at Kinzua (NE) to replace a wrought iron bridge; this viaduct was reportedly the longest and highest in the world, being 2,053 feet long and 301 feet high. (Crist, "Erie Memories," p. 56).

17. **Emporium, Pa.:** The summit of the Pennsylvania Railroad's Buffalo-to-Harrisburg route was located at Keating Summit (SE).

18. **Galeton, Pa.:** Galeton (SE) was the junction of two Baltimore & Ohio Railroad branch lines, which extended north into New York state to interchange with the Erie Railroad at Wellsville and Addison.

19. **Wellsboro, Pa.:** The Erie and New York Central Railroads shared their separate lines with each other.

20. **Sayre, Pa.:** The principal shops of the Lehigh Valley Railroad were located at Sayre (NE), which was also a crew change point.

21. **Towanda, Pa.:** A distinctive Lehigh Valley Railroad main line coaling station was located just west of Towanda station (NW).

22. **Susquehanna, Pa.:** Three famous major viaducts are shown on this map: Kingsley (NW) and Tunkhannock (SW) on the Lackawanna Railroad and Starrucca (NE) on the Erie Railroad.

23. **Honesdale, Pa.:** The principal junction of the New York Ontario and Western was at Cadosia (NW).

24. **Monticello, N.Y.:** The present-day Route 17 (now Interstate 86) was built on or along the abandoned railroad grade of the New York, Ontario and Western Railway from Ferndale (NE) to Cook's Falls (NW) and westward to Hancock (map 23, NW).

25. **Youngstown, Ohio:** An extraordinary tangle of the four major eastern railroads—the B&O, the Erie, the Pennsylvania, and the New York Central—served the mighty steel mills of the Mahoning River Valley (SE). (See also detail map 25A, **Youngstown.**)

26. **Greenville, Pa.:** The heavily and beautifully engineered line of the Bessemer and Lake Erie Railroad, carrying ore south to Pittsburgh and coal north to Conneaut and Erie, crosses this map from NW to SE.

27. **Oil City, Pa.:** The Baltimore & Ohio Railroad (originally the Buffalo, Rochester and Pittsburgh Railway) used a switchback to overcome the grades at Foxburg (SE).

28. **Brookville, Pa.:** The historic county seat of Brookville (SE) was a junction on the New York Central Railroad line connecting the eastern and western Pennsylvania coal fields with the lake port of Ashtabula, Ohio.

29. **Du Bois, Pa.:** Brockway (SW) was the western end of the soon-to-be-abandoned Pittsburgh, Shawmut and Northern Railroad, which ran for 145 steep and twisting miles north to the little New York town of Wayland, and an interchange with the Lackawanna Railroad.

30. **Clearfield, Pa.:** Driftwood (NE) was a lonely Pennsylvania Railroad junction where the "low-grade" branch from Red Bank, Pa., met the Buffalo-to-Harrisburg main line.

31. **Renovo, Pa.:** Renovo (NW), from the Italian for "to renew," was both a major shop and a crew change point on the Buffalo-to-Harrisburg main line of the Pennsylvania Railroad.

32. **Williamsport, Pa.:** Williamsport (SE) was a major interchange between the New York Central Railroad and the Reading Company, providing competition to the Pennsylvania Railroad.

33. **Milton, Pa.:** The main lines of the Pennsylvania Railroad and the Reading Company crossed each other three times in the 25 miles between Milton (SW) and Williamsport (SW).

34. **Berwick, Pa.:** The Pennsylvania Railroad line along the south bank of the Susquehanna River survives today as part of the Canadian Pacific Railway's north-south east coast link between the Norfolk Southern at Sunbury, Pennsylvania, and Canada and via the Guilford Rail System, New England.

35. **Scranton, Pa.:** This map and its four detail maps show the northern heart of the eastern Pennsylvania anthracite coal fields. Filled with mines and breakers, it was served by no fewer

than seven Class I railroads: the Central Railroad of New Jersey, the Delaware & Hudson, the Lackawanna, the Erie, the New York, Ontario & Western, the Lehigh Valley, and the Pennsylvania.

35A. Wilkes-Barre, Pa.: The unique "Ashley Planes" of the Central Railroad of New Jersey were still in operation in 1946, using a cable-operated, inclined-plane railroad to move freight cars between Mountain Top and Ashley.

35B. Pittston, Pa.: The Lehigh Valley Railroad operated a low-grade freight cut-off around Wilkes-Barre to reach its Coxton Yard, west of Pittston, where crews were also changed.

35C. Scranton West, Pa.: Today, many stream locomotives are on display and some operate from the Steamtown Railroad Museum—part of the National Parks system. The museum is located opposite the former Lackawanna Railroad station in Scranton, on the site of that railroad's locomotive shops.

35D. Scranton East, Pa.: Steamtown offers steam-powered excursions east to Elmhurst through the deep gorge of Roaring Brook, along the Lackawanna Railroad main line.

36. Pocono, Pa.: The Lackawanna Railroad main line crossed over the summit of the Pocono mountains, using helper engines based at Gravel Place (SE) which were cut off at Lehigh Summit tower (NW), just west of Gouldsboro.

37. Port Jervis, N.Y.: Port Jervis (NE) was both a crew change point and a classification yard on the Erie Railroad. Here, cars destined for New England were separated from those headed for New York.

38. Middletown, N.Y.: At Maybrook (NE), the New Haven Railroad's hump yard (with hand brakemen for each car!) classified cars at the principal freight gateway to and from New England.

39. Harmon, N.Y.: Piermont Dock (SW) was the original historic starting point of the Erie Railroad, which was built westward to Lake Erie at Dunkirk, N.Y.

40. East Liverpool, Ohio: The Pennsylvania Railroad used a portion of its Rochester, Pa.-to-Cleveland line as a low-grade freight bypass west to just east of Canton at Louisville, to avoid the grades and curves of their main line through Columbiana and Leetonia.

41. New Castle, Pa.: The Pennsylvania Railroad's immense classification yard for the Pittsburgh area was located at Conway, just east (south) of Rochester (SW).

42. Butler, Pa.: The Pennsylvania Railroad's Conpitt Jct. (SE) to Pittsburgh (Federal Street; SW) line provided a low-grade freight northern bypass around Pittsburgh.

43. Indiana, Pa.: Indiana is the hometown of American actor Jimmy Stewart. This map shows the western part of an extensive bituminous coal field served by the Baltimore & Ohio, the New York Central, and the Pennsylvania Railroads.

44. Punxsutawney, Pa.: An extensive tangle of coal mine branches, dominated by the Pennsylvania Railroad operating out of Cresson, located on its main line west of the famous Horse Shoe Curve.

45. Altoona, Pa.: East of the locomotive shops of Altoona (SW), the "heart" of the Pennsylvania Railroad, the seven miles between "Spruce" and "Forge" towers, was the only three- track portion of the four-track main line between Harrisburg and Pittsburgh.

46. Lewistown, Pa.: Lewistown station (SE) is now the archives of the Pennsylvania Railroad Technical & Historical Society.

47. Mifflin, Pa.: Just west of Mifflin (SW), on the Pennsylvania Railroad main line, stood the massive Denholm Coal Dock. Long since retired, its stone abutments still stand.

48. Sunbury, Pa.: Most of the Pennsylvania Railroad's steam locomotives now at the state museum at Strasburg were originally stored for a time at the large roundhouse at Northumberland (NW). For a time in recent history, the site of "Kase" tower (NW) became the western end of the Guilford Rail System freight main line from Mattawaumkeag, Maine, just over 750 miles to the northeast!

49. Pottsville, Pa.: This map dramatically shows the southwestern part of the anthracite coal

fields in the vicinity of Pottsville (SE), Mount Carmel (NW) and Shenandoah (NE). (See also detail maps 49A, **Mt. Carmel**; 49B, **Minersville**; 49C, **Shenandoah**; and 49D, **Haucks**.)

50. **Hazelton, Pa.:** This is the southern part of the anthracite coal fields, dominated by the Central New Jersey, the Lehigh Valley, and the Lehigh & New England Railroads. All three railroads came together to cross Blue Mountain through Lehigh Gap. (See also detail maps 50A, **Hazelton**; 50B, **Ashmore**; and 50C, **Hauto**.)

51. **Allentown, Pa.:** The complex trackage in the center of the map represents the large cement-producing industry, served mainly by the Lackawanna Railroad and the Lehigh & New England Railroad. (See also detail maps 51A, **Easton-Phillipsburg**; 51B, **Allentown**; 51C, **Catasauqua**; 51D, **Coplay-Northampton**; and 51E, **Bethlehem**.)

52. **Somerville, N.J.:** The Lackawanna Railroad built its magnificent "New Jersey Cut-Off" (NW) between Port Morris tower (NE) and Slateford Jct. (map 51, NE). Completed in 1911, it shortened the distance from Hoboken to Buffalo by 11 miles, an achievement that was celebrated by renumbering all the mileposts west of Slateford Jct. to Buffalo.

53. **Newark, N.J.:** This is the most complex map in the atlas, with over 430 named places. Nine major railroads converged in northern New Jersey, the western gateway to New York City. However, only the Pennsylvania Railroad provided direct all-rail service to and through New York City. (See also detail maps 53A, **Newark**; and 53B, **Jersey City**.)

54. **New York City, N.Y.:** The complex rail network east of the Hudson River, with the Long Island Railroad, controlled by the Pennsylvania Railroad, exclusively served Long Island. The New York Connecting Railroad, jointly owned by the Pennsylvania and the New Haven Railroads, provided the key connecting link across the Hell Gate Bridge to New England. (See also detail maps 54A, **The Bronx**; 54B, **Sunnyside**; 54C, **Fresh Pond**; and 54D, **Jamaica**.)

55. **Wheeling, W.Va.:** Wheeling was the historic western terminus of the original Baltimore & Ohio Railroad main line, 379 miles west of Baltimore. (See also detail map 55A, **Wheeling**.)

56. **Carnegie, Pa.:** The hilly western approaches to Pittsburgh were traversed by the Pennsylvania's main line to St. Louis and by the Pittsburgh and West Virginia Railway (part of the "Alphabet Route"), with its many tunnels. In 1946 the P&WV still reached across the Monongahela River to its own stub-end passenger terminal, just east of the so-called "Golden Triangle." (See also detail map 56A, **Carnegie**.)

57. **Pittsburgh, Pa.:** This is the second most complex map in this volume, with nearly 350 place names. The Pennsylvania Railroad is clearly dominant; however, the Pittsburgh & West Virginia Railway, which was a link in the "Alphabet Route," crosses from Bruceton (NW) to Connellsville (SE), where an end-on junction was made with the Western Maryland Railway. (See also detail maps 57A, **Pittsburgh**; 57B, **Port Perry**; and 57C, **Connellsville**.)

58. **Latrobe, Pa.:** The Pennsylvania Railroad main line followed the Conemaugh River through the famous gorge known as the "Packsaddle," just beyond the Conpitt Jct. signal tower.

59. **Johnstown, Pa.:** The Pennsylvania Railroad main line traverses their summit tunnels at Gallitzin (NE). Nearby, the New Portage Branch of the Pennsylvania included a curve called the "Muleshoe," a smaller version of the "Horseshoe Curve" (map 60, NW).

60. **Hollidaysburg, Pa.:** The world-famous "Horseshoe Curve" of the Pennsylvania Railroad is shown in the very NW corner. Two narrow-gauge railroads, the East Broad Top and the Huntingdon & Broad Top Mountain Railroad & Coal Co., still had extensive operations in 1946.

61. **Shippensburg, Pa.:** The East Broad Top Railroad still operates as a tourist railroad at Orbisonia (SW). Another "Alphabet Route" junction was located at Shippensburg (SE), where the Western Maryland connected to the Reading.

62. **Carlisle, Pa.:** The scenic bend in the Susquehanna River at Duncannon (NE) prompted the Pennsylvania Railroad to name the signal tower here "View."

63. **Harrisburg, Pa.:** The major eastern junction of the Pennsylvania Railroad had its major

classification yard at Enola on the west bank of the Susquehanna River, and Harrisburg yard was on the east bank. Crews operated to and from six radiating lines, with both eastward lines having been electrified in the late 1930s.

64. **Lancaster, Pa.:** The name "Cork" for the interlocking station at Lancaster (SW) followed a Pennsylvania Railroad policy of short, clear names for towers. They often honored major nearby shippers—in this case, the Armstrong Cork Company.

65. **Reading, Pa.:** The "heart" of the Reading Company, with its large system locomotive shop and its unique "Reading Outer Station," was set in the middle of a wye junction. (See also detail map 65A, **Frazer**.)

66. **North Philadelphia, Pa.:** The famous Pennsylvania Railroad Main Line ran to Paoli (SW) and on to Pittsburgh, along with the extensive northern Philadelphia suburban lines and branches of the Reading Company (SE, NE). The Trenton Cut-Off of the Pennsylvania crosses from Glen Loch (SW) to Morrisville (SE). (See also detail maps 66A, **Norristown**; and 66B, **Manayunk**.)

67. **Trenton, N.J.:** The first railroad between New York and Philadelphia, the Camden & Amboy Railroad, was still active in 1946 as the Pennsylvania Railroad line from Camden (SW) to South Amboy (NE).

68. **Freehold, N.J.:** The New York & Long Branch Railroad represents the joint operation of the Central of New Jersey and the Pennsylvania Railroads from the Raritan River (NW) to Bay Head Junction (SE). This line saw the final regular steam operations of the famous Pennsylvania K-4s class Pacific steam locomotives.

69. **Long Branch, N.J.:** This is the easternmost point of the six states in this volume, where the Central of New Jersey Railroad ran close to the Atlantic Ocean shore.

70. **Barnesville, Ohio:** The three-foot narrow gauge Ohio River and Western Railway, which ended its days as part of the vast Pennsylvania Railroad, crosses from east to west.

71. **Moundsville, W.Va.:** The western part of the original Baltimore & Ohio Railroad main line turned northwest to Wheeling and missed the Pennsylvania state line by less than one mile.

72. **Waynesburg, W.Va.:** The Waynesburg and Washington Railroad was part of the Pennsylvania system and, until 1944, was three-foot narrow gauge.

73. **Uniontown, Pa.:** Southwestern Pennsylvania bituminous coal fields were served by the Baltimore & Ohio Railroad, the Monongahela Railway and the Pennsylvania Railroad. (See also detail map 73A, **Bowest** [with detail maps 57B and 57C].)

74. **Confluence, Pa.:** For nearly 40 miles, from Sand Patch to Connellsville, the Baltimore & Ohio Railroad and the Western Maryland Railway paralleled each other on opposite banks of the Youghiogheny and Casselman Rivers—B&O trunk line vs. WM "Alphabet Route."

75. **Cumberland, Pa.:** The principal eastern junction city of the Baltimore & Ohio Railroad had a large shop and classification yard; the latter was enlarged in the mid 1950s. (See also detail map 75A, **Frostburg** [with detail map 65A].)

76. **Magnolia, W.Va.:** The Baltimore & Ohio Railroad's "Magnolia Cut-Off," built in 1914, provided a more direct alignment between Orleans Road (SW) and Okonoko (map 75 SE) and was six miles shorter than the main line, which followed the ox-bow curves of the Potomac River.

77. **Hagerstown, Md.:** Here the vast, coal-hauling Norfolk and Western Railway system reached its northeasternmost point, connecting primarily with the Pennsylvania Railroad.

78. **Gettysburg, Pa.:** The Reading Company once had a branch south from Gettysburg to Round Top, the focus of the second day of the three-day Battle of Gettysburg.

79. **York, Pa.:** President Abraham Lincoln rode north from Washington, D.C., to Hanover Junction (NW) on what was then the Northern Central Railway, and then turned west to Gettysburg, where he delivered his simple but magnificent *Gettysburg Address.*

80. **Perryville, Md.:** This is the location of two major rail bridges across the Susquehanna River and the junction of the single-track, electrified "Port Road" of the Pennsylvania Railroad, which runs along the east bank of that river to Harrisburg.

81. **Wilmington, Del.:** The Pennsylvania Railroad overhauled and serviced its electric locomotive and multiple-unit cars at Wilmington Shops, north of Wilmington station near Landlith. (See also detail map 81A, **Wilmington**.)

82. **Philadelphia, Pa.:** The Pennsylvania–Reading Seashore Lines were created in 1933 when both railroads unified their operations between Camden and Atlantic City and between Cape May and Bridgeton. Of special interest (on detail map 82A) are the complex Pennsylvania Railroad's track layout and junction around "Zoo" tower in west Philadelphia. (See also detail maps 82A, **Philadelphia**; and 82B, **Camden**.)

83. **Winslow Jct., N.J.:** Another, though smaller, "flying and burrowing" junction is found at Winslow Jct., where Atlantic City and Cape May lines joined both the ex–West Jersey and Seashore Railroad (PRR) and the ex–Atlantic City Railroad (RDG). Winslow Jct. also had dual connections to the Central of New Jersey (CNJ).

84. **Whitings, N.J.:** Philadelphia seashore vacationers reached the northern New Jersey shore via the Pennsylvania & Atlantic Railroad, controlled by the Pennsylvania Railroad.

85. **Athens, Ohio:** In January 1953, when ex-President Harry S Truman returned to Independence, Missouri, he traveled west on the Baltimore & Ohio Railroad through Athens (NE).

86. **Parkersburg, W.Va.:** At Parkersburg, the Baltimore & Ohio Railroad main line west to Cincinnati and St. Louis crossed the Ohio River at the only railroad bridge in the 175 miles between Wheeling, West Virginia (Bellaire), and Galliopolis, Ohio.

87. **Marietta, W.Va.:** On the Baltimore & Ohio Railroad main line through Cairo (SE), there are no fewer than 14 tunnels in 35 miles!

88. **West Union, W.Va.:** There are 7 tunnels in 32 miles of the Baltimore & Ohio Railroad main line.

89. **Clarksburg, W.Va.:** Nearly all of this complex network of lines are the Baltimore & Ohio Railroad, except for the isolated branches of the Western Maryland (NW), which are reached by trackage rights over the B&O from Bowest Jct. near Connellsville, Pennsylvania.

90. **Kingwood, W.Va.:** In September 1984, on an auto trip all the way from Connecticut to Grafton, West Virginia, I could not find the right dirt road which led to "Q" tower at Hardman, West Virginia, on the Baltimore & Ohio Railroad!

91. **Oakland, Md.:** At Altamont, Md., the Baltimore & Ohio Railroad reached the summit of the Alleghany Mountains, at an elevation of 2,628 feet.

92. **Keyser, W.Va.:** The 50-mile Petersburg Branch of the Baltimore & Ohio Railroad ran deeply into a remote area of West Virginia along the South Branch of the Potomac River.

93. **Winchester, Va.:** Winchester is the south end of what was once the Cumberland Valley Division of the Pennsylvania Railroad, 115 miles south of Harrisburg, Pennsylvania.

94. **Martinsburg, W.Va.:** The Baltimore & Ohio Railroad served the historic town of Harper's Ferry (NE) at the scenic junction of the Potomac and Shenandoah Rivers.

95. **Frederick, Md.:** The original Baltimore & Ohio Railroad main line to Wheeling, West Virginia (NE; NW), crossed through the watersheds of two Chesapeake Bay rivers—the Patapsco and the Patuxent.

96. **Baltimore, Md.:** Baltimore is a city notable for the many tunnels through which both the Baltimore & Ohio and the Pennsylvania Railroad main lines must pass (see detail map 96A)— four separate tunnels on the Pennsylvania Railroad and nine separate tunnels on the Baltimore & Ohio Railroad! (See also detail map 96A, **Baltimore**.)

97. **Middle River, Md.:** Before World War II, travelers to Maryland's Eastern Shore went by steamboat from Baltimore to Love Point, and thence by train to any number of points, well-connected by an extensive Pennsylvania Railroad network.

98. **Dover, Del.:** The Delmarva Division of the Pennsylvania Railroad served not just the state capital at Dover but also the rich agricultural counties of Delaware, Maryland, and Virginia, as well as the oyster-packing ports of Maryland's Eastern Shore.

99. **Bridgeton, N.J.:** This quiet corner of South Jersey, with its natural supply of sand and oyster grounds, was served mainly by the Central of New Jersey Railroad and also by the Pennsylvania–Reading Seashore Lines.

100. **Tuckahoe, N.J.:** The popular southern New Jersey seashore was extensively served, first in redundant duplicate by both the Reading and the Pennsylvania, and finally, after the 1933 unification, by the Pennsylvania–Reading Seashore Lines.

101. **Atlantic City, N.J.:** Before the development of the Interstate Highway System and the commercial airline network, many a trainload of happy convention-goers from all over the United States arrived at Atlantic City by train.

102. **Galliopolis, Ohio:** At Kanauga, the New York Central Railroad crossed the Ohio River into West Virginia to gain access to the coal fields of that state.

103. **Ravenswood, W.Va.:** "RS&G Jct." stands for Ravenswood, Spencer & Glenville Jct., a predecessor of the Baltimore & Ohio Railroad.

104. **Spencer, W.Va.:** Deep in the heart of the West Virginia mountains, this town is hard to reach by rail.

105. **Gassaway, W.Va.:** The abandoned three-foot narrow gauge West Virginia Midland Railway was built to harvest lumber rather than coal, and it once connected not only the Baltimore & Ohio Railroad but also the Western Maryland Railway at Webster Springs (Hilton, *American Narrow Gauge Railroads*, p. 556).

106. **Buckhannon, W.Va.:** Passenger service on the abandoned narrow gauge Valley River Railroad existed until 1931 (Hilton, *American Narrow Gauge Railroads*, p. 556).

107. **Elkins, W.Va.:** Deep in the West Virginia mountains, an interchange existed between the Baltimore-based Western Maryland Railway and the Chesapeake & Ohio Railway, a railroad originating in tidewater Virginia.

108. **Petersburg West, W.Va.:** This is one of the only atlas maps in this volume that is nearly devoid of railroads!

109. **Edinburg, Va.:** Both the Shenandoah River and its North Fork were followed by two major railroads of the south, the Norfolk & Western and the Southern Railways.

110. **Front Royal, Va.:** South of the Baltimore & Ohio Railroad station at Strasburg Jct., the southern line to Harrisonburg, Virginia, was once part of a 157-mile B&O branch that ran from Harper's Ferry to a connection with the Chesapeake & Ohio Railway at Lexington, Virginia.

111. **Warrenton, Va.:** Today, the Southern Railway (now Norfolk Southern) line from Manassas (NE) through Markham (NW) is part of Norfolk Southern's east coast route from Atlanta, Georgia, to Harrisburg, Pennsylvania.

112. **Washington, D.C.:** In 1946, Potomac Yard (NE) was the major rail freight interchange between "north" and "south." Two northern trunk line railroads, the Baltimore & Ohio and the Pennsylvania Railroads, interchanged with three southern railroads—the Chesapeake & Ohio and the Southern Railways and the Richmond, Fredericksburg & Potomac Railroad. The latter, in turn, connected with the Atlantic Coast Line Railroad and the Seaboard Air Line Railway at Richmond, 110 miles farther south. (See also detail map 112A, **Washington, D.C.**)

113. **Anacostia, Md.:** Baltimore & Ohio Railroad freight trains to and from the west, destined to Potomac Yard, had to traverse two successive wye connections near Hyattsville, Maryland (NW). (See also detail map 113A, **Anacostia.**)

114. **Easton, Md.:** This map shows the pre–Bay Bridge, pre–World War II pattern of Eastern Shore tidewater branch lines of the Pennsylvania Railroad, now mostly abandoned.

115. **Harrington, Del.:** The Wilmington-to-Delmar, and thence to Norfolk, Del-Mar-Va Division of the Pennsylvania Railroad once was sufficiently busy to require both double track and interlocking stations with signaled crossovers and passing sidings, spaced every 10 to 15 miles apart.

116. **Lewes, Del.:** The eastern end of the 70-mile Pennsylvania Railroad subsidiary of the Baltimore & Eastern Railroad was at Lewes, where the Delaware Bay meets the Atlantic Ocean.

117. **Cape May, N.J.:** This popular New Jersey seashore resort was, until 1933, served by two parallel railroads—the Atlantic City Railroad (Reading Company) and the West Jersey & Seashore Railroad (Pennsylvania Railroad).

118. **Ashland, Ky.:** The original Norfolk & Western Railway from Norfolk to Cincinnati ran over a hilly, circuitous inland route, called the "Twelve Pole Line," until a new main line, from Naugatuck to Kenova, was built along the east bank of Tug Fork and the Big Sandy River.

119. **Huntington, W.Va.:** The Chesapeake & Ohio Railway's major locomotive shops were at Huntington, West Virginia.

120. **Charleston, W.Va.:** The capital of West Virginia, Charleston, was served by three trunk line railroads—the Baltimore & Ohio Railroad, the Chesapeake & Ohio Railway. and the New York Central Railroad.

121. **Gauley Bridge, W.Va.:** At Deepwater (SW), the New York Central Railroad's farthest extension into the south, there was a direct interchange with the Virginian Railway.

122. **Gauley River, W.Va.:** The Nicholas, Fayette & Greenbrier Railroad was created in 1926 by the Chesapeake & Ohio Railway and the New York Central Railroad—sharing equally service to the coal fields along the Gauley and Meadow Rivers.

123. **Marlinton, W.Va.:** Even though it was not a "common carrier," the Cherry River Boom & Lumber Cos. Railroad is shown; its operations ended in 1948 (Edson, *Railroad Names)*.

124. **Hot Springs, Va.:** The famous tourist railroad, the Cass Scenic Railroad, was still a private logging railroad in 1946. Its steep (over 6 percent) gradient required the use of switchbacks and "Shay"-geared steam locomotives. It connected with the Chesapeake & Ohio Railway at Cass and the Western Maryland Railway at Spruce (Hilton, *American Narrow Gauge Railroads,* p. 84).

125. **Staunton, Va.:** The Chesapeake Western Railway was once part of a 158-mile Baltimore & Ohio Railroad branch from Harper's Ferry, West Virginia, to Lexington, Virginia.

126. **Harrisonburg, Va.:** Harrisonburg is the location of the East Coast office of the Center for American Places.

127. **Charlottesville, Va.:** Chesapeake & Ohio Railway trains to and from Washington, D.C., had trackage rights over the Southern Railway to Orange, Virginia, where C&O rails were reached for the short trip to Gordonsville to join the C&O main line from Fort Monroe, Virginia, to Cincinnati, Ohio.

128. **Culpeper, Va.:** Most of the nearly 40-mile-long Virginia Central Railway was abandoned in 1937, reduced to a little over 1 mile of switching track near Fredericksburg (Hilton, *American Narrow Gauge Railroads,* (p. 545).

129. **Fredericksburg, Va.:** The U.S. Navy branch ran to the Naval Proving Ground, which was named for R/Adm. John A. Dahlgren, USN, a Civil War inventor of scientifically designed, large-caliber naval guns. Information from www.history.navy.mil/sources/va/dahlgr.htm.

130. **Popes Creek, Md.:** The 40-mile U.S. Government railroad branch led to the Naval Air Station–Patuxent River in Virginia/Maryland tidewater country.

131. **Patuxent River, Md.:** See note for map 130.

132. **Salisbury, Md.:** The Pennsylvania Railroad drawbridge interlocking station at Pocomoke was named for Alexander J. Cassatt, who was president of the Pennsylvania Railroad at the beginning of the twentieth century. President Cassatt greatly modernized the Pennsylvania: he boldly tunneled under the Hudson River and built Penn Station in New York City.

133. **Berlin, Md.:** Ocean City, Maryland, was another early seacoast resort served by a Pennsylvania subsidiary, the Baltimore & Eastern Railroad, which, in turn, was connected to Baltimore by a Chesapeake Bay steamboat at Love Point, Maryland (map 97).

134. **Williamson, W.Va.:** The Norfolk & Western Railway main line stayed close to the east bank of Tug Fork, which is also the state line between West Virginia and Kentucky.

135. **Logan, W.Va.:** Three Class I railroads—the Chesapeake & Ohio, the Norfolk & West-

ern, and the Virginian Railways—came together in the mountains of southwestern West Virginia at Gilbert Yard.

136. **Beckley, W.Va.:** The Virginian Railway was electrified from Mullens, W.Va. (SW), 134 miles eastward to Roanoke, Virginia. (See also detail map 136A, **Beckley**.)

137. **Hinton, W.Va.:** The Chesapeake & Ohio Railway main line "Big Bend" tunnel (SW), celebrated in the folk ballad "John Henry," is located just east of milepost 350 from Fort Monroe, Virginia.

138. **Alleghany, W.Va.:** The summit on the Chesapeake & Ohio Railway's crossing of the Alleghany mountains is just east of "Alleghany" tunnel, some 307 miles west of Fort Monroe, Virginia, at 2,072 feet above sea level.

139. **Clifton Forge, Va.:** Just east of Clifton Forge, the main line is rejoined by the James River line, which left the main line at Richmond. The James River Line has a smoother profile but is 38 miles longer.

140. **Lexington, Va.:** Monroe, Virginia, made famous by the song "The Wreck of Old 97," was a crew change point on the Southern Railway.

141. **Tye River, Va.:** In 1946, it was still possible to ride a passenger train on the Nelson & Albemarle Railway between Rockfish on the Southern Railway and Esmont on the Chesapeake & Ohio Railway, six miles north of Warren on the C&O James River Line.

142. **Strathmore, Va.:** Rivanna, the namesake of the Chesapeake & Ohio Railway subdivision that was the eastern half of the James River line, is on the river's north bank (SE).

143. **Maidens, Va.:** The narrow gauge Tidewater & Western Railroad was abandoned in 1917, and its rails were to be sold to France as scrap iron for the war effort. However, the war ended before the metal could be put to its intended use (Hilton, *American Narrow Gauge Railroads*, p. 543).

144. **Richmond, Va.:** The capitol of the Confederacy, Richmond was the zero milepost for every main and branch line on the entire Atlantic Coast Line Railroad and Seaboard Air Line Railway systems, which extended to thousands of cities and small towns in the deep south. (See also detail map 144A, **Richmond**.)

145. **West Point, Va.:** The Southern Railway ran a branch from Danville, Virginia, 179 miles east to West Point at the headwaters of the York River.

146. **Crisfield, Va.:** The 50-mile, nearly dead tangent lower part of the Pennsylvania Railroad's Delmarva line ran for nearly 34 miles across this map.

147. **Pikeville, Ky.:** In Letcher County, southeastern Kentucky, two major railroads—the Louisville & Nashville, 301 miles southeast of Louisville, Kentucky, and the Chesapeake & Ohio Railway, 656 miles west of Fort Monroe, Virginia—came within 2 miles of each other, but did not connect!

148. **Elkhorn City, Ky.:** Here on the Kentucky-Virginia state line is the north end of the 277-mile north-south Clinchfield Railroad (CC&O), which ran from Elkhorn City, Kentucky, to Spartanburg, South Carolina.

149. **Iaeger, W.Va.:** The main line of the Norfolk & Western Railway had no fewer than 15 tunnels in the 25-mile section between Kimball and Clear Fork Branch Jct.

150. **Bluefield, W.Va.:** The summit of the Norfolk & Western Railway's route across the Alleghany Mountains is just west of Bluefield, at an elevation of 2,566 feet.

151. **Radford, Va.:** At Walton, the Norfolk & Western Railway's main connection to the Deep South, the main line to Bristol, heads west-south-west.

152. **Christiansburg, Va.:** O. Winston Link, the photographer who portrayed the Norfolk & Western Railway so well, made many pictures of the small N&W steam locomotive hauling the Blacksburg branch mixed train.

153. **Roanoke, Va.:** Roanoke was the "heart" of the Norfolk & Western Railway, with its famous locomotive erecting shop, which both built and repaired steam locomotives nearly until the

end of steam in the United States. Roanoke also had a large N&W classification yard and the ornate N&W headquarters office building.

154. **Lynchburg, Va.:** A hilly city on the south bank of the James River, where three trunk line railroads—the Chesapeake & Ohio, the Norfolk & Western, and the Southern Railways—crossed and connected.

155. **Appomattox, Va.:** The main line of the Norfolk & Western Railway passes through Appomattox, the county seat of Appomattox County. Just to the northeast stands the Appomattox Court House, where General Robert E. Lee surrendered the Army of Northern Virginia to General Ulysses S. Grant in 1865, ending the American Civil War.

156. **Burkeville, Va.:** The little town of Crewe on the Norfolk & Western Railway main line is one of many railroad crew change points all over the American railroad system.

157. **Camp Pickett, Va.:** One of the early centralized train control installations, constructed with urgency during World War II, was on the single-tracked Seaboard Air Line Railway south from Richmond, Virginia, with its control machine located at Raleigh, North Carolina.

158. **Petersburg, Va.:** An interesting network and rail line pattern, three trunk railroads—the Atlantic Coast Line Railroad and the Norfolk & Western and the Seaboard Air Line Railways—cross and recross each other, alternately under, over, and at grade, at 8 separate places, all within a 15-mile corridor.

159. **Lee Hall, Va.:** Colonial Williamsburg was and is served by a handsome colonial passenger station, built by the Chesapeake & Ohio Railway and presently used by Amtrak.

160. **Fort Monroe, Va.:** The zero milepost of the Chesapeake & Ohio Railway main line to Cincinnati is located in the Chesapeake Bay tidewater area, within an old coastal-harbor defense fort military post, Fort Monroe, Virginia.

161. **Kiptopeke, Va.:** The very southern tip of the remote tidewater and ocean shore Delmarva peninsula was served by the "Standard Railroad of the World," the Pennsylvania Railroad.

162. **Middlesboro, Ky.:** The far western tip represented in this Mid-Atlantic atlas volume—Cumberland Gap, Tennessee—is at the junction of three states.

163. **Harlan, Ky.:** Historic coal country of eastern Kentucky, Harlan County is served by the Louisville & Nashville Railroad system.

164. **Norton, Va.:** Norton is where the Louisville & Nashville Railroad, 290 miles east of Louisville, Kentucky, meets and connects with the Norfolk & Western Railway, 467 miles west of Norfolk, Virginia.

165. **Bristol, Va.:** Here was a historic, end-on junction of two great railroads—the Norfolk & Western and the Southern Railways—where the "Birmingham Special" changed engines on its long journey to and from the Deep South.

166. **Abingdon, Va.:** The 16-mile narrow gauge Laurel Railway (SW), built to harvest lumber, was abandoned in 1924 (Hilton, *American Narrow Gauge Railroads*, p. 520).

167. **Rural Retreat, Va.:** The classic portrait of the southbound Norfolk & Western Railway night train, the "Birmingham Special," was beautifully captured on film by eminent photographer O. Winston Link, at Rural Retreat station (NW), 349 miles west southwest of Norfolk, Virginia.

168. **Galax, Va.:** Virginia's New River "gorge," while not as deep and dramatic as the gorge in West Virginia, nonetheless provides a scenic and rapidly flowing river for the Galax branch of the Norfolk & Western Railway to follow.

169. **Rocky Knob, Va.:** This is another of the very few maps in this atlas nearly devoid of railroads.

170. **Martinsville, Va.:** The north-south line through Martinsville is part of the southernmost extension of the Norfolk & Western Railway, to Winston-Salem, North Carolina.

171. **Danville, Va.:** Danville is the south end of the "mighty rough road" on the Southern Railway, described in the song "The Wreck of Old 97."

172. **South Boston, Va.:** The Atlantic & Danville Railway, operated under lease by the South-

ern Railway, traverses Mecklenburg County, crossing the Seaboard Air Line Railway at La Crosse, Virginia.

173. Victoria, Va.: Victoria, Virginia, was a crew change point on the Virginian Railway, halfway between Norfolk and Roanoke (the next crew change point to the west).

174. Emporia, Va.: The two southern main line railroads—Atlantic Coast Line Railroad and Seaboard Air Line Railway—both of which ran from Richmond, Virginia, to Tampa, Florida, cross this map, headed south.

175. Jarratt, Va.: Two abandoned logging railroad branches—the Atlantic & Danville Railway and the Surry, Sussex & Southampton Railway—once crossed the flat marshland of eastern Virginia.

176. Suffolk, Va.: Suffolk was an interesting railroad junction "nexus;" no fewer than six Class I railroads came together or closely crossed here as each made its western approach to Norfolk, Virginia.

177. Norfolk, Va.: Norfolk and the directly adjacent Newport News have long been the largest rail-to-water port for coal shipments in the United States. (See also detail map 177A, **Norfolk**.)

References

In addition to the listed references, the following types of sources from a variety of railroad company publications were used: employee timetables; signal interlocking diagrams; station lists; system, division, and city/area maps; and track charts.

Atlases and Maps

Hammond's Modern Atlas of the World. 1929. New York: C. S. Hammond & Company, Inc.

Hammond's Universal World Atlas. 1946 (and earlier years). New York: C. S. Hammond & Company, Inc.

United States Geological Survey (USGS)
> United States 1:250,000—Scale Series (base map)
> United States 15-Minute Series
> United States 7.5-Minute Series

Books, Guides, and Manuals

ABC Pathfinder Shipping and Mailing Guide. 1907. Boston: New England Railway Publishing Co.

Crist, Edward J. 1993. *Erie Memories*. New York: Quadrant Press, Inc.

Edson, William D. 1999. *Railroad Names*. Potomac, Maryland: William D. Edson.

Hilton, George W. 1990. *American Narrow Gauge Railroads*. Stanford, California: Stanford University Press.

Jowett, Alan. 1989. *Railway Atlas of Great Britain and Ireland from Pre-Grouping to the Present Day*. Wellingborough, Northamptonshire, England: Patrick Stephens, Ltd.

Moody, John. 1946. *Moody's Steam Railroads*. New York: Moody's Investors Service.

Official Guide of the Railways. 1946. New York: National Railway Publication Co.

Poor's Manual of Railroads. 1923. New York: Poor's Publishing Co.

Preliminary System Plan. Vol. II. 1975. Washington, D.C.: United States Railway Association.

Railway Clearing House. [1915] 1969. *Railway Junction Diagrams*. New York: Augustus M. Kelley.

Saunders, Richard. 2001. *Merging Lines–American Railroads, 1900–1970*. 2001. DeKalb, Illinois: Northern Illinois University Press.

Walker, Mike. 1997. *Steam Powered Video's Comprehensive Railroad Atlas of North America: Great Lakes East*. Dunkirk, Faversham, Kent, England: Ian Andrews.

Walker, Mike. 1997. *Steam Powered Video's Comprehensive Railroad Atlas of North America: Appalachia & Piedmont*. Dunkirk, Faversham, Kent, England: Ian Andrews.

Walker, Mike. 1997. *Steam Powered Video's Comprehensive Railroad Atlas of North America: North East*. Dunkirk, Faversham, Kent, England: Ian Andrews.

Wilner, Frank N. 1997. *Railroad Mergers: History, Analysis, Insight*. Omaha, Nebraska: Simmons-Boardman Books, Inc.

Indexes

INDEX OF COALING STATIONS

Steam locomotives normally were replenished with coal at engine terminals. However, to avoid delay on longer inter-city runs, some coaling stations were located directly on the main track, usually halfway between division or crew change points. The railroad owning the coaling station is signified by their reporting marks shown in parentheses.

INDEX OF INTERLOCKING STATIONS
AND FORMER INTERLOCKING STATIONS

In 1946, the centralized traffic control technology, which is the universal railroad operational standard today, was only beginning to be used on American railroads, and then only to a limited degree in the six mid-Atlantic states of this volume. Then, interlocking stations (or "signal towers," as they are commonly known) were located every five to ten miles along primary main and branch lines and also at nearly every junction or railroad crossing at grade.

These interlocking stations are identified by their name and/or telegraphic call letters (in bold type). In a few instances, tower numbers are shown in bold type in place of call letters. Where needed for clarity, a geographic place name is given in parentheses. Finally the abbreviation of the railroad to whose building style the tower was built is also shown in parentheses, followed by the state, map number, and map quadrant.

The list of interlocking stations is followed by a separate list of former interlocking stations, identified from various historic collections and old railroad employee timetables. The list of former interlocking stations is not offered as a complete list of every interlocking station that ever existed. Such a list remains to be compiled.

Interlocking Stations

5th St., **W** (Reading), (RDG), Pa., 65, NW

11th St., **US** (Pittsburgh), (PRR), Pa., 57A

12th St., **F** (B&O), D.C., 112A, 113, NW

15th St., **UF** (Pittsburgh), (PRR), Pa., 57A

21st St., **BU** (Pittsburgh), (PRR), Pa., 57A

33rd St. Viaduct, **FY** (B&O), Pa., 57A

58th St., **FY** (B&O), Pa., 82A

106th St., **NK** (NYC), N.Y., 54A

Abbott St., **AB** (LV), Pa., 51, SE, 51A

Aberdeen, **A** (B&O), Md., 80, SE

Acca Wye, **AY** (RF&P), Va., 144A

Acre (PRR), Ohio, 55, NW

Aiken, **SA** (B&O), Md., 80, SE

Alan, **A** (PRR), N.J., 82B

Alburtis, **AF** (RDG), Pa., 50, SE

Alexandria Jct., **JD** (B&O), Md., 113, NW

Alleghany, **A** (C&O), Va., 138, SE

Allen Lane, **CW** (PRR), Pa., 66, SE

Allens, **RO** (PRR), Pa., 33, SW

Allentown Yard, **VN** (CNJ), Pa., 51, SW

Altamont, **AM** (B&O), W.Va., 91, NW

Alto, **JK** (PRR), Pa., 45, SW

Amasa, **AM** (NYC), Pa., 26, NW

Anacostia, **JU** (PRR), D.C., 113, NW

Andover Jct., **AJ** (L&HR), N.J., 52, NE

Andover Jct., **BG** (NYC), Ohio, 12, SE

Annville, **NV** (RDG), Pa., 63, NE

Antis, **EF** (PRR), Pa., 45, SW

Apollo, **AP** (PRR), Pa., 42, SE

Arlington (LNE), Pa., 50, NW

Arlington (SIRT), N.Y., 53, SE

Arms, **MQ** (PRR), Del., 98, NE

Arthur (N&W), Va., 152, SW

Arthur Kill Draw, **AK** (SIRT), N.J., 53, SE

Ashford, **AD** (B&O), N.Y., 4, NE

Ashland Yard, **NC** (C&O), Ky., 118, NE

Ashland, **AX** (C&O), Ky., 118, NE

Ashmore, **YA** (LV), Pa., 50B

Ashtabula Harbor, **JM** (NYC), Ohio, 12, NW

Ashtabula Yard, **W** (NYC), Ohio, 12, NW

Ashtabula, **NP** (NYC), Ohio, 12, NW

Ashtabula, **OD** (NYC), Ohio, 12, NW

Glenside, **YM** (RDG), Pa., 66, SE

Glenwood Jct., **WJ** (B&O), Pa., 57A

Glenwood, **GD** (NYC), N.Y., 54, NW

Gordon St. (Allentown), (AT), Pa., 51B

Gordonsville, **G** (C&O), Va., 127, SE

Goshen, **GP** (Erie), N.Y., 38, NW

Gracedale, **F** (LV), Pa., 35, SW

Grafton, **D** (B&O), W.Va., 89, NE

Graham, **FX** (Erie), N.J., 37, SE

Graham, **HQ** (N&W), Va., 150, NW

Graham, **SI** (PRR), Ohio, 25, SE

Grantley, **CO** (PRR), Pa., 79, NE

Granton Jct., **GR** (NYS&W), N.J., 53, NE

Gravel Place, **GR** (DL&W), Pa., 36, SE

Gray, **RM** (PRR), Pa., 45, SW

Great Neck, **G** (LI), N.Y., 54, NE

Great Notch, **GA** (Erie), N.J., 53, NE

Green Spring, **GI** (B&O), W.Va., 75, SE

Greendale, **GN** (RF&P), Va., 144, SW

Greene Jct., **WH** (B&O), Pa., 73, NE

Greene, **BD** (PRR), Pa., 67, SW

Greenville (B&LE), Pa., 26, NW

Greenwich, **28** (NH), Conn., 39, SE

Grosvenor, **AS** (B&O), Ohio, 85, NE

Grove St., **GS** (Erie), N.J., 53B

Grove St., **Z** (DL&W), N.J., 53B

Gwynns Run, **VN** (PRR), Md., 96A

Hack (PRR), N.J., 53B

Hackensack Draw, **HA** (CNJ), N.J., 53A

Hackensack Draw, **HD** (DL&W), N.J., 53B

Hackensack Draw, **HX** (Erie), N.J., 53B

Hackensack River Draw, **DB** (Erie), N.J., 53A

Hager, **HJ** (PRR), Md., 77, SE

Halethorpe, **HX** (B&O), Md., 96, SE

Hall, **JE** (LI), N.Y., 54D

Hamilton St., **HS** (Allentown), (AT), Pa., 51B

Hancock, **HO** (B&O), W.Va., 76, SE

Hardman, **Q** (B&O), W.Va., 90, NW

Harlem River Draw, **DB** (NYC), N.Y., 54A

Harmon, **HM** (NYC), N.Y., 39, SW

Harold, **H** (LI), N.Y., 54B

Harpers Ferry, **HF** (B&O), W.Va., 94, NE

Harrington, **JC** (PRR), Del., 115, NE

Harris, **HG** (PRR), Pa., 63, NW

Harrison Yard, **HR** (DL&W), N.J., 53A

Hart, **SW** (South Wilkes-Barre), (PRR), Pa., 35A

Haselton, **CH** (B&O), Ohio, 25A

Hastings, **HS** (NYC), N.Y., 54, NW

Haucks, **HK** (RDG), Pa., 49D

Haven, **WT** (LI), N.Y., 54, SW

Herrs Island, **CQ** (PRR), Pa., 57A

Hilldale, **MW** (C&O), W.Va., 137, SW

Himrod Jct., **HM** (Erie), Ohio, 25A

Hobbs, **RN** (B&O), W.Va., 94, NW

Hoboken Terminal, **PY** (DL&W), N.J., 53B

Hollis, **IS** (LI), N.Y., 54D

Holly, **DV** (PRR), Del., 98, SE

Holmes, **HG** (PRR), Pa., 66, SE

Homewood Jct., **MD** (PRR), Pa., 41, NW

Homewood, **CM** (PRR), Pa., 57A

Hopewell, **SV** (RDG), N.J., 67, NW

Hornell Yard, **ZY** (Erie), N.Y., 6, NE

Horseheads, **HO** (Erie), N.Y., 8, SW

Houston, **HN** (PRR), Pa., 56, NE

Hudson, **S** (PRR), N.J., 53A

Hudson, **SX** (D&H), Pa., 35B

Hunlock, **RQ** (PRR), Pa., 34, SE

Hunt, **GC** (Huntingdon), (PRR), Pa., 60, NE

Hunter, **RD** (PRR), N.J., 53A

Huntingdon Ave., **HU** (B&O), Md., 96A

Huntington, **DK** (C&O), W.Va., 119, NW

Huntington, **HO** (C&O), W.Va., 119, NW

Hyndman, **Q** (B&O), Pa., 75, SE

Irv, **VA** (Irvineton), (PRR), Pa., 15, NW

Jack (N&W), Va., 157, SE

Jacks Run, **JR** (PRR), Pa., 56A

Jacks, **MU** (PRR), Pa., 61, NW

Jackson, **JA** (Nicolette), (B&O), W.Va., 86, SE

Jacy, **Z** (Jersey City), (PRR), N.J., 53B

Jamesburg, **JG** (PRR), N.J., 68, NW

Jarratt (VGN), Va., 175, NW

Jay, **J** (LI), N.Y., 54D

Jefferson Jct., **JN** (Erie), Pa., 22, NE

Jenkintown Jct., **KI** (RDG), Pa., 66, SE

Jersey City Eng. Term., **B** (CNJ), N.J., 53B

Jersey City, **A** (CNJ), N.J., 53B

Jersey City, **JC** (Erie), N.J., 53B

Jersey City, **JR** (LV), N.J., 53B

Jersey Shore Jct., **JS** (NYC), Pa., 32, SW

Jersey Shore, **SR** (NYC), Pa., 32, SW

Jersey, **PJ** (PRR), N.J., 82B

Journal Square, **WR** (H&M), N.J., 53B

Karny, **GY** (PRR), N.J., 53A

Kase, **DY** (PRR), Pa., 48, NW

Kearny Jct., **KR** (DL&W), N.J., 53A

Keating, **CT** (PRR), Pa., 31, NW

Kendall, **SJ** (PRR), N.Y., 8, SW

Kenova, **KV** (C&O), W.Va., 118, NE

Kenova, **KX** (N&W), W.Va., 118, NE

Keyser Valley, **KV** (DL&W), Pa., 35C

Kilmer, **GR** (PRR), N.J., 53, SW

Kinney, **B** (N&W), Va., 154, NE

Kips, **KD** (PRR), Pa., 48, NE

Kiskiminetas Jct., **AJ** (PRR), Pa., 42, SE

Klapperthal Jct., **KJ** (RDG), Pa., 65, NW

Lackawaxen, **BX** (Erie), Pa., 37, NW

Lake Jct., **L** (CNJ), N.J., 52, NE

Lamokin St., **KN** (PRR), Pa., 82, NW

Landenberg Jct., **WJ** (B&O), Del., 81, SE

Landis, **NV** (PRR), Pa., 64, SW

Landover, **W** (PRR), Md., 113, NW

Lane, **NK** (PRR), N.J., 53A

Lanesboro, **JA** (Erie), Pa., 22, NE

Langhorne, **PV** (RDG), Pa., 67, SW

Lansdale, **MA** (RDG), Pa., 66, SW

Larimer, **CP** (PRR), Pa., 57, NE

Latimer, **MR** (NYC), Ohio, 25, NE

Latrobe, **KR** (PRR), Pa., 58, NW

Laughlin Jct., **GN** (B&O), Pa., 57A

Laurel Hill (PRR), Pa., 56, NE

Laurel Jct., **JX** (LV), Pa., 49D

Lawrence Jct., **JU** (PRR), Pa., 41, NW

Layton, **NS** (B&O), Pa., 57, SE

Lead, **WL** (LI), N.Y., 54, SE

Leavittsburg, **SN** (Erie), Ohio, 25, SW

Lebanon Valley Jct., **VC** (RDG), Pa., 65, NW

Lebanon, **JU** (RDG), Pa., 64, NW

Lee St., **DX** (B&O), Md., 96A

Leetonia, **SB** (PRR), Ohio, 40, NW

Leetsdale, **MY** (PRR), Pa., 41, SE

Lehigh Summit, **GO** (DL&W), Pa., 36, NW

Lehighton, **HI** (LV), Pa., 50, NE

Lemoyne, **J** (PRR), Pa., 63, SW

Lewis, **RW** (PRR), Pa., 46, SE

Lewisburg, **UR** (RDG), Pa., 48, NW

Liberty St., **LR** (Binghamton), (Erie), N.Y., 10, SW, 10A

Linden St., **LS** (Allentown), (AT), Pa., 51B

Linden, **SQ** (PRR), Pa., 32, SE

Little Ferry Draw, **FY** (NYS&W), N.J., 53, NE

Little Ferry, **FY** (NYC), N.J., 53, NE

Lock Haven, **K** (PRR), Pa., 32, SW

Locust Jct. (LV), Pa., 50A

Locust St., **X** (B&O), Pa., 82A

Locust, **OY** (LI), N.Y., 54, NE

Lofty, **OF** (RDG), Pa., 49D

Long Island City, **F** (PRR), N.Y., 54B

Lorraine, **QR** (CNJ), N.J., 53, SW

Lurgan, **GN** (WM), Pa., 61, SE

Lynchburg Union Station., **ND** (C&O), Va., 154, NE

Lynchburg, **X** (N&W), Va., 154, NE

Machias, **CH** (PRR), N.Y., 5, NW

Mahanoy Tunnel, **BF** (RDG), Pa., 49, NE

Main Line Jct., **NQ** (RDG), Pa., 66, SW, 66A

Main St., **JN** (C&O), Va., 144A

Manasquan River Draw, **M** (NY&LB), N.J., 68, SE

Manila, **GR** (B&O), Pa., 75, NW

Manunka Chunk, **U** (DL&W), Pa., 51, NE

Manville Crossing, **XG** (LV), N.J., 52, SE

Marble Hill, **FN** (NYC), N.Y., 54, NW

Marcus Hook, **WO** (PRR), Pa., 82, NW

Martinsburg, **NA** (B&O), W.Va., 94, NW

Matawan, **MR** (NY&LB), N.J., 68, NE

Mattes St., **SR** (Scranton), (DL&W), Pa., 35C

McDougal, **MA** (C&O), W.Va., 121, SE

McKeesport, **MK** (B&O), Pa., 57, NW

McKenzie, **CO** (B&O), Md., 75, SW

Meadow Creek, **MD** (C&O), W.Va., 137, NW

Meyerstown, **MY** (RDG), Pa., 64, NW

Mianus River Draw, **29** (NH), Conn., 39, SE

Midgrade, **MG** (PRR), Pa., 60, NW

Midsteel, **GP** (PRR), Pa., 41, SW

Midway, **MK** (PRR), N.J., 67, NE

Mifflin, **M** (PRR), Pa., 47, SW

Miles, **UR** (PRR), Pa., 46, NW

Milford, **MD** (RF&P), Va., 129, SW

Mill Creek Jct., **CA** (RDG), Pa., 49, SE

Millburn, **MN** (DL&W), N.J., 53, SW

Miller, **R** (B&O), W.Va., 76, SE

Millham, **MO** (PRR), N.J., 67, SE

Milton, **MU** (RDG), Pa., 33, SW

Mineral Springs, **MO** (D&H), Pa., 35A

Miners Mills (D&H), Pa., 35A

Mingo Jct., **MJ** (PRR), Ohio, 55, NE

Minooka Jct., **MJ** (D&H), Pa., 35C

Monaca, **BG** (P&LE), Pa., 41, SW

Monmouth St., **OS** (Erie), N.J., 53B

Monon, **SB** (PRR), Pa., 56A, 57A

Monongahela Jct., **J** (URR), Pa., 57B

Monongahela, **MC** (PRR), Pa., 57, SW

Montandon, **DR** (PRR), Pa., 48, NW

Montclair, **MO** (DL&W), N.J., 53, NE

Montgomery Crossing, **MQ** (Erie), N.Y., 38, NW

Montgomery Crossing, **OG** (RDG), Pa., 33, SW

Montgomery St., **MG** (Trenton), (PRR), N.J., 67, SW

Montour Jct., **MR** (P&LE), Pa., 41, SE

Montview, **MV** (SOU), Va., 154, NE

Moodna Viaduct, **BS** (Erie), N.Y., 38, NE

Morris, **SV** (PRR), Pa., 67, SW

Morristown, **OW** (DL&W), N.J., 53, NW

Mott Haven Jct., **MO** (NYC), N.Y., 54A

Mount Royal, **RM** (B&O), Md., 96A

Mount Vernon, **VO** (NYC), N.Y., 54, NW

Mountain Lake Park, **MK** (B&O), W.Va., 91, NW

Mountain Park, **FQ** (CNJ), Pa., 35A

Mountain View, **MV** (DL&W), N.J., 53, NW

Mt. Airy Jct., **MA** (B&O), Md., 95, NE

Mt. Savage Jct., **J** (B&O), Md., 75, SW

Mullens CTC (VGN), W.Va., 136, SW

Nassau, **CD** (PRR), N.J., 67, NE

Nassau, **MT** (LI), N.Y., 54, SE

National Jct., **NJ** (NYC), N.J., 53B

National, **N** (D&H), Pa., 35C

Nay Aug, **ND** (DL&W), Pa., 35, NE

Nescopeck, **NK** (PRR), Pa., 34, SE

Neshaminy Falls, **JG** (RDG), Pa., 67, SW

Nesquehoning Jct., **PQ** (CNJ), Pa., 50, NE

Neville Island, **FM** (P&LE), Pa., 56, NE, 56A

New Castle Jct., **OA** (B&O), Pa., 41, NW

New Milford, **F** (DL&W), Pa., 22, NE

New Rochelle Jct., **22** (NH), N.Y., 54, NW

New Rochelle Yard, **23** (NH), N.Y., 54, NW

New York Ave., **C** (WATC), D.C., 112A

Newark Bay Draw, **DY** (CNJ), N.J., 53A

Newark Transfer, **BS** (CNJ), N.J., 53A

Newark, **NA** (CNJ), N.J., 53A

Newark, **NK** (LV), N.J., 53A

Newark, **NX** (DL&W), N.J., 53A

Newberry, **NC** (PRR), Pa., 32, SE

Newburgh Jct., **NJ** (Erie), N.Y., 38, NE

Newburgh, **GY** (NYC), N.Y., 38, NE

Newport News, **XA** (C&O), Va., 177, NW

Newton Falls, **HN** (B&O), Ohio, 25, SW

Newtown Jct., **NX** (Tabor), (RDG), Pa., 66, SE

Nicetown Jct., **NI** (RDG), Pa., 82A

Niles Jct., **RS** (B&O), Ohio, 25, SW

Nineveh Jct., **SW** (D&H), N.Y., 10, SE

Niobe Jct., **NE** (Erie), N.Y., 3, SW

Norca, **RT** (PRR), Pa., 49, NW

Norfolk Southern Jct., **NS** (N&W), Va., 177A

Norris, **NQ** (PRR), Pa., 66, SW, 66A

Norristown Jct., **NS** (RDG), Pa., 66, SW, 66A

North Abrams, **PW** (RDG), Pa., 66, SW

North Acca, **NA** (RF&P), Va., 144, SW

North Ave., **NA** (B&O), Md., 96A

North East, **N** (NYC), Pa., 2, SW

North Hawthorne, **NH** (NYS&W), N.J., 53, NE

North Philadelphia, **GD** (PRR), Pa., 82A

North Warren, **WN** (PRR), Ohio, 25, NW

Oak Island Jct., **CY** (CNJ), N.J., 53A

Oak Island Yard, **OK** (LV), N.J., 53A

Oak Point, **4** (NH), N.Y., 54, NW

Oaks, **PK** (PRR), Pa., 66, SW

Ocean Gate Draw, **BG** (PRR), N.J., 84, NE

Oceanport Draw, **OD** (NY&LB), N.J., 68, NE

Odenton, **Z** (PRR), Md., 96, SE

Ohio Jct., **OW** (B&O), Ohio, 25A

Oil City, **RH** (PRR), Pa., 27, NE

Okonoko, **NO** (B&O), W.Va., 75, SE

Olean, **X** (Erie), N.Y., 5, SW

Oley St., **N** (Reading), (RDG), Pa., 65, NW

Oliver, **QN** (B&O), Pa., 73, NE
Orange, **OH** (SOU), Va., 127, NE
Orange, **OR** (DL&W), N.J., 53, NE
Orchard (PRR), Pa., 65, NW
Orleans Road, **AD** (B&O), W.Va., 76, SW
Otts, **DW** (Warren), (PRR), Pa., 15, NE
Overbrook (PRR), Pa., 82, NE
Owego, **OG** (Erie), N.Y., 9, SW
Ozone, **RK** (LI) N.Y., 54, SW

Palmerton, **HX** (CNJ), Pa., 50, NE
Palmyra, **PR** (RDG), Pa., 63, NE
Pamplin, **PA** (N&W), Va., 155, NE
Paoli, **PA** (PRR), Pa., 66, SW
Park Jct., **PK** (RDG), Pa., 82A
Park, **FK** (LI), N.Y., 54, SE
Park, **PG** (PRR), Pa., 81, NW
Park, **SA** (PRR), Pa., 45, SE
Park Place, **PB** (PRR), N.J. 53A
Parkersburg Shops, **SY** (B&O), W.Va., 86, NE
Parkersburg, **OB** (B&O), W.Va., 86, NE
Passaic Draw, **PA** (CNJ), N.J., 53A
Passaic Draw, **PW** (Lyndhurst), (DL&W), N.J., 53, NE
Passaic Draw, **Y** (PRR), N.J., 53A
Passaic Jct., **BT** (Erie), N.J., 53, NE
Passaic River Draw, **BE** (Erie), N.J., 53, NE
Passaic River Draw, **NW** (Erie), N.J., 53A
Passaic River Draw, **WR** (Erie), N.J., 53, NE
Patterson Creek, **FN** (B&O), W.Va., 75, SE
Patterson Jct., **JN** (DL&W), N.J., 53, NE
Patterson Jct., **XW** (Erie), N.J., 53, NE
Peekskill, **37** (NYC), N.Y., 39, NW
Pelham Bay Draw, **14** (NH), N.Y., 54, NW
Penn (East New York), (LI), N.Y., 54, SW
Penn (PRR), Pa., 82A
Penn Haven Jct., **AV** (CNJ), Pa., 50, NE

Penna. Cement, **PC** (LNE), Pa., 51, SW
Pennroad, **SF** (PRR), Pa., 61, SE
Pennsylvania Ave., **VN** (PRR), Pa., 56A
Penrose (PRR), Pa., 82, NE
Perkasie, **PK** (RDG), Pa., 66, NW
Perkiomen Jct., **PK** (RDG), Pa., 66, SW
Perryville, **V** (PRR), Md., 80, SE
Petersburg, **BG** (ACL), Va., 158, NW
PH&P Jct., **HB** (RDG), Pa., 63, NW
Philipsburg, **RG** (NYC), Pa., 45, NE
Phillips St., **F** (CNJ), N.J., 53B
Phillipsburg, **PB** (LV), N.J., 51, SE, 51A
Phillipsburg, **PU** (CNJ), N.J., 51A
Phoebe, **Q** (N&W), Va., 155, NW
Phoenixville, **PN** (PRR), Pa., 65, SE
Phoenixville, **U** (RDG), Pa., 65, SE
Piedmont, **P** (B&O), W.Va., 91, NE
Pike St., **TX** (Reading), (RDG), Pa., 65, NW
Pittsburgh, **DX** (P&LE), Pa., 56A
Pittston Jct., **J** (LV), Pa., 35B
Plainfield, **JA** (CNJ), N.J., 53, SW
Plains Jct., **G** (WBC), Pa., 35A
Plate, **DK** (Dunkirk), (NKP), N.Y., 3, NW
Plymouth Jct., **N** (DL&W), Pa., 35A
Poe (N&W), Va., 158, SW
Pohick, **CW** (RF&P), Va., 112, SE
Point of Rocks, **KG** (B&O), Md., 94, NE
Pond, **DF** (LI), N.Y., 54C
Poplar, **BR** (B&O), Md., 97, NW
Port Clinton, **PN** (RDG), Pa., 49, SE
Port Dickinson, **YO** (D&H), N.Y., 10, SW
Port Morris, **3** (NH), N.Y., 54, NW
Port Morris, **UD** (DL&W), N.J., 52, NE
Port Reading Jct., **PD** (CNJ), N.J., 53, SW

Port, **FY** (LI), N.Y., 54, SE
Port, **MS** (PRR), Pa., 62, NE
Portage, **NY** (PRR), Pa., 59, NE
Portal, **W** (PRR), N.J., 53B
Pottsville Jct., **MJ** (RDG), Pa., 49, SE
Prince, **NI** (C&O), W.Va., 136, NE
Pymatuning, **GH** (Erie), Pa., 26, NW

Queens, **QU** (LI), N.Y., 54, SE
Quinnimont, **QN** (C&O), W.Va., 136, NE

Race St., **RA** (RDG), Pa., 82A
Radebaugh, **RG** (PRR), Pa., 57, NE
Rahway River Draw, **RH** (West Carteret), (CNJ), N.J., 53A
Randolph St. (N&W), Va., 153, NW
Randolph, **RH** (Erie), N.Y., 4, SW
Raritan, **RA** (CNJ), N.J., 52, SE
Raven, **WO** (PRR), Pa., 48, NE
Reading, **RA** (PRR), Pa., 65, NW
Red Bank, **BJ** (PRR), N.J., 42, NE
Red Bank, **RG** (NY&LB), N.J., 68, NE
Redoak, **W** (P-RSL), N.J., 82, NE
Redpen, **AG** (PRR), Pa., 49, NW
Rhode Island Ave., **QN** (B&O), D.C., 112A, 113, NW
Rich, **VO** (McElhattan), (PRR), Pa., 32, SW
Richards, **RS** (LV), Pa., 51, SE, 51A
Ridgewood Jct., **WJ** (Erie), N.J., 53, NE
Riverside Draw, **RC** (Delanco), (PRR), N.J., 67, SW
Riverside Jct., **RJ** (PRR), N.Y., 4, SE
Riverside, **RV** (B&O), Md., 96A
Riverton Jct., **RJ** (N&W), Va., 110, NE
Robesonia, **RA** (RDG), Pa., 64, NE
Rochester, **RC** (PRR), Pa., 41, SW
Rock, **FW** (LI), N.Y., 54, SE
Rockville, **RJ** (PRR), Pa., 63, NW
Rockville, **SI** (PRR), Ohio, 55, NE

Rockwood, **RJ** (WM), Pa., 74, NE
Rodemer, **RO** (B&O), W.Va., 90, NE
Roelofs, **RO** (RDG), Pa., 67, SW
Ronceverte, **RV** (C&O), W.Va., 138, NW
Rose Siding, **GH** (NYC), Pa., 28, SE
Rose, **RV** (PRR), Pa., 45, SW
Roseville Ave., **RO** (DL&W), N.J., 53A
Rowlesburg, **R** (B&O), W.Va., 90, NE
Rutherford, **HX** (RDG), Pa., 63, NW
Rutherford, **VK** (RDG), Pa., 63, NE
Rye, **26** (NH), N.Y., 54, NE

Salamanca, **WJ** (Erie), N.Y., 4, SE
Sand Patch, **SA** (B&O), Pa., 75, NW
Schuylkill Draw, **DR** (B&O), Pa., 82A
Scully, **SY** (PRR), Pa., 56A
Sea Girt, **SG** (NY&LB), N.J., 68, SE
Seaford, **RS** (PRR), Del., 115, SE
Sedgwick Ave., **SK** (NYC), N.Y., 54, NW
Seminary, **AF** (RF&P), Va., 112, NE
Shale (PRR), Ohio, 40, SW
Shamokin, **SJ** (RDG), Pa., 48, NE
Shark River Draw (NY&LB), N.J., 68, SE
Shenango, **XN** (Erie), Pa., 26, NW
Sheridan, **SG** (PRR), Pa., 59, NW
Sheridan, **SR** (RDG), Pa., 64, NE
Shippensburg, **SX** (RDG), Pa., 61, SE
Shire Oaks Yard, **H** (PRR), Pa., 57, SW
Shore, **VN** (PRR), Pa., 66, SE, 82B
Short Line Jct., **J** (B&O), W.Va., 89, NW
Sidney, **GX** (D&H), N.Y., 11, NW
Silver Run, **SR** (B&O), W.Va., 87, SE
Singer, **FH** (Elizabethport), (CNJ), N.J., 53A
Singerly, **SY** (B&O), Md., 81, SW

Sinking Spring, **S** (RDG), Pa., 64, NE

Sixteenth St. Jct., **JU** (RDG), Pa., 82A

Slateford Jct., **SJ** (DL&W), Pa., 51, NE

Slope, **BO** (PRR), Pa., 45, SW

South Amboy, **SA** (NY&LB), N.J., 68, NW

South Branch Drawbridge (VGN), Va., 177A

South End, **RO** (RF&P), Va., 112, NE, 112A

South Fork, **SO** (PRR), Pa., 59, NW

South Mount Vernon, **20** (NH), N.Y., 54, NW

South Norfolk Tower, **MO** (VGN), Va., 177A

South Orange, **J** (DL&W), N.J., 53, SW

South Plainfield, **SP** (LV), N.J., 53, SW

South Wilkes-Barre, **SW** (LV), Pa., 35A

South, **SS** (PRR), Pa., 45, SW

Southwest Jct., **SW** (Greensburg), (PRR), Pa., 57, NE

Sparrowbush, **WX** (Erie), N.Y., 37, NE

Spring St., **GW** (Elizabeth), (CNJ), N.J., 53, SE

Spruce, **SC** (PRR), Pa., 45, SE

Spuyten Duyvil, **DV** (NYC), N.Y., 54, NW

St. George, **TA** (SIRT), N.Y., 53, SE

St. George, **TB** (SIRT), N.Y., 53, SE

Stadium (PRR), Pa., 82, NE

Stamford, **38** (NH), Conn., 39, SE

State (PRR), Pa., 63, NW

State Line, **AN** (LV), N.Y., 8, SE

Stone Cliff, **CS** (C&O), W.Va., 136, NE

Strecker, **HX** (B&O), W.Va., 91, NE

Stroudsburg, **S** (DL&W), Pa., 51, NE

Suffern, **SF** (Erie), N.Y., 38, SE

Suffolk, **S** (N&W), Va., 176, SE

Sugar Notch, **SG** (LV), Pa., 35A

Summit, **ST** (DL&W), N.J., 53, SW

Sunbury, **SF** (RDG), Pa., 48, NW

Sunnyside Yard, **Q** (PRR), N.Y., 54B

Sunnyside Yard, **R** (PRR), N.Y., 54B

Suplee, **W** (RDG), Pa., 65, SW

Susquehanna, **SR** (Erie), Pa., 22, NE

Tabor Jct., **BO** (RDG), Pa., 66, SE

Tannery, **YS** (LV), Pa., 35, SW

Tarrytown, **OW** (NYC), N.Y., 39, SW

Tasker, **AK** (PRR), Del., 81, SE

Terminal, **JR** (RF&P), Va., 144A

Terra Alta, **CA** (B&O), W.Va., 90, NE

Thorn, **HD** (PRR), Pa., 81, NE

Thornton Jct., **RS** (NKP), Pa., 13, NW

Three Bridges, **BR** (LV), N.J., 52, SW

Tidewater Jct., **SJ** (VGN), Va., 177A

Tobyhanna, **DO** (DL&W), Pa., 36, SW

Topton, **FH** (RDG), Pa., 50, SE

Torrance, **BH** (PRR), Pa., 58, NE

Tottenville (SIRT), N.Y., 53, SW

Town, **NC** (WM), Md., 77, SE

Trafford, **SZ** (PRR), Pa., 57, NW

Treichler, **CF** (LV), Pa., 50, SE

Tuckahoe, **KG** (P-RSL), N.J., 100, NW

Tug, **HI** (N&W), W.Va., 149, NE

Tunnel, **TU** (CNJ), Pa., 35, SW

Union Jct., **CS** (PRR), Md., 96A

Union St., **US** (LV), Pa., 51, SW, 51B

Union, **DK** (PRR), N.J., 53, SW

University Heights, **BN** (NYC), N.Y., 54, NW

Valley (LV), N.J., 53, SW

Valley St., **VY** (Erie), Ohio, 25A

Valley, **VA** (LI), N.Y., 54, SE

Van Cortlandt Park Jct., **JS** (NYC), N.Y., 54, NW

Van Nostrand Ave., **HY** (CNJ), N.J., 53B

Van, **VD** (LI), N.Y., 54, SW

Vernon, **HI** (P-RSL), N.J., 82B

Viaduct Jct., **ND** (B&O), Md., 75, SW

Victoria, **HK** (B&O), Pa., 74, NW

View, **JO** (PRR), Pa., 62, NE

Villamont, **US** (N&W), Va., 153, NW

Virginia, **VU** (PRR), D.C., 112, NW, 112A

Vista, **VA** (B&O), Pa., 57, NW

Waldo, **SC** (PRR), N.J., 53B

Waldwick, **WC** (Erie), N.J., 38, SE

Walker, **WA** (B&O), W.Va., 87, SW

Wall, **MI** (PRR), Pa., 47, SW

Walnut St., **JK** (VGN), Va., 153, NW

Walnut St., **S** (Reading), (RDG), Pa., 65, NW

Walton, **BH** (N&W), Va., 151, SE

Wampum Jct., **RK** (PRR), Pa., 41, NW

Ward, **PG** (PRR), Del., 81A

Warren, **BO** (Erie), Ohio, 25, SW

Washington CTC Tower, **ON** (B&O), Pa., 56, SW

Washington St., **PH** (Pittsburgh), (PRR), Pa., 57A

Washington, D.C., **K** (WATC), D.C., 112A

Waterboro, **WO** (Erie), N.Y., 3, SE

Waverly, **SF** (B&O), Md., 96A

Wawa, **J** (PRR), Pa., 82, NW

Wayne Jct., **WS** (RDG), Pa., 82A

Weehawken, **TU** (NYC), N.J., 53B

Weirton Jct., **WC** (PRR), W.Va., 55, NE

Wesleyville, **WV** (NYC), Pa., 1, SE

West Aliquippa, **QA** (P&LE), Pa., 41, SE

West Belt Jct., **BJ** (P&WV), Pa., 56A

West Chester, **WC** (PRR), Pa., 81, NE

West Conway, **WC** (PRR), Pa., 41, SW

West Cumbo, **W** (B&O), W.Va., 77, SW

West End Jct., **WK** (AT), Pa., 51B

West End, **WD** (DL&W), N.J., 53B

West End, **WE** (North Bergen), (NYC), N.J., 53B

West End, **WS** (B&O), W.Va., 90, NW

West Falls No. 1, **FS** (RDG), Pa., 82A

West Falls No. 2, **WV** (RDG), Pa., 82A

West Haverstraw, **HN** (NYC), N.Y., 39, SW

West Hinton, **CW** (C&O), W.Va., 137, SW

West Keyser, **Z** (B&O), W.Va., 92, NW

West Milton, **WM** (RDG), Pa., 33, SW

West Penn Crossing, **VO** (PRR), Pa., 42, NW

West Pittsburgh, **UN** (B&O), Pa., 41, NW

West Secaucus, **SY** (DL&W), N.J., 53B

West Trenton, **CN** (RDG), N.J., 67, NW

Westfield, **WF** (CNJ), N.J., 53, SW

Westfield, **WX** (NYC), N.Y., 2, NE

Weston-Manville, **WX** (RDG), N.J., 52, SE

Weverton, **VO** (B&O), Md., 94, NE

Wheeling, **WR** (B&O), W.Va., 55A

White House, **WH** (CNJ), N.J., 52, SW

White House, **WH** (PRR), N.Y., 5, SW

White Plains North, **NW** (NYC), N.Y., 39, SW

White Sulphur Springs, **WS** (C&O), W.Va., 138, NW

Whitings, **WN** (PRR), N.J., 84, NW

Wildwood, **WD** (B&O), Pa., 42, SW

Williamsport, **WG** (RDG), Pa., 33, SW

Willow Ave., **AV** (NYC), N.J., 53B

Wilmerding, **WG** (PRR), Pa., 57, NW

Wilmington (PRR), Del., 81A

Wilmington Yard, **WY** (PRR), Del., 81A

Win, **WJ** (LI), N.Y., 54, SW

Winslow, **B** (P-RSL), N.J., 83, SW

Woodbridge Jct., **WC** (Perth Amboy), (CNJ), N.J., 53, SW

Woodlawn, **JO** (NYC), N.Y., 54, NW

Works, **RO** (PRR), Pa., 45, SW

Wrights, **NR** (PRR), Pa., 17, SE

Wye, **J** (Allentown), (AT), Pa., 51, SW, 51B

Wye, **SN** (New Portage Jct.), (PRR), Pa., 60, NW

Yellow Creek, **CK** (PRR), Ohio, 40, SE

York, **K** (PRR), Pa., 79, NE

Zane, **PH** (PRR), W.Va., 55, SE, 55A

Zoo (PRR), Pa., 82A

Former Interlocking Stations

14th St., **14** (Altoona), (PRR), Pa., 45, SW

14th St., **BG** (PRR), D.C., 112A

16th St. Crossing, **UR** (PRR), Md., 96A

22nd St., **B** (Philadelphia), (PRR), Pa., 82A

33rd St., **K** (West Philadelphia), (PRR), Pa., 82A

36th St., **D-1** (West Philadelphia), (PRR), Pa., 82A

156th St., **KY** (Melrose), (NYC), N.Y., 54A

180th St., **UP** (NYW&B), N.Y., 54, NW

Adam, **MD** (PRR), Pa., 48, NW

Agnew, **RN** (PRR), Pa., 41, SE

Alford (DL&W), Pa., 22, NW

Alfrata, **AF** (PRR), Pa., 45, SE

Allegheny, **AY** (PRR), Pa., 56A

Allegrippus, **AG** (PRR), Pa., 60, NW

Alston, **SA** (PRR), Del., 98, NE

Altoona, **FG** (PRR), Pa., 45, SW

Amboy, **J** (NYC), Ohio, 12, NE

Analomink, **KN** (DL&W), Pa., 36, SE

Atglen, **NI** (PRR), Pa., 81, NW

Atlantic Highlands Pier, **AH** (CNJ), N.J., 68, NE

Atlantic, **VI** (Erie), Pa., 13, SW

Auburn, **BN** (PRR), Pa., 49, SE

Avondale, **MV** (PRR), Pa., 81, NW

Baker, **FN** (PRR), Pa., 31, NE

Baltic, **BF** (PRR), N.J., 101, NW

Bank, **BA** (PRR), Pa., 48, NW

Bards, **GF** (PRR), Pa., 32, SE

Barnegat Pier, **BN** (PRR), N.J., 84, NE

Bay Head Jct., **HJ** (NY&LB), N.J., 68, SE

Baychester Ave., **BJ** (NYW&B), N.Y., 54, NW

Bell, **MR** (PRR), Pa., 31, SE

Bellewood, **OX** (LV), N.J., 51, SE

Bethel, **BD** (PRR), Del., 115, SE

Betz Jct., **MD** (NYC), Pa., 45, NW

Biddle St., **AC** (PRR), Md., 96A

Big Elk, **BQ** (PRR), Pa., 81, SW

Birmingham, **BR** (PRR), N.J., 83, NE

Black Creek Jct., **BX** (LV), Pa., 50, NW

Bloom, **BM** (PRR), Pa., 49, NW

Bluf, **BC** (PRR), Pa., 48, NE

Boanna, **BO** (PRR), Ohio, 25, SW

Boles, **NB** (PRR), Pa., 41, SW

Boonton, **BN** (DL&W), N.J., 53, NW

Boonton, **BO** (DL&W), N.J., 53, NW

Boston Run Jct., **BJ** (PRR), Pa., 49C

Boyd, **B** (PRR), Pa., 48, NE

Bradenville, **BV** (PRR), Pa., 58, NW

Branchport, **BX** (NY&LB), N.J., 69, NW

Branchton, **UN** (B&LE), Pa., 27, SW

Bridge Siding, **CF** (PRR), Pa., 63, NW

Bridge Three, **CU** (PRR), Del., 81A

Bridgeton Jct., **BJ** (CNJ), N.J., 99, NE

Brigham Road, **CA** (NYC), N.Y., 3, NW

Brilliant "Y," **VI** (PRR), Pa., 57A

Broad Creek, **BC** (PRR), Del., 115, SE

Broadhead, **BH** (PRR), Pa., 56A

Broadway, **BW** (Elizabethport), (CNJ), N.J., 53A

Brockway, **WI** (B&O), Pa., 29, NW

Bronx River Draw, **7** (NH), N.Y., 54A

Buckeye, **OH** (PRR), Ohio, 40, NE

Buffalo St., **BS** (Olean), (PRR), N.Y., 5, SW

Bulger, **HF** (PRR), Pa., 56, NW

Burkeville, **J** (N&W), Va., 156, SE

Burleigh, **BG** (AC), N.J., 100, SW

Bustleton, **BU** (RDG), Pa., 66, SE

Callery, **X** (B&O), Pa., 41, SE

Callicoon, **CO** (Erie), N.Y., 23, NE

Caln, **VA** (PRR), Pa., 81, NW

Calvin, **CN** (B&LE), Pa., 42, NW

Camden Ship Yards, **GJ** (RDG), N.J., 82B

Camden, **A** (PRR), Pa., 82B

Camden, **RM** (PRR), Pa., 82B

Cameron, **CN** (Erie), N.Y., 7, SW

Cameron, **CX** (B&O), W.Va., 71, NE

Campbell Hall, **CP** (NYO&W), N.Y., 38, NW

Canal, **F** (PRR), Pa., 60, NW

Canton Jct., **CJ** (PRR), Md., 96A

Canton, **C** (PRR), Pa., 20, SW

Carrollton, **CT** (Erie), N.Y., 4, SE

Cass St., **CS** (PRR), N.J., 67, SW

Catawissa Jct., **CA** (PRR), Pa., 49, NW

Channel, **BR** (PRR), N.J., 101, NW

Chatsworth Ave., **AT** (Larchmont), (NYW&B), N.Y., 54, NW

Chenango Forks, **CG** (DL&W), N.Y., 10, SW

Chester Jct., **CJ** (DL&W), N.J., 52, NE

Clark, **NS** (PRR), Ohio, 40, SE

Clover Creek Jct., **KZ** (PRR), Pa., 60, NE

Coatesville, **CV** (PRR), Pa., 81, NW

Cold Spring, **45** (NYC), N.Y., 39, NW

Cold Spring, **46** (NYC), N.Y., 39, NW

Coleman, **CZ** (PRR), Pa., 57A

Columbia, **EY** (PRR), Pa., 64, SW

Columbiana, **NA** (PRR), Ohio, 40, NE

Columbus Ave., **CA** (NYW&B), N.Y., 54, NW

Columbus Jct., **CM** (Erie), Pa., 14, NE

Conestoga, **CG** (PRR), Pa., 64, SW

Constable Jct., **CB** (Greenville), (LV), N.J., 53B

Cook's Crossing, **KY** (PRR), Pa., 41, NW

Corning Yard, **YD** (NYC), N.Y., 7, SE

Corning, **WK** (NYC), N.Y., 7, SE

Cornwells Heights, **CO** (PRR), Pa., 67, SW

Crawford Jct., **RF** (NYO&W), N.Y., 38, NW

Creasy, **WR** (PRR), Pa., 34, SE

INDEX OF PASSENGER AND NON-PASSENGER STATIONS

This is an alphabetical list of all the stations shown on the maps of this atlas, with the exception of interlocking stations, former interlocking stations, coaling stations, track pans, tunnels, and viaducts. These are shown in separate lists.

In those locations where two or more railroads serve the same place, the station name of that place is followed, in brackets, by the "reporting marks," or standard abbreviations, of the railroads concerned. In a few instances, adjacent places are given in brackets.

No distinction is made in this index as to the type of station (e.g. passenger stations, remote control interlockings, block stations, or freight-only stations). For that information, the reader may refer to "Map Symbols and Abbreviations" and to the atlas map concerned. Every index entry shows the standard abbreviation of the state within which the station is located.

The last items in each entry are the atlas map number and one of four possible map quadrant designations—NW, NE, SW, or SE—each of which covers one of the four areas of the 15 x 15 minutes each, which, taken together, constitute each atlas sheet. In the case of detail maps, there are no map quadrant designations; instead, the map number is followed by the letter A, B, C, etc., which signifies a particular detail map.

239

Accotink, Va., 112, SE
Accoville, W.Va., 135, NW
Ache Jct., Pa., 73, NW
Ackermanville, Pa., 51, NE
Ackerson, N.J., 37, SE
Ackworth, Pa., 65, SE
Acme, Pa., 56, NW
Acme, Va., 140, SE
Acme, W.Va., 121, SW
Acorn, Pa., 66, SW
Acosta, Pa., 58, SE
Acqueduct Bridge, D.C., 112, NE
Acqueduct, Pa., 62, NE
Acup, W.Va., 121, NW
Ada, W.Va., 150, NE
Adah, Pa., 73, NW
Adam, Pa., 48, NW
Adams, N.J., 68, NW
Adams, Pa., 44, SW
Adams, Pa., 59, SW
Adamsdale, Pa., 49, SE
Adamson, Ky., 147, SE
Adamston, W.Va., 89, NW
Adamstown, Md., 95, NW
Adamsville, Pa., 13, SW
Addington, Va., 164, NE
Addison (B&O; Erie), N.Y., 7, SE
Addison, Ohio, 102, NE
Addison, Va., 158, SW
Adelaide, Pa., 57, SE
Adena, Ohio, 55, SW
Adma, W.Va., 89, SE
Admiralty, W.Va., 136A, NW
Adrian Jct., W.Va., 106, NE
Adrian, N.Y., 6, NE
Adrian, W.Va., 106, NW
Adsit, Va., 174, NE
Advance, Pa., 78, SW
Aetnaville (PRR; W&LE), Ohio, 55, SE, 55A
Affinity (C&O; VGN), W.Va., 136A, SE
Aflex, Ky., 134, SE
Afton, Md., 77, SE
Afton, N.Y., 10, SE
Afton, Va., 126, SW
Agasote, N.J., 67, SW
Ages, Ky., 163, NE
Aggregate, W.Va., 107, NW
Aiken Summit, Va., 170, SE
Aikin, Md., 80, SE
Airey, Md., 115, SW
Ajax Park, N.J., 67, NW

Akeley, Pa., 15, NE
Akron, Pa., 64, SE
Aladdin, Pa., 42, SE
Alba, N.Y., 38, NW
Alba, Pa., 20, SW
Albany, Ohio, 85, SE
Albany, Pa., 50, SW
Alberene, Va., 141, NE
Alberta (SAL; VGN), Va., 174, NW
Albertson, N.Y., 54, NE
Albion (B&LE; PRR), Pa., 13, NW
Albion, N.J., 83, NW
Albright, W.Va., 90, NE
Alden, Ohio, 86, NE
Alden, Pa., 34, SE
Alderson, W.Va., 137, SE
Aldham, Pa., 65, SE
Aldrich, Va., 128, NE
Aldridge, W.Va., 94, NW
Alexander, Ohio, 71, NW
Alexander, W.Va., 106, NE
Alexandria (C&O; RF&P; SOU; W&OD), Va., 112, NE
Alexandria Jct., Va., 112, NE
Alexandria, Pa., 45, SE
Alford, Pa., 22, NW
Alfrata, Pa., 45, SE
Alfred, N.Y., 6, NW
Algoma, W.Va., 150, NW
Algonquin, W.Va., 150, NW
Algren, Va., 177, NW
Alice Mines, Pa., 57, SE
Alice, Ohio, 85, SW
Aliff, W.Va., 135, SE
Aliquippa, Pa., 41, SE
All, Va., 139, SW
Allamuchy, N.J., 52, NW
Alledonia, Ohio, 71, NW
Allegany, Md., 75A
Alleghany, Va., 138, SE
Allegheny (Erie), N.Y., 5, SW
Allegheny (PRR), N.Y., 5, SW
Allen Creek, Va., 141, SW
Allen Jct., Pa., 51, SW
Allen Jct., W.Va., 136, SW
Allen Lane, Pa., 66, SE
Allen, Va., 148, SW
Allen's Mills, Pa., 29, SW
Allendale, N.J., 38, SE
Allenhurst, N.J., 68, SE
Allenport, Pa., 57, SW

Allenport, Pa., 61, NW
Allentown (AT; LV), Pa., 51B
Allentown (LNE), Pa., 51, SW
Allenwood, N.J., 68, SE
Allenwood, Pa., 33, SW
Alley, Va., 149, SW
Alliance Jct., Pa., 49, SE
Allingdale, W.Va., 122, NE
Allison Park, Pa., 42, SW
Allison, Pa., 49B
Allison, Pa., 73, NW
Allisonia, Va., 168, NE
Alloway Jct., N.J., 82, SW
Alloway, N.J., 82, SW
Alloy, W.Va., 121, SW
Allsworth, Pa., 58, NW
Allwood, N.J., 53, NE
Almoden, Pa., 45, NW
Almonate, Va., 149, SE
Almond, N.Y., 6, NE
Almonry, Pa., 35, SW
Alnwick, W.Va., 135, SW
Alpha, N.J., 51, SE
Alpha, Va., 142, SW
Alpine, Va., 139, SE
Alpoca, W.Va., 136, SW
Alston, Del., 98, NE
Altamont, W.Va., 91, NW
Altavista (SOU; VGN), Va., 154, SW
Altenwald, Pa., 77, NE
Alterton, Pa., 62, SW
Althom, Pa., 15, NW
Altman, W.Va., 120, SW
Alton, Va., 171, SE
Alton, W.Va., 106, NE
Altoona, Pa., 45, SW
Alum Creek, W.Va., 120, NW
Alum Rock, Pa., 27, SE
Aluta, Pa., 51, NW
Alva, Ky., 163, SW
Alvan, Pa., 60, SE
Alvarado, Va., 166, SW
Alverton, Pa., 57, SE
Alvon, W.Va., 138, NE
Alwington, Va., 111, SW
Amawalk, N.Y., 39, NW
Ambler, Pa., 66, SE
Amblersburg, W.Va., 90, NE
Amboy, Ohio, 12, NE
Ambridge, Pa., 41, SE
Amcelle, Md., 75, SW
Ameagle, W.Va., 136, NW

Amelia, Va., 157, NW
Amelia, W.Va., 121, NW
Ames, W.Va., 121, SE
Amesville Jct., Pa., 45, NW
Amesville, Ohio, 86, NW
Amherst, Va., 140, SE
Amherstdale, W.Va., 135, NW
Amigo, W.Va., 136, SW
Ammendale, Md., 96, SW
Amoco, Va., 159, SE
Ampere, N.J., 53, NE
Amru, Ky., 162, SE
Amsbry, Pa., 44, SE
Amsterdam, Ohio, 55, NW
Analomink, Pa., 36, SE
Anawalt, W.Va., 150, NW
Anchor, Ky., 162, SW
Ancora, N.J., 83, SW
Andalusia, N.J., 67, SW
Andenried, Pa., 50A
Anderson Road, Pa., 41, SE
Anderson St., N.J., 53, NE
Anderson, Pa., 29, SE
Anderson, Pa., 56, SE
Andersonburg, Pa., 62, NW
Andover, N.J., 52, NE
Andover, N.Y., 6, SW
Andover, Ohio, 12, SE
Andover, Va., 164, NW
Andreas (LNE; LV), Pa., 50, NW
Andreas Siding, Pa., 50, NW
Andreas, Pa., 50, SW
Andrews Ave. (Wildwood), N.J., 117, NW
Andrews Run Jct., Pa., 57, NE
Andrews Settlement, Pa., 18, NW
Andrews, N.J., 83, SW
Angelica, N.Y., 5, NE
Angerona, W.Va., 103, NW
Angle, Va., 170, NE
Anglin, Ky., 118, NW
Angora, Pa., 82, NW
Anita (B&O; PRR), Pa., 44, NW
Anjean, W.Va., 122, SE
Ann St. (Parkersburg), W.Va., 86, NE
Annabelle Branch Connection, W.Va., 89, NW
Annadale, N.Y., 53, SE
Annandale, N.J., 52, SW
Annandale, Pa., 27, SW
Annapolis, Md., 114, NW
Annville, Pa., 63, NE

Anselma, Pa., 65, SE
Ansonia (B&O; NYC), Pa., 19, SW
Anspach, Pa., 74, NW
Anstead, W.Va., 121, SE
Anthony, W.Va., 138, NW
Anthras, Tenn., 162, SW
Antietam, Md., 94, NW
Antioch, W.Va., 89, NE
Antlers, Va., 173, SW
Antrim, Pa., 19, SW
Apalachin, N.Y., 9, SE
Apex, N.Y., 11, SW
Apex, Ohio, 55, NW
Apex, Pa., 35, SW
Apollo, Pa., 42, SE
Appalachia (INT; L&N), Va., 164, NW
Appalachia Coal Co., Pa., 43, SE
Apple Grove, W.Va., 102, SE
Applegate, Va., 141, NE
Appolds, Md., 78, SW
Appomattox, Va., 155, NW
Aquashicola, Pa., 50, NE
Aqueduct, N.Y., 54, SW
Aquia, Va., 129, NW
Ararat, Pa., 22, NE
Arbuckle, W.Va., 103, SW
Arcadia, Md., 79, SW
Arcadia, Pa., 44, NW
Arcadia, Va., 139, SE
Archbald (D&H; NYO&W), Pa., 35, NE
Archer, W.Va., 71, SE
Arcola, Pa., 66, SW
Arcola, W.Va., 122, NE
Ardara, Pa., 57, NE
Ardel, W.Va., 119, NW
Arden, Del., 82, NW
Arden, N.Y., 38, NE
Arden, Pa., 56, SW
Arden, W.Va., 90, SW
Ardmore, Pa., 66, SW
Ardsley, N.Y., 39, SW
Ardsley, Pa., 66, SE
Ardsley-on-Hudson, N.Y., 39, SW
Arensberg, Pa., 73, NW
Argentine, Pa., 27, SW
Argillite, Ky., 118, NW
Argo, Ky., 148, NE
Argonne, W.Va., 134, NE
Arista, W.Va., 150, NW
Aristes Jct., Pa., 49, NW

Aristes, Pa., 49, NW
Arjay, Ky., 162, NE
Arkendale, Va., 129, NW
Arkle, Ky., 162, NW
Arkport (Erie; PS&N), N.Y., 6, NE
Arlington Ave., N.J., 53B
Arlington, Md., 96, NE
Arlington, N.J., 53, NE
Arlington, N.Y., 53, SE
Arlington, Pa., 50, NW
Arlington, Pa., 82, NW
Arlington, W.Va., 150, NW
Armilda, W.Va., 119, SW
Arminius Mines, Va., 128, SW
Armitage, Ohio, 85, NE
Armstrong Mills, Ohio, 71, NW
Arndts, Pa., 51, NW
Arnette, W.Va., 105, NE
Arnettsville, W.Va., 72, SE
Arno, Va., 164, NW
Arnold Hill, W.Va., 107, NW
Arnold, Md., 96, SE
Arnold, Pa., 42, SW
Arnold, W.Va., 105, NE
Arnot, Pa., 19, SE
Arnow, Pa., 59, SW
Arona, Pa., 57, NE
Arrington, Va., 141, SW
Arritt, Va., 138, SE
Arrowchar, N.Y., 53, SE
Arrowhead, Va., 141, NE
Arroyo, Pa., 29, NW
Arroyo, W.Va., 40, SE
Artemus (A-J; L&N), Ky., 162, NW
Arters, Pa., 48, NE
Arthur, Tenn., 162, SE
Arthurs, Pa., 28, NW
Artrip, Va., 165, NE
Arverne, N.Y., 54, SW
Arvondale, W.Va., 106, SW
Arvonia, Va., 142, SW
Asbestos, Pa., 66, SE
Asco, W.Va., 149, NE
Ash Gap, Pa., 35, NE
Ashburn, Va., 95, SW
Ashby, Va., 93, SE
Ashcake, Va., 144, SW
Ashcom, Pa., 60, SW
Ashfield, Pa., 50, NE
Ashford, N.Y., 4, NE
Ashford, W.Va., 120, SE

Ashland (LV; RDG), Pa., 49, NW
Ashland Upper, Pa., 49, NW
Ashland, Del., 81, NE
Ashland, Ky., 118, NE
Ashland, Md., 96, NE
Ashland, N.J., 82, NE
Ashland, Pa., 49, NW
Ashland, Va., 144, NW
Ashland, W.Va., 150, NW
Ashley Planes, Pa., 35A
Ashley, Pa., 35A
Ashmore, Pa., 50B
Ashtabula (NKP; NYC; PRR), Ohio, 12, NW
Ashtabula Harbor (NYC; PRR), Ohio, 12, NW
Ashtola, Pa., 59, SW
Ashton, Pa., 66, SE
Ashton, W.Va., 102, SE
Ashville, N.Y., 3, SW
Ashville, Pa., 44, SE
Aspen, Va., 155, SW
Aspinwall, Pa., 57A
Astor Branch Jct., W.Va., 89, NE
Astor, W.Va., 89, SE
Atcheson, Pa., 56, NW
Atchison, Pa., 73, SW
Atco, N.J., 83, NW
Atenville, W.Va., 119, SE
Atglen, Pa., 81, NW
Athenia (DL&W; Erie), N.J., 53, NE
Athens (B&O; C&O; NYC), Ohio, 85, NE
Atison, N.J., 83, SE
Atkins, Va., 167, NW
Atlantic Ave., N.Y., 54, SE
Atlantic City (P-RSL; PRR-WJ&S; RDG-AC), N.J., 101, NW
Atlantic Highlands, N.J., 68, NE
Atlantic, N.Y., 53, SW
Atlantic, Pa., 13, SW
Atlantic, Pa., 74, NE
Atlasburg, Pa., 56, NW
Atlee, Va., 144, SW
Attoway, Va., 167, NW
Atwell, W.Va., 149, NW
AU Crossover, Md., 75, SW
Auburn (PRR; RDG), Pa., 49, SE
Auburndale, N.Y., 54, NW
Auchenbach, Pa., 49, SW
Aucheys, Pa., 49, SE

Audobon, N.J., 82B
Aughwick, Pa., 61, NW
Augusta Springs, Va., 125, SW
Augusta, Md., 94, NE
Auguston, Pa., 15, SE
Aura, N.J., 82, SE
Austen, W.Va., 90, NW
Austin Jct., Pa., 35B
Austin, Pa., 17, SE
Austinburg, Ohio, 12, NW
Austintown Crossing, Ohio, 25A
Austintown, Ohio, 25, SW
Austinville, Va., 168, NW
Autrey Park, Md., 95, SE
Auville, W.Va., 149, NW
Avalon, Md., 96, SE
Avalon, N.J., 100, SE
Avalon, Pa., 56A
Avella, Pa., 56, NW
Avenel, N.J., 53, SW
Averett, Va., 172, SE
Avery, Pa., 22, SW
Avis Shops, Pa., 32, SW
Avis, Pa., 32, SW
Avis, W.Va., 137, SW
Avoca (DL&W; Erie), N.Y., 7, NW
Avoca Jct., Pa., 35B
Avoca Yard, Pa., 35B
Avoca, Pa., 35B
Avoca, Pa., 35C
Avoca, W.Va., 121, NE
Avon Branch Switch, Pa., 64, NW
Avon, N.J., 68, SE
Avon, Pa., 64, NW
Avondale, Md., 78, SE
Avondale, Ohio, 71, NW
Avondale, Pa., 81, NW
Avonmore, Pa., 43, SW
Awosting, N.J., 38, SW
Axemann, Pa., 46, NW
Axton, Va., 170, SE

B&O Freight House (Erie), Ohio, 25A
B&O Jct., Pa., 30, SW
BA, Ohio, 55, NW
Baber, W.Va., 119, SE
Back Bay, Va., 177, SE
Backbone, Va., 138, NE
Backus, Pa., 16, NE
Bacon Hill, Md., 81, SW
Bacons Neck, N.J., 99, NW
Bacova Jct., Va., 124, NW

Baden, Pa., 41, SE

Baden, W.Va., 102, NE

Bagdad, Pa., 42, SE

Baggaley, Pa., 58, NW

Bailey, Pa., 62, NE

Bailey's Mills, Ohio, 70, NE

Bailey's, Ky., 162, NW

Baileysville, W.Va., 135, SE

Bainbridge, N.Y., 11, NW

Bainbridge, Pa., 63, SE

Bair, Pa., 79, NW

Baird, Pa., 57, SW

Baker Jct., Pa., 79, NW

Baker, Pa., 31, SE

Baker, Pa., 46, NW

Baker, Pa., 56, SE

Baker, Pa., 81, NW

Bakerstown, Pa., 42, SW

Bakersville, N.J., 100, NE

Bakerton, Pa., 44, SW

Bakerton, W.Va., 94, NW

Bala, Pa., 66B

Balcony Falls, Va., 140, SW

Bald Eagle Jct., Pa., 31, SE

Bald Eagle, Pa., 45, SE

Bald Hill, Pa., 30, SW

Bald Pilot, Md., 80, SE

Baldwin Place, N.Y., 39, NW

Baldwin, Md., 97, NW

Baldwin, N.Y., 54, SE

Baldwin, Va., 139, SW

Baleville, N.J., 37, SW

Balkan, Ky., 162, NE

Ballanger, W.Va., 122, SW

Balls Road, Md., 77, SE

Ballsville, Va., 156, NE

Balsden, W.Va., 135, SW

Baltimore (PRR), Md., 96A

Baltimore Heights, Md., 96, SE

Baltusrol, N.J., 53, SW

Bancroft, W.Va., 103, SW

Bandy, Va., 149, SE

Bangor (DL&W; LNE), Pa., 51, NE

Bangor Jct., Pa., 51, NE

Bangor Shops, Pa., 51, NE

Bangor, Va., 165, NW

Banian Jct., Pa., 45, NW

Bank, Pa., 48, NW

Banksville Jct., Pa., 56A

Banksville, Pa., 56A

Banner Mine, W.Va., 90, NE

Banner, Va., 165, NW

Banning (B&O; P&WV), Pa., 57, SE

Bannock, Ohio, 55, SW

Banta, N.J., 53, SW

Barbours Creek, Va., 138, SE

Barboursville, Va., 127, SW

Barboursville, W.Va., 119, NW

Barbourville, Ky., 162, NW

Barclay, Md., 98, SW

Barclay, Pa., 20, SE

Barcroft, Va., 112, NE

Bard, Pa., 75, NE

Bardane, W.Va., 94, NW

Bardonia, N.Y., 38, SE

Bare Hills, Md., 96, NE

Bare Rock, Pa., 74, NE

Baree, Pa., 45, SE

Bareville, Pa., 64, SE

Barker, Ohio, 87, NW

Barking, Pa., 42, SW

Barksdale, Va., 171, SE

Barksdale, W.Va., 137, SW

Barlow, W.Va., 120, NE

Barmouth, Pa., 66B

Barnabas, W.Va., 134, SE

Barnegat (CNJ; SNJ), N.J., 84, NE

Barnegat Pier, N.J., 84, NE

Barnes (S&T; TIV), Pa., 15, SE

Barnes, N.Y., 8, NW

Barnes, Pa., 28, SE

Barnesboro, Pa., 44, SW

Barnesville, Md., 95, SW

Barnesville, Ohio, 70, NE

Barnesville, Pa., 49, NE

Barnett, W.Va., 135, NE

Barneville, Md., 75, SW

Barnitz, Pa., 62, SE

Barnsley, Pa., 81, NW

Barnum, W.Va., 91, NE

Barrackville, W.Va., 72, SE

Barren Creek, W.Va., 121, NW

Barren Springs, Va., 168, NW

Barrett, Pa., 30, SW

Barrington, N.J., 82, NE

Barronvale, Pa., 74, NW

Barrs, W.Va., 104, NW

Barry, Pa., 49, NW

Barryville, Pa., 49, NE

Bartholow, Md., 95, NE

Bartley, N.J., 52, NE

Bartley, W.Va., 149, NE

Bartlick, Va., 148, SW

Barton, Md., 74, SE

Barton, N.Y., 9, SW

Barton, Ohio, 55, SW

Barton, Pa., 43, NE

Barton, W.Va., 123, NW

Bartonsville, Pa., 36, SW

Bartonville, Va., 93, SE

Bartow, N.Y., 54, NW

Bartow, W.Va., 107, SW

Baskerville, Va., 173, SW

Basking Ridge, N.J., 52, SE

Bassett, Va., 170, NW

Bastian, Va., 150, SE

Bastin, Ky., 147, SW

Bath (B&H; DL&W; Erie), N.Y., 7, NW

Bath (DL&W; LNE; NB), Pa., 51, SW

Baumgardner, Pa., 80, NW

Baxter, Ky., 163, NW

Baxter, Pa., 28, SE

Bay Head (PRR), N.J., 68, SE

Bay Head Jct. (NY&LB), N.J., 68, SE

Bay Ridge, N.Y., 53, SE

Bay Terrace, N.Y., 53, SE

Bay View Ave., N.J., 68, NE

Bay View Yard, Md., 96A

Bay View, Va., 161, NW

Bayard, W.Va., 91, NW

Baychester Ave., N.Y., 54, NW

Baychester, N.Y., 54, NW

Baylor, W.Va., 136A, NW

Bayside, N.J., 99, NW

Bayside, N.Y., 54, NW

Bayway (CNJ; SIRT), N.J., 53, SE

BC, Pa., 59, NE

BD, N.J., 67, SE

Beach Bottom, W.Va., 55, SE

Beach Glen, N.J., 53, NW

Beach Haven Crest, N.J., 84, SE

Beach Haven Terrace, N.J., 84, SE

Beach Haven, N.J., 84, SE

Beach Haven, Pa., 34, SE

Beach, Va., 157, NE

Beachdale, Pa., 74, NE

Beacon (NH; NYC), N.Y., 39, NW

Beadling, Pa., 56, NE

Bealeton, Va., 111, SW

Beallsville, Ohio, 70, NE

Beamon, Va., 176, NE

Beans Mill, W.Va., 106, NE

Beans, N.Y., 7, NW

Bear Creek Jct. (B&O; PBR), Md., 97, SW

Bear Creek Jct., Pa., 35, SW

Bear Creek, Pa., 29, NW

Bear Creek, Pa., 35, SW

Bear Creek, W.Va., 105, SW

Bear Lake, Pa., 14, NE

Bear Mountain, N.Y., 39, NW

Bear Mountain, W.Va., 89, SE

Bear Run Jct., Pa., 49C

Bear Swamp, Pa., 31, SE

Bear Valley, Pa., 48, NE

Bear Wallow, W.Va., 150, NW

Bear, Del., 81, SE

Bear, Ky., 162, SE

Bear, Va., 148, NE

Beard, W.Va., 123, SE

Beard's Fork, W.Va., 121, SE

Beards Jct., W.Va., 121, SE

Beartown, W.Va., 149, NW

Beatty, Pa., 58, NW

Beaufort, N.J., 53, NW

Beaupland, Pa., 35, SE

Beaver (P&LE; PRR), Pa., 41, SW

Beaver Creek Jct., W.Va., 122, NE

Beaver Dam, Md., 132, SE

Beaver Dam, N.Y., 8, NW

Beaver Dam, Va., 143, NE

Beaver Falls (PRR), Pa., 41, NW

Beaver Falls & New Brighton, Pa., 41, SW

Beaver Lake, N.J., 37, SE

Beaver Meadow, Pa., 50, NW

Beaver Road, Pa., 56, NE

Beaver Springs, Pa., 47, SE

Beaver Valley, Pa., 49, NW

Beaver, Pa., 45, NE

Beaver, W.Va., 136, SE

Beavertown, Pa., 47, NE

Beccaria, Pa., 45, NW

Becket, Ohio, 71, NW

Beckler, Ohio, 85, NW

Beckley Jct., W.Va., 136A, NE

Beckley, W.Va., 136A, NE

Becks, Pa., 49, SE

Bedford Hills, N.Y., 39, SE

Bedford, Pa., 60, SW

Bedford, Va., 153, NE

Bedington, W.Va., 77, SW

Beech Creek (NYC; PRR), Pa., 31, SE

Beech Farm, Ky., 118, SE

Beech Glen, Pa., 33, NE

Beech Glen, W.Va., 121, SE

Beech Hill, W.Va., 103, NW

Beechmont, Pa., 56, NE

Beechton, Pa., 29, SW

Beechtree, Pa., 29, SW

Beechwood, W.Va., 72, SE

Beechwood, W.Va., 91, SW

Beeler's Summit, Md., 94, NE

Beerston, N.Y., 11, SE

Bel Air, Md., 80, SW

Belair Road, N.Y., 53, SE

Belbois, Pa., 73, NW

Belcamp, Md., 97, NE

Belcher, Ky., 148, NW

Belden, N.Y., 10, SE

Belfast (Erie), N.Y., 5, NE

Belfast (W&B), N.Y., 5, NE

Belfast Jct., Pa., 51, NW

Belford, N.J., 68, NE

Belfry, Ky., 134, SW

Belfry, Pa., 66, SW

Belington (B&O; WM), W.Va., 90, SW

Bell Haven, N.J., 100, NE

Bell Siding, Pa., 56, SW

Bell, Pa., 58, SE

Bell's Valley, Va., 125, SW

Bellaire (B&O; PRR), Ohio, 55, SE, 55A

Bellaire, N.Y., 54, SE

Bellaire, Pa., 63, SE

Bellburn, W.Va., 122, SE

Belle Haven, Va., 146, SW

Belle Mead General Depot, N.J., 67, NE

Belle Valley, Pa., 1, SE

Belle Vernon (P&LE; PRR), Pa., 57, SW

Belle, W.Va., 120, SE

Bellefonte (BFC; CRPA; PRR), Pa., 46, NW

Bellefonte, Ky., 118, NE

Belleplain, N.J., 100, NW

Bellerose, N.Y., 54, SE

Belleville, N.J., 53, NE

Belleville, Pa., 46, SE

Belleville, W.Va., 86, SE

Bellevue Jct., Pa., 82A

Bellevue, Pa., 56A

Bellmawr, N.J., 82, NE

Bellmore, N.Y., 54, SE

Bells Landing, Pa., 44, NE

Bells, Pa., 58, NW

Bellstrace, Ky., 118, SW

Bellwood Park, N.J., 53, NE

Bellwood, Pa., 45, SW

Bellwood, Va., 158, NW

Bellwood, W.Va., 137, NW

Belmar, N.J., 68, SE

Belmar, Pa., 27, NW

Belmont (Erie), N.Y., 5, SE

Belmont (W&B), N.Y., 5, SE

Belmont Park, N.Y., 54, SE

Belmont Park, Va., 94, SE

Belmont St., Ohio, 25, SE

Beloit, Ohio, 40, NW

Belona, Va., 142, SE

Belpre, Ohio, 86, NE

Belsano, Pa., 44, SW

Belsena (NYC; PRR), Pa., 45, NW

Belspring, Va., 151, SE

Belt Jct., Ky., 162, SE

Belt Jct., Va., 177A

Belt Line Crossing (N&W; N&PBL), Va., 177A

Belt Line Jct., Va., 153, NW

Beltsville, Md., 96, SW

Belva (C&O; NYC), W.Va., 121, SE

Belvidere (Erie), N.Y., 5, SE

Belvidere (L&HR; PRR), N.J., 51, NE

Belvidere (PS&N), N.Y., 5, NE

Belvoir, Va., 111, NW

Bemis, W.Va., 107, NE

Bemus Point, N.Y., 3, SW

Ben Avon, Pa., 41, SE

Ben Hur, Va., 163, SE

Ben Lomond, Va., 143, SW

Ben Lomond, W.Va., 102, SE

Ben Roy, Pa., 79, NE

Ben's Run, W.Va., 87, NE

Benbush, W.Va., 90, SE

Bendale, W.Va., 89, SW

Benders Jct., Pa., 51, NW

Bendersville, Pa., 78, NE

Bengies, Md., 97, NW

Benham, Ky., 164, NW

Benham, Va., 165, SE

Bennett Springs, Va., 152, NE

Bennett, N.J., 117, NW

Bennett, W.Va., 105, NE

Bennetts (NY&P), N.Y., 6, SE

Bennetts (PS&N), N.Y., 6, NW

Bennettsville, Ky., 162, NW

Bennezette (B&O; PRR), Pa., 30, NW

Benning (B&O; PRR), D.C., 113, NW, 113A

Benson, Pa., 43, SW

Bentley Springs, Md., 79, SE

Bentley, Ohio, 25, SE

Bentleyville, Pa., 56, SE

Benton, Pa., 34, SW

Bentonville, Va., 110, NW

Bentree, W.Va., 121, NE

Benwood Jct., W.Va., 55, SE, 55A

Benwood Yard, W.Va., 55, SE, 55A

Benwood, W.Va., 55, SE, 55A

Berg Run, Pa., 30, NE

Bergan, Pa., 33, NW

Bergen Hill, N.J., 68, NW

Bergenfield, N.J., 54, NW

Bergholtz, Ohio, 40, SW

Bergoo, W.Va., 123, NW

Berkeley Heights, N.J., 53, SW

Berkeley Run Jct., W.Va., 89, NE

Berkeley Springs, W.Va., 76, SE

Berkeley, N.J., 84, NE

Berkeley, W.Va., 77, SW

Berkey, Pa., 58, SE

Berkley, Pa., 65, NW

Berkley, Va., 177A

Berkley's Mill, Pa., 74, NE

Berkshire, N.Y., 9, NE

Berlin Center, Ohio, 25, SW

Berlin Jct. (EB; WM), Pa., 78, NE

Berlin, Md., 133, NE

Berlin, N.J., 83, NW

Berlin, Pa., 75, NW

Berlinsville No. 2, Pa., 50, NE

Berlinsville, Pa., 50, NE

Bermuda, Va., 158, NW

Bernardsville, N.J., 52, SE

Berne, Pa., 49, SE

Bernice Jct., Pa., 34, NW

Bernice, Pa., 34, NW

Berrton Siding, Pa., 44, SW

Berry Hill, Va., 170, SE

Berryburg Jct., W.Va., 89, SE

Berryburg, W.Va., 89, SE

Berryville, Va., 94, SW

Bertha, Pa., 56, NW

Berton, Va., 151, NE

Berwick (DL&W; PRR), Pa., 34, SE

Berwind, W.Va., 149, NE

Berwinsdale, Pa., 44, NE

Berwyn, Md., 113, NW

Berwyn, Pa., 66, SW

Besco, Pa., 72, NE

Besemer, N.Y., 9, NW

Besoco, W.Va., 136, SE

Bess, Va., 138, SE

Bessemer (B&O; P&LE; PRR; URR), Pa., 57B

Bessemer Jct., Pa., 57B

Best, Pa., 50, SE

Best, Pa., 57, NW

Bethayres, Pa., 66, SE

Bethel, Md., 95, NW

Bethel, Pa., 26, SW

Bethesda, Md., 112, NE

Bethlehem (CNJ; LNE; LV; RDG), Pa., 51, SW, 51E

Bethlehem, Md., 115, SW

Betty Baker, Va., 168, NW

Betz Jct., Pa., 45, NW

Betzwood, Pa., 66, SW

Beuchler, Pa., 49, SW

Beulah Road, Pa., 59, NW

Beulah, W.Va., 107, NW

Beverly, N.J., 67, SW

Beverly, W.Va., 107, NW

BF, Pa., 59, NE

Bias, W.Va., 134, SE

Bickell, W.Va., 136A, NE

Bickford, Pa., 44, NE

Bickmore, W.Va., 121, NE

Bidwell (B&O; WM), Pa., 74, NW

Bidwell, Ohio, 102, NW

Big Bend, Pa., 15, NE

Big Blizzard, W.Va., 123, SW

Big Branch, Ky., 148, NW

Big Chimney, W.Va., 120, NE

Big Creek, W.Va., 119, SE

Big Curve, W.Va., 91, NE

Big Elk, Md., 81, SW

Big Fill, Pa., 16, SW

Big Flats (DL&W; Erie), N.Y., 8, SW

Big Island, Va., 140, SW

Big Meadow Run, Pa., 73, NW

Big Mine Run Jct., Pa., 49, NW

Big Pool, Md., 76, SE

Big Q, W.Va., 137, NW

Big Run, Ohio, 86, NW

Big Run, Pa., 44, NW

Big Run, W.Va., 106, SW

Big Run, W.Va., 124, SW

Big Sandy, W.Va., 149, NE
Big Shanty, Pa., 16, NE
Big Spring, Md., 77, SW
Big Stick, W.Va., 136A, SW
Big Stone Gap (L&N; SOU), Va., 164, NW
Big Towns Creek Br. Jct., Va., 165, NW
Biggs, Ky., 148, NW
Bigler (NYC; PRR), Pa., 45, NW
Biglersville, Pa., 78, NE
Bigley Ave., W.Va., 120, NE
Bigson, W.Va., 135, NE
Bill, Ky., 148, NE
Billings, Pa., 27, NE
Billings, W.Va., 104, NW
Billmeyer, Pa., 63, SE
Bingen, Pa., 51, SW
Bingham, Pa., 16, NE
Bingham, Pa., 18, NE
Binghamton (D&H; DL&W; Erie), N.Y., 10, SW, 10A
Birch Run, W.Va., 121, NE
Birch, Pa., 27, SE
Birchton, W.Va., 135, NE
Bird's Nest, Va., 161, NW
Birdell, Pa., 65, SW
Bird-in-Hand, Pa., 64, SE
Birdsall, N.Y., 6, NW
Birdsboro (PRR; RDG), Pa., 65, NW
Birdwood, Ky., 162, SW
Birge Run, Pa., 31, SE
Birmingham, N.J., 83, NE
Birmingham, Pa., 45, SE
Bishop (PRR; P&WV), Pa., 56, NE
Bishop, Md., 133, NE
Bishop, W.Va., 149, SE
Bishops Bridge, N.J., 83, NW
Bison, W.Va., 105, SE
Bissell, Md., 77, SE
Bitner, Pa., 73, NE
Bittinger, Md., 74, SE
Bittinger, Pa., 78, NE
Bivalve, N.J., 99, SE
Black Betsey, W.Va., 103, SW
Black Cat, W.Va., 121, SW
Black Creek (Erie), N.Y., 5, NE
Black Creek (PRR), N.Y., 5, NE
Black Eagle, W.Va., 136, SW
Black Lick Jct., Pa., 58, NE
Black Oak, W.Va., 92, NW

Black Rock, Pa., 79, SW
Black Siding, N.J., 53, SW
Black Walnut, Pa., 21, SE
Black, Md., 98, NW
Black's Run, Pa., 42, SW
Blackburn, Pa., 57, NE
Blackeley, W.Va., 121, NW
Blackey, Ky., 147, SW
Blackey, Va., 149, NW
Blackford, Va., 166, NW
Blacklog, Pa., 61, SW
Blackmont, Ky., 162, NE
Blacksburg, Va., 152, SW
Blacksnake, Ky., 162, NE
Blackstone, Va., 156, SE
Blacksville, Pa., 72, SE
Blackwell, Pa., 19, SW
Blackwolfe, W.Va., 150, NW
Blackwood (INT; L&N), Va., 164, NE
Blackwood (LV; RDG), Pa., 49, SW
Blackwood Jct., Pa., 49B
Blackwood, N.J., 82, NE
Blain City, Pa., 44, NE
Blain, Pa., 61, NE
Blaine, Ohio, 55, SW
Blaine, W.Va., 91, NE
Blair Four, Pa., 45, SE
Blair, W.Va., 135, NW
Blairs Mills, Pa., 61, NE
Blairs, Va., 171, SW
Blairstown (DL&W; NYS&W), N.J., 52, NW
Blairsville, Pa., 58, NW
Blairton, W.Va., 94, NW
Blake, W.Va., 150, NE
Blanchard, Del., 115, NE
Blanche, Ky., 162, NE
Blandburg, Pa., 45, SW
Blandon, Pa., 65, NW
Blasdell Jct., Pa., 44, NW
Blaser, W.Va., 90, NE
Blauvelts, N.Y., 39, SW
Blaw Nox, Pa., 57, NW
Blenheim, N.J., 82, NE
Blissville, Pa., 16, NW
Bloom, Pa., 29, SE
Bloom, Pa., 49, NW
Bloom, W.Va., 89, NW
Bloomfield (DL&W; Erie), N.J., 53, NE
Bloomfield Jct., Pa., 62, NE

Bloomfield, Md., 114, NE
Bloomfield, N.J., 53, NE
Blooming Grove, N.Y., 38, NE
Bloomingdale, N.J., 38, SW
Bloomington, W.Va., 91, NE
Bloomsburg (DL&W; RDG), Pa., 49, NW
Bloomsbury (CNJ; LV), N.J., 51, SE
Bloomsdorf, Pa., 46, NW
Blossburg, Pa., 19, SE
Blough, Pa., 59, SW
Bloxom, Pa., 146, NE
Blue Anchor, N.J., 83, SW
Blue Ball, Pa., 45, NW
Blue Creek (B&O; NYC), W.Va., 121, NW
Blue Lick, Pa., 74, NE
Blue Mount, Md., 79, SE
Blue Pennant, W.Va., 135, NE
Blue Ridge, Pa., 78, SW
Blue Ridge, Va., 153, NW
Blue Spring Run, Va., 138, SE
Blue Stone, Pa., 32, NW
Blue Sulphur, W.Va., 119, NE
Bluefield, Va., 150, NW
Bluefield, W.Va., 150, NE
Bluemont Jct., Va., 112, NE
Bluemont, Va., 94, SW
Bluestone Jct., W.Va., 150, NW
Bluf, Pa., 48, NE
Bluff, Va., 151, SE
Boanna, Ohio, 25, SW
Boardman, Ohio, 25, SE
Boardman, Pa., 45, NW
Boardville, N.J., 38, SW
Boatswain Branch, Va., 144, SW
Boaz, Va., 176, SW
Bob White, W.Va., 135, NE
Bodine, Pa., 33, NW
Boggs, Pa., 56, NW
Boggsville, Pa., 42, NE
Bogota (NYC; NYS&W), N.J., 53, NE
Bohemia, Va., 158, SW
Boice Siding, Va., 143, SW
Boiling Spring, Va., 141, NE
Boiling Springs, Pa., 62, SE
Boissevain, Va., 150, NW
Bolair, W.Va., 123, NW
Bolivar, N.Y., 5, SE
Bolivar, Pa., 58, NE
Bon Air, Va., 143, SE

Bonair, Pa., 66, SE
Bond, Md., 74, SE
Bonny Blue, Va., 163, NE
Bonny Brook, Pa., 62, SE
Bonsack, Va., 153, NW
Boody (CC&O; N&W), Va., 165, NW
Boomer, W.Va., 121, SW
Boone, Va., 164, SE
Boone, Va., 177, NW
Boones Mill, Va., 153, SW
Boonsboro, Md., 77, SE
Boonton, N.J., 53, NW
Booth, W.Va., 72, SE
Boothwyn, Pa., 82, NW
Borden Crossover, Md., 75A
Borden Shaft, Md., 75, SW
Borden Yard, Md., 75A
Bordentown, N.J., 67, SE
Border, Pa., 59, SW
Borderland, W.Va., 134, SW
Boring, Md., 79, SW
Borland, Pa., 56, NE
Bosco, Ky., 147, NW
Boscobel, Va., 143, SE
Bosher, Va., 143, SE
Boston Run Jct., Pa., 49C
Boston, Pa., 57, NW
Boswell, Pa., 58, SE
Botanical Garden, N.Y., 54, NW
Boudar, Va., 143, SE
Boulder, Va., 164, SE
Boulder, W.Va., 89, SE
Bound Brook, N.J., 52, SE
Boush Jct., Va., 177, NW
Bowden, W.Va., 107, NE
Bowenstown, N.J., 99, NW
Bower Hill, Pa., 56, NE
Bower, W.Va., 105, NE
Bowers, Pa., 65, NE
Bowers, Va., 177, NW
Bowersville, Pa., 44, NW
Bowest (WM), Pa., 73, NE, 73A
Bowest Jct. (B&O; WM), Pa., 73, NE, 73A
Bowie, Md., 96, SW
Bowman, Va., 109, NE
Bowmansdale, Pa., 63, SW
Bowmanstown (CNJ; LV), Pa., 50, NE
Bowmanstown, Pa., 50, NE
Bowood, Pa., 73, NW

Bowyer Creek Jct., W.Va., 136A, SE

Bowyer, W.Va., 136A, SE

Boxley, Va., 151, NW

Boxwood, Va., 170, SE

Boyce, Pa., 56, NE

Boyce, Va., 93, SE

Boyd, Md., 77, SW

Boyd, Md., 95, SW

Boyd, Pa., 48, NE

Boydton, Va., 173, SW

Boyer, Pa., 16, NE

Boyer, W.Va., 107, SW

Boyertown, Pa., 65, NE

Boykins, Va., 175, SE

Boyles, Pa., 48, NW

Boynton, Pa., 74, NE

Boys Home, Va., 138, NE

Braceville (Erie; NYC), Ohio, 25, SW

Bracey, Va., 173, SE

Braddock Heights, Md., 94, NE

Braddock, N.J., 83, SW

Braddock, Pa., 57, NW

Bradenville, Pa., 58, NW

Braders, Pa., 35, SW

Bradevelt, N.J., 68, NE

Bradford (B&O; Erie), Pa., 16, NE

Bradford Jct., Pa., 57, SE

Bradford Shops, Pa., 16, NE

Bradley Beach, N.J., 68, SE

Bradley Jct., Pa., 44, SE

Bradshaw, Md., 97, NW

Bradshaw, W.Va., 149, NW

Brady, W.Va., 119, SE

Brady's Bend, Pa., 42, NE

Braeburn, Pa., 42, SE

Braeholm, W.Va., 135, NW

Brallier, Pa., 60, SW

Brambaugh, Pa., 60, NE

Bramwell, W.Va., 150, NW

Branch Jct., N.J., 53, SW

Branch, Ohio, 40, SE

Branch, Pa., 26, NE

Branchdale, Pa., 49, SW

Branchland, W.Va., 119, SE

Branchport, N.J., 69, NW

Branchton, Pa., 27, SW

Branchville Jct., N.J., 37, SE

Branchville, Md., 96, SW

Branchville, N.J., 37, SE

Branchville, Va., 175, SE

Brand, Va., 125, SE

Brandamore, Pa., 65, SW

Brandon, Pa., 36, NE

Brandon, Pa., 77, NE

Brandonville, Pa., 49, NE

Brandt, Pa., 22, NE

Brandtsville, Pa., 62, SE

Brandy Camp, Pa., 29, NE

Brandy, Va., 111, SW

Brandywine (PRR; US GOV'T), Md., 113, SW

Brandywine Summit, Pa., 81, NE

Brandywine, Del., 81A

Braucher, W.Va., 107, SW

Brave, Pa., 72, SW

Braxton, W.Va., 105, NE

Bray, Va., 154, NE

Bream, W.Va., 120, NE

Breatheds, Md., 77, SE

Breece, W.Va., 120, SW

Breeden, W.Va., 134, NW

Breesport, N.Y., 8, SE

Breinigsville, Pa., 50, SE

Bremo, Va., 142, SW

Brent, Pa., 26, SE

Brentwood, Md., 113, NW

Bretz, W.Va., 73, SW

Briar Creek, Pa., 34, SW

Briarcliff Manor, N.Y., 39, SW

Brickyard, Del., 116, NW

Bridesburg, Pa., 66, SE

Bridewell, Md., 96, SW

Bridge Jct., W.Va., 120, NE

Bridge Jct., W.Va., 121, SE

Bridge No. 2, Va., 151, NW

Bridge Siding, Pa., 63, NW

Bridge St. (Port Washington), N.Y., 54, NW

Bridge St., Pa., 18, SE

Bridge, Pa., 35A

Bridgeburg, Pa., 43, NW

Bridgeport (B&O; PRR), Ohio, 55, SE, 55A

Bridgeport B&W Jct., Pa., 44, NE

Bridgeport Draw, N.J., 82, NW

Bridgeport, N.J., 82, NW

Bridgeport, Pa., 66A

Bridgeport, Va., 142, SW

Bridgeport, W.Va., 89, NW

Bridgeton (CNJ; P-RSL), N.J., 99, NE

Bridgeton Jct. (CNJ), N.J., 99, NE

Bridgeton Port (CNJ), N.J., 99, NE

Bridgeton, Pa., 80, NW

Bridgeville (PRR; P&WV), Pa., 56, NE

Bridgeville, Del., 115, SE

Bridgeville, N.J., 51, NE

Bridgewater, Pa., 41, SW

Bridgewater, Pa., 82, NW

Bridgewater, Va., 126, NW

Brier Hill, Ohio, 25A

Briery, Va., 156, SW

Brigantine Beach, N.J., 101, NW

Brigantine Jct., N.J., 100, NE

Briggs, Ohio, 86, NE

Brighton Ave., N.J., 53, NE

Brilliant, Ohio, 55, NE

Brinker, Pa., 42, NW

Brisben, N.Y., 10, NE

Brisbin, Pa., 45, NW

Brister, Ohio, 70, NW

Bristol (N&W), Va., 165, SE

Bristol (SOU), Tenn., 165, SE

Bristol, N.J., 67, SW

Bristol, W.Va., 88, NE

Bristolville, Ohio, 25, NW

Briston, Md., 97, SE

Bristow, Va., 111, SE

Broad Channel, N.Y., 54, SW

Broad Ford, Pa., 57, SE

Broad Mountain, Pa., 49, SE

Broad Run, Va., 111, NE

Broad St. (Richmond), (RF&P), Va., 144A

Broad St. (Trenton), N.J., 67, SW

Broad Top City, Pa., 60, SE

Broadacre, Ohio, 55, NW

Broadbeck, Pa., 79, NW

Broadford Jct., Pa., 57, SE

Broadkill, Del., 116, SW

Broadnax, Va., 173, SE

Broadway (Fairlawn), N.J., 53, NE

Broadway, N.J., 51, SE

Broadway, N.Y., 54, NW

Broadway, Va., 109, SW

Broadwell, Ohio, 86, NW

Brock Road, Va., 128, NE

Brockport, Pa., 29, NE

Brockton (NKP), N.Y., 3, NW

Brockton (NYC), N.Y., 3, NW

Brockton (PRR), N.Y., 3, NW

Brockton, Pa., 49, SE

Brockway (B&O; Erie; PRR; PS), Pa., 29, SW

Brodhead, Pa., 51, SW

Brogueville, Pa., 79, NE

Bronx Park 180th St., N.Y., 54, NW

Bronxville, N.Y., 54, NW

Brook Hill, W.Va., 91, SE

Brook Park, Pa., 48, NW

Brook Road, Pa., 66A

Brook, W.Va., 55, NE

Brooke, Va., 129, NW

Brookes Mills, Pa., 60, NW

Brookewood, Va., 125, SE

Brookfield, Ohio, 25, SE

Brookland, Pa., 18, NW

Brooklandville, Md., 96, NE

Brooklawn, N.J., 82, NE

Brooklyn Manor, N.Y., 54, SW

Brooklyn, Md., 96, SE

Brooklyn, W.Va., 136, NE

Brookneal (N&W; VGN), Va., 155, SW

Brooks, W.Va., 137, SW

Brookside Colliery, Pa., 48, SE

Brookside, N.J., 52, NE

Brookston, Pa., 16, SW

Brookton, N.Y., 9, NW

Brookville (NYC; PRR; PS), Pa., 28, SE

Brosia, W.Va., 102, NE

Brounland (C&O; KC), W.Va., 120, SW

Brown St. (Richmond), Va., 144A

Brown, Md., 95, SE

Brown, N.J., 83, NW

Brown, W.Va., 89, NW

Brown's Ferry, Pa., 73, NW

Brownfield, Pa., 73, NE

Brownlee, Pa., 19, SW

Browns Mills, N.J., 83, NE

Brownsdale, Pa., 57, SW

Brownstone (B&M; RDG), Pa., 63, NE

Brownsville (P&LE), Pa., 57, SW

Brownsville, Md., 94, NE

Brownton, Pa., 79, NE

Browntown, W.Va., 89, SE

Brownwood, W.Va., 137, NW

Browser (NH; NYC), N.Y., 39, NE

Bruce, Pa., 80, NW

Bruce, Va., 177, NW

Bruceton (B&O; P&WV), Pa., 57, NW

Bruckharts, Pa., 64, SW

Bruin, Pa., 27, SE

Brundage, N.Y., 7, NW
Brunswick, Md., 94, NE
Brush Fork Jct., Ohio, 85, NE
Brush Run, Pa., 78, NE
Brushton, W.Va., 120, SE
Bryansville, Pa., 80, NW
Bryn Athyn, Pa., 66, SE
Bryn Mawr Park, N.Y., 54, NW
Bryn Mawr, Pa., 66, SW
Bryners, Pa., 28, SW
Bryson Mountain, Tenn., 162, SW
Buchanan (C&O; N&W), Va., 139, SE
Buchanan Branch Jct., W.Va., 134, SE
Buchanan, Ky., 118, SE
Buchtel, Ohio, 85, NE
Buck Lodge, Md., 95, SW
Buck Mountain, Pa., 49, NE
Buck Ridge, Pa., 48, NE
Buck Run Jct., Pa., 49, SW
Buck Run, Pa., 81, NW
Buck Siding, Pa., 49, NE
Buckeye Furnace, Ohio, 85, SW
Buckeye, Ohio, 40, NE
Buckeye, W.Va., 123, SE
Buckeye, W.Va., 136, SW
Buckeystown, Md., 95, NW
Buckhannon, W.Va., 106, NE
Buckhorn, Pa., 34, SW
Buckingham, Ky., 147, NE
Buckingham, Pa., 66, NE
Buckley, Pa., 49, SE
Bucknell, Pa., 48, NW
Buckner, Va., 143, NW
Buckton, Va., 110, NW
Bud, Pa., 32, SE
Bud, W.Va., 136, SW
Buell, Va., 177, NW, 177A
Buena Vista (C&O; N&W), Va., 140, SW
Buena Vista, Pa., 42, NW
Buena Vista, Pa., 57, NW
Buena, N.J., 83, SW
Buena, Va., 127, NE
Buffalo Br., Pa., 62, NE
Buffalo Creek, Pa., 42, NE
Buffalo Creek, W.Va., 118, NE
Buffalo Forge, Va., 140, SW
Buffalo Jct., Va., 172, SE
Buffalo Lithia Springs, Va., 172, SE
Buffalo Mills, Pa., 75, NE

Buffalo, Va., 141, SW
Buffalo, W.Va., 103, SW
Buhls, Pa., 41, NE
Bulger, Pa., 56, NW
Bull Creek Jct., Va., 148, NE
Bullock, N.J., 84, NW
Bumpass, Va., 143, NE
Bundy, Va., 164, NW
Bunker Hill, Pa., 64, NW
Bunker Hill, Va., 93, NE
Bunola, Pa., 57, SW
Burdett, N.Y., 8, NW
Burdette, Va., 176, NW
Burdetts Creek, W.Va., 122, SW
Burdine, Ky., 147, SE
Burdine, Pa., 56, NE
Burgess, Va., 157, SE
Burgettstown, Pa., 56, NW
Burghill, Ohio, 25, NE
Burk, W.Va., 73, SW
Burke, Va., 112, NW
Burkeville (N&W; SOU), Va., 156, SE
Burkhelder, Pa., 74, NE
Burks Garden, Va., 150, SW
Burks, W.Va., 136A, NE
Burlington, N.J., 67, SW
Burly, Pa., 45, NW
Burma, W.Va., 136, SE
Burnaugh, Ky., 118, NE
Burnham, Pa., 46, SE
Burnley, Va., 127, SW
Burns, N.Y., 6, NE
Burnside Jct., Pa., 48, NE
Burnside, N.Y., 38, NE
Burnside, Pa., 44, NW
Burnsville Jct., W.Va., 105, NE
Burnsville, W.Va., 105, NE
Burnt House, N.J., 83, NE
Burnwell, W.Va., 121, SW
Burnwood, Pa., 22, NE
Burrell, Va., 177, SW
Burrows, Pa., 18, SE
Burtner, Md., 77, SE
Burton, Ky., 147, NE
Burton, W.Va., 72, SW
Burtons Ford, Va., 165, NW
Burtville, Pa., 17, NE
Bush Hill, Va., 112, NE
Bush River, Md., 97, NE
Bush, Md., 97, NE
Bushkill Center, Pa., 51, NW
Bushkill Jct., Pa., 51, NW

Bushwick Jct., N.Y., 54, SW
Bushwick, N.Y., 54, SW
Buskill, Pa., 36, SE
Bustleton (PRR; RDG), Pa., 66, SE
Bute, Pa., 73, NE
Butler (B&LE; B&O; PRR), Pa., 42, NW
Butler Jct. (B&O), Pa., 42, NW
Butler Jct. (WNF), Pa., 42, SE
Butler, N.J., 38, SW
Butterworths Cor., N.J., 83, NE
Buttonwood, Pa., 35A
Butts, Va., 177, SE
Butzville, N.J., 51, NE
Byberry, Pa., 66, SE
Bycot, Pa., 66, NE
Byers, Pa., 65, SE
Byers, W.Va., 106, SE
Bynum, Md., 80, SW
Byram, Pa., 66, NE
Byrne, W.Va., 72, SE
Byrnedale, Pa., 29, NE
Byromtown, Pa., 15, SE
Byron, W.Va., 89, SW

C&C Jct., W.Va., 107, NW
C.J.I. Camps, N.Y., 37, NE
CA Jct., Pa., 49, NW
Cabell, W.Va., 136A, NE
Cabin Creek Jct., W.Va., 121, SW
Cabin Run, Pa., 20, SE
Cabot, Pa., 42, NW
Cabot, W.Va., 120, SE
Cacapon Lake, W.Va., 92, SE
Caddell, W.Va., 90, NE
Cadiz Jct., Ohio, 55, NW
Cadiz, N.Y., 5, NW
Cadogan, Pa., 42, NE
Cadosia, N.Y., 23, NW
Cady, Va., 144, SW
Cairnbrook, Pa., 59, SW
Cairo (B&O; C&KV), W.Va., 87, SE
Calcite, Pa., 60, NE
Caldonia (B&O; PRR), Pa., 30, NW
Caldwell, N.J., 53, NW
Caldwell, W.Va., 138, NW
Califon, N.J., 52, SW
California, Md., 130, NE
California, Pa., 57, SW
Calla, Ohio, 40, NW

Callaghan, Va., 138, NE
Callery, Pa., 41, SE
Callicoon, N.Y., 23, NE
Caln, Pa., 81, NW
Calumet, Pa., 58, SW
Calvert (Baltimore), (PRR), Md., 96A
Calvert, W.Va., 120, NW
Calverton, Va., 111, SE
Calvin, Ky., 162, SE
Calvin, Pa., 42, NW
Cambridge Springs, Pa., 13, NE
Cambridge, Md., 114, SE
Cambridge, N.J., 67, SW
Camden (AC), N.J., 82B
Camden (Baltimore), (B&O), Md., 96A
Camden Broadway (PRR), N.J., 82B
Camden Heights, Va., 177, NE
Camden Market St. (PRR), N.J., 82B
Camden-on-Gauley, W.Va., 122, NE
Cameron Mills, N.Y., 7, SW
Cameron, N.Y., 7, SW
Cameron, Pa., 30, NE
Cameron, W.Va., 71, NE
Cammal, Pa., 32, NW
Camp Alleghany, W.Va., 138, NW
Camp Appalachia, Va., 139, NW
Camp Ground, Pa., 62, SE
Camp Hill (PRR), Pa., 63, SW
Camp Hill (RDG), Pa., 63, SW
Camp Hill, Pa., 56A
Camp Holabird, Md., 96A
Camp Meade (PRR), Md., 96, SE
Camp Pickett Jct., Va., 157, SW
Camp Pickett, Va., 157, SW
Camp Ritchie, Md., 78, SW
Camp Run, W.Va., 105, SE
Camp Wickham, W.Va., 92, SW
Camp, Pa., 63, SW
Camp, Pa., 78, NE
Camp, W.Va., 121, NE
Campbell (DL&W; Erie), N.Y., 7, SE
Campbell Hall (NYO&W), N.Y., 38, NW
Campbell, Va., 127, SW
Campbells, W.Va., 76, SW
Campell Hall Jct. (Erie), N.Y., 38, NW

Campgaw, N.J., 38, SE
Campville, N.Y., 9, SE
Camworth, Pa., 44, NE
Canaanville, Ohio, 86, NW
Canadohta, Pa., 14, NW
Canal, N.J., 117, NW
Canandea (PRR), N.Y., 5, NE
Canandea (W&B), N.Y., 5, NE
Canaseraga (Erie; PS&N), N.Y., 6, NW
Candor, N.Y., 9, SW
Cane Fork, W.Va., 121, SW
Canebrake, W.Va., 149, NE
Canfield, N.Y., 8, NW
Canfield, Ohio, 25, SW
Canfield, W.Va., 107, NW
Canisteo (Erie; NY&P), N.Y., 6, NE
Cannelton, Pa., 41, NW
Cannelton, W.Va., 121, SW
Cannon, Del., 115, SE
Cannonsburg, Pa., 56, NE
Canody, Va., 140, SE
Canoe Camp, Pa., 19, NE
Canoe Creek Jct., Pa., 60, NW
Canterbury, W.Va., 134, NW
Canton, Md., 96A
Canton, Pa., 20, SW
Canyon Mine, Pa., 73, SW
Cape Charles, Va., 160, NE
Cape May (PRR; P-RSL), N.J., 117, NW
Cape May Courthouse (P-RSL; WJ&S), N.J., 100, SW
Caperton, W.Va., 121, SE
Capeville, Va., 161, SW
Capito, Ky., 162, SW
Capitol St. (Charleston), (B&O; C&O; NYC), W.Va., 120, NE
Capitol View, Md., 95, SE
Capon Road, Va., 93, SW
Capon Springs, Va., 93, SW
Capouse Jct., Pa., 35D
Capron, Va., 175, SE
Captina, Ohio, 71, NW
Captiva Jct., Ohio, 71, NW
Carbide, W.Va., 71, SE
Carbo (C&O; N&W), Va., 165, NE
Carbon Centre, Pa., 42, NW
Carbon Glow, Ky., 147, SW
Carbon, Pa., 26, SW
Carbon, Pa., 49, SE

Carbondale (D&H; NYO&W), Pa., 22, SE
Carbondale (Erie), Pa., 23, SW
Carbondale, Ohio, 85, NW
Cardiff, Md., 80, SW
Cardiff, N.J., 100, NE
Cardiff, Pa., 29, NE
Cardiff, Pa., 59, NW
Cardinal, Ky., 162, NE
Caretta Branch Jct., W.Va., 149, NE
Caretta, W.Va., 149, NE
Carfax, Va., 165, NW
Carl, Pa., 27, NE
Carle Place, N.Y., 54, SE
Carlisle (PRR; RDG), Pa., 62, SE
Carlisle, W.Va., 136, NE
Carlos Jct., Md., 75, SW
Carlos, W.Va., 149, NW
Carlson, Pa., 16, SW
Carlstadt, N.J., 53, NE
Carlton Hill, N.J., 53, NE
Carlton, Pa., 26, NE
Carman, Pa., 29, NW
Carmel, N.Y., 39, NE
Carmen, Ohio, 55, NW
Carmine, Va., 140, NW
Carmona, Pa., 26, SE
Carnegie, Pa., 56A
Carney's Point, N.J., 82, SW
Carnifex Ferry, W.Va., 122, SW
Carnifex, W.Va., 122, SW
Carnog, Pa., 65, SE
Carolina Jct. Yard (NS), Va., 177A
Carolina, Va., 177A
Carolina, W.Va., 89, NW
Caroline, N.Y., 9, NW
Carpenter Coal & Coke, Pa., 57, SE
Carpenter, Del., 82, NW
Carpenter, Ohio, 85, SE
Carpenter, Pa., 66, SE
Carpentersville, N.J., 51, SE
Carranza, N.J., 83, NE
Carrier, Pa., 28, SE
Carroll, N.Y., 5, SW
Carrollton (NYC; PRR), Pa., 44, SE
Carrollton, Md., 79, SW
Carrollton, N.Y., 4, SE
Carrollton, W.Va., 89, SE
Carrsville, Va., 176, SW
Carson, Va., 110, NE

Carson, Va., 158, SW
Carswell, W.Va., 149, NE
Carterton, Va., 165, NE
Cartwright, Pa., 29, NE
Cartwright, Va., 148, NE
Cartwrights, Pa., 16, SE
Carville, Md., 97, SE
Cary, Ky., 162, NE
Caryl, N.Y., 54, NW
Carysbrook, Va., 142, NE
Casanova, Va., 111, SE
Cascade, Pa., 1, SE
Cascade, W.Va., 73, SW
Caskie, Va., 141, SW
Casparis, Pa., 73, NE
Cass, W.Va., 124, NW
Cassandra, Pa., 59, NE
Casselman (B&O; WM), Pa., 74, NE
Casselman, Md., 74, SE
Cassity, W.Va., 106, NE
Cassville, W.Va., 72, SE
Castle Fin, Pa., 80, NW
Castle Shannon, Pa., 56, NE
Castlewood, Va., 165, NW
Cat's Run Jct., Pa., 73, NW
Catalpa, Ky., 118, SE
Cataract, Pa., 30, SE
Catasauqua (CNJ; LNE; LV), Pa., 51, SW, 51C
Catatonk, N.Y., 9, SW
Catawba Crossing, Va., 152, NE
Catawba Jct., W.Va., 72, SE
Catawba Sanitorium, Va., 152, NE
Catawba, W.Va., 72, SE
Catawissa (DL&W; PRR; RDG), Pa., 49, NW
Catlett, Va., 111, SE
Catlettsburg, Ky., 118, NE
Cato, Ky., 163, NE
Catoctin, Md., 78, SW
Catoctin, Md., 94, NE
Catonsville, Md., 96, NE
Catron, Va., 167, NE
Cattarangus, N.Y., 4, NW
Cavetown, Md., 77, SE
Cavode, Pa., 43, NE
Caylor, Va., 162, SE
Cayuga Jct., Pa., 35C
Cayuta, N.Y., 8, NE
Caywood, Ohio, 87, NW
CD Crossover, Pa., 14, NE
Cecil (PRR; P&WV), Pa., 56, NE

Cecil, Va., 165, SE
Cecil, W.Va., 90, NW
Cedar Ave., N.Y., 53, SE
Cedar Bluff, Va., 149, SW
Cedar Brook, N.J., 83, SW
Cedar Creek, Pa., 57, SW
Cedar Creek, Va., 93, SW
Cedar Crest, N.J., 84, NW
Cedar Grove (KC&NW; KEC; NYC), W.Va., 121, SW
Cedar Grove, N.J., 53, NE
Cedar Hill, Pa., 46, SE
Cedar Hollow, Pa., 66, SW
Cedar Knoll, Pa., 65, SW
Cedar Lake, N.J., 83, SW
Cedar Lane, Pa., 64, SE
Cedar Manor, N.Y., 54D
Cedar Point, Va., 143, SW
Cedar Rock, Pa., 61, SW
Cedar Run, Pa., 19, SW
Cedar Street, Ohio, 25A
Cedar, W.Va., 134, SE
Cedarhurst, Md., 79, SW
Cedarhurst, N.Y., 54, SE
Cedarville, N.J., 99, NE
Cedarville, Va., 110, NE
Celco, Va., 151, NW
Cementon, Pa., 50, SE
Center Ave., N.Y., 54, SE
Center Ave., Pa., 56, NW
Center Hill, Pa., 78, NE
Center Road, Pa., 13, SW
Center St., (B&O), Ohio, 25A
Center Valley, Pa., 51, SW
Center Village, N.Y., 10, SE
Center, N.J., 53, SW
Center, N.Y., 6, NW
Center, Pa., 62, NW
Centerville, N.Y., 5, NE
Centerville, Pa., 14, SW
Central Ave., N.J., 100, SE
Central Ave., N.J., 53, NE
Central City, Pa., 59, SW
Central Valley, N.Y., 38, NE
Central, Pa., 34, NW
Central, W.Va., 88, NW
Centralia, Pa., 49, NW
Centralia, W.Va., 105, SE
Centre Hall, Pa., 46, NE
Centre Square, N.J., 82, NW
Centreville, Md., 97, SE
Century Jct., W.Va., 89, SE
Century, W.Va., 89, SE

Cereal, Pa., 57, NE
Ceredo, W.Va., 118, NE
Ceres, N.Y., 5, SW
Cessna, Pa., 59, SE
CG Sidings, Pa., 13, NE
Chad, Ky., 163, NE
Chadd's Ford (PRR), Pa., 81, NE
Chadd's Ford Jct. (RDG), Pa., 81, NE
Chadwick, N.J., 84, NE
Chafee, W.Va., 91, NE
Chaffee's, Pa., 16, SW
Chain, Pa., 50, SW
Chaintown, Pa., 57, SE
Chalfont, Pa., 66, NE
Cham, W.Va., 135, NW
Chambers, N.Y., 8, NW
Chambersburg (PRR; WM), Pa., 77, NE
Chambersville, Pa., 43, SE
Chambersville, Va., 93, SE
Champion, Ohio, 25, NW
Champion, Pa., 56, NW
Chancellor, Va., 128, NE
Chandler, Ohio, 55, NW
Chapel Road, Pa., 1, SE
Chapel, Md., 114, NE
Chapel, Va., 151, NW
Chapelle, W.Va., 105, SE
Chapin, Va., 125, SW
Chaplin, W.Va., 72, SE
Chapline, W.Va., 55A
Chapman Quarries, Pa., 51, NW
Chapman, Ky., 118, SE
Chapman, Pa., 50, SE
Chapman, W.Va., 105, NE
Chapmanville, W.Va., 134, NE
Chappaqua, N.Y., 39, SW
Charleroi, Pa., 57, SW
Charles Town (B&O; N&W), W.Va., 94, NW
Charleston, Md., 81, SW
Charlie Hope, Va., 174, NW
Charlotte Hall, Md., 130, NW
Charlotteburg, N.J., 38, SW
Charlottesville (C&O), Va., 127, SW
Charlottesville Union Station (SOU), Va., 127, SW
Charlton, Md., 77, SW
Charlton, Va., 138, SE
Charmian, Pa., 78, SW
Chase City, Va., 173, NW

Chase, Md., 97, NW
Chase, Pa., 45, NW
Chatham, N.J., 53, SW
Chatham, Va., 171, NW
Chatsworth, N.J., 83, NE
Chattaroy, W.Va., 134, SW
Chattolanee, Md., 96, NE
Chauncey, N.Y., 54, NW
Chauncey, Ohio, 85, NE
Chauncey, W.Va., 135, NW
Cheat Bridge, W.Va., 107, SW
Cheat Jct., W.Va., 107, NW
Cheat River, Pa., 73, SW
Chelayan, W.Va., 121, SW
Chelsea, N.J., 101, NW
Chelten Ave., Pa., 66, SE
Cheltenham, Md., 113, SW
Cheltenham, Pa., 66, SE
Chemung Sta., Pa., 20, NE
Chemung, N.Y., 8, SE
Chenango Bridge, N.Y., 10, SW
Chenango Forks, N.Y., 10, SW
Chenoa, Ky., 162, SW
Cheriton, Va., 161, NW
Cherokee, W.Va., 121, SW
Cherry Creek, N.Y., 3, SE
Cherry Hill, Va., 112, SW
Cherry Run, Pa., 16, SW
Cherry Run, W.Va., 76, SE
Cherry Springs, Pa., 18, SW
Cherry St. (New Castle), (PRR), Pa., 41, NW
Cherry Tree (CT&D; NYC; PRR), Pa., 44, SW
Chesaco Park, Md., 97, NW
Chesapeake, W.Va., 120, SE
Cheshire, Ohio, 102, NE
Chesilhurst, N.J., 83, SW
Chester (ACL; SAL; TWRR), Va., 158, NW
Chester (B&O; PRR), Pa., 82, NW
Chester (CNJ; DL&W), N.J., 52, NE
Chester (PRR; RDG), Pa., 82, NW
Chester Heights, N.Y., 54, NW
Chester Heights, Pa., 82, NW
Chester Jct., N.J., 52, NE
Chester Springs, Pa., 65, SE
Chester, Md., 114, NW
Chester, N.Y., 38, NW
Chester, W.Va., 40, SE
Chesterbrook, Pa., 66, SW

Chesterfield, Pa., 45, NW
Chestertown, Md., 97, SE
Chestnut Hill (PRR; RDG), Pa., 66, SE
Chestnut St., Pa., 56, SW
Chestnut Yard, Va., 168, SW
Cheswick, Pa., 42, SW
Cheswold, Del., 98, SE
Cheverly, Md., 113, NW
Chevy Chase, Md., 112, NE
Chew Road, N.J., 83, SW
Chewsville, Md., 77, SE
Chewton, Pa., 41, NW
Cheyney, Pa., 81, NE
Chickahominy, Va., 144, SW
Chickies, Pa., 63, SE
Chicora, Pa., 42, NE
Chiefton (B&O; WM), W.Va., 89, NW
Childs, Md., 81, SW
Childs, Pa., 14, NE
Chilhowie, Va., 166, NE
Chipmunk, N.Y., 4, SE
Chippewa, Pa., 33, SW
Christian, Va., 125, SE
Christian, W.Va., 135, SW
Christiana, Pa., 81, NW
Christiansburg, Va., 152, SW
Christie, Va., 172, SW
Chula, Va., 157, NW
Church Hill, Tenn., 164, SE
Church Road, Va., 157, SE
Churchland, Va., 177, NW
Churchville, Pa., 66, SE
Cinco, W.Va., 121, NW
Cinderella, W.Va., 134, SE
Cirtsville, W.Va., 136, NW
Cisna Run, Pa., 62, NW
Clader, Pa., 49, NE
Claghorn Siding, Pa., 58, NE
Claghorn, Pa., 58, NE
Claiborne, Md., 114, NW
Claiborne, Va., 129, SW
Clairfield, Tenn., 162, SW
Clairton (PRR; URR), Pa., 57, NW
Clairton Jct., Pa., 57, NW
Clandennin, W.Va., 121, NW
Clara, Pa., 17, NE
Claremont Park, N.Y., 54, NW
Claremont Wharf, Va., 159, SW
Claremont, Va., 159, SW
Claremont, W.Va., 136, NE

Claren, W.Va., 149, NE
Clarence, Pa., 31, SW
Clarendon, Pa., 15, NE
Claridge, Pa., 57, NE
Clarington (B&O), W.Va., 71, NW
Clarington (PRR), Ohio, 71, NW
Clarion Jct. (B&O; Erie), Pa., 29, NE
Clarion, Ohio, 85, SW
Clarion, Pa., 28, SW
Clarion, Va., 154, SW
Clark, N.J., 53, SW
Clark, Ohio, 40, SE
Clark, Pa., 60, SE
Clark, Va., 151, SW
Clarkes Gap, Va., 94, SE
Clarks Ferry, Pa., 63, NW
Clarks Gap, W.Va., 150, NW
Clarks Mills, Pa., 26, NE
Clarks Summit, Pa., 35, NE
Clarksboro, N.J., 82, NE
Clarksburg Branch Jct., W.Va., 89, NW
Clarksburg, Pa., 43, SW
Clarksburg, W.Va., 89, NW
Clarksville (A&D; SOU), Va., 172, SE
Clarksville Jct., Va., 172, SE
Clarksville, Pa., 72, NE
Clarkton, Va., 172, NW
Clarmatan, W.Va., 135, SW
Clawson, W.Va., 123, NE
Clay Rock, Pa., 62, SW
Clay, Va., 154, NE
Clay, W.Va., 121, NE
Clayford (PRR), Pa., 73, NW
Claymont, Del., 82, NW
Claypool, W.Va., 137, NW
Clayport, Ohio, 40, SE
Claysburg, Pa., 60, NW
Claysville, Pa., 56, SW
Clayton, Del., 98, NE
Clayton, Md., 97, NW
Clayton, N.J., 82, SE
Claytonia, Pa., 27, SW
Clayville, Va., 157, NW
Clear Brook, Va., 93, NE
Clear Creek Jct., Ky., 147, NE
Clear Fork Branch Jct., W.Va., 149, NE
Clearco, W.Va., 122, SE
Clearfield (B&O; PRR), Pa., 30, SW

Clearfield Jct., Pa., 30, SW
Clearwater Park, Va., 139, NW
Clement, Pa., 48, NW
Clementon, N.J., 83, NW
Clements, W.Va., 89, SE
Clemo, Pa., 23, SW
Cleonia, Pa., 64, NW
Clermont (PS&N; PRR), Pa., 17, SW
Clermont Jct., Pa., 17, NW
Clermont, Md., 96A
Cleveland, Va., 165, NE
Clevenger, W.Va., 136, SW
Clevers, Pa., 56A
Cleversburg Jct., Pa., 62, SW
Cleversburg, Pa., 62, SW
Click, Tenn., 164, SE
Cliff Mine, Pa., 56, NE
Cliffield, Va., 149, SE
Clifford, Md., 96, SE, 96A
Clifford, Pa., 48, NW
Cliffview, Va., 168, SW
Cliffwood, N.J., 68, NE
Clift Yard, W.Va., 150, NW
Clifton Forge, Va., 139, NW
Clifton, N.J., 53, NE
Clifton, N.Y., 53, SE
Clifton, Pa., 35, SE
Clifton, Pa., 41, SE
Clifton, Pa., 63, SE
Clifton, Pa., 82, NW
Clifton, Va., 112, NW
Clifton, W.Va., 85, SE
Climax, Pa., 43, NW
Clinchburg, Va., 166, NW
Clinchco, Va., 148, SW
Clinchfield, Va., 165, NE
Clinchport, Va., 164, SW
Clinton Ave., N.J., 53, SW
Clinton Road, N.Y., 54, SE
Clinton St. (DL&W; LV), N.Y., 8, NE
Clinton St., Md., 96A
Clinton, N.J., 52, SW
Clintondale, Pa., 31, SE
Cloe (B&O; PRR), Pa., 44, NW
Clokey, Pa., 56, SE
Clonmel, Pa., 81, NW
Clopper, Md., 95, SE
Closplint, Ky., 163, NE
Closter, N.J., 54, NW
Cloth, W.Va., 135, NW
Clover Lick, W.Va., 124, NW

Clover, Va., 172, NE
Clover, W.Va., 102, SE
Cloverdale, Va., 153, NW
Cluster Springs, Va., 172, SW
Cly, Pa., 63, SE
Clyde, N.J., 67, NE
Clyde, W.Va., 120, NE
Clyffeside, Ky., 118, NE
Clymer, N.Y., 2, SE
Clymer, Pa., 43, SE
CM Jct., Pa., 14, NE
CO, W.Va., 55, NE
Coal Centre, Pa., 57, SW
Coal Fork, W.Va., 120, NE
Coal Glen, Pa., 29, SW
Coal Grove, Ohio, 118, NE
Coal Hollow, Pa., 29, NE
Coal Jct. (B&O; WM), Pa., 58, SE
Coal Lick Run Branch Jct., Pa., 73, NE
Coal Mountain, W.Va., 135, SE
Coal Run Jct., Pa., 45, NW
Coal Run, Pa., 43, SW
Coal Run, Pa., 45, NW
Coal Run, Pa., 74, NE
Coal Siding, W.Va., 122, NE
Coal, W.Va., 121, SW
Coalboro, Va., 157, NE
Coalburg, Ohio, 25, SE
Coalburg, W.Va., 121, SW
Coaldale, Pa., 50C
Coaldale, Pa., 60, SE
Coaldale, W.Va., 150, NW
Coalfield, W.Va., 121, SW
Coalmont, Pa., 60, SE
Coalport, Pa., 44, NE
Coalridge, W.Va., 121, NW
Coalton, Ky., 118, NW
Coalton, W.Va., 107, NW
Coalwood, W.Va., 149, NE
Coatesville (PRR; RDG), Pa., 81, NW
Coats, Ohio, 70, NE
Cobb, W.Va., 122, SE
Cobbs, Va., 161, NW
Cobham, Pa., 15, SW
Cobham, Va., 127, SW
Coburn, Pa., 47, NW
Cochecton, N.Y., 23, SE
Cochran, Va., 174, NW
Cochranton, Pa., 13, SE
Cockeysville, Md., 96, NE
Coco, W.Va., 121, NW

Coder, Pa., 28, SE
Codorns, Pa., 79, NW
Coeburn, Va., 165, NW
Coffman, Pa., 58, SW
Coffman, W.Va., 89, NE
Coger, W.Va., 105, NE
Cogley, W.Va., 71, NE
Coheva, Pa., 64, NW
Cohill, Md., 76, SE
Cokeburg Jct., Pa., 56, SE
Cokeburg, Pa., 56, SE
Coketon, W.Va., 90, SE
Colburns, N.Y., 3, SW
Colchester, N.Y., 11, SE
Colchester, Va., 112, SE
Colcord, W.Va., 136, NW
Cold Harbor, Va., 144, SW
Cold Run, Pa., 65, SW
Cold Spring Harbor, N.J., 117, NW
Cold Spring, N.J., 117, NW
Cold Spring, N.Y., 4, SW
Cold Spring, N.Y., 39, NW
Cold Spring, Pa., 20, SE
Cold Spring, Pa., 63, NE
Cold Spring, Va., 140, NE
Cold Springs, Pa., 79, NW
Coldiron, Ky., 163, NW
Cole, Ohio, 86, NW
Colebrook, Pa., 63, SE
Colebrookdale, Pa., 65, NE
Colegrove, Pa., 17, SW
Coleman, Pa., 57A
Coleman, Pa., 59, SW
Colemans Mill, Va., 144, NW
Coles, Ohio, 25, SE
Coles, Pa., 34, SW
Coles, Pa., 60, SE
Colesburg, Pa., 18, NW
Coleville, Pa., 46, NW
Colfax, Pa., 42, SW
Colfax, W.Va., 89, NE
Colgate Creek, Md., 96A
Colledgeville, Pa., 66, SW
College Park, Md., 113, SW
College Point, N.Y., 54, NW
College, Pa., 41, NW
Colley Ave., Va., 177A
Collier, Pa., 45, SW
Collier, Pa., 73, NW
Collier, W.Va., 55, NE
Collingdale, Pa., 82, NW
Collingswood, N.J., 82B
Collington, Md., 113, NW

Collins Siding, N.J., 83, SW
Collins, Ky., 147, NE
Collins, Pa., 63, SE
Collins, Va., 129, SW
Collinsburg, Pa., 57, SW
Collinsdale, W.Va., 121, SW
Collison, Va., 139, NW
Colmar, Ky., 162, SE
Colmar, Pa., 66, NW
Cologne, N.J., 100, NE
Colona, Pa., 41, SW
Colonia, N.J., 53, SW
Colonial Colliery (PRR; RDG), Pa., 49, NW
Colora, Md., 80, SE
Colosse, Va., 176, NW
Columbia (PRR; RDG), Pa., 64, SW
Columbia Ave., Pa., 82A
Columbia Sulphur Springs, W.Va., 123, SE
Columbia X Roads, Pa., 20, NW
Columbia, N.J., 51, NE
Columbia, Va., 142, NE
Columbia, W.Va., 121, SW
Columbiana (PRR; Y&S), Ohio, 40, NE
Columbus Ave. (NH; NYW&B), N.Y., 54, NW
Columbus Jct., Pa., 14, NE
Columbus, N.J., 67, SE
Columbus, Pa., 14, NE
Colver, Pa., 44, SW
Colwell, Pa., 43, NW
Colza, Pa., 14, NE
Comfort Run, Pa., 44, NE
Commerce St. (Bridgeton), (P-RSL), N.J., 99, NE
Commodore, Pa., 44, SW
Communipaw (CNJ; LV), N.J., 53B
Como, N.J., 68, SE
Compton, Va., 110, NW
Conboy, Pa., 77, NE
Concord, Del., 81A
Concord, Va., 155, NW
Concordville, Pa., 81, NE
Concrete Bridge, Pa., 44, SW
Concrete, Va., 126, NE
Condron, Pa., 44, SE
Conemaugh, Pa., 59, NW
Conestoga East, Pa., 64, SW
Conestoga, Pa., 65, SW
Coneville, Pa., 17, NE

Conewago, Pa., 63, SE
Conewango, N.Y., 3, SE
Confluence (B&O; WM), Pa., 74, NW
Congers, N.Y., 39, SW
Congo, W.Va., 40, SE
Conifer, Pa., 28, SE
Conklin, N.Y., 10, SW
Conneaut (B&LE; NKP; NYC), Ohio, 12, NE
Conneaut Jct., Pa., 13, NW
Conneaut Lake Park, Pa., 13, SW
Conneaut Lake, Pa., 13, SW
Conneaut Yard (NKP), Ohio, 12, NE
Conneautville (B&LE; PRR), Pa., 13, NW
Connellsville (B&O; P&LE; PRR), Pa., 57C
Connor, Ohio, 55, SE
Conoway, Va., 148, NE
Conowingo, Md., 80, SE
Conshohocken (PRR; RDG), Pa., 66A
Constitution, Ohio, 86, NE
Contee, Md., 96, SW
Convent, N.J., 53, NW
Conway, Pa., 35, SW
Conway, Pa., 41, SE
Cooch, Del., 81, SE
Cook's Falls, N.Y., 24, NW
Cook's Mills, Pa., 75, NE
Cooks, Pa., 60, SE
Cookstown, N.J., 67, SE
Cool Spring, Del., 116, SE
Cool Spring, Pa., 26, NE
Cool Spring, Va., 129, NW
Cool Spring, W.Va., 87, SW
Coolbaugh, Pa., 36, SE
Cooleys, Md., 132, SE
Coolville, Ohio, 86, SW
Coolwell, Va., 140, SE
Coonville, Ohio, 85, NW
Coopers (DL&W; Erie), N.Y., 7, SE
Coopersburg, Pa., 51, SW
Coopersville, Pa., 80, NE
Copeland, Pa., 57, NW
Copeland, Va., 139, NE
Copen, W.Va., 105, NE
Copenhaver, Va., 166, NE
Copenhaver, W.Va., 121, NW
Coplay, Pa., 50, SE

Coplay, Pa., 51D
Copper Hill, N.J., 67, NW
Coral, Pa., 43, SE
Coral, Pa., 58, NE
Corapeke, N.C., 176, SE
Corapolis, Pa., 41, SE
Corbett, Md., 77, SE
Corbett, Md., 79, SE
Corbett, N.Y., 11, SE
Corbit, Del., 81, SE
Cordelia, Pa., 64, SW
Cordova, Md., 115, NW
Core, W.Va., 72, SE
Corinth, W.Va., 91, NW
Corliss, Pa., 56A
Cornelia, Ohio, 85, SW
Corning (DL&W; Erie; NYC), N.Y., 7, SE
Corning Yard (NYC), N.Y., 7, SE
Corning, Pa., 65, NE
Cornwall (CWL; PRR), Pa., 64, NW
Cornwall, N.Y., 38, NE
Cornwall, Va., 140, NW
Cornwallis (B&O; HS), W.Va., 87, SE
Cornwells Heights, N.J., 67, SW
Corona, N.Y., 54, SW
Corry (Erie; PRR), Pa., 14, NE
Corsons, Pa., 66, SW
Cortland Jct., N.Y., 9, NW
Cortland, Ohio, 25, NE
Corts, Ky., 162, NW
Corydon, Pa., 16, NW
Coryville (PS&N; PRR), Pa., 17, NW
Cos Cob, Conn., 39, SE
Cossart, Pa., 81, NE
Costello, Pa., 17, SE
Costen, Md., 132, SE
Cottageville, W.Va., 103, NW
Cotton Hill, W.Va., 121, SE
Coudersport, Pa., 17, NE
Coulson, Pa., 26, NE
Coulter, Pa., 28, SE
Coulter, Pa., 57, NW
Country Life Press, N.Y., 54, SE
County Farm, W.Va., 119, SW
County Home, N.Y., 7, NW
County Home, Pa., 29, NE
County House, N.Y., 6, NW
County Jct., Pa., 57, NE
County Line, Pa., 66, SE

Courtland, Va., 175, SE
Courtney, Pa., 57, SW
Courtney, W.Va., 103, SW
Courtneys Mills, Pa., 26, SE
Cove Run, W.Va., 90, SW
Cove, Pa., 60, NW
Cove, Pa., 63, NW
Covedale, Pa., 60, NE
Covel, W.Va., 150, NW
Coventry, N.Y., 10, NE
Coverts (P&LE; PRR), Pa., 41, NW
Covesville, Va., 141, NE
Covington, Pa., 19, SE
Covington, Va., 139, NW
Cowan, Pa., 42, NE
Cowan, Va., 151, SE
Cowan, Va., 164, SE
Cowanesque, Pa., 18, NE
Cowansburg, Pa., 57, NW
Cowden, Pa., 56, NE
Cowen, W.Va., 122, NE
Cowl, W.Va., 40, SE
Cowley, Pa., 20, SW
Cox Creek, Md., 96, SE
Cox Landing, W.Va., 119, NW
Cox, Md., 130, NW
Coxton Yard, Pa., 35B
Coxton, Ky., 163, NW
Coy Jct., Pa., 43, SE
CR Jct., W.Va., 86, NE
Crab Creek, Ohio, 25A
Crabapple, Ohio, 71, NW
Craborchard, W.Va., 136A, SE
Crabtree, Pa., 58, NW
Craddock, W.Va., 106, SW
Crafton, Pa., 56A
Crafts, N.Y., 39, NE
Crafts, Va., 151, NE
Craigheads, Pa., 62, SE
Craigs Corners, Pa., 36, SE
Craigsville, Pa., 42, NE
Craigsville, Va., 125, SW
Cramer, Pa., 29, SW
Cranberry Jct., Pa., 50A
Cranberry Lake, N.J., 52, NE
Cranberry, N.J., 68, NW
Cranberry, W.Va., 136A, NE
Crandalltown, Pa., 33, NW
Crane Jct., Pa., 50, SE
Crane Jct., Pa., 51, SW
Crane, Va., 139, NE
Craneco, W.Va., 135, NE

Cranesville, Pa., 13, NW
Cranford (CNJ; LV; RV), N.J., 53, SW
Crawford Jct., N.Y., 38, NW
Crawford Jct., Pa., 16, NE
Crawford, N.Y., 5, NE
Crawford, W.Va., 106, NW
Crawford, W.Va., 107, NW
Crayton, Pa., 13, NW
Cream Ridge, N.J., 67, SE
Creasy, Pa., 34, SW
Creed, W.Va., 120, NE
Creeds, Va., 177, SE
Creek Jct., Va., 166, SE
Creek Siding, Pa., 48, NW
Creekside, Pa., 43, SE
Creels, W.Va., 86, SE
Creigh, Pa., 77, NW
Creighton, Pa., 42, SW
Creighton, Va., 144, SW
Crellin, Md., 91, NW
Crenshaw, Pa., 29, NE
Creola, Ohio, 85, NW
Cres, Pa., 80, NW
Crescent, Ohio, 55, SW
Crescentville, Pa., 66, SE
Cresco, Pa., 36, SW
Creslo, Pa., 59, NE
Cresmont, Pa., 66, SE
Cresmont, W.Va., 122, NW
Cresskill, N.J., 54, NW
Cresson, Pa., 59, NE
Cressona, Pa., 49, SE
Crest, Va., 164, NW
Crestwood, N.Y., 54, NW
Crewe, Va., 156, SE
Cribb, Pa., 57, NE
Crichton, W.Va., 122, SE
Crimora, Va., 126, SW
Cripple Creek, Va., 167, NE
Crisfield, Md., 146, NW
Crisp, Md., 96, SE
Crites, W.Va., 135, NW
Critz, Va., 169, SE
Crockett, Ky., 162, NE
Crockett, Va., 167, NE
Croft, Pa., 30, SW
Cromby, Pa., 65, SE
Cromer, W.Va., 107, SW
Crook Horn Draw, N.J., 100, SE
Croome, Md., 113, NW
Cropwell, N.J., 83, NW
Crosby, Pa., 17, SW

Crosiers Run Branch Jct., Pa., 73, NW

Cross Fork Jct., Pa., 18, SW

Crossland, Pa., 57C

Crossley, N.J., 84, NW

Croton Falls, N.Y., 39, NE

Croton Heights, N.Y., 39, NW

Croton Lake, N.Y., 39, SW

Croton-on-Hudson, N.Y., 39, SW

Crow Summit, W.Va., 103, NE

Crow, Ohio, 71, NW

Crown City, W.Va., 102, SW

Crown Hill, W.Va., 121, SW

Crown, Pa., 28, NW

Croxton Yard, N.J., 53B

Croxton, Ohio, 55, NE

Croydon, N.J., 67, SW

Croyland, Pa., 29, NW

Crozet, Va., 126, SE

Crucible, Pa., 73, NW

Crugers, N.Y., 39, SW

Cruickshanks, W.Va., 122, SW

Crum Lynne, Pa., 82, NW

Crum, W.Va., 134, NW

Crummies, Ky., 163, NE

Crumpler, W.Va., 150, NW

Crystal Hill, Va., 172, NW

Crystal Lake, N.J., 38, SE

Crystal Lake, N.Y., 5, NW

Crystal Run, N.Y., 38, NW

Crystal, W.Va., 150, NW

Cub Creek Jct., W.Va., 135, SW

Cuba (Erie), N.Y., 5, SW

Cuba (PRR), N.Y., 5, SW

Cucumber, W.Va., 149, NE

Culbertson, Pa., 61, SE

Cullen, Va., 155, SE

Culloden, W.Va., 119, NE

Culmerville, Pa., 42, SW

Culpeper, Va., 128, NW

Cumberland (B&O; WM), Md., 75, SW

Cumberland Gap (L&N; SOU), Tenn., 162, SE

Cumberland, Ky., 164, NW

Cumberland, Va., 156, NE

Cumbo Yard, W.Va., 77, SW

Cumbola, Pa., 49, SE

Cummings, Pa., 43, SE

Cumru Jct., Pa., 65, NW

Cunningham, N.C., 171, SE

Cunningham, Pa., 42, SW

Cunningham, W.Va., 92, SW

Curb, Ky., 148, NW

Curry Run, Pa., 44, NE

Curry, Pa., 60, NW

Curtin, Pa., 46, NE

Curtin, W.Va., 122, NE

Curtis Bay Jct., Md., 96A

Curtis Bay, Md., 96, SE

Curtis Jct. (FV; MC&C), Ohio, 86, NW

Curtis, N.Y., 7, SE

Curve, Va., 151, NE

Curwensville, Pa., 44, NE

Cush Cushion, Pa., 44, SW

Cushings, W.Va., 103, NW

Custer City (B&O; Erie), Pa., 16, NE

Custer, Pa., 66, SW

Cuthbert, N.J., 82B

Cutler Summit, Pa., 18, NW

Cutler, Ohio, 86, NW

CV 87, W.Va., 77, SW

CV Jct., Pa., 19, NE

CV Jct., Va., 93, SE

CW Rly. Jct., Va., 126, NW

Cynwyd, Pa., 66B

Cypher, Pa., 60, SW

Cyrus, W.Va., 118, NE

D&M Jct. (PRR; RDG), Pa., 62, SE

Dabney, Va., 112, SE

DaCosta, N.J., 83, SW

Daffan, Va., 129, NW

Dagsboro, Del., 116, SE

Dagus, Pa., 29, NE

Daguscahonda, Pa., 29, NE

Dahlgren Jct., Va., 129, NW

Dahlgren, Va., 129, NE

Dahlia, Va., 174, SE

Dahoga Jct., Pa., 16, SW

Dahoga, Pa., 16, SE

Dailey, W.Va., 107, NW

Dain, W.Va., 122, SE

Dale Summit, Pa., 46, NW

Dale, Pa., 49, SE

Dale, Pa., 65, SE, 65A

Dallas, Pa., 35, NW

Dallastown Jct., Pa., 79, NE

Dallastown, Pa., 79, NE

Dalmatia, Pa., 48, SW

Dalton, Pa., 22, SE

Damascus (L; N&W), Va., 166, SW

Damon, Va., 141, NE

Dan's Run, W.Va., 75, SE

Daniels, Md., 96, NW

Danielsville, Pa., 50, NE

Danieltown, Va., 174, NW

Dankel, Pa., 65, NE

Danstown, Pa., 58, SW

Dante, Va., 165, NW

Danville (DL&W; RDG), Pa., 48, NE

Danville, Va., 171, SW

Danville, W.Va., 120, SW

Daphna, Va., 109, SW

Darby (B&O; PRR), Pa., 82, NW

Darby, W.Va., 120, SE

Darbyville, Va., 163, NE

Darent (B&O; PRR), Pa., 73, NE

Daretown, N.J., 82, SW

Darkwater, Pa., 49, SE

Darlington, Pa., 41, NW

Darlington, Pa., 58, NW

Darlington, Pa., 82, NW

Dartmont, W.Va., 120, SE

Dartmoor, W.Va., 107, NW

Daub, Md., 95, NW

Dauphin (PRR; RDG), Pa., 63, NW

Davis, N.J., 67, SE

Davis, Pa., 18, NE

Davis, Va., 140, NW

Davis, W.Va., 91, SW

Davisville, W.Va., 87, SW

Davy, W.Va., 149, NE

Daw, Va., 149, SW

Dawes, W.Va., 121, SW

Dawson Mill, Va., 141, NE

Dawson Mine, W.Va., 89, NW

Dawson Run, Pa., 15, SW

Dawson, Md., 74, SE

Dawson, Pa., 57, SE

Dawson, W.Va., 92, NW

Dawson, W.Va., 149, NE

Day Jct., Pa., 16, NW

Day, Va., 151, NW

Daylesford, Pa., 66, SW

Dayton, N.J., 67, NE

Dayton, N.Y., 4, NW

Dayton, Pa., 43, NE

Dayton, Pa., 48, SE

Dayton, Va., 126, NW

DB Tower, W.Va., 121, SW

Deacon, Va., 154, NE

Deal, N.J., 68, SE

Deal, Pa., 75, NW

Dean, Pa., 1, SE

Dean, Pa., 44, SE

Dean, W.Va., 119, SW

Deans, N.J., 67, NE

Deanwood, Md., 113, NW

Decatur, Pa., 45, NE

Decatur, Va., 140, NW

Decota, W.Va., 121, SW

Deegans, W.Va., 122, SW

Deep Cut (PRR), N.J., 68, NW

Deep Hollow, Pa., 35, SW

Deepwater, W.Va., 121, SW

Deer Lick, Pa., 72, NW

Deer Park, W.Va., 91, NW

Deer, Pa., 30, SE

Deer, Pa., 45, SE

Deerfield, Md., 78, SW

Deerfield, N.J., 82, SE

Deeter, Pa., 74, NW

DeForest Jct., Ohio, 25, SW

DeGolia, Pa., 16, NE

Dehn, W.Va., 135, NW

Deibler, Pa., 48, NE

DeKalb St. (Norristown), (RDG), Pa., 66A

DeKays, N.J., 38, SW

Delair, N.J., 82B

Delanco, N.J., 67, SW

Delano Jct., Pa., 49D

Delano, Pa., 49D

Delano, Va., 148, SW

Delaplane, Va., 111, NW

Delawanna, N.J., 53, NE

Delaware City, Del., 81, SE

Delaware River Pier, Del., 81A

Delaware Water Gap, Pa., 51, NE

Delaware, N.J., 51, NE

Delbarton, W.Va., 134, SE

Delevan, N.Y., 5, NW

Dellslow, W.Va., 73, SW

Dellwood, Pa., 29, SW

Delmar, Del., 132, NE

Delmar, Va., 166, SW

Delmont, Pa., 57, NE

Delong, Pa., 50, SE

Delorme, W.Va., 134, SE

Delphi, W.Va., 122, NE

Delps, Pa., 51, NW

Delta, Pa., 80, SW

Delton, Va., 168, NE

Delvale, Va., 164, NW

Demarest, N.J., 54, NW

Demmler Transfer (B&O; P&LE), Pa., 57B
Demuth, Va., 153, NW
Den Mar, W.Va., 123, SE
Denbo, Pa., 57, SW
Dendron, Va., 159, SW
Denniston (A&D; N&W), Va., 172, SW
Denniston, Pa., 57A
Dennisville, N.J., 100, SW
Denny, Pa., 42, SW
Dent, Pa., 16, NE
Denton Br. Jct., Md., 115, NW
Denton, Ky., 118, NW
Denton, Md., 115, NW
Dents Run (B&O; PRR), Pa., 30, NW
Denver, Pa., 64, SE
Denver, W.Va., 71, NE
Deposit, N.Y., 11, SW
Depue, W.Va., 104, NW
Derby, Va., 164, NW
Derry Hale, W.Va., 136, NE
Derry, Pa., 58, NW
Derwood, Md., 95, SE
Detour, Md., 78, SW
Devault, Pa., 65, SE
Devol, Ohio, 87, NW
Devon, Pa., 66, SW
Devon, W.Va., 134, SE
Dewart, Pa., 33, SW
Dewing Mill Branch Jct., Pa., 73, SW
Dewitt Siding, Va., 157, SE
Dexter City, Ohio, 70, SW
Dexter, Ohio, 85, SE
Diamond Town, Pa., 49A
Diamond, Va., 148, NE
Diamondville, Pa., 44, SW
Diana, W.Va., 106, SW
Dias, Pa., 58, NE
Diascund, Va., 159, NW
Dick, Pa., 57, NE
Dickerson Run, Pa., 57, SE
Dickerson, Md., 95, SW
Dickey, Pa., 43, NW
Dickinson, W.Va., 121, SW
Dickson City, Pa., 35D
Dickson, Pa., 35D
Dicksonburg, Pa., 13, SW
Dilks, Pa., 42, NW
Dille, Ohio, 71, NW
Dillersville, Pa., 64, SW

Dilley, Tenn., 162, SW
Dillner, Pa., 73, SW
Dillon, Ky., 163, NE
Dillon, Va., 139, SE
Dillonvale (NYC; W&LE), Ohio, 55, SW
Dills, Pa., 62, SE
Dillsburg Jct., Pa., 62, SE
Dillsburg, Pa., 62, SE
Dilltown, Pa., 59, NW
Dillwyn, Va., 142, SW
Dimeling, Pa., 45, NW
Dimock, Pa., 22, SW
Dineharts, N.Y., 7, NW
Dingess, W.Va., 134, NE
Dinwiddie, Va., 157, SE
Dione, Ky., 163, NE
Dishart, Pa., 44, SE
Dismal Creek Jct., Va., 148, SE
Disputanta, Va., 158, SE
Divener, Pa., 42, NW
Divide, N.J., 82B
Dividing Creek, N.J., 99, NE
Division St., (PRR), Ohio, 25A
Division St., Pa., 35A
Dix, Pa., 45, SE
Dixie, W.Va., 89, NW
Dixie, W.Va., 121, NE
Dixmont, Pa., 41, SE
Dixonville, Pa., 43, SE
Dobbs Ferry, N.Y., 39, SW
Dobra, W.Va., 135, NW
Dock, N.Y., 39, SW
Dock, W.Va., 120, NW
Dodamead, Va., 143, SE
Dodges, Pa., 35C
Doe Gully (B&O; WM), W.Va., 76, SW
Doe Run, Pa., 81, NW
Doherty, Ohio, 70, NW
Dola, W.Va., 89, NW
Dolphin, N.J., 100, NE
Dolphin, Va., 174, NW
Donaldson Jct. (B&O; CRB&L), W.Va., 122, NE
Donaldson, Pa., 15, SE
Donaldson, Pa., 49, SW
Donaldson, W.Va., 92, NE
Dondale, W.Va., 137, NW
Dongan Hills, N.Y., 53, SE
Donhegan No. 1, W.Va., 122, SE
Donhegan No. 10, W.Va., 122, NE
Donnelly, Pa., 27, SE

Donohoe, Pa., 58, NW
Donora (PRR; P&WV), Pa., 57, SW
Dooms, Va., 126, SW
Dora Jct., Va., 151, SW
Dora, Pa., 43, NE
Dora, Va., 151, SE
Doran, Va., 149, SW
Dorchester Jct. (INT; L&N), Va., 164, NE
Dorchester, N.J., 100, NW
Dorchester, Va., 164, NE
Dorfee, W.Va., 121, NE
Dornsife, Pa., 48, SW
Dorothy, N.J., 100, NW
Dorothy, W.Va., 136, NW
Dorset, Ohio, 12, SE
Dorset, Va., 157, NW
Dorsey, Md., 96, SE
Dorsey, Ohio, 71, NW
Dorsey, W.Va., 136, NE
Dorsey's, Pa., 80, NE
Dorton, Ky., 147, NE
Dory, Va., 175, NE
Doswell (C&O; RF&P), Va., 144, NW
Dothan, W.Va., 136, NE
Doty, Ohio, 85, NE
Dotyville, Pa., 14, SE
Douglas, Ky., 147, NE
Douglas, Pa., 57, SW
Douglas, W.Va., 90, SE
Douglassville (PRR; RDG), Pa., 65, NE
Douglaston, N.Y., 54, NW
Dover (CNJ; DL&W), N.J., 52, NE
Dover Farm, Va., 143, SW
Dover, Del., 98, SE
Dow, Va., 159, SE
Dowler Jct., Pa., 44, NW
Downardsville, Ohio, 85, SW
Downer Jct., Pa., 57, SW
Downer, N.J., 82, SE
Downes, Md., 115, NW
Downingtown (PRR; RDG), Pa., 65, SE
Downs, Pa., 65, SE
Doylestown, Pa., 66, NE
Draffin, Ky., 148, NW
Drake, Pa., 26, SE
Drake, Pa., 27, NW
Drakes Branch, Va., 172, NE
Draper, N.C., 170, SE

Draper, Va., 151, SE
Dravosburg, Pa., 57, NW
Drawbridge Jct., Del., 81A
Drehersville, Pa., 49, SE
Dreibelbis, Pa., 50, SW
Dresher, Pa., 66, SE
Dressen, Ky., 163, NW
Drewryville, Va., 175, SW
Drift, Ky., 147, NW
Drifton Jct., Pa., 35, SE
Drifton, Pa., 35, SW
Driftwood, Pa., 30, NE
Drill, Va., 149, SW
Drivers, Va., 176, NE
Droop Mountain, W.Va., 123, SW
Drowning Ford, Va., 166, SW
Druckers, Pa., 29, SE
Drum Hill, N.C., 176, SE
Drummond, Pa., 29, NE
Drums, Pa., 35, SE
Dry Branch, Va., 151, SE
Dry Branch, W.Va., 121, SW
Dry Creek, W.Va., 122, SW
Dry Fork Branch Jct., Va., 149, SW
Dry Fork Jct., W.Va., 149, NW
Dry Fork, Va., 171, NW
Dry Run, Ohio, 25A
Dry Run, Ohio, 40, SE
Dryden, N.Y., 9, NW
Dryden, Va., 164, NW
Dublin, Va., 151, SE
DuBois (B&O; PRR), Pa., 29, SW
Duckworth, W.Va., 88, NW
Dudley, Pa., 60, SE
Dudley, Va., 172, NW
Duffield, Va., 164, SW
Duffields, W.Va., 94, NW
Duhring, Pa., 16, SW
Dukes, W.Va., 104, NW
Dulevy Siding, Pa., 57, SW
Duley, Md., 113, SW
Dumas, Pa., 74, NW
Dumbarton, Va., 144, SW
Dumont, N.J., 54, NW
Dumring, W.Va., 150, NW
Dunbar (B&O; PRR), Pa., 73, NE
Dunbar, Ohio, 86, NE
Dunbar, Va., 164, NE
Dunbar, W.Va., 120, NE
Dunbarton, N.J., 83, SW
Duncan, W.Va., 103, NE

Duncannon (PRR; SR&W), Pa., 62, NE
Duncansville, Pa., 60, NW
Dundale, Pa., 58, NW
Dundalk (B&O; PRR), Md., 96, NE
Dundas (B&O; C&O), Ohio, 85, SW
Dundas, Va., 173, NE
Dundon (B&O; BC&G), W.Va., 121, NE
Dunellen, N.J., 53, SW
Dunfee, Ohio, 71, NW
Dunfee, W.Va., 136, NE
Dungannon, Va., 165, NW
Dungarvin, Pa., 45, SE
Dungriff, W.Va., 120, SE
Dunham, Ky., 147, SE
Dunkard, W.Va., 72, SE
Dunkelburgers, Pa., 48, NE
Dunkirk (Erie), N.Y., 3, NW
Dunkirk (NKP), N.Y., 3, NW
Dunkirk (NYC), N.Y., 3, NW
Dunkirk (PRR), N.Y., 3, NW
Dunlap, Pa., 67, SW
Dunlay (C&O; N&W), Ky., 148, NE
Dunleavy Jct., Ky., 148, NW
Dunleavy, Ky., 148, NW
Dunlo, Pa., 59, NE
Dunlow, W.Va., 119, SW
Dunmore (DL&W), Pa., 35D
Dunmore (Erie), Pa., 35D
Dunmore Shops, Pa., 35D
Dunn Loring, Va., 112, NE
Dunn, Va., 139, SW
Dunnfield, N.J., 51, NE
Dunnings Creek Jct., Pa., 60, SW
Dunwoodie, N.Y., 54, NW
Duo, W.Va., 122, SE
Durand, Va., 174, SE
Durant City Jct., Pa., 16, SW
Durant, N.Y., 39, SW
Durbin (C&O; WM), W.Va., 107, SW
Durea Jct., Pa., 35B
Durgon, W.Va., 108, NE
Durham, Pa., 51, SE
Durrett, Va., 141, NE
Duryea, Pa., 35B
Dushore, Pa., 21, SW
Dusquene, Pa., 57B
Dutchess Jct., N.Y., 39, NW

Duvall, Pa., 56, SW
Dwina, Va., 165, NW
Dyes, N.Y., 10, SE
Dyesville, Ohio, 85, SE
Dykemans, N.Y., 39, NE
Dyre Ave., N.Y., 54, NW
Dysart, Pa., 44, SE

E&W Jct., Pa., 51, SW
E. 3rd St., N.Y., 54, NW
E. 6th St., N.Y., 54, NW
Eagle Hill Jct., Pa., 49, SE
Eagle Rock, Pa., 27, NE
Eagle Rock, Va., 139, SW
Eagle, Pa., 45, SE
Eagle, W.Va., 121, SW
Eagles Mere, Pa., 33, NE
Eagleville, Ohio, 12, SW
Eagleville, Pa., 31, SE
Eakle, W.Va., 122, NW
Eakle's Mill, Md., 94, NE
Earle, N.J., 68, NE
Earleigh Heights, Md., 96, SE
Earling, W.Va., 135, NW
Earnest, Pa., 43, SE
Earnest, Pa., 66A
Easley, W.Va., 120, SE
East 22nd St., N.J., 53A
East 45th St., N.J., 53A
East Altoona, Pa., 45, SW
East Bangor (DL&W), Pa., 51, NE
East Bangor (LNE), Pa., 51, NE
East Bank, W.Va., 121, SW
East Beech, Pa., 31, SE
East Bell, Pa., 56, SW
East Berlin, Pa., 79, NW
East Bloomsburg, Pa., 49, NW
East Bradford (B&O; Erie), Pa., 16, NE
East Brady, Pa., 42, NE
East Branch, N.Y., 23, NE
East Branch, Pa., 28, SE
East Brookside, Pa., 48, SE
East Burlington, N.J., 67, SW
East Butler, Pa., 42, NW
East Cadiz, Ohio, 55, NW
East California, Pa., 57, SW
East Chester, N.Y., 38, NW
East Claysville, Pa., 56, SW
East Clokey, Pa., 56, SE
East Dry Run, Pa., 40, SE
East DuBois Jct., Pa., 29, SW

East Duvall, Pa., 56, SW
East Earl, Pa., 64, SE
East End, N.J., 68, NW
East End, Pa., 50B
East Falls, Pa., 82A
East Fayetteville, Pa., 77, NE
East Ferry St., N.J., 53A
East Fredericktown, Pa., 57, SW
East Freedom, Pa., 60, NW
East Freehold, N.J., 68, NW
East Greensburg, Pa., 57, NE
East Greenville, Pa., 65, NE
East Gulf, W.Va., 136, SW
East Hickory, Pa., 15, SW
East Honesdale, Pa., 23, SE
East Ithaca, N.Y., 9, NW
East Julian, Pa., 46, NW
East Lane, Pa., 66, SE
East Lexington (B&O; C&O), Va., 140, NW
East Liberty, Pa., 57A
East Lincoln Ave. (Mt. Vernon), N.Y., 54, NW
East Liverpool, Ohio, 40, SE
East Long Branch, N.J., 69, NW
East Lynn, W.Va., 119, SW
East Middletown, Pa., 63, SE
East Millsboro, Pa., 73, NW
East Millstone, N.J., 52, SE
East Musgrove, Pa., 43, NW
East New Market, Md., 115, SW
East New York, N.Y., 54, SW
East Orange (DL&W; Erie), N.J., 53, NE
East Orwell, Ohio, 12, SW
East Palestine, Ohio, 40, NE
East Patterson, N.J., 53, NE
East Petersburg, Pa., 64, SW
East Pittsburgh, Pa., 57, NW
East Plumville, Pa., 43, NE
East Rices Landing, Pa., 73, NW
East Richmond, Va., 144, SW
East Riverside, Pa., 73, NW
East Rockaway, N.Y., 54, SE
East Roscoe, Pa., 57, SW
East Salamanca, N.Y., 4, SE
East Salisbury, Pa., 74, NE
East Sewell, W.Va., 136, NE
East Spotswood, N.J., 68, NW
East Steubenville, W.Va., 55, NE
East Summit, Pa., 18, NE
East Taylorstown, Pa., 56, SW
East Toronto, W.Va., 55, NE

East Vandergrift, Pa., 42, SE
East View, N.Y., 39, SW
East Vindex, Md., 91, NE
East Wade, Pa., 56, SE
East Warwood, W.Va., 55, SE, 55A
East Waterford, Pa., 61, NE
East Waverly, N.Y., 8, SE
East Weeksbury, Ky., 147, NE
East Weirton, W.Va., 55, NE
East Williamson, W.Va., 134, SW
East Williston, N.Y., 54, NE
East Windsor, N.Y., 10, SE
East Wye Switch, Pa., 51, SW
East Yard, Pa., 58, SE
Eastbrook, Pa., 26, SW
Easton (CNJ; LV), Pa., 51, SE
Easton Jct., Md., 114, NE
Easton, Md., 114, NE
Eastville, Va., 161, NW
Eaton, W.Va., 87, SW
Eatontown, N.J., 68, NE
Ebbvale, Md., 79, SW
Ebensburg Jct., Pa., 59, NE
Ebensburg, Pa., 59, NE
Ebervale, Pa., 50, NW
Eccles Jct., W.Va., 136A, NW
Eccles, W.Va., 136A, NW
Eccleston, Md., 96, NE
Echo, Pa., 43, NW
Echo, W.Va., 119, SW
Eckenrode Mall, Pa., 44, SE
Eckhart Jct. (B&O; WM), Md., 75, SW
Eckhart, Md., 75, SW
Eckley Jct., Pa., 50B
Eclipse, Pa., 27, NW
Eclipse, Pa., 56, SE
Economy, Pa., 41, SE
Eddington, N.J., 67, SW
Eddystone (B&O; PRR), Pa., 82, NW
Eddyville, Pa., 43, NW
Edelman, Pa., 51, NW
Eden, Md., 132, NE
Eder, Md., 81, SW
Edgar, N.J., 53, SW
Edge Moor Yard, Del., 81A
Edge Moor, Del., 81A
Edgecliff, Pa., 42, SE
Edgemere, N.Y., 54, SW
Edgemont, Md., 77, SE
Edgemont, Pa., 50, NE

Edgerton, Va., 174, NW
Edgewater Park, N.J., 67, SW
Edgewater, Pa., 42, SW
Edgewood Grove, Pa., 74, NE
Edgewood, Ky., 162, SW
Edgewood, Md., 97, NW
Edgewood, Pa., 57A
Edgeworth, Pa., 41, SE
Edgley, N.J., 67, SW
Edinburg (B&O; P&LE; PRR),
    Pa., 26, SW
Edinburg, Va., 109, NE
Edison, N.J., 37, SE
Edisonville, Pa., 64, NW
Edmondson, Md., 96A
Edna, W.Va., 72, SE
Edri, Pa., 43, SW
Edsall, Va., 112, NE
Edward, Pa., 72, NE
Edwards, N.Y., 6, SE
Edwards, Pa., 44, SW
Edwardsville, Pa., 35, NW
Edwright, W.Va., 135, NE
Egg Harbor (AC; P-RSL), N.J., 83,
    SE
Eggleston (N&W; VGN), Va., 151,
    NE
Egypt, Pa., 50, SE
Ehrenfield, Pa., 59, NW
Eight Mile, W.Va., 121, NW
Eighty Four, Pa., 56, SE
Eisaman, Pa., 57, NE
Ekman, W.Va., 150, NW
Elam, Va., 155, NE
Elba, Ohio, 70, SW
Elbel, Pa., 44, NW
Elberon, N.J., 68, NE
Elbert, W.Va., 149, NE
Elbon, Pa., 29, NE
Eldo, Pa., 60, NW
Eldon, Ohio, 70, NW
Eldora, Pa., 80, NE
Eldorado, Pa., 60, NW
Eldred, Pa., 17, NW
Eleanora Mines, Pa., 29, SW
Elgin, Pa., 14, NE
Elgin, Va., 110, SW
Elimer, Ky., 147, NE
Elizabeth (CNJ; PRR), N.J., 53,
    SE
Elizabeth (P&LE; PRR), Pa., 57,
    NW
Elizabeth Ave. (CNJ), N.J., 53, SE

Elizabeth, W.Va., 87, SW
Elizabethport, N.J., 53A
Elizabethtown, Pa., 63, SE
Elizabethville, Pa., 48, SW
Elk City, Pa., 27, SE
Elk Garden, W.Va., 91, NE
Elk Hill, Va., 142, SE
Elk Mills, Md., 81, SW
Elk River Bridge, W.Va., 123, NE
Elk River Jct., W.Va., 107, NW
Elk Run Jct. (PRR), Pa., 44, NW
Elk, W.Va., 120, NE
Elkdale Jct., Pa., 44, SW
Elkhorn City (CC&O; C&O), Ky.,
    148, NW
Elkhorn City Jct., Ky., 148, NW
Elkhorn, W.Va., 150, NW
Elkhurst, W.Va., 121, NE
Elkins Jct., W.Va., 107, NW
Elkins Park, Pa., 66, SE
Elkins, W.Va., 107, NW
Elkland (B&O; NYC), Pa., 19, NW
Elko, Va., 158, NE
Elkram, Pa., 50, NW
Elkridge Jct., W.Va., 121, SW
Elkridge, W.Va., 121, SW
Elkton (CHW; N&W), Va., 126,
    NE
Elkton, Md., 81, SW
Elkton, Ohio, 40, NE
Elkview, Pa., 81, NW
Elkview, W.Va., 121, NW
Elkwater, W.Va., 106, SE
Elkwood, Va., 111, SW
Ellendale Forge, Pa., 63, NW
Ellendale, Del., 116, NW
Ellenton, Pa., 20, SW
Ellerslie, Md., 75, SW
Ellerson, Va., 144, SW
Ellett, Va., 144, NW
Ellett, Va., 152, SW
Ellicott City, Md., 96, NW
Ellicottville, N.Y., 4, NE
Ellinboro, W.Va., 87, NE
Elliottsburg, Pa., 62, NW
Ellisburg, Pa., 18, NW
Elliston, Va., 152, SE
Ellsworth, Ohio, 25, SW
Ellsworth, Pa., 56, SE
Ellwood City, Pa., 41, NW
Ellwood Jct. (PRR), Pa., 41, NW
Ellwood, Ky., 147, NE
Ellwood, Md., 115, SW

Elm Grove, W.Va., 55, SE
Elm Park, N.Y., 53, SE
Elm St. (Norristown), (RDG), Pa.,
    66A
Elm, N.J., 83, SW
Elma, Va., 141, NW
Elmer, N.J., 82, SE
Elmhurst (DL&W; Erie), Pa., 35,
    NE
Elmhurst, N.Y., 54, SW
Elmira (DL&W; Erie; LV; PRR),
    N.Y., 8, SW
Elmira Heights (DL&W; LV),
    N.Y., 8, SW
Elmo, Pa., 27, NE
Elmont, Va., 144, SW
Elmora Ave., N.J., 53, SE
Elmore, W.Va., 136, SW
Elmsford, N.Y., 39, SW
Elmwood (AC; P-RSL), N.J., 83,
    SE
Elmwood Road, N.J., 83, NW
Elmwood, Pa., 82, NW
Elmwood, W.Va., 119, SW
Elrama, Pa., 57, NW
Elrico, Pa., 57, NE
Elsmere, N.J., 82, SE
Elswick, Ky., 147, NE
Eltingville, N.Y., 53, SE
Elton, N.Y., 5, NW
Elton, Pa., 59, NE
Elvaton, Md., 96, SE
Elverson, Pa., 65, SW
Elverton, Va., 164, NE
Elverton, W.Va., 121, SE
Elwell, W.Va., 102, SE
Elwyn, Pa., 82, NW
Elys, Ky., 162, NW
Emblem, Pa., 57, NW
Embreeville, Pa., 81, NE
Emerald (LV; RDG), Pa., 50, SE
Emerling, Ky., 163, NW
Emerson, N.J., 53, NE
Emigsville, Pa., 63, SE
Emil, Va., 140, SW
Emlenton, Pa., 27, NW
Emlenton, Pa., 27, SE
Emmanuel, Ky., 162, NW
Emmart, W.Va., 106, NW
Emmaus, Pa., 51, SW
Emmett, W.Va., 135, SW
Emmitsburg, Md., 78, SW
Emmons, W.Va., 120, SW

Emory, Va., 166, NW
Emoryville, W.Va., 91, NE
Empire Quarries, N.Y., 38, NW
Empire, Ohio, 40, SE
Emporia (ACL; A&D), Va., 174,
    SE
Emporium, Pa., 17, SE
Emsworth, Pa., 41, SE
End of Branch, Ohio, 71, NW
End of Branch, Pa., 59, SW
End of Branch, Pa., 75, NW
End of Tipton R.R., Pa., 45, SW
Endeavor, Pa., 15, SW
Endicott, N.Y., 9, SE
Enfield, Pa., 66, SE
Engle, W.Va., 94, NW
Englewood, N.J., 54, NW
English, W.Va., 149, NE
Englishtown, N.J., 68, NW
Enlow, Pa., 56, NE
Enoch, Pa., 58, SE
Enola Yard, Pa., 63, NW
Enon, Pa., 41, NW
Enterprise, Pa., 49, NW
Enterprise, Pa., 79, NE
Enterprise, W.Va., 89, NW
Entriken, Pa., 60, NE
Ephrata, Pa., 64, SE
Eraw, W.Va., 135, SW
Erbacon, W.Va., 105, SE
Erie (B&LE), Pa., 1, SE
Erie (NKP), Pa., 1, SE
Erie (NYC), Pa., 1, SE
Erie (PRR), Pa., 1, SE
Erie Jct., Pa., 29, SW
Erie, W.Va., 89, NW
Erin, N.Y., 8, SE
Erin, W.Va., 149, NE
Erma, N.J., 117, NW
Ermine, Ky., 147, SW
Erskine, N.J., 38, SW
Erwin, W.Va., 90, NE
Erwins Crossing, N.Y., 7,
    SE
Erwins, N.Y., 7, SE
Erwinton, Pa., 29, NW
Esco, Ky., 147, NE
Eskdale, W.Va., 121, SW
Esmer, Pa., 1, SE
Esmont (C&O; NEA), Va., 141,
    NE
Espanong, N.J., 52, NE
Espy, Pa., 34, SW

Espyville, Pa., 13, SW
Essen, Pa., 56, NE
Esserville, Va., 164, NE
Essex Fells (Erie; MT&E), N.J., 53, NW
Essex St., N.J., 53, NE
Essick, Pa., 33, NE
Essington (PRR; RDG), Pa., 82, NW
Estar, W.Va., 103, NW
Esthill, Ky., 147, NW
Ethel, Ohio, 70, NW
Ethel, W.Va., 135, NW
Etna (B&O; PRR), Pa., 57A
Etna, N.Y., 9, NW
Etna, Pa., 42, SW
Etna, Pa., 57A
Etowah, W.Va., 120, NE
Euclid, Pa., 42, NW
Euclid, Va., 177, NE
Eugene, Pa., 59, SW
Eulaine, Va., 156, NW
Eunice, W.Va., 135, NE
Eureka, Ohio, 40, NE
Eureka, W.Va., 87, NW
Evans City, Pa., 41, NE
Evans, Pa., 73, NE
Evans, W.Va., 103, NW
Evanston, Pa., 60, NE
Evansville, N.J., 83, NE
Evansville, Pa., 65, NW
Evarts, Ky., 163, NE
Everett, Pa., 60, SW
Evergreen, Md., 96, NE
Evergreen, Va., 155, NW
Everson, Pa., 57, SE
Everson, W.Va., 89, NE
Evington, Va., 154, SW
Ewell, Va., 159, NE
Ewing, N.J., 67, NW
Ewing, Pa., 63, SE
Ewing, Va., 163, SW
Ewings, Pa., 56, NE
Excelcior, W.Va., 149, NE
Excelsior, Pa., 45, NW
Excelsior, Pa., 49, NW
Exchange, W.Va., 105, NE
Exmoor, Pa., 49, SW
Exmore, Va., 146, SW
Export, Pa., 57, NE
Export, W.Va., 137, NW
Exton, Pa., 65, SE
Eyer, Pa., 45, SE

Eyersgrove Jct., Pa., 33, SE
Eyersgrove, Pa., 33, SE

Faber, Va., 141, NE
Factoryville, Pa., 22, SW
Fagg, Va., 152, SW
Fair Oaks, Va., 144, SW
Fairbank, Pa., 73, NW
Fairbrook, Pa., 46, SW
Fairchance (B&O; PRR), Pa., 73, NE
Fairfax, Va., 112, NW
Fairfield, Pa., 33, SW
Fairfield, Pa., 78, NW
Fairfield, Va., 140, NW
Fairgrounds, Pa., 65, NW
Fairhope, Pa., 75, NW
Fairmont (B&O; MON), W.Va., 89, NE
Fairmont Ave., N.J., 53, NE
Fairmont Belt Line Jct., W.Va., 89, NE
Fairmont, Pa., 42, NE
Fairmont, Pa., 80, NE
Fairplay, W.Va., 76, SW
Fairpoint, Ohio, 55, SW
Fairton, N.J., 99, NE
Fairview (NKP), Pa., 1, SW
Fairview (NYC), Pa., 1, SW
Fairview Pit, Pa., 1, SW
Fairview, N.J., 53, NE
Fairwood, Va., 167, SW
Falconer, N.Y., 3, SE
Fall Creek, Va., 171, SW
Fallen Timber Jct., Pa., 45, SW
Fallen Timber, Pa., 44, SE
Fallen Timber, W.Va., 71, SE
Falling Rock, W.Va., 121, NW
Falling Spring, Va., 139, NW
Falling Waters, W.Va., 77, SW
Falls Church, Va., 112, NE
Falls Creek, Pa., 29, SW
Falls Mills, Va., 150, NW
Falls View, W.Va., 121, SE
Falls, Md., 96, NW
Falls, Pa., 35, NW
Fallsburgh Tunnel, N.Y., 24, SE
Fallsburgh, N.Y., 24, SE
Fallston, Pa., 41, SW
Falmouth, Pa., 63, SE
Fanshaw, Va., 158, NW
Fanwood, N.J., 53, SW
Far Hills-Bedminster, N.J., 52, SE

Far Rockaway, N.Y., 54, SE
Farm School, Pa., 66, NE
Farm, W.Va., 120, NW
Farm, W.Va., 149, NE
Farmer, Va., 171, NW
Farmers Valley (PS&N; PRR), Pa., 17, NW
Farmersville, N.Y., 5, NW
Farmingdale (CNJ; PRR), N.J., 68, SE
Farmington, Del., 115, NE
Farmington, Pa., 50, SE
Farmington, Va., 126, SE
Farmington, W.Va., 72, SE
Farmville (N&W; TWRR), Va., 156, NW
Farnhurst, Del., 81, SE
Farrandsville, Pa., 31, SE
Farrell (NYC; PRR), Ohio, 25, SE
Farview, Pa., 23, SW
Farwell, Pa., 31, NE
Fassett, Pa., 20, NW
Faulkner, W.Va., 107, NE
Faunce, Pa., 45, NW
Fayette City, Pa., 57, SW
Fayette, W.Va., 121, SE
Fayetteville, Pa., 77, NE
Featherstone, Va., 112, SE
Federal St., Pa., 56A
Federal, Ohio, 86, NW
Federal, Pa., 56, NE
Federalsburg, Md., 115, SW
Feds Creek Spur Jct., Ky., 148, NE
Fellwick, Pa., 66, SE
Felton, Del., 98, SE
Felton, Pa., 79, NE
Fendley, Va., 157, NE
Fenelton, Pa., 42, NE
Fenmore, Pa., 79, NE
Fentress, Va., 177, SE
Fenwick, N.J., 82, SW
Fenwick, W.Va., 122, SE
Ferenbaugh, N.Y., 7, SE
Ferguson, Pa., 73, NE
Ferguson, W.Va., 119, SW
Fern Glen, Pa., 49, NE
Fern Hill, Pa., 81, NE
Fern Rock, Pa., 66, SE
Fernbank, Pa., 27, NE
Ferncliff, Md., 80, SW
Ferncliff, Pa., 45, NW
Ferndale, Ky., 162, SE

Ferndale, Md., 96, SE
Ferndale, N.Y., 24, NE
Ferndale, Pa., 15, SE
Ferndale, Pa., 49, NE
Ferndale, Pa., 59, NW
Fernwood, Ohio, 55, NE
Fernwood, Pa., 45, NW
Fernwood, Pa., 82, NW
Ferrell, W.Va., 120, NW
Ferrellsburg, W.Va., 119, SE
Ferro, Ohio, 71, NW
Ferrol, Va., 125, SW
Ferrona, Pa., 25, SE
Ferrum, Va., 169, NE
Ferry St., N.J., 53A
Fetterman, Pa., 41, NW
Fetterman, W.Va., 89, NE
Ficht, Pa., 56A
Field, Va., 149, SE
Fieldale (D&W; N&W), Va., 170, SW
Fieldmore Springs, Pa., 14, SE
Fieldsboro, N.J., 67, SE
Fiery Siding, Md., 77, SE
Fife, Pa., 56, NE
Filbert, W.Va., 149, NE
Filer, Pa., 26, SE
Fillmore (Erie), N.Y., 5, NE
Fillmore (PRR), N.Y., 5, NE
Fillmore, Pa., 46, NW
Finchley, Va., 173, SW
Finley, N.J., 99, NE
Finleyville, Pa., 56, NE
Finleyville, Pa., 60, SE
Finney, Va., 165, NE
Finneywood, Va., 173, NW
Fire Creek, W.Va., 136, NE
Fireco, W.Va., 136, SE
First Ford, Va., 151, NW
First Fork, Pa., 30, NE
First St., W.Va., 40, SE
Firthcliffe, N.Y., 38, NE
Fish Camp, W.Va., 89, SE
Fish, Pa., 79, NW
Fish's Eddy, N.Y., 23, NE
Fisher, N.J., 83, SW
Fisher, Ohio, 85, NE
Fisher's Hill, Va., 110, NW
Fishers Crossing, W.Va., 106, SW
Fishers Siding, Pa., 43, SW
Fishers, Pa., 66, SE
Fishersville, Va., 126, SW
Fishertown, Pa., 59, SE

Fishing Creek, Pa., 80, NW
Fitz Henry, Pa., 57, SW
Fitzpatrick, W.Va., 136A, SE
Five Block, W.Va., 135, NW
Five Forks, Pa., 77, NE
Five Oaks, Va., 150, SW
Five Points, Pa., 44, NE
Flaggy Meadow, W.Va., 72, SE
Flagtown (CNJ; LV), N.J., 52, SE
Flanders, N.J., 52, NE
Flat Lick, Ky., 162, NW
Flat Rock, Va., 143, SW
Flat Top, W.Va., 150, NW
Flatbush Ave., N.Y., 54, SW
Flatwoods, W.Va., 105, SE
Fleetwood, N.Y., 54, NW
Fleetwood, Pa., 65, NW
Fleming Summit, Pa., 44, SW
Fleming, Ky., 147, SE
Fleming, Ohio, 86, NE
Flemington (CNJ; LV; PRR), N.J., 52, SW
Flemington Jct., N.J., 52, SW
Flemington, W.Va., 89, NE
Flemingville, N.Y., 9, SE
Flicksville, Pa., 51, NE
Flint, W.Va., 107, NE
Flinton, Pa., 44, SE
Flipping, W.Va., 150, NW
Floodwood, Ohio, 85, NE
Floral Park, N.Y., 54, SE
Floreffe, Pa., 57, NW
Florence Yard, Pa., 51, SW
Florence, N.J., 67, SW
Florence, N.J., 83, SW
Florida, N.Y., 38, NW
Florin, Pa., 63, SE
Florinel, Pa., 63, SE
Flourtown, Pa., 66, SE
Floyd, W.Va., 71, SE
Flushing Main St., N.Y., 54, NW
Foch, W.Va., 120, SW
Fogle Mine, Pa., 74, NE
Folcroft, Pa., 82, NW
Foley, W.Va., 152, NW
Follansbee, W.Va., 55, NE
Folsom, N.J., 83, SW
Folsom, W.Va., 88, NE
Fombell, Pa., 41, NE
Fonde, Ky., 162, SW
Fontaine, Pa., 65, SW
Fontaine, Va., 170, SW

Foot of Mountain, Pa., 49, SE, 49B
Foot of President St., Md., 96A
Footedale, Pa., 73, NW
Forbes Road, Pa., 57, NE
Force, Pa., 29, NE
Ford City, Pa., 42, NE
Ford, N.J., 37, SE
Ford, Va., 157, SE
Fordham, N.Y., 54, NW
Fordham, Pa., 43, NE
Fords Branch, Ky., 147, NE
Fordwick, Va., 125, SW
Forest City (Erie; NYO&W), Pa., 23, SW
Forest Glen, Md., 95, SE
Forest Grove, N.J., 83, SW
Forest Hill, Md., 80, SW
Forest Hill, N.J., 53, NE
Forest Hills, N.Y., 54, SW
Forest Hills, Pa., 66, SE
Forest, Va., 154, NW
Forest, W.Va., 136A, SE
Fork Ridge Jct., Tenn., 162, SW
Fork Union, Va., 142, NW
Fork, Va., 156, NW
Forked River, N.J., 84, NE
Forks, Pa., 34, SW
Forrest Hall, Md., 130, NE
Forrest, Pa., 59, NW
Forsyth, N.Y., 2, NE
Fort Blackmore, Va., 164, NE
Fort Branch, W.Va., 135, NW
Fort Defiance, Va., 126, SW
Fort Dix, N.J., 67, SE
Fort Gay, W.Va., 118, SE
Fort George G. Meade (B&O), Md., 96, SE
Fort Hill, Pa., 66, SE
Fort Hill, Pa., 74, NW
Fort Lee, Va., 158, NW
Fort Louden, Pa., 77, NW
Fort Miles, Del., 116, NE
Fort Mitchell, Va., 173, NW
Fort Monroe, Va., 160, SW
Fort Montgomery, N.Y., 39, NW
Fort Pitt, Pa., 56A
Fort Richmond, N.Y., 53, SE
Fort Robinson, Pa., 62, NW
Fort Spring, W.Va., 137, SE
Fort Wadsworth, N.Y., 53, SE
Fort Washington, N.Y., 54, NW
Fortuna, Pa., 66, NW

Forty Fort, Pa., 35A
Fossilville, Pa., 75, NE
Foster Falls, Va., 168, NW
Foster, Pa., 22, SW
Foster, W.Va., 71, NW
Foul Rift, N.J., 51, NE
Fount, Ky., 162, NW
Fountain Rock, Md., 95, NW
Four Mile, Ky., 162, NE
Four Pole Spur Jct., W.Va., 135, SW
Fourth Ave. (Pittsburgh), Pa., 57A
Foustwell, Pa., 59, SW
Fowblesburg, Md., 79, SW
Fowler, Ohio, 25, NE
Fowlerville, Pa., 34, SW
Fox Chase, Pa., 66, SE
Fox Hall, Va., 177, NE
Foxburg (B&O; PRR), Pa., 27, SE
Foys Hill, Md., 81, SW
Frackville, Pa., 49, NE
Frametown, W.Va., 105, SW
Francis Mine No. 1, W.Va., 89, NE
Francis Mine, W.Va., 91, SW
Francis, W.Va., 136, SW
Franconia, Va., 112, NE
Frank Ave., N.Y., 54, SW
Frank, W.Va., 107, SW
Frankford (PRR; RDG), Pa., 66, SE
Frankford Jct., Pa., 66, SE, 82B
Frankford, Del., 116, SE
Franklin (A&D; F&C; SAL), Va., 176, SW
Franklin (DL&W; NYS&W), N.J., 37, SE
Franklin (Erie; NYC; PRR), Pa., 27, NE
Franklin Ave. (Nutley), N.J., 53, NE
Franklin City, Va., 133, SW
Franklin Jct., Va., 176, SE
Franklin St. (Reading), (RDG), Pa., 65, NW
Franklin, N.Y., 11, NE
Franklinville, N.J., 82, SE
Franklinville, N.Y., 5, NW
Frankstown, Pa., 60, NW
Frankville, Md., 74, SE
Franz, Pa., 43, NE
Frazer, Pa., 65, SE, 65A

Frazier, W.Va., 137, SE
Fred, W.Va., 120, SE
Frederick (B&O; PECo.; PRR), Md., 95, NW
Frederick Hall, Va., 143, NW
Frederick Jct., Md., 95, NW
Frederick Road, Md., 96, NE
Fredericksburg (RF&P; VC), Va., 129, NW
Fredericktown, Pa., 57, SW
Freedom, N.Y., 5, NW
Freedom, Pa., 41, SW
Freedonia, Pa., 26, NW
Freehold (CNJ; PRR), N.J., 68, NW
Freeland Crossing, Pa., 35, SW
Freeland, Md., 79, SE
Freeland, Pa., 35, SW
Freeman, N.J., 82, NE
Freeman, N.Y., 7, SW
Freeman, Pa., 16, SE
Freeman, Va., 174, NE
Freemansburg (CNJ; LV), Pa., 51, SW
Freeport (PRR), Pa., 42, SE
Freeport Jct. (PS), Pa., 42, SE
Freeport, N.Y., 54, SE
Freeze Fork, W.Va., 135, NW
Fremont, Va., 148, SW
French Creek Jct., Pa., 65, SW
French Creek, Pa., 13, SE
French, Va., 151, NW
French, W.Va., 75, SE
Frenchton, W.Va., 106, NW
Frenchtown, N.J., 51, SE
Frenchville, Pa., 30, SE
Freneau, N.J., 68, NE
Fresh Pond Jct. (LI; NYCN), N.Y., 54C
Fresh Pond, N.Y., 54C
Frey's Grove, Pa., 63, SE
Frick, Pa., 57C
Fricks Lock, Pa., 65, SE
Friedens, Pa., 59, SW
Friendly, W.Va., 70, SE
Friendship (Erie), N.Y., 5, SE
Friendship (PS&N), N.Y., 5, SE
Friendship, Md., 133, NE
Friendship, N.J., 82, SW
Friendsville, Md., 74, SW
Fries Jct., Va., 168, NW
Fries, Va., 168, SW
Frisco, Pa., 41, NW

Frisco, Tenn., 164, SE
Fritz, Pa., 33, SW
Fritztown, Pa., 64, NE
Froman, Pa., 57, SW
Front Royal (N&W; SOU), Va., 110, NE
Front Royal Jct. (SOU), Va., 110, NE
Frostburg (C&PA; WM), Md., 75A
Frostburg (WM), Md., 75, SW, 75A
Frostburg, Pa., 43, NE
Frosts, Ohio, 86, NW
Frugality, Pa., 45, SW
Fruitland, Md., 132, NE
Frutchey, Pa., 36, SE
Fry, W.Va., 119, SE
Frye, Pa., 57, SW
Ft. Meade Jct., Md., 96, SW
Fuccy, W.Va., 105, SE
Fuller, Pa., 29, SW
Fuller, Pa., 57, SE
Fullers, Ky., 118, SE
Fullerton, Pa., 51C
Fulmor, Pa., 66, SE
Fulton Brick Yard, Va., 144, SW
Fulton, Pa., 49, NW
Fulton, Va., 144A
Funkstown, Md., 77, SE
Fuqua, W.Va., 120, NW

G&B Jct., W.Va., 89, SE
G&E Jct., W.Va., 137, NW
Gabe, Va., 149, SE
Gage, W.Va., 107, NW
Gaines Jct., Pa., 18, SE
Gaines Mill, Va., 144, SW
Gaines, Pa., 18, NE
Gainesboro, Va., 93, NW
Gainesville, Va., 111, NE
Gaither, Md., 96, NW
Gaithersburg, Md., 95, SE
Gala, Va., 139, SW
Galax, Va., 168, SW
Gale, W.Va., 106, NE
Galeton, Pa., 18, SE
Galilee, N.J., 69, NW
Gallagher, W.Va., 121, SW
Gallant Green, Md., 113, SW
Gallatin, Pa., 57, SW
Galliopolis (B&O), W.Va., 102, NE

Galliopolis (C&O), Ohio, 102, NE
Gallitzin, Pa., 59, NE
Galloway Jct., W.Va., 89, SE
Galloway, W.Va., 89, SE
Gallup, Ky., 118, SE
Galmish, W.Va., 71, SE
Galt, Md., 78, SE
Galts Mill, Va., 154, NE
Gamble, Pa., 56, SE
Gamoca, W.Va., 121, SE
Gang Mills, N.Y., 7, SE
Ganister, Pa., 60, NE
Ganoga Lake, Pa., 34, NW
Gans, Pa., 73, SW
Gap Jct., Pa., 50, NE
Gap Road, Pa., 77, NW
Gap, Pa., 80, NE
Gapland, Md., 94, NE
Gara Jct., Pa., 50, NW
Garden City, N.Y., 54, SE
Garden Lake, N.J., 83, NW
Garden, Va., 149, SW
Gardener, Pa., 65, NE
Gardiner Ave. (New Castle), (Erie), Pa., 41, NW
Gardiner, Pa., 45, NE
Gardiner, Va., 149, SW
Gardiners Farm, Pa., 62, SE
Gardiners Switch, Pa., 35A
Gardners, Pa., 62, SE
Garfield, N.J., 53, NE
Garfield, Ohio, 40, NW
Garland (NYC; PRR), Pa., 15, NW
Garland, W.Va., 149, NW
Garlick Mine, Pa., 60, SE
Garman, Pa., 44, SW
Garner, W.Va., 120, SW
Garrett (B&O; WM), Pa., 74, NE
Garrett Park, Md., 95, SE
Garrett Road, Pa., 82, NW
Garrett, Ky., 147, NW
Garretts Mill, Md., 94, NE
Garrison, N.Y., 39, NW
Garvers Ferry, Pa., 42, SE
Garwood, N.J., 53, SW
Garwood, W.Va., 150, NW
Garwoods (Erie; PS&N), N.Y., 6, NW
Gary, W.Va., 149, NE
Gascola, Pa., 57, NW
Gaskill, Ky., 147, SE
Gassaway Branch Switch, W.Va., 105, SW

Gassaway, W.Va., 105, SW
Gaston, W.Va., 89, SW
Gastonville (B&O; P&WV), Pa., 57, NW
Gate City, Va., 164, SE
Gate, N.J., 83, SE
Gates, N.C., 176, SW
Gates, Pa., 16, NW
Gates, Pa., 73, NW
Gauley Bridge, W.Va., 121, SE
Gauley Mills, W.Va., 122, NE
Gauley River Jct., W.Va., 122, NE
Gauley, W.Va., 121, SE
Gawango, Pa., 16, NW
Gaylord, Va., 94, SW
Gazzam, Pa., 44, NE
Gedney Way, N.Y., 39, SW
Gehman, Pa., 65, NE
Geiger, Pa., 58, SE
Geigertown, Pa., 65, SW
Geisingers, Pa., 51, SW
Genasco, N.J., 53, SW
Genesee (B&O; NY&P), Pa., 18, NW
Genesee Springs, Pa., 18, NW
Geneva (NKP; NYC), Ohio, 12, NW
Geneva, Pa., 13, SE
Genoa, W.Va., 119, SW
George School, Pa., 67, SW
George, Pa., 56, NW
Georgetown Jct., Md., 95, SE
Georgetown, D.C., 112, NE
Georgetown, Del., 116, SW
Gerhards, Pa., 50, NW
Germania, N.J., 83, SE
Germania, Pa., 18, SE
Germans, Pa., 50, NE
Germansville, Pa., 50, SE
Germantown Road, Pa., 66, SE
Germantown, Md., 95, SW
Germantown, Pa., 66, SE
Germonds, N.Y., 39, SW
Gerstell, W.Va., 92, NW
Getty Square (Yonkers), N.Y., 54, NW
Gettysburg (RDG; WM), Pa., 78, NE
Gettysburg Jct., Pa., 62, SE
Geyelin Park, Pa., 33, NE
Giatto, W.Va., 150, NW
Gibbstown, N.J., 82, NW
Gibraltar, Pa., 65, NW

Gibson "B" Yard, Pa., 73A
Gibson Colliery, Pa., 57, SW
Gibson, Ky., 147, NE
Gibson, N.Y., 54, SE
Gibson, Ohio, 70, NW
Gibsonia, Pa., 42, SW
Gibsons Point, Pa., 82A
Gibsons, Va., 162, SE
Gilbert (N&W; VGN), W.Va., 135, SW
Gilbert Yard (N&W; VGN), W.Va., 135, SW
Gilbert, Va., 127, SW
Gilbert, W.Va., 72, SE
Gilberton Branch Jct., Pa., 49C
Gilberton, Pa., 49C
Giles, W.Va., 121, SE
Gilkeson, Pa., 56, SE
Gill, W.Va., 119, SE
Gillespie, N.J., 68, NW
Gillespie, Va., 149, SE
Gillespie, W.Va., 105, SE
Gillespie, W.Va., 106, SW
Gillett, Pa., 20, NW
Gillette, N.J., 53, SW
Gilliam, W.Va., 150, NW
Gillintown, Pa., 31, SW
Gilman, W.Va., 107, NW
Gilmer, W.Va., 105, NE
Gilmerton, Va., 177, NW, 177A
Gilmore Jct., Pa., 56, NE
Gilmore Mills, Va., 139, SE
Gilson, Pa., 15, SE
Ginter, Pa., 45, NW
Gipsy, Pa., 44, NW
Girard (Erie) Ohio, 25, SE
Girard Jct. (B&O; LE&E), Ohio, 25A
Girard Jct., Pa., 49C
Girard Manor, Pa., 49, NE
Girard, Pa., 1, SW
Girard, Pa., 13, NW
Girardville (LV; RDG), Pa., 49, NW
Girdle Tree, Md., 133, SW
Girdler, Ky., 162, NW
Gist Siding, Pa., 73, NE
Given, Va., 138, SE
Gladden (PRR; P&WV), Pa., 56, NE
Glade Hill, Va., 170, NW
Glade Spring, Va., 166, NW
Glade, Pa., 15, NE

Glade, W.Va., 136, NE
Gladhill, Pa., 78, SW
Gladstone, N.J., 52, SE
Gladstone, Pa., 82, NW
Gladstone, Va., 141, SW
Gladwin, W.Va., 90, SE
Gladwyne, Pa., 66B
Glady, W.Va., 107, NE
Gladys, Va., 154, SE
Glamorgan, Va., 164, NE
Glasgow, Del., 81, SE
Glasgow, Pa., 45, SW
Glasgow, Pa., 65, NE
Glasgow, Va., 140, SW
Glasgow, W.Va., 121, SW
Glass Works, N.J., 82, SW
Glassboro, N.J., 82, SE
Glassmere, Pa., 42, SW
Glassport, Pa., 57, NW
Glassrock, W.Va., 72, SE
Glatfelter, Pa., 79, NE
Gleason, W.Va., 91, NE
Glebe, W.Va., 92, SW
Glen Allen, Va., 143, SE
Glen Alum, W.Va., 135, SW
Glen Burnie, Md., 96, SE
Glen Campbell (NYC; PRR), Pa.,
    44, NW
Glen Carlyn, Va., 112, NE
Glen Cove, N.Y., 54, NE
Glen Easton, W.Va., 71, NE
Glen Eyre, Pa., 36, NE
Glen Falls, W.Va., 89, NW
Glen Ferris, W.Va., 121, SE
Glen Gardner, N.J., 52, SW
Glen Hayes, W.Va., 118, SE
Glen Hazel, Pa., 16, SE
Glen Head, N.Y., 54, NE
Glen Hope, Pa., 44, NE
Glen Huddy, W.Va., 121, SW
Glen Iron, Pa., 47, NE
Glen Irwin, Pa., 42, SE
Glen Jct., Pa., 49, NE
Glen Jean, W.Va., 136, NE
Glen Loch, Pa., 65, SE, 65A
Glen Lyn (N&W; VGN), Va., 151,
    NW
Glen Lyon (CNJ; PRR), Pa., 34,
    SE
Glen Manor, Pa., 64, SW
Glen Mawr, Pa., 33, NE
Glen Mills, Pa., 82, NW
Glen Moore, N.J., 67, NW

Glen Moore, Pa., 65, SW
Glen Morgan, W.Va., 136, SE
Glen Morris, Md., 96, NW
Glen Onoko, Pa., 50, NW
Glen Osborne, Pa., 41, SE
Glen Richey, Pa., 45, NW
Glen Riddle, Pa., 82, NW
Glen Ridge (DL&W; Erie), N.J.,
    53, NE
Glen Rock, N.J., 53, NE
Glen Rock, Pa., 79, NE
Glen Rogers, W.Va., 136, SW
Glen Rose, Pa., 81, NW
Glen St. (Glen Cove), N.Y., 54, NE
Glen Union, Pa., 31, NE
Glen White Jct. (C&O; VGN),
    W.Va., 136A, NW
Glen White, W.Va., 136A, NW
Glen Wilton, Va., 139, NW
Glen, Pa., 65, SE, 65A
Glenarm, Md., 97, NW
Glenartney, Md., 96, SE
Glenbrook, Conn., 39, SE
Glenburn, Pa., 22, SE
Glencoe, Md., 79, SE
Glencoe, Ohio, 55, SW
Glencoe, Pa., 75, NW
Glendale, N.Y., 54, SW
Glendale, Pa., 29, SE
Glendale, W.Va., 71, NW
Glendon, N.J., 82, NE
Glendon, Pa., 51, SE
Glendon, W.Va., 105, SW
Glenfield, Pa., 41, SE
Glenita, Va., 164, SW
Glenlake, Pa., 67, SW
Glenn, Pa., 56A
Glenndale, Md., 113, NW
Glenns, Ohio, 102, NW
Glenolden (B&O; PRR), Pa., 82,
    NW
Glenover, Pa., 57, NW
Glenray, W.Va., 137, SE
Glenrock, Va., 177, NE
Glens, N.J., 38, SW
Glenshaw, Pa., 42, SW
Glenside, Pa., 66, SE
Glenvar (N&W; VGN), Va., 152,
    NE
Glenville, Pa., 79, NW
Glenwillard, Pa., 41, SE
Glenwood Jct., N.Y., 38, NW
Glenwood, N.J., 38, NW

Glenwood, N.Y., 54, NW
Glenwood, Pa., 57A
Glenwood, W.Va., 102, SE
Glidden, Ky., 163, NW
Globe, W.Va., 40, SE
Glory, Pa., 44, SW
Gloucester, N.J., 82, NE
Glover Gap, W.Va., 72, SW
Glyn Heath, N.J., 100, NE
Glynden, Pa., 14, NE
Glyndon, Md., 96, NW
Godby, W.Va., 134, NE
Godeffroy's, N.Y., 37, NE
Godfrey, Pa., 42, SE
Goehring, Pa., 41, NE
Goff, Pa., 27, SW
Gold Dale, Va., 128, NW
Gold Mine, Pa., 48, SE
Gold, Pa., 18, NW
Golden Ring, Md., 97, NW
Golden, W.Va., 123, SW
Golden's Bridge, N.Y., 39, NE
Goldenville, Pa., 78, NE
Goldsboro, Md., 98, SW
Goldsboro, Pa., 63, SW
Golt, Md., 98, NW
Good Spring Colliery, Pa., 49, SW
Good Spring, Pa., 49, SW
Good, Pa., 44, NE
Goode, Va., 154, NW
Goodman, Pa., 45, SE
Goodman, W.Va., 134, SW
Goodmans, N.J., 53, SW
Goods Corner, Pa., 59, NW
Goodview, Va., 153, SE
Goodwill, W.Va., 150, NW
Goodwin, W.Va., 106, NE
Goodwins Ferry, Va., 151, NE
Goodyear, Pa., 62, SE
Gorbush, Md., 96, NW
Gordon St., Pa., 51B
Gordon, Pa., 49, NW
Gordon, Va., 141, SW
Gordonsville, Va., 127, SE
Gordonville, Pa., 64, SE
Gore, Va., 93, NW
Gorman, W.Va., 91, NW
Gorton, Pa., 45, NE
Goshen, N.J., 100, SW
Goshen, N.Y., 38, NW
Goshen, Pa., 80, NE
Goshen, Va., 140, NW
Goss Run Jct., Pa., 45, NW

Gould, Pa., 57, NW
Gouldsboro, Pa., 36, SW
Gowen, Pa., 49, NE
Grace Siding, Pa., 43, SE
Grace, W.Va., 92, NE
Graceham, Md., 78, SW
Graceton, Pa., 43, SE
Grafton, W.Va., 89, NE
Graham, N.Y., 37, NE
Graham, N.Y., 39, SW
Graham, W.Va., 103, NW
Grand Central Terminal (New
    York), (NH; NYC), N.Y., 54, NW
Grand Central, Pa., 51, NW
Grand Valley, Pa., 14, SE
Grand View, N.Y., 39, SW
Grandy, Va., 174, NW
Granite Springs, N.Y., 39, NW
Granite, Pa., 78, NE
Granite, Va., 143, SE
Granoque, Del., 81, NE
Grant Ave., N.J., 53, SW
Grant City, N.Y., 53, SE
Grant, Ky., 118, NW
Grantsville, Md., 74, SE
Granville, Pa., 46, SE
Grape Run Jct., Pa., 50A
Grapeville, Pa., 57, NE
Grapevine, W.Va., 134, SE
Grasmere, N.Y., 53, SE
Grasonville, Md., 114, NE
Grass Flat, Pa., 30, SE
Grasselli Branch Jct., W.Va., 89,
    NW
Grassland, Pa., 82, NW
Grassy Sound Draw, N.J., 100,
    SW
Graters Ford, Pa., 66, SW
Gratham, Pa., 63, SW
Gratztown (B&O; PRR), Pa., 57,
    SW
Grave Creek, W.Va., 71, NE
Gravel Bank, Ohio, 86, NE
Gravel Jct., Ohio, 71, NW
Gravel Place, Pa., 36, SE
Gravers, Pa., 66, SE
Graves, Va., 140, SW
Gravity, Pa., 36, NW
Gray (A&D; VGN), Va., 175, NW
Gray (PRR), Pa., 58, NE
Gray (WM), Pa., 58, SE
Gray Jct., Pa., 58, SE
Gray, Md., 96, NW

Gray, Pa., 30, SW

Graybill, Pa., 79, NW

Grays Ferry Draw, Pa., 82A

Grays Ferry, Pa., 82A

Grays Landing, Pa., 73, NW

Grayson, Ky., 118, NW

Graystone, Md., 79, SE

Greason, Pa., 62, SW

Great Belt, Pa., 42, NW

Great Bend, Pa., 22, NE

Great Cacapon, W.Va., 76, SW

Great Kills, N.Y., 53, SE

Great Meadows, N.J., 52, NW

Great Neck, N.Y., 54, NE

Great Notch, N.J., 53, NE

Great Valley, N.Y., 4, SE

Green Bank, Del., 81, SE

Green Bay (N&W; SOU), Va., 156, SW

Green Bottom, W.Va., 102, SW

Green Bush, Va., 146, SE

Green Cove, Va., 166, SE

Green Hill, Pa., 81, NE

Green Lane, Pa., 66, NW

Green Lawn, Pa., 81, NW

Green Park, Pa., 62, NW

Green Point, Pa., 63, NE

Green Pond Jct., N.J., 38, SW

Green Ridge (B&O), W.Va., 76, SW

Green Ridge (WM), Md., 76, SW

Green Ridge, Pa., 35D

Green Spring, W.Va., 75, SE

Green Springs Jct. (PRR; WM), Md., 96, NW

Green Springs, Pa., 57A

Green Springs, Va., 127, SE

Green Valley, Pa., 28, SE

Greenawald, Pa., 50, SW

Greenbrier Jct., W.Va., 107, NW

Greencastle, Pa., 77, NE

Greendale, Pa., 16, SW

Greendale, Pa., 77, NE

Greendale, Va., 144, SW

Greendale, W.Va., 121, NE

Greendell, N.J., 52, NW

Greene, N.Y., 10, NW

Greenfield, Pa., 64, SE

Greenfields, Pa., 48, SE

Greenford, Ohio, 40, NW

Greenhurst, N.Y., 3, SW

Greenland, Tenn., 164, SW

Greenmount, Md., 79, SW

Greenock, Pa., 57, NW

Greenough, Ky., 148, NW

Greens Bridge, N.J., 51, SE

Greensboro, Md., 115, NW

Greensburg, Pa., 57, NE

Greenvale, N.Y., 54, NE

Greenview, W.Va., 135, NW

Greenville (B&LE; Erie; PRR), Pa., 26, NW

Greenville (CNJ; PRR), N.J., 53B

Greenville, Del., 81, NE, 81A

Greenville, Va., 125, SE

Greenwald, Pa., 58, NW

Greenway, Va., 141, SW

Greenwich Pier, N.J., 99, NW

Greenwich, Conn., 39, SE

Greenwich, N.J., 99, NW

Greenwich, Va., 177, NE

Greenwood Jct., Pa., 50, NW

Greenwood, Del., 115, NE

Greenwood, N.Y., 6, SE

Greenwood, Pa., 73A

Greenwood, Va., 126, SW

Greenwood, W.Va., 88, NW

Greer, W.Va., 73, SW

Gregg, Pa., 56, NE

Gregorytown, N.Y., 11, SE

Greismore, Pa., 44, SW

Grenloch, N.J., 82, NE

Grenoble, Pa., 66, SE

Gress, Pa., 77, SE

Gretna (F&P; SOU), Va., 171, NW

Grey Oaks, N.Y., 54, NW

Greycourt, N.Y., 38, NW

Greystone, N.Y., 54, NW

Greythorne, Pa., 62, SW

Griffith Siding, W.Va., 137, NW

Griffith, Va., 139, NE

Griffiths, Pa., 16, SE

Griggs, Ky., 162, NE

Grim, W.Va., 103, SW

Grimes, Md., 77, SW

Grindstone, Pa., 57, SW

Grippe, W.Va., 120, SE

Grizzard, Va., 175, SW

Groffdale, Pa., 64, SE

Grogg, Va., 158, NW

Grosclose, Va., 167, NW

Grottoes, Va., 126, NW

Grovania, Pa., 48, NE

Grove City, Pa., 26, SE

Grove Hill, Va., 109, SE

Grove St., N.J., 53, NE

Grove, Md., 95, NW

Grove, Va., 159, SE

Grover, Pa., 20, SW

Groves, W.Va., 105, SW

Groveton, Pa., 41, SE

Grundy, Va., 148, NE

Guernsey, Pa., 78, NE

Guffey, Pa., 16, NE

Guffey, Pa., 57, NW

Guilford Springs, Pa., 77, NE

Guilford, Md., 96, SW

Guilford, N.Y., 11, NW

Guilford, Pa., 77, NE

Guilfoyl, Pa., 28, NE

Guinea, Va., 129, SW

Guldens, Pa., 78, NE

Gulf Jct., W.Va., 136, SW

Gulf Summit, N.Y., 10, SE

Gulf Switch, W.Va., 136A, SE

Gulf, Ohio, 12, NW

Gulfport, N.Y., 53, SE

Gum Run, Pa., 49, NE

Gum Stump, Pa., 46, NW

Gumm, W.Va., 122, SE

Gun Hill Road, N.Y., 54, NW

Gund, Va., 148, SE

Gunpow, Md., 97, NW

Gunpowder, Md., 79, SW

Gunpowder, Md., 97, NW

Gunton Park, Va., 151, SW

Gustavus, Ohio, 25, NE

Guthrie Spur Jct., Pa., 43, SE

Guthrie, Va., 141, NE

Guyan, W.Va., 135, SW

Guyandot Jct., W.Va., 136, SW

Guyandotte (B&O; C&O), W.Va., 119, NW

Guyencourt, Del., 81, NE

Guysville, Ohio, 86, NW

Gwynbrook, Md., 96, NW

Gwynedd Valley, Pa., 66, SW

Gypsy, W.Va., 89, NW

Haberman, N.Y., 54, SW

Hackensack, N.J., 53, NE

Hacker Valley, W.Va., 106, SW

Hackett, Pa., 56, SE

Hackettstown, N.J., 52, NW

Haddix, W.Va., 90, SW

Haddleton, W.Va., 120, SW

Haddonfield, N.J., 82B

Haden, Va., 139, SW

Hadley, Pa., 26, NE

Hagans, Va., 163, SW

Hagerstown (B&O; N&W; PECo.; PRR; WM), Md., 77, SE

Hagood, Va., 173, SE

Hahntown, Pa., 57, NE

Haines Jct., Pa., 35D

Haines, Pa., 50, SE

Haines, Pa., 80, SE

Hainesburg (DL&W; NYS&W), N.J., 51, NE

Hainesburg Jct., N.J., 51, NE

Hainesport, N.J., 83, NW

Hale, Va., 150, NW

Hales Eddy, N.Y., 11, SW

Hales Gap, W.Va., 151, NW

Halethorpe, Md., 96, SE

Half Moon, Pa., 36, SW

Halfway Siding, W.Va., 77, SW

Halfway, Md., 77, SW

Halfway, W.Va., 137, SE

Halifax, Pa., 63, NW

Halifax, Va., 172, NW

Hall, Md., 113, NE

Hall, Pa., 57, NW

Hallen, W.Va., 86, SE

Hallon, N.Y., 7, SE

Halls (RDG; W&NB), Pa., 33, SW

Halls, Ohio, 55, SW

Hallsboro, Va., 157, NE

Hallstead, Pa., 22, NE

Hallston, Pa., 27, SW

Hallton (CLAR; TIV), Pa., 29, NW

Halltown, W.Va., 94, NW

Hallwood, Va., 146, NE

Hallwood, W.Va., 102, NE

Halo, W.Va., 122, NE

Halpine, Md., 95, SE

Halsey, N.J., 37, SW

Halsey, Va., 154, NE

Hambleton, W.Va., 90, SE

Hamburg (PRR; RDG), Pa., 50, SW

Hamburg, N.J., 37, SE

Hamburg, Pa., 50, SW

Hamilton Beach, N.Y., 54, SW

Hamilton, N.J., 52, SE

Hamilton, Va., 94, SE

Hamilton, Va., 129, NW

Hamilton, W.Va., 121, SE

Hamler, Pa., 28, SW

Hamley Run, Ohio, 85, NE

Hamlin, Pa., 17, SW

Hamlin, Va., 165, NW

Hammel, N.Y., 54, SW
Hammond, Pa., 19, NE
Hammond, W.Va., 89, NE
Hammondsport, N.Y., 7, NE
Hammondsville, Ohio, 40, SE
Hammondsville, Pa., 57, SE
Hammondton (AC; P-RSL), N.J.,
   83, SW
Hampshire Club, W.Va., 92, NW
Hampshire, W.Va., 91, NE
Hampstead, Md., 79, SW
Hampton (CNJ; DL&W), N.J., 52,
   SW
Hampton Jct., W.Va., 106, NE
Hampton Yard, Pa., 35C
Hampton, Va., 160, SW
Hampton, W.Va., 106, NE
Hamptonburgh, N.Y., 38, NE
Hancock (B&O), W.Va., 76, SE
Hancock (Erie; NYO&W), N.Y.,
   23, NW
Hancock (WM), Md., 76, SE
Hancock, Pa., 50, SE
Hand, Va., 175, SE
Handley, W.Va., 121, SW
Hanford, N.J., 37, NE
Hanging Rock, Va., 152, NE
Hankins, N.Y., 23, NE
Hanlin, Pa., 56, NW
Hanna, Ohio, 55, SW
Hannah, Pa., 45, NE
Hannastown, Pa., 58, NW
Hannibal, W.Va., 71, SW
Hanover (PRR; WM), Pa., 79, NW
Hanover Jct. (PRR; WM), Pa., 79,
   NE
Hanover Yard, Pa., 35, SW
Hanover, Md., 96, SE
Hanover, N.J., 53, NW
Hanover, Va., 144, NW
Hansford, W.Va., 121, SW
Hansrote, W.Va., 76, SW
Happy Creek, Va., 110, NE
Harbell, Ky., 162, SE
Harbor Bridge, Pa., 26, SW
Harbor Creek (NKP), Pa., 2, SW
Harbor Creek (NYC), Pa., 2, SW
Harbor, Pa., 80, NW
Harding, W.Va., 107, NW
Hardman, W.Va., 90, NW
Hardware, Va., 142, SW
Hardwood, Va., 165, NW
Hardwood, W.Va., 122, NE

Hardy, Ky., 134, SE
Hardy, Va., 153, SW
Harewood Park, Md., 97, NW
Harewood, W.Va., 121, SW
Harlan Jct., Ky., 163, NW
Harlan, Ky., 163, NW
Harlan, Pa., 28, SE
Harleigh Jct., Pa., 49, NE
Harlem River, N.Y., 54A
Harlingen, N.J., 67, NE
Harman, Md., 96, SE
Harman, W.Va., 107, NE
Harmarville, Pa., 42, SW
Harmer, Ohio, 87, NW
Harmon, N.Y., 39, SW
Harmon, Va., 148, NE
Harmonsburg, Pa., 13, SW
Harmony Grove, Md., 95, NW
Harmony, N.J., 51, NE
Harmony, Pa., 41, NE
Harper, W.Va., 107, NW
Harper, W.Va., 136A, NW
Harpers Ferry, W.Va., 94, NE
Harpursville, N.Y., 10, SE
Harriman, N.Y., 38, NE
Harrington Park, N.J., 54, NW
Harrington, Del., 115, NE
Harrington, Va., 139, NW
Harris Ferry, W.Va., 86, SE
Harris Siding, Va., 143, SE
Harrisburg (PRR; RDG), Pa., 63,
   NW
Harrisburg Yard, Pa., 63, NW
Harrison (DL&W; Erie; H&M),
   N.J., 53A
Harrison (NH; NYW&B), N.Y., 54,
   NE
Harrison City, Pa., 57, NE
Harrison St., N.J., 53, NE
Harrison Valley, Pa., 18, NE
Harrison, Ky., 162, SW
Harrison, Pa., 58, SE
Harrison, W.Va., 91, NE
Harrisonburg (CHW; SOU), Va.,
   126, NW
Harriston, Va., 126, SW
Harrisville, Pa., 26, SE
Harrisville, W.Va., 87, SE
Hart's Run, W.Va., 138, NW
Harter, W.Va., 123, NE
Hartford, N.J., 83, NW
Hartford, Ohio, 25, NE
Hartford, W.Va., 86, SW

Hartland (B&O; MCK), W.Va.,
   121, NE
Hartly, Del., 98, SE
Hartman, Pa., 64, SE
Hartranft, Pa., 66, SW
Harts, W.Va., 119, SE
Hartsdale, N.Y., 39, SW
Hartstown, Pa., 13, SW
Hartwood, N.Y., 24, SE
Hartz, Pa., 50B
Hartzel, W.Va., 88, NE
Hartzell, Pa., 50, SE
Harvard, N.Y., 11, SE
Harvey Jct., Pa., 35A
Harvey, Va., 164, NW
Harvey's Lake, Pa., 34, NE
Harwood Jct., Pa., 49, NE
Harwood, Md., 96, SE
Hasbrouck Heights, N.J., 53, NE
Haskell, N.J., 38, SW
Haskell, Va., 165, SE
Hastings Branch Jct., Pa., 44, SE
Hastings, N.Y., 54, NW
Hastings, Pa., 44, SE
Hastings, W.Va., 71, SE
Hatboro, Pa., 66, SE
Hatch, N.J., 82B
Hatcher, W.Va., 135, SE
Hatfield, Ky., 134, SW
Hatfield, Pa., 45, SE
Hatfield, Pa., 66, NW
Hauto (CNJ; LNE), Pa., 50C
Hauto Scale, Pa., 50C
Havaco, W.Va., 149, NE
Haven, N.Y., 24, SE
Haverford, Pa., 66, SW
Haverstraw (Erie; NYC), N.Y., 39,
   SW
Havre-de-Grace, Md., 80, SE
Hawk Run, Pa., 45, NE
Hawk's Mountain, N.Y., 23, NE
Hawk's Nest, W.Va., 121, SE
Hawkins Point, Md., 96, SE
Hawkins, Pa., 57A
Hawks, Ohio, 85, SW
Hawley, Pa., 36, NE
Hawley, W.Va., 137, NW
Haworth, N.J., 54, NW
Haws Ave., Pa., 66A
Haws, Pa., 66A
Hawstone, Pa., 46, SE
Hawthorn, Pa., 28, SW
Hawthorne, N.J., 53, NE

Hawthorne, N.Y., 39, SW
Hayden, Md., 98, SW
Haydenville, Ohio, 85, NW
Haymarket, Va., 111, NE
Haymond, Ky., 147, SE
Haynes, N.Y., 10, NE
Hays (P&LE; PRR), Pa., 57A
Hays Grove, Pa., 62, SW
Hays, Ohio, 55, NW
Hays, Pa., 73, NW
Haysi, Va., 148, SW
Haysville, Pa., 41, SE
Hayter, Va., 166, SW
Hayville, N.J., 83, SW
Haywood, W.Va., 89, NW
Hazel Creek Jct., Pa., 50, NW
Hazel Kirk, Pa., 57, SW
Hazelbrook Jct., Pa., 49, SW
Hazelbrook, Pa., 50B
Hazelhurst, Pa., 16, SE
Hazelton Jct., Pa., 49D
Hazelton Jct., Pa., 50A
Hazelton Mills, Pa., 16, NE
Hazelton, Pa., 50A
Hazen, Pa., 41, NE
Hazlet, N.J., 68, NE
Hazy Creek, W.Va., 135, NE
HD, Ohio, 55, NE
Head of Grade, Pa., 49, NE
Hearn, Del., 132, NE
Heart Lake, Pa., 22, NW
Heaters, W.Va., 105, NE
Heathcote, N.Y., 54, NW
Heathville, Pa., 28, SE
Heaton (PRR; RDG), Pa., 66, SE
Hebron, Md., 132, NE
Hebron, Va., 157, SW
Heckmer, W.Va., 106, SW
Hecks (PRR; RDG), Pa., 63, NW
Hecla, Pa., 50, SW
Hecla, Pa., 57, SE
Heda Park, Pa., 46, NE
Heidrick, Ky., 162, NW
Heilman Dale, Pa., 64, NW
Heilwood, Pa., 44, SW
Heinley, Pa., 65, NE
Heislerville, N.J., 100, SW
Heldrick, Pa., 28, SE
Helen (C&O; VGN), W.Va., 136,
   SW
Helen, Pa., 73, NW
Hellam, Pa., 79, NE
Hellen Mills, Pa., 29, NE

Hellertown, Pa., 51, SW
Hellier, Ky., 148, NW
Helmetta, N.J., 68, NW
Helmick, W.Va., 107, SW
Helvetia, Pa., 29, SW
Hemfield, Pa., 64, SW
Hemlock Hollow, W.Va., 137, NW
Hemlock, Pa., 15, NE
Hempfield Br. Jct., Pa., 57, NE
Hemphill, Ky., 147, SE
Hemphill, W.Va., 149, NE
Hempstead Gardens, N.Y., 54, SE
Hempstead, N.Y., 54, SE
Henderson (PRR; RDG), Pa.,
    66A
Henderson, Md., 98, SW
Henderson, Pa., 28, SW
Henderson, W.Va., 102, NE
Hendersonville, Pa., 56, NE
Hendricks, Pa., 66, NW
Hendricks, W.Va., 90, SE
Henlawson, W.Va., 135, NW
Henrico, Va., 144, SW, 144A
Henrietta, Pa., 60, NW
Henry Clay, Ky., 148, NW
Henry Clay, Pa., 62, SE
Henry, Ohio, 86, NE
Henry, Va., 170, NW
Henry, W.Va., 91, SW
Henry's Bend, Pa., 27, NE
Henry's Mill, Pa., 15, SE
Henryton, Md., 96, NW
Henryville, Pa., 36, SE
Hepburnville, Pa., 32, NE
Hepzibah, W.Va., 89, NW
Hercules, Pa., 51, NW
Heritage, N.J., 82, SE
Herlan, Ohio, 70, NW
Herman Siding, Ky., 147, NE
Herman, Pa., 42, NW
Herminie, Pa., 57, NE
Hermitage (RF&P; SAL), Va.,
    144A
Hermitage, N.Y., 7, NW
Hermitage, Pa., 26, SW
Herndon (PRR; RDG), Pa., 48,
    SW
Herndon Branch Jct., Pa., 48, NE
Herndon, Va., 112, NW
Herndon, W.Va., 136, SW
Hernshaw, W.Va., 120, SE
Herrick Center, Pa., 23, SW
Herrick, Ohio, 55, SW

Herring Run, Md., 96A
Hershey, Pa., 63, NE
Hesbon, Pa., 58, NE
Hesston, Pa., 60, NE
Heston, Ohio, 40, NE
Hetzel, W.Va., 135, NW
Heverly, Pa., 44, SE
Hewitt, N.J., 38, SW
Hewitt, Ohio, 85, NW
Hewlett, N.Y., 54, SE
Hewlett, Va., 143, NE
Hewlett, W.Va., 118, SE
Hi Hat, Ky., 147, NE
Hiawassee, Va., 168, NE
Hiawatha, W.Va., 150, NE
Hibernia, N.J., 53, NW
Hick's Ferry, Pa., 34, SE
Hickman, Del., 115, NE
Hickman, Pa., 56, NE
Hickory Ground, Va., 177, SE
Hickory Grove, Pa., 22, NE
Hickory Hall, Pa., 60, SE
Hickory Hill, Va., 126, SE
Hickory Lick, W.Va., 106, SE
Hickory, Pa., 56, NW
Hickox, Pa., 18, NW
Hicksville, N.Y., 54, NE
Hicksville, Va., 150, SE
Higbie Ave., N.Y., 54, SW
Higgins, W.Va., 72, SE
Higginsville, N.J., 52, SW
High Bridge, N.J., 52, SW
High Bridge, N.Y., 54, NW
High Bridge, Va., 156, NW
High Coal, W.Va., 136, NW
High House, Pa., 73, NW
High Rock, Pa., 80, NW
Highfield, Md., 78, SW
Highland Ave., N.J., 53, NE
Highland Ave., Pa., 82, NW
Highland Beach, N.J., 69, NW
Highland Falls, N.Y., 39, NW
Highland Mills, N.Y., 38, NE
Highland Park, Va., 144A
Highland, Pa., 41, NW
Highland, Pa., 66, SE
Highland, Va., 141, SE
Highland, W.Va., 106, NE
Highlands, N.J., 69, NW
Highlandtown Jct., Md., 96A
Highlandtown, Md., 96A
Highspire, Pa., 63, SW
Highsplint, Ky., 163, NE

Hightstown Jct. (PRR; UTR), N.J.,
    67, NE
Hightstown, N.J., 67, NE
Hilda, Va., 175, NW
Hildebrand, W.Va., 73, SW
Hill Crest, Pa., 66, SE
Hill Top, N.J., 82, NE
Hill Yard, Pa., 51, NE
Hill, Md., 113, NE
Hill, Pa., 42, SE
Hill, Va., 164, SE
Hillburn, N.Y., 38, SE
Hillcrest, N.J., 67, SW
Hillen (Baltimore), (WM), Md.,
    96A
Hilliard, N.J., 84, SE
Hilliards, Pa., 27, SW
Hillman Jct., Pa., 44, NW
Hillman, Pa., 44, NW
Hills (MTR; PRR), Pa., 56, NE
Hills Crossing, W.Va., 103, NW
Hillsboro, Md., 115, NW
Hillsboro, Pa., 59, SW
Hillsdale, N.J., 38, SE
Hillside Jct., Pa., 35C
Hillside, Md., 81, SW
Hillside, N.J., 53, SE
Hillside, N.Y., 54D
Hillside, Pa., 58, NW
Hillsville, Pa., 26, SW
Hilltop Mine, W.Va., 89, NW
Hilltop, Pa., 66, NW
Hilltop, Va., 170, SW
Hillville, Md., 130, NE
Hilton, Va., 165, SW
Hiltons, N.J., 68, NE
Himbaugh, Pa., 14, NW
Himyar, Ky., 162, NW
Hinkels, Pa., 50, NW
Hinman, N.Y., 8, NE
Hinsdale (Erie), N.Y., 5, SW
Hinsdale (PRR), N.Y., 5, SW
Hinterleiter, Pa., 50, SE
Hinton, W.Va., 137, SW
Hiram, Ky., 163, NE
Hiram, Va., 166, NW
Hitchcock Quarry, Va., 174, SE
Hitchins (C&O; EK), Ky., 118,
    NW
Hite, W.Va., 72, SE
Hitop, W.Va., 121, NW
Hix, W.Va., 149, SE
Hoadley's, Pa., 23, SW

Hoard, W.Va., 73, SW
Hobbs, Md., 115, NW
Hoblitzell, Pa., 75, NE
Hoboken, N.J., 53B
Hobson (NYC), Ohio, 102, NE
Hobson Jct. (C&O), Ohio, 102, NE
Hobson Yard (C&O), Ohio, 102,
    NE
Hockendauqua, Pa., 51C
Hockessin, Del., 81, NE
Hocking Mine, Ohio, 85, NE
Hockman, Va., 150, SW
Hoddon Heights, N.J., 82, NE
Hoffman, Pa., 62, NE
Hoffmans, N.J., 52, SW
Hoffmanville, Md., 79, SW
Hog Island, Pa., 82, NW
Hogsett, W.Va., 102, SE
Ho-Ho-Kus, N.J., 53, NE
Hokes, Pa., 79, SW
Holban Yard, N.Y., 54D
Holcomb Rock, Va., 140, SW
Holcomb, W.Va., 122, NE
Holden, Pa., 28, SE
Holden, W.Va., 134, NE
Holiday, Pa., 19, NE
Holland, N.J., 51, SE
Holland, N.Y., 54, SW
Holland, Pa., 67, SW
Holland, Va., 176, SW
Holland, W.Va., 107, NE
Holliday, Va., 143, NE
Hollidaysburg, Pa., 60, NW
Hollins, Md., 96, NE
Hollins, Va., 153, NW
Hollis, N.Y., 54D
Hollowfield, Md., 96, NW
Holly Dak, Del., 82, NW
Holly Grove, Md., 133, NE
Holly Grove, W.Va., 121, SW
Holly Jct. (B&O; WVM), W.Va.,
    105, SE
Holly, W.Va., 105, SE
Holly, W.Va., 121, SW
Hollyhurst, W.Va., 120, SW
Hollywood, Md., 130, NE
Hollywood-West End, N.J., 69,
    NW
Holmes, Pa., 82, NW
Holmesburg Jct., Pa., 66, SE
Holsopple, Pa., 59, SW
Holston, Tenn., 164, SE
Holt, Ky., 118, SE

Holt, Va., 144, SW

Holtwood, Pa., 80, NW

Home Creek, Va., 148, NE

Home, Pa., 43, SE

Homeland, Md., 96, NE

Homer City (B&O; PRR), Pa., 43, SE

Homer City (PRR), Pa., 43, SE

Homestead (P&LE; PRR), Pa., 57A

Homestead, W.Va., 102, SE

Hometown, Pa., 50, NW

Homeville, Va., 175, NE

Homewood, Pa., 57A

Honaker, Va., 149, SW

Honesdale (D&H; Erie), 23, SW

Honey Brook, Pa., 65, SW

Honey Creek, Pa., 46, SE

Honey Grove, Pa., 61, NE

Honey Pot, Pa., 34, SE

Hoo Hoo, W.Va., 136A, SW

Hood, Md., 76, SE

Hood, Pa., 67, NW

Hoods Mill, Md., 95, NE

Hooker, Pa., 42, NW

Hooks, Pa., 43, NW

Hoover, Pa., 73, NW

Hooverhurst, Pa., 44, NW

Hooversville, Pa., 59, SW

Hopatcong Jct. (CNJ), N.J., 52, NE

Hope, Ohio, 85, NW

Hopedale (NYC; P&WV), Ohio, 55, NW

Hopemont, W.Va., 90, NE

Hopeton, Va., 146, NE

Hopewell, Ky., 118, NW

Hopewell, Md., 132, SW

Hopewell, N.J., 67, NW

Hopewell, Pa., 60, SW

Hopewell, Va., 158, NW

Hopkins Fork, W.Va., 120, SE

Hopkins, W.Va., 107, SW

Hopkins, W.Va., 120, SW

Hopper, W.Va., 138, NW

Hopping Jct., N.J., 68, NE

Hopping, N.J., 68, NE

Horn Springs, Pa., 51, NW

Horn, Pa., 48, NW

Hornberger's Siding, Md., 80, SW

Hornell (Erie; PS&N), N.Y., 6, NE

Horner, W.Va., 89, SW

Hornerstown, N.J., 67, SE

Horning, Pa., 57, NW

Horrell, Pa., 60, NW

Horrock, W.Va., 123, SW

Horseheads (DL&W; Erie; LV; PRR), N.Y., 8, SW

Horton, N.J., 52, NE

Horton, Ohio, 55, SW

Horton, Va., 139, SW

Horton, W.Va., 107, NE

Horton's, N.Y., 23, NE

Hosensack, Pa., 65, NE

Hosterman, W.Va., 124, NW

Hostetter, Pa., 58, NW

Hostler, Pa., 45, SE

Hot Spot, Ky., 147, SW

Hot Springs, Va., 124, SW

Hotchkiss, W.Va., 136, SW

Hotcoal, W.Va., 136A, SE

Hother, Pa., 77, NE

Houchins, Va., 152, SW

Houghton Farm, N.Y., 38, NE

Houghton, N.Y., 5, NE

Hoult, W.Va., 72, SE

Houseville, Pa., 42, SW

Houston Jct. (B&LE; PRR), Pa., 26, SE

Houston, Del., 115, NE

Houtzdale, Pa., 45, NW

Hovers, Pa., 15, SE

Howard (B&O; Erie), Pa., 16, NE

Howard Beach, N.Y., 54, SW

Howard, Pa., 31, SE

Howards, Pa., 30, NW

Howardsville, Va., 141, SE

Howardville, Md., 96, NE

Howell, N.J., 68, SE

Howells, N.Y., 38, NW

Howellville (PRR; RDG), Pa., 66, SW

Howesville, W.Va., 90, NW

Hoyt, W.Va., 105, NE

Hoytville, Pa., 19, SW

Hubball, W.Va., 119, SE

Hubbard (Erie; NYC), Ohio, 25, SE

Hubbard Jct., Va., 166, NW

Hubbard Springs, Va., 163, SE

Hubbardstown, W.Va., 118, SE

Hublersburg, Pa., 46, NE

Huddleston, Va., 154, SW

Huddy, Ky., 134, SW

Hudson Ave., N.J., 54, NW

Hudson Jct., N.Y., 38, NW

Hudson Yard, N.J., 51, SE

Hudsondale, Pa., 50, NW

Huey, Pa., 27, SE

Huff Creek Jct., W.Va., 135, SE

Huff Jct., W.Va., 135, SW

Huffsville, W.Va., 135, SW

Hugenot Park, N.Y., 53, SE

Hugheston, W.Va., 121, SW

Hughesville, Md., 113, SW

Hughesville, Pa., 33, SE

Huguenot, N.Y., 37, NE

Hull St. (Richmond), (SOU), Va., 144A

Hull, W.Va., 149, NW

Hulmes, Pa., 34, SW

Hulton Ferry, Pa., 42, SW

Hulton, Pa., 42, SW

Humbert, Pa., 74, NW

Hume, N.Y., 5, NE

Hummel, Pa., 60, NE

Hummelstown, Pa., 63, NE

Humphrey, Pa., 58, SW

Hundred, W.Va., 72, SW

Hunker, Pa., 57, SE

Hunlock Creek, Pa., 34, SE

Hunnewell, Ky., 118, NW

Hunt, W.Va., 135, SW

Hunter Run Jct., Pa., 57, SE

Hunter, Pa., 28, NW

Hunter, Pa., 48, NW

Hunter, Va., 112, NW

Hunters Park, Pa., 46, NW

Hunters Point Ave., N.Y., 54B

Hunters Run, Pa., 62, SE

Huntingdon (H&BTM; PRR), Pa., 60, NE

Huntingdon, Pa., 66, SE

Huntington (B&O; C&O), W.Va., 119, NW

Huntley, Pa., 30, SW

Hunton, Va., 143, SE

Huntsdale, Pa., 62, SW

Huntsville, N.J., 52, NW

Hurd, N.J., 52, NE

Hurford, Ohio, 55, SW

Hurley, Va., 148, NE

Hurlock (B&E; PRR), Md., 115, SW

Huron, Pa., 73, NW

Hurricane, Va., 165, NE

Hurricane, W.Va., 119, NE

Hursley, Md., 133, SW

Hurt, Va., 154, SW

Husband, Pa., 58, SE

Huskin Run, Pa., 59, SW

Husted, N.J., 82, SE

Hutchins, Pa., 16, SE

Hutchinson (B&O; WM), W.Va., 89, NW

Hutchinson, N.J., 51, NE

Hutchinson, W.Va., 135, NW

Hutchison (B&O; PRR), Pa., 73, NE

Hutton (B&O; PR), W.Va., 91, NW

Huttonsville (VYRR; WM), W.Va., 107, SW

Huttonsville Jct., W.Va., 107, NW

HX Tower, Pa., 26, SE

HX Tower, Pa., 36, NE

HY Cabin, Va., 139, NW

Hyattsville, Md., 113, NW

Hyde Park, Pa., 42, SE

Hyde, Md., 97, NW

Hyde, Pa., 30, SW

Hydes (Erie; PS&N), Pa., 29, NE

Hydetown, Pa., 14, SE

Hyers, W.Va., 105, NE

Hyndman (B&O; PRR), Pa., 75, NE

Hyner, Pa., 31, NE

Hyper-Humus, N.J., 37, SE

Iaeger, W.Va., 149, NW

Ice, Ky., 147, SW

Icedale, Pa., 65, SW

Idamar, Pa., 43, SE

Idamay, W.Va., 89, NW

Idaville, Pa., 62, SE

Idlewild, Pa., 58, NW

Idlewood, Pa., 56A

Ifield, Pa., 73, NW

Ijamsville, Md., 95, NW

Ilchester, Md., 96, NW

Imboden, Va., 164, NW

Imlaystown, N.J., 67, SE

Imler, Pa., 59, SE

Immaculata, Pa., 65A

Imperial, Pa., 56, NW

Imperial, W.Va., 106, NE

Independence, W.Va., 90, NW

Indian Camp, Pa., 28, SE

Indian Creek (B&O; WM), Pa., 73, NE

Indian Crossing, Pa., 17, NW

Indian Head Jct. (PRR; US GOV'T), Md., 113, SW

Indian Head, Md., 112, SE
Indian Head, Pa., 58, SW
Indian Rock, Va., 139, SE
Indian Yard, Va., 149, SE
Indian, Va., 149, SW
Indian, W.Va., 120, NW
Indiana (B&O; PRR), Pa., 43, SE
Indiana Jct., Pa., 44, NW
Indiantown Gap, Pa., 63, NE
Industrial School, W.Va., 88, NE
Industry, Pa., 41, SW
Inez, W.Va., 119, NE
Ingham, Va., 109, SE
Inghams, Ohio, 85, NW
Ingleby, Pa., 47, NW
Inglenook, Pa., 63, NW
Ingleside (N&W; VGN), W.Va., 150, NE
Ingleside, Va., 177, NE
Ingram Branch, W.Va., 121, SE
Ingram, Pa., 56A
Inlet, Va., 128, NW
Inman, Va., 164, NW
Instanter, Pa., 16, SE
Institute, W.Va., 120, NW
Insull, Ky., 163, NW
Interior, Va., 151, NE
Intervale, Va., 139, NW
Inwood (LI), N.Y., 54, SE
Inwood (NYC), N.Y., 54, NW
Inwood, Pa., 63, NE
Inwood, Va., 93, NE
Iona Island, N.Y., 39, NW
Iona, N.J., 82, SE
Iowa, Pa., 28, SE
Ira, W.Va., 105, SW
Irene, W.Va., 120, SW
Iron Bridge, Pa., 57, SE
Iron Gate, Va., 139, NW
Iron Hill, Md., 81, SW
Irondale, Ohio, 40, SE
Irondale, Va., 164, NW
Ironia, N.J., 52, NE
Ironshire, Md., 133, NW
Ironstone, Pa., 65, NE
Ironto, Va., 152, SW
Ironton, Pa., 50, SE
Iroquois, Pa., 62, NE
Irvin, W.Va., 105, SE
Irvineton (NYC; PRR), Pa., 15, NW
Irving Ave. (Bridgeton), (P-RSL), N.J., 99, NE

Irving, Pa., 49, SW
Irving, Va., 153, NE
Irving, W.Va., 89, NW
Irvington, N.J., 53, SE
Irvington, N.Y., 39, SW
Irvona (NYC; PRR), Pa., 44, NE
Irvona Jct. (NYC), Pa., 44, NE
Irvona, W.Va., 90, NE
Irwin, Pa., 57, NE
Irwin, Va., 143, SW
Isaban, W.Va., 135, SW
Isabella, Pa., 65, SW
Isabella, Pa., 73, NW
Ischna, N.Y., 5, SW
Iselin, N.J., 53, SW
Iselin, Pa., 43, SW
Island Creek, Ky., 147, NE
Island Ford, Va., 126, NE
Island Heights, N.J., 84, NE
Island Park, N.Y., 54, SE
Island, Va., 142, SE
Isle, Pa., 41, NE
Itaska, N.Y., 10, NW
Ithaca (DL&W; LV), N.Y., 8, NE
Itmann, W.Va., 136, SW
Ivanhoe, Va., 168, NW
Ivor, Va., 176, NW
Ivy Rock, Pa., 66A
Ivy, Va., 126, SE
Ivyland, Pa., 66, SE
Ivywood, Pa., 42, SW

J&B Jct., Pa., 16, SE
J, Ohio, 55, NE
Jack's Mountain, Pa., 78, NW
Jacks Creek, Ky., 147, NE
Jackson Ave., N.J., 53B
Jackson Center, Pa., 26, NE
Jackson Summit, Pa., 19, NE
Jackson, Md., 80, SE
Jackson, Pa., 1, SE
Jacksonburg, W.Va., 71, SE
Jacksons Mill, W.Va., 89, SW
Jacksonville, Ohio, 85, NE
Jacksonville, Pa., 50, SW
Jacksonville, Pa., 51, SW
Jacobs Creek (B&O; P&LE), Pa., 57, SE
Jacobs Ferry, Pa., 73, NW
Jacobsburg, Ohio, 71, NW
Jacobsburg, Pa., 51, NW
Jaffa, Va., 171, SW
Jamacia, N.Y., 54D

James City, Pa., 16, SW
James River Jct., Va., 175, SW
James, Va., 154, NE
Jamesburg, N.J., 68, NW
Jamestown (Erie), N.Y., 3, SE
Jamestown (JW&NW), N.Y., 3, SE
Jamestown (NYC; PRR), Pa., 26, NW
Jamison City, Pa., 34, NW
Jamison, Pa., 15, SW
Jamison, Pa., 18, SW
Jamisonville, Pa., 42, NW
Jane Lew, W.Va., 89, SW
Janesville, Pa., 45, NW
Janie, W.Va., 135, NE
Jarman Gap, Va., 126, SE
Jarratt (ACL; VGN), Va., 175, NW
Jarretts Ford, W.Va., 121, NW
Jasper, Va., 164, NW
Jazbo, W.Va., 136, SW
Jeanette, Pa., 57, NE
Jeddo, Pa., 50B
Jeff, Pa., 82A
Jefferson Ave., N.Y., 53, SE
Jefferson, Md., 94, NE
Jefferson, N.J., 82, SE
Jefferson, Ohio, 12, SW
Jefferson, Ohio, 55, NE
Jefferson, Pa., 49, SE
Jeffress, Va., 172, SE
Jeffrey, W.Va., 135, NW
Jeffreytown, Pa., 56, NE
Jenkin Jones, W.Va., 150, NW
Jenkins Ford, Va., 139, NW
Jenkins Jct., Pa., 35B
Jenkins, Ky., 147, SE
Jenkintown, Pa., 66, SE
Jenner, Pa., 58, SE
Jennings Ordinary, Va., 156, SE
Jennings, Md., 74, SE
Jenningston, W.Va., 107, NE
Jenny Gap, W.Va., 136A, SW
Jere, W.Va., 72, SE
Jericho, Va., 176, SE
Jermyn (D&H; NYO&W), Pa., 22, SE
Jerome Jct., Pa., 59, SW
Jerome, Pa., 59, SW
Jerome, W.Va., 76, SW
Jerry's Run, Va., 138, NE
Jerryville, W.Va., 123, NW
Jersey City (CNJ; Erie; LV; PRR), N.J., 53B

Jersey Mills, Pa., 32, NW
Jersey Shore (NYC), Pa., 32, SW
Jersey Shore (PRR), Pa., 32, SE
Jerusalem, Ohio, 70, NE
Jessup, Md., 96, SW
Jessup, W.Va., 90, NW
Jessup-Peckville, Pa., 35D
Jeter, W.Va., 137, NE
Jetersville, Va., 156, NE
Jethro, Ohio, 40, SE
Jetsville, W.Va., 122, SE
Jewell Valley, Va., 149, SW
Jewell, Va., 149, SW
Jewett (PRR; W&LE), Ohio, 55, NW
JM Jct., Pa., 43, SW
JN Cabin, Va., 139, SE
JO Sidings, Pa., 14, NW
Job, W.Va., 107, NE
Jobstown, N.J., 67, SE
Johanna Heights, Pa., 65, SW
Johanna, Pa., 65, SW
John Veith Jct., Pa., 49, SW
Johnetta, Pa., 42, SE
Johns, W.Va., 120, SE
Johnson City (DL&W; Erie), N.Y., 10, SW
Johnson, N.Y., 37, NE
Johnsonburg (B&O; PRR), Pa., 29, NE
Johnsonburg, N.J., 52, NW
Johnsons, Ohio, 25, NE
Johnsons, Ohio, 71, NW
Johnsons, W.Va., 86, SE
Johnston, Va., 142, SW
Johnstown (B&O; C&BL; PRR), Pa., 59, NW
Johnstown, W.Va., 122, SE
Johnsville, Pa., 66, SE
Joint Line Jct., Pa., 64, SW
JoJo Jct., Pa., 16, SW
Jones Creek, Va., 170, SW
Jones Crossing, W.Va., 103, NE
Jones Mills, Pa., 58, SW
Jones Point, N.Y., 39, NW
Jones, Md., 96, SE
Jones, W.Va., 150, NW
Jonestown, Pa., 64, NW
Joppa, Md., 97, NW
Jordan, N.J., 82B
Jordan, Pa., 50, SE

Jordan, W.Va., 72, SE
Josephine (B&O; PRR), Pa., 58, NE
Josephine, Va., 164, NE
Joshua Falls, Va., 154, NE
Journal Sq. (H&M), N.J., 53B
Joyce, Md., 96, SE
Joyner, Va., 175, NW
Judge, W.Va., 135, SE
Julian, Pa., 46, NW
Juliustown, N.J., 67, SE
Jumbo Draw, N.J., 82, NW
Jumbo, N.J., 82, NW
Jumbo, W.Va., 106, SW
Juneau, Pa., 44, NW
Juniata Furnace, Pa., 62, NE
Juniata, Pa., 73, NE
Junior, W.Va., 107, NW
Juniper, Va., 177, SW
Juno, W.Va., 149, NE
Justice, W.Va., 135, SW
Jutland, N.J., 52, SW

Kale, W.Va., 150, NE
Kanauga Jct., Ohio, 102, NE
Kanauga, Ohio, 102, NE
Kanawa, W.Va., 87, SW
Kanawha Falls, W.Va., 121, SE
Kane Jct., Pa., 16, SW
Kane, Pa., 16, SW
Kanes Creek, W.Va., 73, SW
Kanesholm, Pa., 16, SE
Kanona (DL&W; Erie; PRAT), N.Y., 7, NW
Kansas City, Pa., 42, NE
Kanty, Pa., 1, SE
Kapp, Pa., 48, NW
Karns, Pa., 42, SE
Karo, Va., 110, NW
Karthaus, Pa., 30, SE
Kasson, Pa., 16, SE
Kato, Pa., 31, SW
Katonah, N.Y., 39, NE
Kaufman Run, Pa., 59, SW
Kaufman, Pa., 74, NW
Kaulmont, Pa., 29, NE
Kayford, W.Va., 121, SW
Kayjay, Ky., 162, SW
Kaylor, Pa., 42, NE
Kaymoor, W.Va., 121, SE
Kayoulah, Va., 168, NW
Keane, Pa., 23, SW
Keansburg, N.J., 68, NE

Kearney, Pa., 60, SE
Kearneysville, W.Va., 94, NW
Kearny, N.J., 53A
Keating Jct., Pa., 31, NW
Keating Summit (B&O; PRR), Pa., 17, SE
Keating, Pa., 31, NW
Keedysville, Md., 94, NE
Keeneys Creek, W.Va., 121, SE
Keezletown, Va., 126, NW
Keffers, Pa., 49, SW
Kegley, W.Va., 150, NE
Keifer, W.Va., 76, SW
Keister Jct., Pa., 73, NW
Keister, W.Va., 138, NW
Keisters, Pa., 27, SW
Keith, W.Va., 120, SE
Kellar, W.Va., 87, NW
Keller, Va., 146, SW
Kellettville, Pa., 15, SW
Kelley, Pa., 56A
Kelley, W.Va., 151, NW
Kelleysville, W.Va., 151, NW
Kellogg, Ky., 118, NW
Kellogg, Pa., 21, SW
Kellogg, W.Va., 118, NE
Kelly, Pa., 42, SE
Kelly, Va., 154, NE
Kelsa, Va., 148, NE
Kelsey, Ohio, 71, NW
Kelton, Pa., 81, NW
Kemper St. (Lynchburg), (SOU), Va., 154, NE
Kemps, Md., 77, SW
Kempton Jct., W.Va., 91, SW
Kempton, Md., 91, SW
Kempton, Pa., 50, SW
Kenbridge, Va., 173, NE
Kendalia, W.Va., 121, NW
Kendall Grove, Va., 161, NW
Kendall, Md., 74, SW
Kendell, Pa., 41, SE
Kenilworth, N.J., 53, SW
Kenilworth, W.Va., 40, SE
Kennard, Ohio, 85, NW
Kennard, Pa., 26, NW
Kennedy, N.Y., 3, SE
Kennedyville, Md., 98, NW
Kennerdell, Pa., 27, NW
Kennett Square, Pa., 81, NE
Kenneys, Pa., 65, SW
Kenniston, W.Va., 123, SE
Kenny, Pa., 57, NW

Kenova (B&O; C&O; N&W), W.Va., 118, NE
Kenrock, Pa., 60, SE
Kensico Cemetery, N.Y., 39, SW
Kensington, Md., 95, SE
Kensington, Ohio, 40, SW
Kent Jct., Va., 164, NE
Kent, Del., 115, NE
Kent, Va., 167, NE
Kentmare Passing (PRR), Del., 81A
Kentmere (RDG), Del., 81A
Kentmere Jct., Del., 81A
Kenton, Del., 98, SE
Kenvil (CNJ; DL&W), N.J., 52, NE
Kenvir, Ky., 163, NE
Kenwood, Md., 96, NE
Kenwood, Ohio, 55, NW
Kenyon, Va., 176, SE
Keokee, Va., 164, NW
Kephart, Pa., 74, NW
Kepler, W.Va., 135, SE
Kerens, W.Va., 90, SW
Kermit, Va., 164, SE
Kermit, W.Va., 134, NW
Kerns, Va., 151, NE
Kernstown, Va., 93, SE
Kerr, Ohio, 70, NE
Kerrmoor, Pa., 44, NE
Kerrs, Ohio, 102, NW
Kerry's, N.Y., 11, SW
Kersey, Pa., 29, NE
Kessler, W.Va., 137, NE
Keswick Grove, N.J., 84, NW
Keswick, Va., 127, SW
Ketner, Pa., 16, SE
Kettle Island, Ky., 162, NE
Kew Gardens, N.Y., 54, SW
Kewanee, Ky., 147, NE
Key, Ohio, 71, NW
Keymar (PRR; WM), Md., 78, SE
Keymar Passing, Md., 78, SE
Keyport, N.J., 68, NE
Keyser Valley Shops, Pa., 35C
Keyser, W.Va., 92, NW
Keystone, Pa., 40, NE
Keystone, Pa., 75, NW
Keystone, W.Va., 150, NW
Keysville, Va., 156, SW
Kilarm Jct., W.Va., 89, NW
Kilbourne, Pa., 18, NE
Kilby, Va., 176, SE
Killarney, W.Va., 136, SW

Killawog, N.Y., 9, NE
Kimball, W.Va., 149, NE
Kimberling, W.Va., 135, SW
Kimberly, Ohio, 85, NE
Kimberton, Pa., 65, SE
Kimbles, Pa., 36, NE
Kimbleton, Va., 151, NE
Kimmel, Pa., 43, SE
Kimmel, Pa., 60, SE
Kimmelton, Pa., 59, SW
Kincaid, Va., 139, NW
King of Prussia, Pa., 66, SW
King Ridge, Pa., 74, NW
King, Pa., 66, SW
King's Bridge, Pa., 80, NE
King's Mill, Pa., 62, NE
Kingmont, W.Va., 89, NE
Kings Bridge, N.Y., 54, NW
Kings Creek (Peninsula Jct.), Md., 132, SE
Kings Creek, W.Va., 55, NE
Kings, Ohio, 55, NE
Kingsbridge Road, N.Y., 54, NW
Kingsdale, Pa., 78, SE
Kingsland, N.J., 53, NE
Kingsley, Pa., 22, NW
Kingsport, Tenn., 164, SE
Kingston (DL&W; LV), Pa., 35A
Kingston Forty-Fort, Pa., 35A
Kingston, Md., 132, SE
Kingston, N.J., 67, NE
Kingston, Pa., 58, NW
Kingston, W.Va., 136, NW
Kingsville (NKP; NYC), Ohio, 12, NE
Kingsville, Pa., 28, SW
Kingsville, W.Va., 106, NE
Kingwood (B&O; WVN), W.Va., 90, NE
Kingwood Jct., W.Va., 90, NE
Kinkora, N.J., 67, SW
Kinney, Va., 154, NE
Kino, Va., 139, SE
Kinsman, Ohio, 25, NE
Kinter, Pa., 44, SW
Kinzers, Pa., 80, NE
Kinzua Viaduct, Pa., 16, NE
Kinzua, Pa., 16, NW
Kipling, Ohio, 70, NW
Kipps, N.Y., 38, NW
Kiptopeke, Va., 161, SW
Kirbyton, W.Va., 120, SE
Kire, Va., 151, NE

Lawndale, Md., 79, SW
Lawndale, Pa., 66, SE
Lawrence, N.J., 67, NE
Lawrence, N.Y., 54, SE
Lawrenceville (Erie; NYC), Pa., 19, NE
Lawrenceville, Va., 174, NW
Lawsonham, Pa., 43, NW
Lawyer, Va., 154, NE
Lax, W.Va., 135, NW
Layland, W.Va., 137, NW
Layman, Ky., 163, NW
Laymans, Va., 126, NW
Layton, Pa., 57, SE
Lazeaville, W.Va., 55, NE
Le Gore, Md., 78, SW
Leach, Ky., 118, NE
Leadsville, W.Va., 107, NW
Leadville, Ohio, 25A
Leahigh, Md., 96, NE
Leak Run, Pa., 57, NW
Leaksville Jct., Va., 170, SE
Leaman Place (PRR; SRC), Pa., 64, SE
Leatherwood, Pa., 28, SW
Leavittsburg (B&O; Erie), Ohio, 25, SW
Lebanon (PRR; RDG), Pa., 64, NW
Lebanon, N.J., 52, SW
Lecato, Va., 146, NE
Leckrone (B&O; MON), Pa., 73, NW
Ledgewood, N.J., 52, NE
Ledy, Pa., 77, NE
Lee Bell, W.Va., 106, SE
Lee Hall, Va., 159, SE
Lee Town, Va., 148, NE
Lee Vale, W.Va., 135, NE
Lee, Pa., 34, SE
Lee, Va., 143, SW
Lee's Cross Roads, Pa., 62, SW
Leechburg, Pa., 42, SE
Leeland, Md., 113, NE
Leeland, Va., 129, NW
Lees, Pa., 65, SE
Leesburg, N.J., 100, NW
Leesburg, Pa., 26, SE
Leesburg, Va., 94, SE
Leesport (PRR; RDG), Pa., 65, NW
Leesville, Va., 154, SW
Leetonia (Erie; PRR), Ohio, 40, NW

Leewood, W.Va., 121, SW
LEF&C Jct., Pa., 28, SE
Leftwich, Va., 154, NE
Lehigh Gap (CNJ; LV), Pa., 50, NE
Lehighton (CNJ; LV), Pa., 50, NE
Lehman, Pa., 35, NW
Lehmasters, Pa., 77, NW
Leidighs, Pa., 62, SE
Leiter, W.Va., 107, NW
Leith (B&O; PRR), Pa., 73, NE
Lemley, W.Va., 72, SE
Lemon, Va., 139, SW
Lemont, Pa., 46, NW
Lenape, Pa., 81, NE
Lenhartsville, Pa., 50, SW
Lenni, Pa., 82, NW
Lennig, Va., 172, NW
Lennodo, Pa., 58, NW
Lenola, N.J., 83, NW
Lenore, W.Va., 134, NW
Leola, Pa., 64, SE
Leolyn, Pa., 20, SW
Leon, Ky., 118, NW
Leon, Ohio, 12, SE
Leon, W.Va., 103, NW
Leonard, N.J., 68, NE
Leonard, Va., 165, SE
Leonardo (USN), N.J., 68, NE
Leonia, N.J., 54, NW
Lerchs, Pa., 51, SW
Leroy, W.Va., 103, NE
Lesage, W.Va., 102, SW
Leslie, Md., 81, SW
Leslie, W.Va., 122, SE
Lesmalinston, W.Va., 90, NW
Lester (C&O; VGN), W.Va., 136A, SW
Lester Manor, Va., 145, SW
Letart, W.Va., 103, NW
Letcher, Ky., 147, SW
Levi, W.Va., 120, NE
Levisa Jct., Ky., 148, NW
Lewes, Del., 116, NE
Lewis (PRR; UTR), N.J., 67, SE
Lewis Run (B&O; Erie), Pa., 16, NE
Lewisburg (PRR; RDG), Pa., 48, NW
Lewisburg Siding, Pa., 48, NW
Lewisburg, N.J., 37, SE
Lewistown, Md., 78, SW
Lewistown, Pa., 46, SE
Lewisville, Ohio, 70, NE

Lexington, Va., 140, NW
LG-14, Pa., 80, NE
LG-21, Pa., 80, NW
LG-30 Buzzard Rock, Pa., 80, NW
LG-54, Pa., 63, SE
Liberty Corners, N.Y., 37, NE
Liberty Grove, Md., 80, SE
Liberty, N.Y., 24, NW
Liberty, Pa., 17, SE
Library Jct., Pa., 56, NE
Library, Pa., 56, NE
Lichty, Pa., 50, SE
Lick Branch, W.Va., 150, NW
Lick Creek, Ky., 148, NW
Lick Island, Pa., 30, NE
Lick Run Jct., Pa., 73, NE
Lick Run, Va., 139, NW
Lico, W.Va., 120, SE
Liggett, Ky., 163, NW
Light Street (PRR; RDG), Pa., 34, SW
Light, W.Va., 135, SW
Lightfoot, Va., 159, NW
Ligon, Ky., 147, NE
Ligonier, Pa., 58, SE
Lillian, W.Va., 89, SE
Lilly, Pa., 59, NE
Lillybrook, W.Va., 136, SE
Lima, Va., 171, SW
Lime Bluff, Pa., 33, SW
Lime Ridge, Pa., 34, SW
Lime Rock, Pa., 64, SW
Lime Rock, W.Va., 90, SE
Limekiln, Md., 95, NW
Limestone Jct., Pa., 42, NE
Limestone, N.Y., 4, SE
Limestone, Pa., 28, SW
Limestone, W.Va., 89, NW
Limeton, Va., 110, NW
Linan, W.Va., 107, SW
Lincoln, Ohio, 85, SW
Lincoln Park, N.J., 53, NW
Lincoln University, Pa., 81, NW
Lincoln Yard, Pa., 79, NW
Lincoln, N.Y., 54, NW
Lincoln, W.Va., 120, NW
Lincolndale, N.Y., 39, NE
Lincolnville, Pa., 14, NW
Lincoln City, Del., 116, NW
Linden (NYC), Pa., 32, SE
Linden Hall, Pa., 46, NW
Linden Jct., N.J., 53, SE
Linden, N.J., 53, SW

Linden, Va., 110, NE
Linden, Va., 164, NW
Lindenwold, N.J., 83, NW
Lindley, N.Y., 7, SE
Lindsay, Va., 127, SE
Lindsey, W.Va., 134, SE
Lineboro, Md., 79, SW
Linesville (B&LE; PRR), Pa., 13, SW
Linfield, Pa., 65, SE
Linkwood, Md., 115, SW
Linn, Pa., 57, SW
Linore, Pa., 17, SE
Linville, Va., 109, SW
Linwood Country Club, N.J., 100, NE
Linwood, Md., 78, SE
Linwood, N.J., 100, NE
Lipco, Va., 157, SW
Lipscomb, Va., 126, SW
Lisbon (Erie; PL&W), Ohio, 40, NW
Lisle, N.Y., 9, NE
Listie, Pa., 58, SE
Listonburg, Pa., 74, NW
Litchfield, N.Y., 9, SW
Lithia, Va., 153, NE
Lithicum, Md., 96, SE
Lititz, Pa., 64, SW
Little Cacapon, W.Va., 76, SW
Little Creek, Va., 177, NE
Little Falls (DL&W; Erie), N.J., 53, NE
Little Falls, W.Va., 72, SE
Little Ferry (NYC; NYS&W), N.J., 53, NE
Little Gap, Pa., 50, NE
Little Genesee, N.Y., 5, SE
Little Gunpowder, Md., 80, SW
Little Hocking, Ohio, 86, NE
Little Neck, N.Y., 54, NE
Little Orleans, Md., 76, SW
Little Run Jct., Pa., 50, SE
Little Silver, N.J., 68, NE
Little Valley, N.Y., 4, NW
Littlestown, Pa., 78, SE
Littleton, W.Va., 71, SE
Lively, W.Va., 136, NW
Livermore, Pa., 58, NW
Liverpool, W.Va., 103, NE
Livingston Manor, N.Y., 24, NW
Livingston, N.Y., 53, SE
Livingston, W.Va., 121, SW

Lizard Creek Jct., Pa., 50, NE
Llandaff, Md., 114, SE
Llanerch, Pa., 82, NW
Llewellyn Crossing, Pa., 49B
Llewellyn, N.J., 53, NE
Llewellyn, Pa., 49, SW
Lloydell, Pa., 59, NE
Lloydville, Pa., 45, SW
Loch Laird Jct., Va., 140, SW
Loch Lomond Jct. (NYC; PRR), Pa., 45, NE
Loch Raven, Md., 96, NE
Locher, Va., 140, SW
Lochgelly, W.Va., 121, SE
Lochland, Pa., 50, SE
Lock 2, W.Va., 121, SW
Lock 3, Pa., 57, NW
Lock 11, W.Va., 73, SW
Lock 12, W.Va., 72, SE
Lock 13, W.Va., 72, SE
Lock Haven (NYC; PRR), Pa., 32, SW
Lock Ridge, Pa., 50, SE
Locke Valley, Pa., 61, SW
Lockport, Pa., 50, SE
Lockport, Pa., 58, NE
Locksley, Pa., 81, NE
Lockvale, Pa., 44, NW
Lockwood Crossing, Ohio, 12, NE
Lockwood, Ky., 118, NE
Lockwood, N.Y., 8, SE
Lockwood, Ohio, 25, NW
Locust Dale, Pa., 49, NW
Locust Gap, Pa., 49, NW
Locust Grove, N.J., 83, NW
Locust Manor, N.Y., 54, SW
Locust Point, Md., 96A
Locust Summit Jct., Pa., 49, NW
Locust Summit, Pa., 49, NW
Locust Valley, N.Y., 54, NE
Locust, Pa., 44, NW
Locust, Pa., 49, NW
Locust, W.Va., 123, SE
Lodgeville, W.Va., 89, NW
Lofton, Va., 140, NE
Lofty Jct., Pa., 49D
Lofty, Pa., 49D
Logan Ferry, Pa., 42, SW
Logan, Pa., 66, SE
Logan, Va., 140, SW
Logan, W.Va., 135, NW
Logansport, Pa., 42, SE
Logue, Md., 77, SE

Logue, Pa., 18, SW
Lomax, W.Va., 149, NE
Lonaconing, Md., 75, SW
London Bridge, Va., 177, NE
London, W.Va., 121, SW
Lone Cedar, W.Va., 86, SW
Lone Jack, Va., 154, NE
Lone, Md., 76, SE
Long Beach, N.Y., 54, SE
Long Branch, N.J., 69, NW
Long Branch, W.Va., 136, NW
Long Bridge, N.J., 52, NW
Long Bridge, Pa., 58, NW
Long Eddy, N.Y., 23, NE
Long Flat, N.Y., 11, SE
Long Green, Md., 97, NW
Long Island, Va., 154, SE
Long Park, Pa., 64, SW
Long Run, Ohio, 55, SW
Long Run, Pa., 49, NE
Long Run, W.Va., 88, NE
Long Valley Station, Pa., 20, SE
Long Valley, N.J., 52, NW
Long Valley, Pa., 20, SE
Longacre, W.Va., 121, SW
Longdale, Va., 139, NW
Longdale, W.Va., 103, NW
Longfellow, Pa., 46, SE
Longport, N.J., 100, NE
Longs, Md., 78, SW
Longsdorf, Pa., 62, SW
Longstreth, Ohio, 85, NE
Longview, Pa., 56, NE
Lookout Jct., Pa., 22, SE
Lookout, Ky., 148, NW
Lookout, W.Va., 122, SW
Looney's Creek, Va., 148, NE
Loop Line Switches, W.Va., 55A
Loop Run, Pa., 30, SE
Loop, Pa., 43, NE
Loop, Pa., 60, NW
Loop, W.Va., 106, NE
Loopemont, W.Va., 138, NW
Lopez, Pa., 34, NW
Lorado, W.Va., 135, NE
Loraine, Pa., 65, NW
Lorberry Jct., Pa., 49, SW
Lorberry, Pa., 49, SW
Lordville, N.Y., 23, NE
Lore City, Ohio, 70, NW
Loreley, Md., 97, NW
Lorentz, W.Va., 89, SW
Loretto Ave., N.J., 100, SE

Loretto Road, Pa., 44, SE
Lorraine, N.J., 53, SE
Lorraine, Va., 143, SE
Lorton, Va., 112, SE
Lory, W.Va., 120, SW
Losh's Run, Pa., 62, NE
Lost Creek, Pa., 49, NE
Lost Creek, W.Va., 89, SW
Lothair, Md., 130, NW
Lottsville, Pa., 15, NW
Lotus, Pa., 56, NE
Loucks, Pa., 79, NE
Louden Hill, Pa., 22, NW
Loudenville, W.Va., 71, NE
Loudon Park, Md., 96, NE
Louellen, Ky., 163, NE
Louisa, Ky., 118, SE
Louisa, Va., 128, SW
Louise, W.Va., 55, NE
Lounsberry, N.Y., 9, SW
Love Point, Md., 97, SW
Lovell, Pa., 14, NE
Lovett, Pa., 59, NE
Low Gap, W.Va., 120, SW
Low Moor, N.J., 69, NW
Low Moor, Va., 139, NW
Low Phos, Pa., 73, NW
Low, Ky., 163, NW
Lowber, Pa., 57, SW
Lowell Ave., Ohio, 25, SE
Lowell, W.Va., 137, SE
Lowellville, Ohio, 25, SE
Lower Ferry, W.Va., 55, NE
Lower Mill, Del., 81, NE
Lower Switch, N.Y., 9, NW
Lowerre, N.Y., 54, NW
Lowesville, Va., 140, SE
Lowman, N.Y., 8, SE
Lowney, W.Va., 134, NW
Lowrey, Pa., 20, SW
Lowry, Va., 154, NW
Lowrys, Pa., 43, SW
Lowville, W.Va., 72, SE
Loyall, Ky., 163, NW
Loyalton, Pa., 48, SW
Loys, Md., 78, SW
Loysville, Pa., 62, NW
Lucas, Pa., 57A
Lucaston, N.J., 83, NW
Lucasville, Pa., 42, NW
Lucerne Jct., Pa., 43, SE
Lucinda, Pa., 28, NW
Luckett, Pa., 59, NE

Ludlow, N.Y., 54, NW
Ludlow, Pa., 16, SW
Ludlow-Asbury, N.J., 51, SE
Luhrig, Ohio, 85, NE
Luke, Va., 148, NE
Luke, W.Va., 91, NE
Lumber, Pa., 44, NE
Lumber, W.Va., 107, NW
Lumberport, W.Va., 89, NW
Lumberton, N.J., 83, NW
Lumberton, Va., 175, NE
Lunt, Va., 112, NE
Luray, Va., 110, SW
Lurich, Va., 151, NW
Lushbaugh, Pa., 30, NE
Luthersburg, Pa., 29, SE
Lutherville, Md., 96, NE
Lutzville, Pa., 60, SW
Luzerne, Pa., 35A
Luzon, N.Y., 24, SE
LV Jct., Pa., 48, SW
Lyburn, W.Va., 135, NW
Lykens (PRR; RDG), Pa., 48, SE
Lyle, Va., 139, SE
Lynbrook, N.Y., 54, SE
Lynces Jct., Pa., 13, SW
Lynch, Ky., 164, NW
Lynch, Md., 97, NE
Lynch, Pa., 15, SE
Lynch, Va., 158, NW
Lynchburg (C&O; N&W), Va., 154, NE
Lynco, W.Va., 135, SE
Lyndell, Pa., 65, SE
Lyndhurst, N.J., 53, NE
Lyndhurst, Va., 126, SW
Lynnhaven, Va., 177, NE
Lynnport, Pa., 50, SW
Lynnwood, Va., 126, NW
Lyons Run Branch Jct., Pa., 57, NE
Lyons, N.J., 52, SE
Lyons, Pa., 26, NE
Lyons, Pa., 65, NW
Lytle, Pa., 49, SW

M&K Junction, W.Va., 90, NE
M&NE Jct., Ohio, 85, SE
M&U Jct., N.J., 37, NE
MA, Pa., 67, SW
Maben, W.Va., 136, SW
Mabie, W.Va., 107, NW
Mabscott, W.Va., 136A, NE

MacCorkle, W.Va., 120, SW
Mace Spring, Va., 165, SW
MacFarlan, W.Va., 87, SE
Machias (B&O), N.Y., 5, NW
Machias (PRR), N.Y., 5, NW
Machipongo, Va., 161, NW
Mackeyville, Pa., 32, SW
Mackin, Pa., 42, NW
Macksburg, Ohio, 70, SW
Macoby Siding, Pa., 66, NW
Macon, Va., 143, SW
Macpelah Jct., W.Va., 89, SW
Macungie, Pa., 50, SE
Mada, W.Va., 135, SE
Madera (NYC; PRR), Pa., 45, NW
Madge, Pa., 16, SW
Madiera Hill, Pa., 44, NW
Madison College, Va., 126, NW
Madison Run, Va., 127, SE
Madison Run, W.Va., 90, NE
Madison, N.J., 53, NW
Madison, Pa., 27, SE
Madison, Pa., 57, NE
Madison, W.Va., 120, SW
Madley, Pa., 75, NE
Magazine, N.J., 82, NW
Magee, Pa., 15, SW
Maggie, W.Va., 102, NE
Magnolia, Md., 97, NW
Magnolia, N.J., 82, NE
Magnolia, Va., 176, SE
Magnolia, W.Va., 76, SW
Magruder, Va., 159, NE
Mahaffey Jct. (NYC; PRR), Pa., 44, NE
Mahan, W.Va., 121, SW
Mahanoy City (LV; RDG), Pa., 49, NE
Mahanoy Plane, Pa., 49, NE
Mahanoy Plane, Pa., 49, NW
Mahanoy, Pa., 62, NE
Maher, W.Va., 134, NW
Mahoning (PRR; PS), Pa., 43, NW
Mahopac, N.Y., 39, NE
Mahwah, N.Y., 38, SE
Maiden Creek, Pa., 65, NW
Maidens, Va., 143, SW
Maidsville, W.Va., 73, SW
Main Ave., N.J., 83, SW
Main St. (Norristown), (RDG), Pa., 66A

Main St. (Richmond), (C&O), Va., 144A
Main St. (Washington), (WAW), Pa., 56, SE
Mainville (PRR; RDG), Pa., 49, NW
Maitland, Pa., 46, SE
Maitland, W.Va., 149, NE
Maizeville, Pa., 49C
Majestic, Ky., 134, SE
Major, Pa., 30, NW
Major, Va., 140, SW
Maken, W.Va., 89, NW
Makmie Park, Va., 146, NE
Malaga, N.J., 82, SE
Malba, N.Y., 54, NW
Malden, W.Va., 120, NE
Mallory, W.Va., 135, SW
Mallow, Va., 139, NW
Maloney, Va., 155, SE
Maltby Jct., Pa., 35A
Malvern, N.Y., 54, SE
Malvern, Pa., 65, SE
Mamaroneck (NH; NYW&B), N.Y., 54, NE
Mamaroneck Ave., N.Y., 39, SW
Mammoth, Pa., 58, SW
Mammoth, W.Va., 121, NW
Mampa, Ky., 134, SE
Man, W.Va., 135, SW
Manahawken, N.J., 84, SW
Manasquan (NY&LB; PRR), N.J., 68, SE
Manassas, Va., 112, NW
Manayunk (PRR; RDG), Pa., 66B
Manbur, Va., 144, SW
Mance, Pa., 75, NW
Manco, Ky., 148, NW
Manhassett, N.Y., 54, NE
Manhattan Beach, N.Y., 54, SW
Manheim, Pa., 64, SW
Manheim, W.Va., 90, NE
Manitou, N.Y., 39, NW
Mann, Ohio, 12, SE
Mann's Choice, Pa., 59, SE
Manning Jct., Pa., 36, NW
Manning, Tenn., 162, SW
Mannington, W.Va., 72, SW
Manor, Pa., 57, NE
Manor, Pa., 64, SW
Manor, Va., 110, NE
Manorville, Pa., 42, NE
Manown, Pa., 57, SW

Manry, Va., 175, NE
Mansfield, Ky., 162, SE
Mansfield, Pa., 19, NE
Mansion, Va., 154, SE
Manteo, Va., 141, SE
Mantolocking, N.J., 68, SE
Mantown, W.Va., 90, NE
Manumuskin, N.J., 100, NW
Manus, W.Va., 134, NE
Manver (C&I; CT&D), Pa., 44, SW
Manville (LV; RDG), N.J., 52, SE
Manville-Finderne, N.J., 52, SE
Mapes, N.Y., 6, SW
Maple Grove, Md., 79, SW
Maple Shade, N.J., 82, NE
Maple Springs, N.Y., 3, SW
Maple, Pa., 57, SW
Maple, Pa., 66, SW
Mapleton, Ohio, 71, NW
Mapleton, Pa., 61, NW
Mapleton, W.Va., 105, SE
Mapleville, Md., 77, SE
Maplewood (DL&W; RV), N.J., 53, SW
Maplewood, Pa., 36, NW
Maplewood, Va., 156, NE
Marathon, N.Y., 9, NE
Marble Hill, N.Y., 54, NW
Marcoal, Ky., 134, SW
Marcus Hook (PRR; RDG), Pa., 82, NW
Mardale Springs, Md., 132, NW
Marengo, Pa., 45, SE
Marfork, W.Va., 135, NE
Marfrance, W.Va., 122, SE
Margaret, N.C., 175, SW
Margaret, Pa., 43, NW
Margate, N.J., 100, NE
Marguerite, Pa., 58, NW
Marianna, Pa., 56, SE
Marianna, W.Va., 135, SE
Marienville, Pa., 28, NE
Marietta (B&O; MC&C; PRR), Ohio, 87, NW
Marietta Jct., Pa., 64, SW
Mariner's Harbor, N.Y., 53, SE
Marion (M&RV; N&W), Va., 166, NE
Marion Center, Pa., 43, NE
Marion Jct., N.J., 53B
Marion, Md., 132, SW
Marion, Pa., 77, NE

Marion, W.Va., 90, NE
Markham, Pa., 81, NE
Markham, Va., 111, NW
Markle, Pa., 43, NE
Marklesburg, Pa., 60, NE
Markleton (B&O; WM), Pa., 74, NE
Marksboro, N.J., 52, NW
Marlboro, Md., 113, NE
Marlboro, N.J., 68, NE
Marlbrook, Va., 140, NW
Marley Neck Br. Jct., Md., 96, SE
Marlinton, W.Va., 123, SE
Marlow, Md., 96, SE
Marlton, N.J., 83, NW
Marmet (C&O; WVS), W.Va., 120, NE
Marne, W.Va., 121, NE
Marnie, W.Va., 135, NE
Marquis, Ohio, 40, NW
Marriotsville, Md., 96, NW
Marrowbone Jct., Ky., 148, NW
Marrowbone, Ky., 148, NW
Mars Hill Summit, Pa., 80, NE
Mars, Pa., 41, SE
Marsh Creek, Pa., 19, NW
Marsh Fork Jct., W.Va., 136A, NW
Marsh Hill Jct. (PRR; S&NY), Pa., 33, NW
Marsh Run, Pa., 63, SW
Marshall St. (Norristown), (RDG), Pa., 66A
Marshall, Va., 111, NW
Marshalls Creek, Pa., 36, SE
Marshalls, N.Y., 7, NW
Marshalltown, Va., 138, SE
Martha, Pa., 45, NE
Martha, W.Va., 119, NW
Martic Forge, Pa., 80, NW
Martin, Pa., 73, NW
Martins Creek (DL&W; LNE; PRR), Pa., 51, NE
Martins Creek Jct., Pa., 51, NE
Martins Ferry (B&O; PRR; W&LE), W.Va., 55, SE, 55A
Martins, N.J., 37, SE
Martinsburg (B&O; PRR), W.Va., 94, NW
Martinsburg Jct., Pa., 60, NW
Martinsburg, Pa., 60, NW
Martinsville (D&W; N&W), Va., 170, SW
Marvindale, Pa., 16, SE

Marwood, Pa., 42, NW

Maryd, Pa., 49, NE

Marydel, Md., 98, SE

Maryland School, Md., 96, NE

Marysville, Pa., 63, NW

Mason, Va., 175, NW

Mason, W.Va., 85, SE

Mason-Dixon, Pa., 77, SW

Masontown, Pa., 73, NW

Masontown, W.Va., 73, SW

Masonville, N.J., 83, NW

Maspeth, N.Y., 54, SW

Massaponax, Va., 129, SW

Massey, Md., 98, NW

Massies Mill, Va., 140, NE

Masten, Pa., 20, SW

Masters, W.Va., 122, SW

Masthope, Pa., 23, SE

Matawan, N.J., 68, NE

Matewan, W.Va., 134, SE

Mather, Pa., 72, NE

Matney, Va., 149, NW

Matoaka, W.Va., 150, NE

Mattern Jct., Pa., 46, NW

Mattingly, W.Va., 90, NE

Mauch Chunk (CNJ; LV), Pa., 50, NE

Maugansville, Md., 77, SE

Mauk, Pa., 43, NE

Maurertown, Va., 110, NW

Maurice River, N.J., 99, SE

Mauricetown, N.J., 99, NE

Mauricetown, N.J., 100, NW

Mausdale, Pa., 48, NE

Mauzy, Va., 126, NE

Max Meadows, Va., 168, NW

Maxim, N.J., 68, SE

Maxwell, Pa., 57, SW

Maxwell, Va., 149, SE

May King, Ky., 147, SW

May, W.Va., 107, SW

Maybeury, W.Va., 150, NW

Maybrook (NH), N.Y., 38, NE

Maybrook Jct. (L&HR), N.Y., 38, NE

Mayburg, Pa., 15, SE

Mayes, Pa., 44, NE

Mayfield Yard, Pa., 22, SE

Mayfield, Pa., 22, SE

Maynard (B&O; W&LE), Ohio, 55, SW

Mayport, Pa., 28, SE

Mays Landing, N.J., 100, NE

Maytown, Va., 165, NW

Mayview, Pa., 56, NE

Mayville (JW&NW), N.Y., 3, NW

Mayville (PRR), N.Y., 3, NW

Mayville Jct., N.Y., 3, NW

Mayville, N.J., 100, SW

Maywood, N.J., 53, NE

Maywood, N.Y., 11, NW

MB, Pa., 67, SW

McAdams, Pa., 56, NW

McAdoo, Pa., 50A

McAfee, N.J., 37, SE

McAlpin (C&O; VGN), W.Va., 136A, SW

McArthur, Ohio, 85, SW

McAuley, Pa., 49, NW

McAvan, Ohio, 86, NE

McBride, Pa., 42, NW

McCalls Ferry, Pa., 80, NW

McCalls, Pa., 80, NW

McCalmont, Pa., 42, NW

McCarr, W.Va., 134, SE

McCarty, W.Va., 122, NE

McClainville, Ohio, 55, SW

McClure, Pa., 47, SW

McComas, W.Va., 150, NW

McConnell, W.Va., 135, NW

McConnellstown, Pa., 60, NE

McCoole, Md., 92, NW

McCormick, Pa., 43, NE

McCoy, Pa., 56, NW

McCoy, Va., 151, SE

McCreery, W.Va., 136, NE

McCullough, Ohio, 40, SE

McCullough, Pa., 57, NE

McDaniel, Md., 114, NW

McDonald (MTR; PRR), Pa., 56, NW

McDonald's "Y," Pa., 15, SW

McDonaldton, Pa., 75, NW

McDonough, Md., 96, NW

McDowell, Ky., 147, NE

McDowell, W.Va., 150, NW

McElhattan, Pa., 32, SW

McGaheysville, Va., 126, NE

McGareys (McGeary), Pa., 28, SE

McGees (NYC; PRR), Pa., 44, NW

McGuire Park, W.Va., 89, SW

McIntyre, Pa., 43, SW

McKean, Pa., 42, SE

McKee City, N.J., 100, NE

McKee, Pa., 60, NW

McKee, Pa., 62, NE

McKeefney, W.Va., 71, NW

McKeesport, Pa., 57, NW

McKendree, W.Va., 136, NE

McKenney, Va., 174, NE

McKinley, Pa., 16, SW

McKinneys, N.Y., 8, NE

McLaughlin, Pa., 57, NW

McLean, W.Va., 90, SW

McLeans, Pa., 66, NW

McLyn, W.Va., 135, SE

McMechen, W.Va., 71, NE

McMillan, W.Va., 90, NE

McMinns Summit (B&O; PRR), Pa., 29, SW

McMurray, Pa., 56, NE

McNeill, W.Va., 92, SW

McNutt, W.Va., 105, SE

McQuaid, W.Va., 136, NE

McRae's, Va., 156, NW

McRobert, Ky., 147, SE

McRoss, W.Va., 137, NW

McVeigh, Ky., 134, SW

McVey, W.Va., 136A, SE

McVeytown, Pa., 61, NE

McWhorter, W.Va., 89, SW

McWilliams, Pa., 43, NW

Mead, W.Va., 136, SW

Meador, Va., 153, SE

Meadow (ACL), Va., 144A

Meadow (SOU), Va., 144, SE

Meadow Bridge, W.Va., 137, NW

Meadow Brook, N.Y., 38, NE

Meadow Creek (C&O; NF&G), W.Va., 137, NW

Meadow Fork, W.Va., 136, NE

Meadow View, Va., 166, NW

Meadowbrook, N.Y., 54, SE

Meadowbrook, Pa., 66, SE

Meadowbrook, W.Va., 89, NW

Meadowdale, W.Va., 103, NE

Meadowlands, Pa., 56, SE

Meadows Yard, N.J., 53, A, B

Meads, Ky., 118, NE

Meadville (B&LE; Erie), Pa., 13, SE

Meadville Jct., Pa., 13, SW

Mears, Va., 146, NE

Mease, Pa., 45, NW

Mechanicsburg, Pa., 62, SE

Mechanicstown, Ohio, 40, SW

Mechanicsville, Md., 130, NE

Medford, Md., 78, SE

Medford, N.J., 83, NW

Media, Ohio, 70, NE

Media, Pa., 82, NW

Medix Run (B&O; PRR), Pa., 30, NW

Meechum's River, Va., 126, SE

Meetz, Va., 111, SW

Meherrin (SOU; VGN), Va., 156, SW

Mehoopany, Pa., 21, SE

Meigs, Ohio, 85, SE

Meiser, Pa., 48, NW

Melcroft, Pa., 58, SW

Melfa, Va., 146, SE

Mellin, W.Va., 87, SE

Mellinger, Pa., 80, NW

Melrose, Md., 79, SW

Melrose, N.J., 83, NW

Melrose, N.Y., 54A

Melrose, Pa., 22, NE

Melton, Va., 127, SE

Melville, W.Va., 135, NW

Melvin, Ky., 147, NE

Melvin, Va., 164, SE

Menantico, N.J., 100, NW

Mendenhall, Pa., 81, NE

Mendham, N.J., 52, NE

Mendota, Va., 165, SW

Menlo Park, N.J., 53, SW

Mercer (B&LE; PRR), Pa., 26, SE

Mercer Jct., Pa., 26, NE

Mercer Pike, Pa., 13, SE

Mercer Road, Pa., 42, NW

Mercers Bottom, W.Va., 102, SE

Mercersburg Jct., Pa., 77, NW

Mercersburg, Pa., 77, NW

Merchantville, N.J., 82B

Meredith, Pa., 29, SW

Meriden, W.Va., 89, SE

Merion, Pa., 82, NW

Merkle, Pa., 50, SE

Merna, Ky., 163, NW

Merrick, N.Y., 54, SE

Merrickville, N.Y., 11, NE

Merrill, Pa., 29, NE

Merrill, Pa., 41, SW

Merrilon Ave., N.Y., 54, SE

Merrimac Mines, Va., 152, SW

Merrimac, Va., 152, SW

Merrimac, W.Va., 134, SE

Merritt, Ohio, 70, NW

Merrittstown, Pa., 73, NW

Mertztown, Pa., 50, SE

Meshoppen, Pa., 21, SE
Messengerville, N.Y., 9, NE
Messmore, Pa., 73, NW
Mesta, Pa., 57A
Metalton, W.Va., 136A, NW
Metamora, Pa., 50, SE
Metuchen (LV; PRR; RDG), N.J., 53, SW
Metz, W.Va., 72, SW
Metzler, Pa., 74, NW
Meyers Crossing, Ohio, 85, NE
Meyersdale (B&O; WM), Pa., 74, NE
Miami, W.Va., 121, SW
Micajah, W.Va., 150, NW
Micco, W.Va., 135, NW
Mickleton, N.J., 82, NE
Mickley's, Pa., 50, SE
Middle Branch, N.J., 84, NW
Middle Fork, W.Va., 121, NW
Middle River, Md., 97, NW
Middle Siding, Md., 75, SW
Middle Valley, N.J., 52, NW
Middleburg, Md., 78, SE
Middleburg, Pa., 47, NE
Middlebury, Pa., 19, NW
Middlebush, N.J., 67, NE
Middleport, Ohio, 85, SE
Middleport, Pa., 49, SE
Middlesboro, Ky., 162, SE
Middlesex, N.J., 53, SW
Middlesex, Pa., 62, SE
Middletown (Erie; M&U; NYO&W), N.Y., 38, NW
Middletown (PRR; RDG), Pa., 63, SE
Middletown, Del., 98, NE
Middletown, Md., 94, NE
Middletown, N.J., 100, NE
Middletown, N.J., 68, NE
Middletown, Va., 93, SW
Midkiff, W.Va., 119, SE
Midland Mine, W.Va., 137, NE
Midland Park, N.J., 53, NE
Midland, Md., 75, SW
Midland, Pa., 41, SW
Midland, Va., 111, SE
Midlothian, Va., 143, SE
Midmont, Pa., 16, SE
Midvale (B&O; MF), W.Va., 106, NE
Midvale, Pa., 77, SE
Midvale, Va., 140, NW

Midway, Pa., 56, NW
Midway, Pa., 80, NW
Midway, Va., 141, SE
Midway, W.Va., 121, NW
Midwest, W.Va., 121, SW
Mifflin Cross Roads, Pa., 49, NW
Mifflin Jct. (MTR; URR), Pa., 57, NW
Mifflin, Pa., 34, SW
Mifflin, Pa., 47, SW
Mifflin, W.Va., 135, NW
Mifflinburg, Pa., 47, NE
Mikegrady, Ky., 148, NW
Milam Jct., W.Va., 136, SW
Milam, W.Va., 136, SW
Milan, Pa., 20, NE
Milboro, Va., 139, NE
Milburn, W.Va., 121, SW
Mile Branch, W.Va., 149, NW
Milesburg, Pa., 46, NW
Milesville, Pa., 57, SW
Milford, Del., 116, NW
Milford, N.J., 51, SE
Milford, Va., 129, SW
Miliken, W.Va., 120, NE
Mill Creek (VYRR; WM), W.Va., 107, SW
Mill Creek Road, W.Va., 137, NE
Mill Creek, Del., 81, NE
Mill Creek, Pa., 29, NW
Mill Creek, Pa., 61, NW
Mill Hall (NYC; PRR), Pa., 32, SW
Mill Hill Gap, Pa., 49, SE
Mill Hollow, W.Va., 121, NW
Mill Lane, Pa., 65, SE
Mill Neck, N.Y., 54, NE
Mill Park, Pa., 65, NE
Mill Plain, Conn., 39, NE
Mill Rift, Pa., 37, NE
Mill Rock, Ohio, 40, NE
Mill Run Jct., Pa., 74, NW
Mill St. (Salisbury), (B&E), Md., 132, NE
Mill Village, Pa., 14, NW
Millbank, Pa., 58, NW
Millbourne Mills, Pa., 82, NW
Millbrook, Va., 125, SE
Millburn, N.J., 53, SW
Millburn, Pa., 26, SE
Miller Farm, Pa., 14, SE
Miller Run, Pa., 59, SW

Miller Yard (CC&O; INT), Va., 165, NW
Miller, Md., 79, SW
Miller, Ohio, 55, NW
Miller, Pa., 51, NW
Millers Camp, W.Va., 136, NE
Millers, Pa., 14, NW
Millers, Pa., 15, SE
Millers, Pa., 49, SE
Millers, Pa., 50, SW
Millersburg, Pa., 48, SW
Millerton, Pa., 20, NW
Millfield, Ohio, 85, NE
Millham (RDG), N.J., 67, SE
Millikens, Pa., 44, NE
Millington, Md., 98, NW
Millington, N.J., 52, SE
Millmont, Pa., 47, NE
Millport, N.Y., 8, NW
Millport, Pa., 17, NE
Mills, N.Y., 9, NE
Mills, Ohio, 102, NE
Mills, Pa., 18, NE
Mills, Va., 149, SW
Millsboro (MON; PRR), Pa., 73, NW
Millsboro, Del., 116, SW
Millstone, Ky., 147, SW
Millvale (B&O; PRR), Pa., 57A
Millville, N.J., 99, NE
Millville, Pa., 33, SE
Millville, Pa., 34, SW
Millville, W.Va., 94, NW
Millway, Pa., 64, SE
Millwood, N.Y., 39, SW
Millwood, Pa., 58, NW
Millwood, W.Va., 103, NW
Milmay, N.J., 100, NW
Milnor, Pa., 77, NE
Milroy, Pa., 46, SE
Milton, Del., 116, NW
Milton, N.C., 171, SE
Milton, Pa., 33, SW
Milton, W.Va., 119, NE
Mina, Pa., 17, SE
Minden, W.Va., 136, NE
Mine Hill Crossing, Pa., 49, SE
Mine Run, Va., 128, NW
Minebrook, N.J., 52, SE
Minefield, Md., 80, SW
Mineola, N.Y., 54, SE
Mineral Point, Pa., 59, NW
Mineral Ridge, Ohio, 25, SW

Mineral Spring, Pa., 74, NE
Mineral, Ohio, 85, NW
Mineral, Va., 128, SW
Miners Mills (CNJ), Pa., 35A
Minersville (PRR; RDG), Pa., 49B
Minerton, Ohio, 85, SW
Mingo Yard (PRR; W&LE), Ohio, 55, NE
Minisink, N.J., 52, NE
Minister, Pa., 15, SE
Mink Shoals, W.Va., 120, NE
Minnick, Md., 80, SE
Minnie, W.Va., 71, SW
Minolta, N.J., 83, SW
Minson, N.J., 82B
Mint Spring, Va., 125, SE
Miquon (PRR; RDG), Pa., 66A, B
Miracle, Ky., 162, NE
Mitchell Field, N.Y., 54, SE
Mitchell, Va., 127, NE
Mitchell's, Pa., 45, NW
Mizpah, N.J., 100, NW
Moatsville, W.Va., 90, SW
Mocanaqua, Pa., 34, SE
Moccasin Gap, Va., 164, SE
Modena, Pa., 81, NW
Modoc, Ohio, 85, NE
Moffett, Md., 76, SE
Mogees, Pa., 66A
Mohican, Pa., 42, NE
Mohola, N.J., 37, SE
Mohrsville, Pa., 65, NW
Molino, Pa., 49, SE
Molly, Pa., 48, NW
Molus, Ky., 163, NW
Monaca, Pa., 41, SW
Monarch, W.Va., 121, SW
Monaville, W.Va., 135, NW
Monclo, W.Va., 135, NW
Monday, Ohio, 85, NE
Monessen (P&LE; P&WV), Pa., 57, SW
Moneta, Va., 153, SE
Monitor Jct., W.Va., 134, NE
Monkton, Md., 79, SE
Monmouth Beach, N.J., 69, NW
Monmouth Jct., N.J., 67, NE
Monocacy (PRR; RDG), Pa., 65, NW
Monongah, W.Va., 89, NE
Monongahela (P&LE; PRR), Pa., 57, SW

Monroe, N.J., 37, SE

Monroe, N.Y., 38, NE

Monroe, Pa., 42, SE

Monroe, Va., 140, SE

Monroe, W.Va., 107, NW

Monroe-Cedar Knoll, N.J., 53, NW

Monroeton (LV; S&NY), Pa., 21, SW

Monroeville, N.J., 82, SE

Monrovia St. (New Castle), (PRR), Pa., 41, NW

Monrovia, Md., 95, NW

Monsey Heights, N.Y., 38, SE

Monsey, N.Y., 38, SE

Mont Clare, Pa., 65, SE

Montana, W.Va., 72, SE

Montandon, Pa., 48, NW

Montcalm, W.Va., 150, NW

Montchanin, Del., 81, NE

Montclair (DL&W; Erie), N.J., 53, NE

Montclair Heights, N.J., 53, NE

Montcoal, W.Va., 135, NE

Montello, Pa., 64, NE

Montessori School, Pa., 66, NE

Monteview, Md., 95, NW

Montgomery (PRR; RDG), Pa., 33, SW

Montgomery, Va., 152, SW

Montgomery, W.Va., 121, SW

Montgomeryville, Pa., 42, NE

Monticello, N.Y., 24, SE

Montivideo, Md., 96, SW

Montivideo, Va., 126, NW

Montour Falls, N.Y., 8, NW

Montour Jct. (B&O), Pa., 57, NW

Montour Jct. (MTR; P&LE), Pa., 41, SE

Montoursville, Pa., 33, SW

Montpelier, Va., 127, SE

Montrose (DL&W; LV), Pa., 22, NW

Montrose, N.Y., 39, NW

Montrose, W.Va., 90, SW

Montvale, N.J., 38, SE

Montvale, Va., 153, NE

Monument, Pa., 31, SE

Moon Run, Pa., 56A

Moonachie, N.J., 53, NE

Moonville, Ohio, 85, NW

Moore, Pa., 82, NW

Moore, W.Va., 90, SE

Moorefield, W.Va., 92, SW

Mooreland, Va., 143, SE

Moores Jct., Ohio, 87, NW

Moores Mill, Pa., 60, NW

Moores, N.J., 99, NE

Moores, W.Va., 104, NW

Mooresburg, Pa., 48, NE

Moorestown, N.J., 83, NW

Moorhead (NYC), Pa., 2, SW

Moorheads (NKP), Pa., 2, SW

Moorings, Va., 159, SW

Moors Mill, Pa., 62, SW

Moosic, Pa., 35C

Morado (PRR), Pa., 41, NW

Moraine, N.Y., 6, NE

Moran, Va., 156, SW

Morann, Pa., 45, NW

Morantown, Md., 75, SW

Morea, Pa., 49, NE

Morewood, Pa., 57, SE

Morgan (PRR; P&WV), Pa., 56, NE

Morgan, Md., 95, NE

Morgan, N.J., 68, NW

Morgan, Pa., 57, SE

Morgan, Pa., 82, NW

Morgan, W.Va., 86, SW

Morgan, W.Va., 150, NW

Morgansville, W.Va., 88, NE

Morgantown (B&O; MON), W.Va., 73, SW

Morganville, N.J., 68, NE

Morganza, Pa., 56, NE

Moritz, N.Y., 4, SE

Morrell, Pa., 60, NW

Morris County Jct., N.J., 52, NE

Morris Creek Jct., W.Va., 121, SW

Morris Heights, N.Y., 54, NW

Morris Park (NH; NYW&B), N.Y., 54, NW

Morris Park Shops, N.Y., 54D

Morris Park, N.Y., 54D

Morris Plains, N.J., 53, NW

Morris Run (Erie), Ohio, 25A

Morris Run, Pa., 19, SE

Morris, N.J., 82B

Morris, Ohio, 86, NW

Morris, Pa., 19, SW

Morris, Pa., 49, SE

Morrisdale (NYC; PRR), Pa., 45, NE

Morrison, Pa., 16, NW

Morrison, Va., 144A

Morrison, Va., 160, SW

Morrison, W.Va., 105, SE

Morrissania, N.Y., 54, NW

Morristown (DL&W; MT&E), N.J., 53, NW

Morrisville, N.J., 67, SW

Morsemere, N.J., 53, NE

Morstein, Pa., 65A

Morton, Pa., 82, NW

Mortonville, Pa., 81, NW

Moscow, Pa., 35, NE

Moscow, W.Va., 40, SE

Moselem, Pa., 50, SW

Moseley (SOU; TVRR), Va., 157, NW

Moser Run Jct., Pa., 73, NW

Moses, W.Va., 150, NW

Moss Creek Jct., Pa., 44, SW

Moss Run, Va., 138, NE

Mossy Creek, Va., 125, NE

Mostoller, Pa., 59, SW

Motley, Va., 154, SW

Motters, Md., 78, SW

Moundsville, W.Va., 71, NE

Mount Airy Jct., Md., 95, NE

Mount Airy, Md., 95, NE

Mount Alto Park, Pa., 77, NE

Mount Alto, Pa., 77, NE

Mount Arlington, N.J., 52, NE

Mount Braddock, Pa., 73, NE

Mount Carbon, W.Va., 121, SW

Mount Carmel Jct., Pa., 49, NW

Mount Clare Jct., Md., 96A

Mount Crawford, Va., 126, NW

Mount Dallas (H&BTM; PRR), Pa., 60, SW

Mount Eagle, Pa., 46, NE

Mount Etna, Pa., 45, SE

Mount Holly Springs, Pa., 62, SE

Mount Hope (CWL), Pa., 64, SW

Mount Hope (RDG), Pa., 64, SW

Mount Hope, N.J., 52, NE

Mount Hope, N.Y., 54, NW

Mount Hope, W.Va., 136, NE

Mount Jackson, Va., 109, NE

Mount Jewett (B&O; Erie), Pa., 16, SE

Mount Joy, Pa., 64, SW

Mount Kisco, N.Y., 39, SE

Mount Pleasant (B&O; PRR), Pa., 57, SE

Mount Pleasant, Del., 81, SE

Mount Pleasant, N.J., 52, NE

Mount Pleasant, N.J., 100, SW

Mount Pleasant, N.Y., 39, SW

Mount Pocono, Pa., 36, SW

Mount Royal (Baltimore), (B&O), Md., 96A

Mount Royal, N.J., 82, NE

Mount Savage (C&PA; WM), Md., 75, SW

Mount Savage, Ky., 118, NW

Mount Sidney, Va., 126, NW

Mount Solon, Va., 125, NE

Mount St. Vincent, N.Y., 54, NW

Mount Tabor, N.J., 53, NW

Mount Union (EBT; PRR), Pa., 61, NW

Mount Upton, N.Y., 11, NW

Mount Vernon (NH; NYC), N.Y., 54, NW

Mount Vernon, Pa., 63, SE

Mount Wilson, Md., 96, NW

Mount Wolf, Pa., 63, SE

Mountain Ave., N.J., 53, NE

Mountain Dale, N.Y., 24, SE

Mountain Grove, Pa., 49, NE

Mountain Lake Park, W.Va., 91, NW

Mountain Lakes, N.J., 53, NW

Mountain Spring, N.Y., 37, NE

Mountain Springs, Pa., 34, NE

Mountain Top, Pa., 35, SW

Mountain View (DL&W; Erie), N.J., 53, NW

Mountain View, Va., 140, NW

Mountain View, W.Va., 90, NW

Mountain, N.J., 53, NW

Mountaindale, Pa., 45, SW

Mountainville, N.Y., 38, NE

Mountcastle, Va., 158, NE

Mountville, N.J., 53, NW

Mountville, Pa., 64, SW

Mouthcard, Ky., 148, NW

Mower & Reaper, Ohio, 25A

Mowhawk, W.Va., 135, SW

Mowry, Pa., 30, SE

Moycock, N.C., 177, SE

Moyer, Pa., 57C

Moyers, Pa., 49, SE

Moylan-Rose Valley, Pa., 82, NW

Mrya, Ky., 147, NE

Mt. Airy, N.C., 168, SE

Mt. Airy, Pa., 66, SE

Mt. Airy, W.Va., 123, NE

Mt. Alton, Pa., 16, NE

Mt. Bernard, Va., 143, SW

Mt. Bethel, Pa., 51, NE
Mt. Calvary, N.J., 100, NE
Mt. Carmel (LV; PRR), Pa., 49A
Mt. Carmel, Va., 167, NW
Mt. Clare, Md., 96A
Mt. Cuba, Del., 81, NE
Mt. Ephraim, N.J., 82, NE
Mt. Ephraim, Ohio, 70, NW
Mt. Gretna, Pa., 64, NW
Mt. Holly, N.J., 83, NW
Mt. Hope, Md., 96, NE
Mt. Ivy, N.Y., 38, SE
Mt. Washington, Md., 96, NE
Mt. Winans, Md., 96A
Mud Jct., W.Va., 134, NE
Muddlety Falls, W.Va., 122, NE
Muddlety, W.Va., 122, NW
Muddy Creek Forks, Pa., 80, NW
Mueller, Pa., 74, NW
Muhlenberg, Pa., 65, NW
Muirkirk, Md., 96, SW
Mukden, Pa., 58, SE
Muleshoe, Pa., 59, NE
Mulford, N.J., 37, SE
Mullens, W.Va., 136, SW
Mullica Hill, N.J., 82, SE
Mullikin, Md., 113, NE
Muncy (PRR; RDG), Pa., 33, SW
Muncy Valley, Pa., 33, NE
Munden, Va., 177, SE
Munhall, Pa., 57A
Munson, Pa., 45, NE
Munster, Pa., 59, NE
Murdock, Pa., 74, NE
Murphy, W.Va., 89, SE
Murray Hill, N.J., 53, SW
Murray Hill, N.Y., 54, NW
Murray, Pa., 63, NE
Murray, W.Va., 103, NE
Murraysville, Pa., 57, NE
Murrayville, W.Va., 86, SW
Musconetcong Jct., N.J., 51, SE
Muse Jct., Pa., 56, NE
Musgrove, Pa., 43, NW
Music, Ky., 118, NW
Musser, Pa., 46, SW
Mutual, Pa., 58, SW
MW, Pa., 67, SW
MX Crossover, Ohio, 12, NE
MY, Md., 75, SW
MY, Pa., 67, SW
Myersville, Md., 77, SE
Myles, W.Va., 107, NW

Myobeach, Pa., 21, SE
Myoma, Pa., 41, SE
Myrick, Ky., 162, NW
Myrtle, Va., 176, NE
Mystic, Pa., 17, NE

N.J. & N.Y. Jct., N.J., 53, NE
N.Y. State Fish Hatchery, N.Y., 7,
   NW
NA Tower, Pa., 13, NW
Naaman, Del., 82, NW
Nace, Va., 153, NW
Nadine, Pa., 57, NW
Naginey, Pa., 46, SE
Nallen, W.Va., 122, SW
Nance, Va., 158, NE
Nancy's Run, W.Va., 104, NW
Nansen, Pa., 16, SW
Nantyglo (C&I; PRR), Pa., 59,
   NW
Nanuet, N.Y., 38, SE
Naomi Pines, Pa., 36, SW
Napier, Ohio, 86, NE
Napier, Pa., 59, SE
Napiers, N.Y., 5, NW
Narberth, Pa., 66, SW
Narrows (N&W; VGN), Va., 151,
   NW
Narrows, Md., 114, NE
Narrows, W.Va., 71, NE
Narrowsburg, N.Y., 23, SE
Naruna, Va., 154, SE
Narva, Ohio, 55, NW
Narvon, Pa., 65, SW
Nash, Va., 157, NE
Nashua, Pa., 26, SW
Nashville, Pa., 79, NW
Nasons, Va., 127, NE
Nassau Boulevard, N.Y., 54, SE
Nassau, Del., 116, NE
Nassau, N.Y., 53, SE
Nassawadox, Va., 161, NW
Natalie Jct., Pa., 49C
Nathalie, Va., 172, NW
Naticoke, Pa., 34, SE
National, Pa., 56, NE
National, W.Va., 72, SE
Natrona, Pa., 42, SE
Natural Bridge (C&O; N&W), Va.,
   139, SE
Natural Well, Va., 139, NW
Naugatuck, W.Va., 134, NW
Naughright, N.J., 52, NE

Naval Academy, Md., 114, NW
Naval Proving Ground, Va., 129,
   NE
Naval Weapons Center, Va., 159,
   SE
Navarro, Pa., 51, SW
Navesink Beach, N.J., 69, NW
Nay Aug, Pa., 35, NE
Nazareth (DL&W; LNE), Pa., 51,
   SW
NC, W.Va., 55, NE
NE Junction, N.Y., 3, SW
Neal, W.Va., 118, NE
Nealy, Pa., 41, NE
Nebasco, Va., 112, SW
Nebo, Pa., 58, SW
Nebraska, Pa., 28, NW
Neelyton, Pa., 61, SW
Neelytown, N.Y., 38, NE
Neff (W&LE), Ohio, 55, SW
Neffs (B&O), Ohio, 55, SW
Neffs, W.Va., 91, NE
Negley, Ohio, 40, NE
Negro Arm, Va., 143, SW
Neibert, W.Va., 135, NW
Nella, N.C., 166, SE
Nellis, W.Va., 120, SE
Nelson, Pa., 19, NE
Nelson, W.Va., 120, SE
Nelsons, Pa., 27, SW
Nelsonville Yard, Ohio, 85, NE
Nelsonville, Ohio, 85, NE
Nemacolin, Pa., 73, NW
Nemours, W.Va., 150, NW
Neola, W.Va., 138, NE
Neon Jct., Ky., 147, SE
Neon, Ky., 147, SE
Nepera Park, N.Y., 54, NW
Nepperhan, N.Y., 54, NW
Neptune, W.Va., 86, SW
Nescopeck, Pa., 34, SE
Neshaminy Falls, Pa., 67, SW
Neshannock Falls, Pa., 26, SW
Neshantic (CNJ; LV), N.J., 52, SE
Nesquehoning (CNJ; LNE), Pa.,
   50, NW
Netcong, N.J., 52, NE
Netherwood, N.J., 53, SW
Nettie, W.Va., 122, SE
Neversink, Pa., 65, NW
Nevin, Pa., 81, NW
New Albany, Pa., 21, SW
New Alexandria, Ohio, 55, NE

New Alexandria, Pa., 58, NW
New Berlin Jct. (NYO&W; UV),
   N.Y., 11, NW
New Berlinville, Pa., 65, NE
New Bethlehem, Pa., 28, SW
New Bloomfield, Pa., 62, NE
New Boston Jct. (LV; PRR), Pa.,
   49, NE
New Brighton, N.Y., 53, SE
New Brighton, Pa., 41, SW
New Britain, Pa., 66, NE
New Brunswick (PRR), N.J., 68,
   NW
New Canton, Tenn., 164, SW
New Canton, Va., 142, SW
New Castle (B&O; P&LE; PRR),
   Pa., 41, NW
New Castle Colliery, Pa., 49, SE
New Castle Road, Del., 81A
New Castle, Del., 81, SE
New Castle, Va., 138, SE
New Centerville, Pa., 66, SW
New Church, Va., 146, NE
New City, N.Y., 39, SW
New Columbia, Pa., 33, SW
New Cumberland, Pa., 63, SW
New Cumberland, W.Va., 55, NE
New Dorp, N.Y., 53, SE
New Egypt, N.J., 67, SE
New England, W.Va., 86, SE
New Era, W.Va., 103, NE
New Florence, Pa., 58, NE
New Franklin, Pa., 77, NE
New Freedom (PRR; STRT), Pa.,
   79, SE
New Galilee (PRR; Y&S), Pa., 41,
   NW
New Garden, Pa., 81, NW
New Geneva, Pa., 73, NW
New Germantown, Pa., 61, NE
New Glasgow, Va., 141, SW
New Hampton, N.Y., 38, NW
New Haven, W.Va., 103, NW
New Hempstead, N.Y., 38, SE
New Holland, Pa., 64, SE
New Hope, Pa., 67, NW
New Hyde Park, N.Y., 54, SE
New Kensington, Pa., 42, SW
New Kingston, Pa., 62, SE
New Lisbon, N.J., 83, NE
New Market, N.J., 53, SW
New Market, Va., 109, SE
New Marshfield, Ohio, 85, NE

New Martinsville, W.Va., 71, SW
New Midway, Md., 78, SW
New Milford, N.J., 53, NE
New Milford, N.Y., 38, SW
New Milford, Pa., 22, NE
New Millport, Pa., 44, NE
New Oxford, Pa., 78, NE
New Philadelphia, Pa., 49, SE
New Plymouth, Ohio, 85, NW
New Providence, N.J., 53, SW
New Providence, Pa., 80, NE
New Ringgold, Pa., 50, SW
New River, Va., 151, SE
New Rochelle, N.Y., 54, NW
New Salem, Pa., 73, NW
New Salisbury, Ohio, 40, SE
New Sewickley, Pa., 41, NW
New Siding, Va., 124, SW
New Stanton, Pa., 57, SE
New Thacker, W.Va., 134, SE
New Village, N.J., 51, SE
New Waterford, Ohio, 40, NE
New Wilmington (PRR; SH), Pa., 26, SW
New Windsor, Md., 78, SE
New York Ave., N.J., 100, NE
New's Ferry, Va., 171, SE
Newark (B&O; PRR), Del., 81, SW
Newark (CNJ; DL&W; Erie; H&M; PRR), N.J., 53A
Newark Airport, N.J., 53A
Newark Transfer, N.J., 53A
Newark Valley, N.Y., 9, SE
Newark, W.Va., 87, SW
Newberry (RDG), Pa., 32, SE
Newberry Jct. (NYC), Pa., 32, SE
Newbridge, Del., 81A
Newburg, W.Va., 90, NW
Newburgh (Erie), N.Y., 38, NE
Newcomer, Pa., 73, NW
Newell, Pa., 57, SW
Newell, W.Va., 40, SE
Newfield Jct. (B&O; CPA), Pa., 18, NW
Newfield, N.J., 82, SE
Newfield, N.Y., 8, NE
Newfield, Pa., 18, NW
Newfoundland, N.J., 38, SW
Newgate, Md., 77, SW
Newhall, W.Va., 149, NE
Newhouse, Ohio, 40, NE
Newkirk, Pa., 50, NW

Newlinsburg, Pa., 57, NE
Newport (PRR; SR&W) Pa., 62, NE
Newport News, Va., 177, NW
Newport, Del., 81, SE
Newport, N.J., 99, NE
Newport, Pa., 41, NW
Newsoms, Va., 175, SE
Newton Falls (B&O; NYC), Ohio, 25, SW
Newton Hamilton, Pa., 61, NW
Newton, N.J., 37, SE
Newton, Pa., 15, NW
Newton, Pa., 49, NE
Newton, Va., 142, SW
Newton, W.Va., 106, SE
Newtonville, N.J., 83, SW
Newtown Square, Pa., 82, NW
Newtown, Pa., 49, SW
Newtown, Pa., 57, NW
Newtown, Pa., 67, SW
Newville, Pa., 62, SW
Niagara, Va., 153, NW
Nicetown, Pa., 82A
Nicholas, Va., 142, NW
Nichols, N.Y., 7, SW
Nichols, N.Y., 9, SW
Nicholson, Pa., 22, SW
Nick Haven, Pa., 57, SW
Nickel, Pa., 67, SW
Nicolette, W.Va., 86, SE
Niday, Va., 150, NE
Nigh, Ky., 148, NW
Nilan, Pa., 73, SW
Nile, N.Y., 5, SE
Niles (Erie; PRR), Ohio, 25, SW
Niles Valley, Pa., 19, NW
Niles, N.Y., 11, NW
Niles, Pa., 27, NW
Nineveh, N.Y., 10, SE
Niobe, N.Y., 3, SW
Nipton, Pa., 44, SW
Nitro, Pa., 43, NW
Nitro, W.Va., 120, NW
Nittany, Pa., 46, NE
Niver Jct., Pa., 75, NW
Niverton, Pa., 74, SE
NK Target (Erie), Ohio, 25A
No. 6 Jct., Pa., 35D
Noble, Pa., 66, SE
Nobles, Pa., 14, SW
Noblestown, Pa., 56, NE
Noel, Pa., 59, NE

Noel, Va., 143, NE
Nokesville, Va., 111, SE
Nolan, W.Va., 134, SW
Nolting, Va., 128, NW
Nora, Va., 148, SW
Norcross, Va., 151, NE
Nordmont, Pa., 34, NW
Norfolk (NS; N&W; VGN), 177A
Norge, Va., 159, NW
Norma, N.J., 99, NE
Normal, Ky., 118, NE
Norman, Pa., 28, SE
Norman, W.Va., 136, SW
Normandie, N.J., 69, NW
Norristown (PRR; P&WN), Pa., 66A
North Alberta, Va., 174, NW
North Anna, Va., 143, NE
North Apollo, Pa., 42, SE
North Asbury Park, N.J., 68, SE
North Ave. (Baltimore), (M&PA), Md., 96A
North Bangor, Pa., 51, NE
North Beach Haven, N.J., 84, SE
North Bellwood, Va., 158, NW
North Bend, Pa., 31, NE
North Bergen (NNJ; NYS&W), N.J., 53, NE
North Bergen (NYC; NYS&W), N.J., 53B
North Bessemer (B&LE; URR), Pa., 57, NW
North Branch (B&O; WM), Md., 75, SE
North Branch, N.J., 52, SE
North Broad St., Pa., 82A
North Burgess, Va., 157, SE
North Caldwell, W.Va., 138, NW
North Chester, Va., 158, NW
North Clarion Jct., Pa., 28, SW
North Cochran, Va., 174, NW
North Danville, Va., 171, SW
North Dewitt, Va., 157, SE
North Dinwiddie, Va., 157, SE
North East (NKP), Pa., 2, SW
North East (NYC), Pa., 2, SW
North East, Md., 81, SW
North Elizabeth, N.J., 53, SE
North End, Pa., 32, SE
North Fairmont, W.Va., 89, NE
North Fork (PRR; SCO), Pa., 31, SE
North Fork, W.Va., 150, NW

North Garden, Va., 141, NE
North Girard, Pa., 1, SW
North Hackensack, N.J., 53, NE
North Harford, N.Y., 9, NE
North Hawthorne, N.J., 53, NE
North Hills, Pa., 66, SE
North Jackson, Ohio, 25, SW
North Jct. (PRR), Va., 177A
North Lima, Ohio, 40, NE
North Lithicum, Md., 96, SE
North Long Branch, N.J., 69, NW
North Lynch, Va., 158, NW
North McKenney, Va., 174, NE
North Mountain, Va., 125, SW
North Mountain, W.Va., 77, SW
North Newport News, Va., 160, SW
North Oakland, Pa., 42, NW
North Pelham, N.Y., 54, NW
North Petersburg, Va., 158, SW
North Philadelphia (PRR), Pa., 82A
North Pinch, W.Va., 121, NW
North Pine Grove, Pa., 49, SW
North Point, Pa., 43, NE
North Rahway, N.J., 53, SW
North Rawlings, Va., 174, NW
North River Gap, Va., 125, NE
North Roanoke, Va., 153, NW
North Ryan, Va., 158, SW
North Spencer, N.Y., 8, NE
North Sulger, Pa., 29, SW
North Vineland, N.J., 82, SE
North Wales, Pa., 66, SW
North Warfield, Va., 174, NW
North Warren, Pa., 15, NE
North Wildwood, N.J., 117, NW
North Wilkes-Barre, Pa., 35A
North Woodbury, N.J., 82, NE
Northampton Jct., Pa., 51, SW
Northampton, Pa., 51, SW, 51D
Northbrook, Pa., 81, NE
Northfield, N.J., 100, NE
Northrups, N.J., 37, SE
Northumberland (DL&W; PRR), Pa., 48, NW
Northvale, N.J., 39, SW
Northvale, N.J., 54, NW
Northwest, Va., 177, SE
Norton (INT; L&N; N&W), Va., 164, NE
Norton Branch, Ky., 118, NW
Norton Hollow, N.Y., 6, SE

Norton, W.Va., 107, NW
Norvell, W.Va., 136, NE
Norwood, N.J., 54, NW
Norwood, Pa., 82, NW
Norwood, Va., 141, SW
Nostrand Ave., N.Y., 54, SW
Notch Cliff, Md., 96, NE
Notre Dame, Md., 96, NE
Nottingham, Pa., 80, SE
Nottoway, Va., 156, SE
Novelty, Va., 170, NE
Noxen, Pa., 34, NE
Nuckolls, W.Va., 121, SW
Number Nine, Md., 75A
Numine, Pa., 43, NW
Nunnery, Pa., 77, NE
Nurney, Va., 176, SE
Nutbush, Va., 156, SW
Nutley, N.J., 53, NE
Nuttall, W.Va., 121, SE
NY Cabin, Va., 160, SW
Nyack, N.Y., 39, SW
NYC Jct., Ohio, 25A
Nypen, N.Y., 8, SW

O&B Short Line Jct. (B&O), Pa., 73A
O'Brien, W.Va., 89, SE
O'Donnell Mine, W.Va., 89, NW
O'Donnell, Pa., 29, SW
O'Toole, W.Va., 150, NW
Oak Crest, Md., 96, SW
Oak Grove, Del., 115, SE
Oak Hall, Pa., 46, NW
Oak Hall, Va., 146, NE
Oak Hill Jct., W.Va., 136, NE
Oak Hill, Pa., 57, NW
Oak Hill, W.Va., 136, NE
Oak Lane, Pa., 66, SE
Oak Park, W.Va., 73, SW
Oak Point Yard, N.Y., 54A
Oak Ridge, Pa., 28, SW
Oak Tree, N.J., 53, SW
Oak Tree, Pa., 43, NE
Oakbourne, Pa., 81, NE
Oakcrest, N.J., 100, NE
Oakdale, Pa., 56, NE
Oakfield, Ohio, 25, NW
Oakington, Md., 80, SE
Oakland (PRR), Pa., 41, NW
Oakland, N.J., 38, SE
Oakland, N.J., 82, SW
Oakland, Pa., 26, SW

Oakland, Pa., 43, NW
Oakland, Pa., 65, SE
Oakland, Va., 126, SE
Oakland, W.Va., 91, NW
Oakleigh, Md., 96, NE
Oakley, N.Y., 6, NE
Oaklyn, N.J., 82B
Oakmont, Pa., 42, SW
Oakmont, W.Va., 91, NE
Oakridge, N.J., 38, SW
Oaks (PRR; RDG), Pa., 66, SW
Oakvale (N&W; VGN), W.Va., 151, NW
Oakville, Pa., 58, NW
Oakville, Pa., 62, SW
Oakwood Heights, N.Y., 53, SE
Oakwood, Md., 96, SE
Oakwood, Va., 148, SE
Occoquan, Va., 112, SE
Ocean City 10th St., N.J., 100, NE
Ocean City Gardens, N.J., 100, NE
Ocean Gate, N.J., 84, NE
Ocean Heights, N.J., 100, NE
Ocean, Md., 75, SW
Oceana, Va., 177, NE
Oceana, W.Va., 135, SE
Oceanport, N.J., 68, NE
Oceanside, N.Y., 54, SE
Oceanville, N.J., 101, NW
Ochra, Va., 158, NW
Ocoonita, Va., 163, SE
Octoraro, Md., 80, SE
Odenton, Md., 96, SE
Odessa, N.Y., 8, NW
Oella, Md., 96, NW
Ohio Jct., Ohio, 55, SE, 55A
Ohio Pyle (B&O; WM), Pa., 74, NW
Ohio Steel Jct., Ohio, 25A
Ohio Works Jct. (Erie), Ohio, 25A
Ohio Works, Ohio, 25A
Ohley, W.Va., 121, SW
Ohlton, Ohio, 25, SW
Oil City (Erie; NYC; PRR), Pa., 27, NE
Oil Siding, W.Va., 121, NW
Okonoko, W.Va., 75, SE
Olanta, Pa., 44, NE
Olcott, Ky., 162, SW
Old Bridge, N.J., 68, NW
Old Church, Va., 144, SE
Old Clarendon, Pa., 15, NE
Old Forge, Pa., 35C

Old Fort, Pa., 47, SW
Old Furnace, Pa., 41, NE
Old Gauley, W.Va., 121, SE
Old Greenwich, Conn., 39, SE
Old Maids Lane, N.J., 83, NE
Old Town, Md., 75, SE
Old Town, N.Y., 53, SE
Oldman, N.J., 82, NW
Oldwick, N.J., 52, SE
Olean (Erie), N.Y., 5, SW
Olean (PRR), N.Y., 5, SW
Olean (PS&N), N.Y., 5, SW
Oleika, Ky., 162, SE
Oleopolis, Pa., 27, NE
Olinger, Va., 164, NW
Olive, Va., 129, SW
Olive, W.Va., 107, SW
Oliver, Pa., 73, NE
Olivers Mills, Pa., 35, SW
Olmstead, Pa., 17, NE
Olney, Pa., 66, SE
Olyphant (D&H; NYO&W), Pa., 35D
Olyphant Yard, Pa., 35D
Omal, Ohio, 71, SW
Omar, W.Va., 135, NW
Omrod, Pa., 50, SE
Ona, W.Va., 119, NE
Oneida Jct., Pa., 50A
Oneida, Pa., 42, NW
Oneida, Pa., 49, NE
Oneonta, N.Y., 11, NE
Onley, Va., 146, SE
Onoville, N.Y., 4, SW
Ontario, Va., 173, NW
Ontelaunee Park, Pa., 50, SW
Opekiska, W.Va., 72, SE
Open Fork Jct., W.Va., 121, SE
Ophir, Pa., 45, NE
Opossum Run Jct., Pa., 57C
Oquaga, N.Y., 11, SW
Oradell, N.J., 53, NE
Oral, W.Va., 89, NE
Oramel, N.Y., 5, NE
Orange (DL&W; Erie), N.J., 53, NE
Orange (SOU; VC), Va., 127, SE
Orange Grove, Md., 96, SW
Orangeburg (Erie; NYC), N.Y., 39, SW
Orangeburg, N.J., 37, SE

Orangeville, Ohio, 25, NE
Orangeville, Pa., 34, SW
Orbison, Ohio, 85, NE
Orbisonia, Pa., 61, SW
Orchard St., N.J., 53, NE
Orchard, N.J., 82, NE
Ore Branch Jct., Pa., 62, SE
Ore Hill, Pa., 60, NW
Ore Valley, Pa., 79, NE
Oreland Jct., N.J., 52, NE
Oreland, Ohio, 85, NW
Oreland, Pa., 66, SE
Oreminea, Pa., 60, NW
Oreton, Ohio, 85, SW
Oreton, Va., 164, NW
Orgas, W.Va., 120, SE
Orianna, Va., 159, SE
Orient, Pa., 73, NW
Oriskany, Va., 139, SW
Orkney, Ky., 147, NE
Orlando Jct., W.Va., 105, NE
Ormsbee Road, Pa., 14, NE
Ormsby, Pa., 16, NE
Orniston, Pa., 31, SW
Orphanage, Pa., 48, NE
Orrtanna, Pa., 78, NW
Orson, Pa., 23, NW
Orston, N.J., 82, NE
Ortley, N.J., 84, NE
Orvilla, Pa., 66, NW
Orwig, Pa., 79, NE
Orwigsburg, Pa., 49, SE
Osage, N.J., 82, NE
Osage, W.Va., 72, SE
Osaka, Va., 164, NW
Osborn St., Pa., 59, NW
Osborne, Md., 113, SW
Osborne, Va., 166, NE
Osbornes, Pa., 27, SW
Oscar, Pa., 43, NW
Oscawanna, N.Y., 39, SW
Osceola (B&O; NYC), Pa., 19, NW
Osceola Mills, Pa., 45, NW
Osgood (B&LE; NYC), Pa., 26, NW
Ossining, N.Y., 39, SW
Ostend, Pa., 44, NE
Osterburg, Pa., 59, SE
Oswald, W.Va., 136, NE
Oswayo, Pa., 17, NE
Otego, N.Y., 11, NE
Otisville, N.Y., 37, NE
Otsego, W.Va., 136, SW

Ottawa, Pa., 33, SE
Ottawa, W.Va., 135, NW
Otter Point, Md., 97, NW
Otter, W.Va., 104, SE
Otto Colliery, Pa., 49, SW
Otto, Pa., 48, SW
Ours Hill, W.Va., 106, NE
Outcalt, N.J., 68, NW
Outcrop, Pa., 73, NW
Outwood, Pa., 49, SW
Overall, Va., 110, NW
Overbrook, Del., 116, NE
Overbrook, Pa., 82, NE
Overton Br. Jct., Pa., 57, SE
Owego (DL&W; Erie; LV), N.Y., 9, SW
Owens, Del., 115, NE
Owens, N.J., 37, NE
Owens, W.Va., 120, NE
Owensport, W.Va., 87, SW
Owings Mills, Md., 96, NW
Owings, W.Va., 89, NW
Oxford (DL&W; NYO&W), N.Y., 10, NE
Oxford (LO&S; PRR), Pa., 81, NW
Oxford Furnace, N.J., 52, NW
Oxford, Md., 114, SE
Oxford, N.Y., 38, NE
Oxley, W.Va., 107, SW
Oyster Bay, N.Y., 54, NE
Oyster Point, Va., 159, SE
Ozark, Ohio, 70, NE
Ozone Park, N.Y., 54, SW

P&LE Yard, Ohio, 25A
Pace, Va., 171, SE
Pacific Ave., N.J., 53B
Packer No. 5, Pa., 49, NW
Packerton Yard, Pa., 50, NE
Packsaddle, Pa., 58, NE
Pactolius, Ky., 118, NW
Paden City, W.Va., 71, SW
Padonia, Md., 96, NE
Paeonian Springs, Va., 94, SE
Page, Ky., 162, SE
Page, Va., 149, SW
Page, W.Va., 121, SW
Pageton, W.Va., 150, NW
Paine, Pa., 29, NE
Paint Bank, Va., 138, SW
Paint Creek, Pa., 59, SW
Paint Mills, Pa., 28, NW

Painted Post (DL&W; Erie), N.Y., 7, SE
Painter, Va., 146, SW
Painterville, Pa., 47, SW
Palanka Jct., Pa., 56, SE
Palanka, Pa., 56, NE
Palatine, N.J., 82, SE
Palermo, N.J., 100, SE
Palestine, W.Va., 87, SW
Palisades Park, N.J., 53, NE
Palm, Pa., 65, NE
Palmer, Pa., 58, NE
Palmer, Pa., 58, NW
Palmer, W.Va., 105, SE
Palmerton (CHR; CNJ; LNE), Pa., 50, NE
Palmerton East, Pa., 50, NE
Palmyra, N.J., 66, SE
Palmyra, Pa., 63, NE
Palmyra, Va., 142, NW
Pamplin, Va., 155, NE
Pamunkey, Va., 144, SE
Pan, Ohio, 55, NW
Panama, N.Y., 2, SE
Pancoast, N.J., 83, SW
Pancoast, Pa., 29, SW
Panther, W.Va., 149, NW
Paoli, Pa., 66, SW
Paper Mill (PRR; RDG), Pa., 34, SW
Paper Mill, Pa., 79, NE
Paper Mills, N.J., 51, SE
Paper Mills, Pa., 66, SE
Paramount, Md., 77, SE
Parchment, W.Va., 103, NW
Pardee, Pa., 47, NW
Pardee, Va., 147, SW
Pardee, W.Va., 135, NE
Pardoe, Pa., 26, SE
Park Hill, N.Y., 54, NW
Park Place, Pa., 35D
Park Ridge, N.J., 38, SE
Park, N.Y., 8, SE
Park, W.Va., 89, NE
Parkdale, N.J., 83, SW
Parker, N.Y., 11, NW
Parker, Pa., 27, SE
Parker, Va., 128, NW
Parker's Glen, Pa., 37, NW
Parkerford, Pa., 65, SE
Parkers Landing (B&O; PRR), Pa., 27, SE
Parkersburg, W.Va., 86, NE

Parkesburg, Pa., 81, NW
Parkhead, Md., 76, SE
Parkland, Pa., 67, SW
Parkside, N.Y., 54, SW
Parksley, Va., 146, NE
Parksville, N.Y., 24, NW
Parkton, Md., 79, SE
Parkview, Pa., 57, NW
Parkville, N.J., 82, NE
Parkwood, Pa., 43, SW
Parlin, N.J., 68, NW
Parnassus, Pa., 42, SW
Parr, Va., 139, SW
Parrish, Pa., 29, NW
Parrot, Va., 151, SE
Parryville, Pa., 50, NE
Parsons, Pa., 35A
Parsons, W.Va., 90, SE
Parsonsburg, Md., 133, NW
Pasadena, Md., 96, SE
Pasadena, N.J., 84, NE
Paschall, N.C., 173, SE
Passaic (DL&W; Erie), N.J., 53, NE
Passaic Jct., N.J., 53, NE
Passaic Park, N.J., 53, NE
Patapsco, Md., 79, SW
Patchen, Pa., 44, NW
Pathfork, Ky., 163, NW
Patrick Henry, Va., 155, SW
Patrick Springs, Va., 169, SE
Patterson (DL&W; Erie), N.J., 53, NE
Patterson Ave., N.J., 53B
Patterson Broadway, N.J., 53, NE
Patterson City, N.J., 53, NE
Patterson Creek, W.Va., 75, SE
Patterson Jct., Va., 168, NW
Patterson St., (St. Clair), Pa., 49, SE
Patton Colliery (NYC; PRR), Pa., 44, SE
Patton, Md., 132, NE
Patton, Pa., 44, SE
Patuxent River, Md., 131, NW
Patuxent, Md., 96, SE
Paul Green, W.Va., 136, SW
Paulding, N.J., 82, SW
Pauley, Ky., 147, NE
Paulsboro Draw, N.J., 82, NE
Paulsboro, N.J., 82, NE
Paulson, Ky., 162, SE
Pavonia, N.J., 82B

Paw Paw, W.Va., 76, SW
Pax (C&O; VGN), W.Va., 136, NW
Paxinos (PRR; RDG), Pa., 48, NE
Paxtang, Pa., 63, NW
Paxton Siding, Pa., 48, SW
Paxtonville, Pa., 47, NE
Payne, Pa., 41, NE
Payne, Va., 170, SW
Paynes, Va., 142, NW
Peabody, Ky., 163, NE
Peach Bottom, Pa., 80, NE
Peach Creek, W.Va., 135, NW
Peach Glen, Pa., 62, SE
Peacock, Pa., 56, NW
Peahala, N.J., 84, SE
Peake, Va., 144, SW
Peale, Pa., 45, NE
Peapack (DL&W; P&NJ), N.J., 52, SE
Pearch, Va., 140, SW
Pearisburg, Va., 151, NW
Pearl River, N.Y., 38, SE
Pearre, Md., 76, SW
Pearson, W.Va., 107, NW
Pebble Dell, Pa., 15, SE
Pecan, Pa., 27, NW
Pechin, Pa., 73, NE
Peck's Run, W.Va., 89, SE
Pecks Mill, W.Va., 135, NW
Peckville, Pa., 35D
Pedricktown, N.J., 82, NW
Peekskill, N.Y., 39, NW
Peermont, N.J., 100, SE
Peg, Ky., 134, SW
Pekin, W.Va., 135, SW
Pelham Manor, N.Y., 54, NW
Pelham Parkway, N.Y., 54, NW
Pelham, N.C., 171, SW
Pelham, N.Y., 54, NW
Pelhamwood, N.Y., 54, NW
Pellettown, N.J., 37, SE
Pemberton (C&O; VGN), W.Va., 136A, SE
Pemberton (PRR; UTR), N.J., 83, NE
Pemberton, Va., 142, SE
Pembroke (N&W; VGN), Va., 151, NE
Pembroke, Pa., 82, NW
Pembroke, W.Va., 105, SE
Penbryn, N.J., 83, NW
Pence Springs, W.Va., 137, SE
Pencoyd, Pa., 66B

Pendleton, Va., 143, NW
Penfield (B&O; PRR), Pa., 29, SE
Peninsula, W.Va., 55A
Penitentiary Switch, Pa., 48, NW
Penlan, Va., 142, SW
Penllyn, Pa., 66, SE
Penn Argyl Shops, Pa., 51, NW
Penn Argyl, Pa., 51, NW
Penn Hook, Va., 170, NE
Penn Mary Jct., Md., 96A
Penn, Pa., 57, NE
Pennbrook, Pa., 66, SW
Pennbrook, W.Va., 136, NE
Penncaird, Va., 126, NW
Pennhurst, Pa., 65, SE
Penniman Jct., Va., 159, NE
Pennine, Pa., 57, NE
Pennington, N.J., 67, NW
Pennington, Pa., 45, SE
Pennington, Va., 163, NE
Penn-Mar, Md., 77, SE
Penns Grove, N.J., 82, SW
Penns Neck, N.J., 67, NE
Pennsauken, N.J., 82B
Pennsboro, W.Va., 88, NW
Pennsburg, Pa., 66, NW
Pennsdale, Pa., 33, SW
Pennside, Pa., 13, NW
Pennsylvania Col., Pa., 44, SW
Pennsylvania Furnace, Pa., 45,
SE
Pennsylvania State Asylum, Pa.,
27, NW
Pennsylvania Station (New York),
(LI; LV; NH; PRR), N.Y., 54,
SW
Pennwood Park, Md., 97, SW
Penny Bridge, N.Y., 54, SW
Penny, Ky., 147, NE
Penobscot, Pa., 35, SW
Penola, Va., 144, NW
Penowa, Pa., 55, NE
Penred (P-RSL), N.J., 101, NW
Pentacre, W.Va., 121, NW
Pentland, Pa., 79, NW
Penton, N.J., 82, SW
Pentress, W.Va., 72, SE
Penvir, Va., 151, NW
Peoria, W.Va., 89, NW
Pepper (N&W; VGN), Va., 151,
SE
Pequannock, N.J., 53, NW
Pequea, Pa., 80, NW

Pequest, N.J., 52, NW
Per Se, Pa., 30, SE
Perdix, Pa., 63, NW
Perdue, Va., 157, NE
Perkasie, Pa., 66, NW
Perkins, N.J., 67, SW
Perkintown, N.J., 82, SW
Perkiomen Jct., Pa., 66, SW
Perkiomen, Pa., 66, NW
Perry, Pa., 49, SE
Perryman, Md., 97, NE
Perrysburg, N.Y., 4, NW
Perryville, Md., 80, SE
Perth Amboy (CNJ; LV), N.J., 53,
SW
Perulac, Pa., 61, NE
Pete, Pa., 45, SE
Peters Creek Jct., Pa., 57, NW
Petersburg (ACL; N&W; SAL),
Va., 158, SW
Petersburg, N.J., 100, NE
Petersburg, Pa., 45, SE
Petersburg, W.Va., 108, NE
Peterson, W.Va., 105, NE
Petit, W.Va., 152, NW
Petroleum Center, Pa., 14, SE
Petroleum, W.Va., 87, SW
Petrolia, Pa., 27, SE
Pettus, W.Va., 135, NE
Petty Island Draw, N.J., 82B
Pew, Pa., 28, SW
Peyton, Va., 125, SE
Peytona, W.Va., 120, SE
PF-2, Pa., 63, NW
Phalanx (Erie; NYC), Ohio, 25,
NW
Pheasant Ridge, Va., 177, SE
Phenix, Va., 155, SE
Phico, W.Va., 134, NE
Philadelphia (B&O), Pa., 82A
Philadelphia Broad St. Station
(PRR), Pa., 82A
Philadelphia Reading Terminal
(RDG), Pa., 82A
Philadelphia Suburban Station
(PRR), Pa., 82A
Philipsburg (NYC; PRR), Pa., 45,
NE
Phillip, Va., 165, SW
Phillippi, W.Va., 89, SE
Phillips (B&O; NYC), Pa., 19, NW
Phillipsburg (CNJ; DL&W; PRR),
N.J., 51, SE, 51A

Phillipse Manor, N.Y., 39, SW
Phillipston, Pa., 42, NE
Philmont, Pa., 66, SE
Philpott, Va., 169, NE
Philson, Pa., 75, NW
Phoebus, Va., 160, SW
Phoenix Park, Pa., 49, SW
Phoenix, Md., 79, SE
Phoenix, N.J., 68, NW
Phoenixville (PRR; RDG), Pa.,
65, SE
Picatinny, N.J., 52, NE
Pickens, W.Va., 106, SE
Pickering, Pa., 65, SE
Picture Rocks, Pa., 33, NE
Piedmont, W.Va., 91, NE
Pierce, Pa., 57, NW
Pierce, W.Va., 90, SE
Piermont, N.Y., 39, SW
Pig Point, Va., 177, NW
Pikeland, Pa., 65, SE
Pikesville, Md., 96, NE
Pikeville, Ky., 147, NE
Pilkington, Va., 157, NW
Pilot, Md., 80, SE
Pinch, W.Va., 121, NW
Pine Beach, N.J., 84, NE
Pine Bluff, W.Va., 89, NW
Pine Brook, N.Y., 54, NW
Pine City, N.Y., 8, SW
Pine Creek, Pa., 57A
Pine Creek, W.Va., 134, NE
Pine Crest, N.J., 83, NE
Pine Flats (C&I; CT&D), Pa., 44,
SW
Pine Forest Jct., Pa., 49, SE
Pine Forge, Pa., 65, NE
Pine Furnace, Pa., 43, NW
Pine Grove Furnace, Pa., 62, SW
Pine Grove, Pa., 49, SW
Pine Grove, W.Va., 71, SE
Pine Hill Jct., Pa., 75, NW
Pine Hill, Pa., 75, NW
Pine Island Jct. (Erie; LNE), N.Y.,
38, NW
Pine Island, N.Y., 38, NW
Pine Jct., Pa., 49D
Pine Oak, Va., 148, NE
Pine Run, Pa., 42, SE
Pine St. (Crisfield), Md., 146, NW
Pine Top, Va., 138, SE
Pine Valley, N.J., 83, NW
Pine Valley, N.Y., 8, SW

Pine Valley, Ohio, 55, SW
Pine, Pa., 19, SW
Pine, Pa., 32, SW
Pine, Va., 165, NW
Pinepoca, W.Va., 136, NE
Pinesburg, Md., 77, SW
Pineville, W.Va., 135, SE
Pinewald, N.J., 84, NE
Piney Fork, Ohio, 55, NW
Piney River, Va., 140, SE
Piney, Pa., 28, SW
Pineyville, Ky., 162, NE
Pink Ash Jct., Pa., 50B
Pinners Point, Va., 177A
Pinola, Pa., 61, SE
Pinson Fork Jct., Ky., 134, SW
Pinson, Ky., 134, SW
Pisgah, Va., 149, SE
Pitcairn, Pa., 57, NW
Pitman, N.J., 82, SE
Pittsburg Jct. (P&WV; W&LE),
Ohio, 55, NW
Pittsburgh (B&O; PRR), Pa., 57A
Pittsburgh (P&LE; P&WV), Pa.,
56A
Pittsburgh Allegheny, Pa., 56A
Pittsfield (NYC; PRR), Pa., 15,
NW
Pittston (D&H), Pa., 35B
Pittston (DL&W), Pa., 35B
Pittston (LV), Pa., 35B
Pittstown, N.J., 52, SW
Pittsville, Md., 133, NW
Pittsville, Va., 171, NW
Pixley, N.Y., 5, NW
Pkin, Va., 140, NE
Placid, W.Va., 87, SW
Plainfield, N.J., 53, SW
Plainfield, Pa., 77, NE
Plains (WBE), Pa., 35A
Plains Jct. (CNJ; Erie), Pa., 35B
Plains, N.J., 37, SE
Plainsboro, N.J., 67, NE
Plainsville, Pa., 35B
Plandome, N.Y., 54, NE
Plane 4, Md., 95, NE
Planebrook, Pa., 65, SE
Plank Road, Pa., 79, NE
Plank, Pa., 79, NW
Plantation, Va., 161, SW
Plasterco, Va., 166, NW
Platea, Pa., 13, NW
Plauderville, N.J., 53, NE

Pleasant Creek, W.Va., 89, SE
Pleasant Gap, Pa., 46, NE
Pleasant Hill, N.C., 174, SE
Pleasant Hill, Va., 126, NW
Pleasant Mount, Pa., 23, SW
Pleasant Plains, N.Y., 53, SE
Pleasant Shade, Va., 174, SE
Pleasant Stream, Pa., 33, NW
Pleasant Unity, Pa., 58, SW
Pleasant Valley, N.Y., 7, NW
Pleasant Valley, Pa., 51, SW
Pleasant Valley, Va., 126, NW
Pleasant View, W.Va., 103, NW
Pleasantview, Pa., 62, NW
Pleasantville (AC; P-RSL), N.J., 100, NE
Pleasantville, N.Y., 39, SW
Plum Creek, Pa., 57, NE
Plumville, Pa., 43, NE
Plunkett, W.Va., 135, SE
Plymouth (D&H), Pa., 35A
Plymouth (DL&W), Pa., 35A
Plymouth Jct. (WBC), Pa., 35A
Plymouth Jct., Pa., 66A
Plymouth Meeting (PRR; RDG), Pa., 66, SW
Plymouth, W.Va., 103, SW
Pocahontas, Va., 150, NW
Pocantico Hills, N.Y., 39, SW
Pocket (L&N; SOU), Va., 163, NE
Pocomoke, Md., 132, SE
Pocono Lake, Pa., 36, SW
Pocono Summit (DL&W; WBE), Pa., 36, SW
Pocopson, Pa., 81, NE
Pogue, Pa., 61, SW
Pohick, Va., 112, SE
Point Bridge, Pa., 56A
Point Lick Jct., W.Va., 120, NE
Point Marion, Pa., 73, SW
Point Mills, W.Va., 55, SE
Point of Rocks, Md., 94, NE
Point Pleasant (B&O; NYC), W.Va., 102, NE
Point Pleasant, N.J., 68, SE
Point, W.Va., 107, NW
Poland, Pa., 73, NW
Poland, W.Va., 91, NE
Polk Gap, W.Va., 136, SW
Polk Jct., Pa., 27, NW
Polk, Pa., 27, NW
Polk, W.Va., 86, SW
Pollock Mills, Pa., 72, NE

Polon, W.Va., 149, NW
Pomeroy, Ohio, 85, SE
Pomeroy, Pa., 81, NW
Pomona, N.J., 100, NE
Pomona, N.Y., 38, SE
Pompton Jct., N.J., 38, SW
Pompton Lakes, N.J., 38, SW
Pompton Plains, N.J., 53, NW
Pond Creek Jct., Pa., 35, SW
Pond Eddy, Pa., 37, NW
Pond Fork, W.Va., 121, NW
Pond Hill, Pa., 34, SE
Pond Jct., W.Va., 120, SW
Ponders, Del., 116, NW
Pontoon, Pa., 48, NE
Ponza, Ky., 162, SE
Pool Point, Ky., 148, NW
Poole, Va., 157, SE
Popes Creek, Md., 130, NW
Popeville, Ky., 163, NE
Poplar Island, W.Va., 89, NE
Poplar Springs, Va., 158, NW
Poplar, Md., 97, NW
Port Alleghany (CPA; PRR), Pa., 17, NW
Port Bowkley, Pa., 35A
Port Carbon, Pa., 49, SE
Port Chautauqua, N.Y., 3, SW
Port Chester (NH; NYW&B), N.Y., 39, SE
Port Clinton, Pa., 49, SE
Port Covington, Md., 96A
Port Crane, N.Y., 10, SW
Port Deposit, Md., 80, SE
Port Elizabeth, N.J., 100, NW
Port Hunter, Ohio, 40, SE
Port Indian, Pa., 66, SW
Port Jervis (Erie; NYO&W), N.Y., 37, NE
Port Kennedy, Pa., 66, SW
Port Matilda, Pa., 45, NE
Port Monmouth, N.J., 68, NE
Port Morris, N.Y., 54A
Port Murray, N.J., 52, NW
Port Norris, N.J., 99, SE
Port Orange, N.Y., 37, NE
Port Perry Br. Jct., Pa., 57, NW
Port Perry, Pa., 57B
Port Providence, Pa., 66, SW
Port Reading (CNJ; RDG), N.J., 53, SW
Port Republic, Va., 126, NW

Port Richmond (Philadelphia), Pa., 82B
Port Royal (PRR; TV), Pa., 47, SW
Port Royal, Pa., 57, SW
Port Tobacco, Md., 113, SW
Port Vue, Pa., 57, NW
Port Washington, N.Y., 54, NE
Port Washington, Pa., 66, SE
Port, Pa., 34, SE
Port, Pa., 80, NW
Portage, Pa., 59, NE
Porter Jct., Ky., 147, NW
Porter, Del., 81, SE
Porterfield, Ohio, 86, NE
Porters Falls, W.Va., 71, SW
Porters, Pa., 79, NW
Porters, W.Va., 121, NW
Portersville, Pa., 41, NE
Porterwood, W.Va., 90, SE
Portland (DL&W; LNE), Pa., 51, NE
Portland (NKP), N.Y., 3, NW
Portland (NYC), N.Y., 3, NW
Portland Mills, Pa., 29, NW
Portlock Yard (N&W), Va., 177A
Portlock, Va., 177, NW, 177A
Portsmouth (ACL; A&D; SAL), Va., 177A
Portville (PRR), N.Y., 5, SW
Portville (PS&N), N.Y., 5, SW
Posm, Va., 154, NE
Possum Point, Va., 112, SW
Post Creek, N.Y., 8, SW
Post Mill, W.Va., 89, SE
Post Siding, Pa., 32, SW
Poston, Ohio, 85, NE
Potomac Ave., Md., 77, SE
Potter Brook, Pa., 18, NE
Potter, N.J., 53, SW
Potters, Ky., 118, SE
Pottersville, N.J., 52, SE
Potts Run Jct., Pa., 45, NW
Potts Valley Jct., Va., 151, NE
Pottsgrove, Pa., 48, NW
Pottstown (PRR; RDG), Pa., 65, SE
Pottsville (PRR; RDG), Pa., 49, SE
Pottsville 12th St., Pa., 49B
Pounding Mill, Va., 149, SE
Powell, W.Va., 89, NE
Powellton, W.Va., 121, SW

Powhatan, Ohio, 71, NW
Powhatan, Va., 143, SW
Powhatan, W.Va., 150, NW
Poyntelle, Pa., 23, NW
Pratt, W.Va., 121, SW
Preisser, Pa., 44, SW
Premier, W.Va., 149, NE
Prenter, W.Va., 120, SE
Prescott, Pa., 64, NW
Presho, N.Y., 7, SE
President Station, Md., 96A
President, Pa., 27, NE
Presto, Pa., 56, NE
Preston Park, Pa., 23, NW
Preston, Md., 115, SW
Preston, Va., 170, SW
Preston, W.Va., 90, NE
Prestonia, W.Va., 105, SE
Price Hill Jct., W.Va., 136, NE
Price Hill, W.Va., 136, NE
Price, Ky., 147, NE
Price, Md., 98, SW
Price, N.C., 170, SW
Price, W.Va., 72, SE
Prickett Creek Jct., W.Va., 72, SE
Prickett, W.Va., 72, SE
Prilliman, Va., 169, NE
Primos, Pa., 82, NW
Primrose, Pa., 56, NW
Prince, W.Va., 136, NE
Princes Bay, N.Y., 53, SE
Princess Anne, Md., 132, SE
Princess Anne, Va., 177, NE
Princess, Ky., 118, NE
Princeton Jct., N.J., 67, NE
Princeton, N.J., 67, NE
Princeton, Pa., 41, NE
Princeton, W.Va., 150, NE
Principio, Md., 80, SE
Pritchard, Pa., 19, NE
Pritchard, W.Va., 118, SE
Probst's Mill, Pa., 31, SE
Proctor, W.Va., 71, SW
Profitt, Va., 127, SW
Prompton, Pa., 23, SW
Prospect Ave., N.J., 53, NE
Prospect Plains, N.J., 68, NW
Prospect St. (Trenton), (RDG), N.J., 67, SW
Prospect St., N.J., 100, SE
Prospect, N.J., 53, NE
Prospect, N.J., 82, NW
Prospect, N.Y., 2, SE

Prospect, Va., 155, NE

Prosser, N.Y., 5, SW

Protectory, Pa., 66, SW

Providence (NYO&W), Pa., 35D

Providence Forge, Va., 158, NE

Providence Jct., Va., 177, NW, 177A

Providence-Market St., Pa., 35D

Providence-Provident Mills, Md., 81, SW

PRR Transfer, Pa., 43, SE

Pruden, Ky., 162, SW

Pulaski (Erie; PRR), Pa., 26, SW

Pulaski, Va., 151, SW

Pullen, Pa., 66, NW

Pump House (Erie), Ohio, 25A

Pump Station, Pa., 61, NW

Pumphreys, Md., 96, SE

Pungo, Va., 177, SE

Punxsutawney (B&O; PRR), Pa., 44, NW

Purcell, Va., 164, NW

Purcellville, Va., 94, SE

Purchase Line, Pa., 44, SW

Purdy, Va., 174, NE

Purdy's, N.Y., 39, NE

Purnell, Md., 96, NE

Purseglove, W.Va., 72, SE

Purvis, Va., 176, SE

Putnam, Va., 149, SW

Putnam, W.Va., 120, NW

Putney, Ky., 163, NE

Putney, W.Va., 121, NW

Putneyville, Pa., 43, NW

Putt, W.Va., 136, SW

PV Jct., Md., 77, SW

PW&S Jct., Pa., 58, SE

Pylesville, Md., 80, SW

Quail Run, N.J., 84, NE

Quakake (LV; RDG), Pa., 49D

Quaker City, Ohio, 70, NW

Quaker Ridge, N.Y., 4, SW

Quaker Ridge, N.Y., 54, NW

Quakertown (QB; RDG), Pa., 66, NW

Qualey, Ohio, 86, NW

Quantico, Va., 112, SW

Quarrier, W.Va., 121, SW

Quarry Jct., Pa., 51, NE

Quarry, Md., 80, SE

Quarry, Va., 165, NW

Quarryville (LO&S; PRR), Pa., 80, NE

Quarryville, N.J., 37, NE

Queen Anne, Md., 115, NW

Queen Jct., Pa., 42, NW

Queen Lane, Pa., 66, SE

Queen Shoals, W.Va., 121, NW

Queen, Pa., 60, NW

Queens Village, N.Y., 54, SE

Queenstown, Md., 114, NE

Quemahoning Jct., Pa., 59, SW

Queponco, Md., 133, NW

Quick, W.Va., 121, NW

Quickle, W.Va., 105, SW

Quicksburg, Va., 109, SE

Quincy, Pa., 77, NE

Quinland, W.Va., 120, SW

Quinnimont, W.Va., 136, NE

Quinnwood, Pa., 16, SE

Quinton, N.J., 82, SW

Quinton, Va., 144, SE

Quinwood, W.Va., 122, SE

Raccoon, Pa., 56, NW

Race, N.J., 82B

Rachel, W.Va., 72, SW

Racine, W.Va., 120, SE

Radburn (Fairlawn), N.J., 53, NE

Radcliff, Ohio, 85, SW

Radcliffe, W.Va., 72, SE

Radebaugh, Pa., 57, NE

Radford, Va., 151, SE

Radix, N.J., 83, SW

Radnor, Pa., 66, SW

Radnor, W.Va., 119, SW

Ragan, Del., 81A

Rahns, Pa., 66, SW

Rahway, N.J., 53, SW

Rainbow, Ohio, 87, NW

Rainelle Jct., W.Va., 137, NW

Rainelle, W.Va., 137, NW

Raines, Va., 156, NW

Rainey Branch Jct., Pa., 73, NE

Raitt, Va., 148, NE

Ralco, Va., 173, NE

Raleigh, W.Va., 136, NE

Rallston, Md., 80, SW

Ralphton, Pa., 58, SE

Ralston, Pa., 20, SW

Ramage, W.Va., 135, NW

Ramapo, N.Y., 38, SE

Rambo, Pa., 66, SW, 66A

Ramey, Pa., 45, NW

Ramp, Pa., 41, SW

Ramsay, Ohio, 71, NW

Ramsaytown, Pa., 28, SE

Ramsey (INT; N&W), Va., 164, NE

Ramsey, N.J., 38, SE

Ramsey, Pa., 32, NW

Rand, Pa., 57A

Randall, W.Va., 73, SW

Randolph, Md., 95, SE

Randolph, N.Y., 4, SW

Randolph, Pa., 58, SE

Randolph, Va., 172, NE

Ranger, W.Va., 119, SE

Ranicks, Pa., 49, NE

Rankin (B&O; P&LE), Pa., 57A

Ransom, Pa., 35, NW

Ranson, W.Va., 94, NW

Raphine, Va., 140, NE

Rapidan, Va., 127, NE

Raritan River, N.J., 68, NW

Raritan, N.J., 52, SE

Rasselas, Pa., 16, SE

Ratchford, Ohio, 85, SW

Ratcliff, Ky., 148, NW

Rathbone, N.Y., 7, SW

Rathbun, Pa., 30, NW

Rattling Run, Pa., 63, NE

Rauchs, Pa., 50, SW

Raughts, Del., 116, NW

Rausch Creek, Pa., 49, SW

Rausch Gap, Pa., 48, SE

Rausch Gap, Pa., 63, NE

Raven Rock, Pa., 66, NE

Raven Rock, W.Va., 87, NE

Raven, Va., 149, SW

Ravencliff, W.Va., 136, SW

Ravenswood, W.Va., 103, NW

Ravensworth, Va., 112, NE

Rawl, W.Va., 134, SE

Rawlings (B&O; WM), Md., 75, SW

Rawlings, Va., 174, NW

Ray, Pa., 43, SW

Ray, Pa., 48, NE

Ray, W.Va., 152, NW

Rayland, Ohio, 55, NE

Raymitton, Pa., 27, NW

Raymond City, W.Va., 120, NW

Raymond, Pa., 18, NW

Raymond, Pa., 28, SE

Rays, Ky., 162, NW

Raywood, W.Va., 124, NW

RC Jct., Ky., 148, NW

Rea, Pa., 56, NW

Read, N.J., 83, SW

Reader, W.Va., 71, SE

Reading (PRR), Pa., 65, NW

Reading Center, N.Y., 8, NW

Reading Jct., Pa., 59, SW

Reading Outer Station (RDG), Pa., 65, NW

Readville, Ky., 118, SW

Ready, Pa., 43, NW

Ream, W.Va., 149, NE

Reamer, W.Va., 121, NW

Rectortown, Va., 111, NW

Red Ash, Va., 149, SW

Red Ash, W.Va., 136, NE

Red Bank, N.J., 68, NE

Red Bank, Pa., 42, NE

Red Creek Jct., W.Va., 107, NE

Red Diamond, Ohio, 85, NW

Red Hill, Pa., 66, NW

Red Hill, Va., 141, NE

Red House (Erie), N.Y., 4, SW

Red House (PRR), N.Y., 4, SW

Red House, W.Va., 103, SW

Red Jacket, W.Va., 134, SE

Red Lion, Del., 81, SE

Red Lion, Pa., 79, NE

Red Onion Switch, N.Y., 38, NW

Red Run, W.Va., 90, SE

Red Star, W.Va., 136, NE

Red Warrior Jct., W.Va., 121, SW

Redden, Del., 116, SW

Redmond, Pa., 26, SE

Redstone Jct., Pa., 73, NE

Reduction, Pa., 57, SW

Redwood, Va., 153, SW

Reed (CCK; NYC), W.Va., 120, NE

Reed Crossing, N.J., 83, NW

Reed Jct., Pa., 43, SW

Reed, Ohio, 55, NW

Reed, Pa., 48, NE

Reed, Va., 168, NW

Reed, W.Va., 120, NE

Reeder, Pa., 67, NW

Reeders, Pa., 36, SW

Reedsville (KV; PRR), Pa., 46, SE

Reedsville, W.Va., 73, SW

Reedy, W.Va., 104, NW

Reega, N.J., 100, NE

Reels Mill, Md., 95, NW

Rees Mill, Pa., 72, NE

Reese, W.Va., 137, NW

Reesedale, Pa., 43, NW
Reeves, N.J., 83, NW
Refton, Pa., 80, NE
Regan Jct., Pa., 44, SW
Rego Park, N.Y., 54, SW
Rehoboth, Del., 116, SE
Reid, Md., 77, SE
Reids Grove, Md., 115, SW
Reidsburg, Pa., 28, SW
Reiner, Pa., 48, SE
Reinholds, Pa., 64, NE
Reisers, Pa., 48, NE
Reitz, Pa., 29, SW
Relay, Md., 96, SE
Relay, Pa., 79, NE
Rembrandt, Pa., 43, SE
Remington, Va., 111, SW
Remoh, Pa., 43, SE
Renchans, N.Y., 7, NW
Renfrew, Pa., 42, NW
Renick, W.Va., 138, NW
Rennerdale, Pa., 56, NE
Reno (Erie; NYC), Pa., 27, NE
Renovo, Pa., 31, NW
Rensford, W.Va., 121, NW
Renton, Pa., 42, SE
Repaupo, N.J., 82, NW
Republic Hill, Pa., 51, SW
Republic, Ky., 148, NW
Republic, Pa., 73, NW
Republic, W.Va., 136, NW
Reservoir, Pa., 60, NW
Reservoir, Va., 159, SE
Reston, W.Va., 120, SW
Retort, Pa., 45, NE
Retreat, Pa., 34, SE
Reusens, Va., 154, NE
Revell, Md., 96, SE
Revere Colliery, Pa., 73, NW
Revloc (C&I; PRR), Pa., 59, NW
Rex, Ohio, 118, NE
Rexford, Ohio, 55, NW
Rexis Branch Jct., Pa., 59, NW
Rexis, Pa., 59, NW
Rexville, N.Y., 6, SE
Reybold, Del., 81, SE
Reynolds, Pa., 50, SW
Reynoldsdale, Pa., 59, SE
Reynoldsville, Pa., 29, SW
Reynoldsville, W.Va., 89, NW
Rhea, Ky., 163, NW
Rheems, Pa., 63, SE
Rheims, N.Y., 7, NW

Rhoads, Pa., 31, SW
Rhodell, W.Va., 136, SW
Rhodesdale, Md., 115, SW
Rhodesville, Va., 128, NW
Ribold, Pa., 42, NW
Rice, Va., 156, NW
Rice's Landing, Pa., 72, NE
Riceville, Pa., 14, NW
Rich Creek, Va., 151, NW
Rich Ford, W.Va., 90, SE
Richam, Ky., 147, NE
Richard, W.Va., 73, SW
Richardson Jct., Pa., 49, SW
Richburg, N.Y., 5, SE
Richenbach, Pa., 65, NW
Richfol, Pa., 56, NE
Richford, N.Y., 9, NE
Richland, N.J., 100, NW
Richland, Pa., 64, NW
Richlands, Va., 149, SW
Richmond (RARR), Va., 144, SW,
    144A
Richmond Hill, N.Y., 54D
Richmond Valley, N.Y., 53, SE
Richmond, N.J., 82, SE
Richmond, Pa., 77, NW
Richwood (B&O; CRB&L), W.Va.,
    122, SE
Rick Creek Jct., W.Va., 121, SE
Ricketts, Pa., 34, NW
Riddlesburg, Pa., 60, SE
Riddleton, N.J., 82, SW
Ridenour, W.Va., 121, SW
Riders, N.J., 83, NE
Riderville, Pa., 16, NE
Riderwood, Md., 96, NE
Ridge Branch Jct., Pa., 43, SE
Ridge Road, Pa., 66A
Ridgedale, W.Va., 92, NE
Ridgefield Park (NYC; NYS&W),
    N.J., 53, NE
Ridgefield, N.J., 53, NE
Ridgely, Md., 115, NW
Ridgeview, W.Va., 120, SW
Ridgeway, N.Y., 39, SW
Ridgeway, Va., 93, NE
Ridgeway, Va., 170, SW
Ridgway (B&O; PRR), Pa., 29,
    NE
Ridley Park (B&O; PRR), Pa.,
    82, NW
Riegelsville (PRR; QB), N.J., 51,
    SE

Rife, Va., 148, NE
Riffe, W.Va., 137, SE
Rift, W.Va., 149, NE
Rigby, Ohio, 70, NW
Riker, Pa., 44, NW
Rileyville, Va., 110, NW
Rillton, Pa., 57, NE
Rimel, W.Va., 124, SW
Rimersburg, Pa., 28, SW
Rimerton, Pa., 42, NE
Rinehart, W.Va., 88, NE
Ringdale, Pa., 34, NW
Ringgold, Pa., 43, NE
Ringgold, Va., 171, SW
Ringoes, N.J., 67, NW
Ringtown, Pa., 49, NE
Ringwood Jct., N.J., 38, SW
Ringwood, N.J., 38, SW
Rinker's, Pa., 36, SW
Rinn, Pa., 43, NE
Rio Grande (P-RSL; WJ&S), N.J.,
    100, SW
Rio, Va., 127, SW
Ripley Landing, W.Va., 103, NW
Ripley, N.Y., 2, NE
Ripley, W.Va., 103, NE
Ripplemead (N&W; VGN), Va.,
    151, NE
Rippon, W.Va., 94, SW
Rising Springs, Pa., 46, NE
Rising Sun, Md., 80, SE
Risley, N.J., 100, NW
Ritchie, Pa., 31, NE
Ritter, W.Va., 149, NW
Ritts, Pa., 27, SE
Rivanna Jct. (C&O), Va., 144A
Rivanna, Va., 142, SE
River Edge, N.J., 53, NE
River St., N.J., 53, NE
River Valley, Pa., 42, SW
River, Ohio, 40, SE
River, W.Va., 107, SW
Riverdale, Md., 113, NW
Riverdale, N.J., 53, NW
Riverside Jct. (NYO&W), Pa.,
    35D
Riverside Jct. (NYS&W), Pa., 35D
Riverside Jct., N.Y., 4, SE
Riverside Jct., Pa., 35D
Riverside, Conn., 39, SE
Riverside, N.J., 53, NE
Riverside, N.J., 67, SW
Riverside, N.Y., 10, SW

Riverside, N.Y., 54, NW
Riverside, Va., 140, NW
Riverton (N&W; SOU), Va., 110,
    NE
Riverton, N.J., 66, SE
Riverton, Pa., 57, NW
Riverville, Va., 141, SW
Rivesville Jct. (B&O; WM), W.Va.,
    72, SE
Rixley, Va., 129, SW
Roach, W.Va., 119, NE
Roanoke (N&W; VGN), Va., 153,
    NW
Roanville, W.Va., 106, NW
Roaring Branch, Pa., 20, SW
Roaring Creek Connection, W.Va.,
    107, NW
Roaring Fork, Va., 164, NE
Roaring Run, Pa., 58, SW
Roaring Spring, Pa., 60, NW
Robanna, N.J., 82, SE
Robbins, Del., 116, NW
Robbins, Pa., 57, NW
Robbinsville, N.J., 67, SE
Robbs, Pa., 56, NE
Roberts, Md., 98, SW
Roberts, Pa., 26, SW
Roberts, Pa., 74, NE
Roberts, W.Va., 90, SW
Robertsburg, W.Va., 103, SW
Robertsdale, Pa., 60, SE
Robesonia, Pa., 64, NE
Robey, Va., 128, NE
Robey, W.Va., 89, NW
Robeyville, Ohio, 55, SW
Robinette, W.Va., 135, NW
Robinhood, W.Va., 135, NE
Robinson Creek, Ky., 147, NE
Robinson, Md., 96, SE
Robius, Va., 143, SE
Robson, W.Va., 121, SE
Rochelle Park, N.J., 53, NE
Rochester, Pa., 41, SW
Rock Castle, Va., 143, SW
Rock Cliff, Ky., 162, NW
Rock Creek, Ohio, 12, SW
Rock Creek, W.Va., 120, SW
Rock Cut, Pa., 51, SW
Rock Enon Springs, Va., 93, SW
Rock Forge, W.Va., 73, SW
Rock Glen, Pa., 49, NE
Rock House, W.Va., 135, NE
Rock Jct., Pa., 35D

Rock Point, Pa., 41, NW
Rock Rift, N.Y., 11, SE
Rock Run, Pa., 45, NE
Rock Run, W.Va., 88, NW
Rock Tavern, N.Y., 38, NE
Rock, Md., 80, SE
Rock, Pa., 49, NE
Rock, Pa., 49, SW
Rock, W.Va., 150, NE
Rock-a-Walkin, Md., 132, NE
Rockaway (CNJ; DL&W), N.J., 52, NE
Rockaway Jct., N.J., 52, NE
Rockaway Park, N.Y., 54, SW
Rockburn, Pa., 79, NE
Rockdale Jct., Md., 79, SW
Rockdale, N.Y., 11, NW
Rockdale, Pa., 42, NW
Rockdale, Pa., 50, SE
Rockdale, Pa., 82, NW
Rockdale, W.Va., 55, NE
Rocketts Jct., Va., 144A
Rockfish (NEA), Va., 141, NE
Rockfish (SOU), Va., 141, NW
Rockhill, Pa., 66, NW
Rockhouse, Ky., 148, NW
Rockland, Del., 81, NE
Rockland, Ohio, 86, NE
Rockland, Pa., 27, SW
Rockland, W.Va., 137, SE
Rocklyn, Pa., 65, SW
Rockmere, Pa., 27, NE
Rockport, Pa., 50, NW
Rocks, Md., 80, SW
Rocks, W.Va., 92, NE
Rockstream, N.Y., 8, NW
Rockton, Pa., 29, SE
Rockton, W.Va., 105, SW
Rockview, Pa., 46, NW
Rockville (PRR; RDG), Pa., 63, NW
Rockville Center, N.Y., 54, SE
Rockville, Md., 95, SE
Rockville, N.Y., 5, NE
Rockwells Mills, N.Y., 11, NW
Rockwood, Pa., 74, NE
Rocky Gap, Va., 150, SE
Rocky Hill, N.J., 67, NE
Rocky Mount (F&P; N&W), Va., 153, SW
Rocky Point, Va., 139, SE
Rocky Ridge (EMIT; WM), Md., 78, SW

Rocky Ridge, Pa., 60, SE
Roda, Va., 164, NW
Rodbourne, N.Y., 8, SE
Roddy, Pa., 62, NE
Rodemer, W.Va., 90, NE
Roderfield, W.Va., 149, NE
Rodes, W.Va., 136, NE
Roebling, N.J., 67, SW
Roeders, Pa., 49, SE
Roelofs, Pa., 67, SW
Roemer, Pa., 26, NW
Rogers (PRR), Ohio, 40, SW
Rogers (Y&S), Ohio, 40, NE
Rogers Mills, Pa., 74, NW
Rohrbough, W.Va., 106, NW
Rohrerstown, Pa., 64, SW
Rohrersville, Md., 94, NE
Rolfe, Pa., 29, NE
Rolfe, W.Va., 150, NW
Rollyson, W.Va., 105, NE
Rome, Ohio, 12, SW
Romney Jct., W.Va., 92, NW
Romney, W.Va., 92, NW
Ronceverte, W.Va., 138, SW
Ronco, Pa., 73, NW
Ronda, W.Va., 121, SW
Roneys Point, W.Va., 55, SE
Ronk, Pa., 64, SE
Rook, Pa., 56A
Roosevelt, N.J., 68, NE
Roots, Pa., 45, SW
Rorer, W.Va., 123, SW
Roscoe, N.Y., 24, NW
Roscoe, Pa., 57, SW
Rose Hill, Va., 163, SW
Rose Lake, Pa., 18, NW
Rose Point, Pa., 41, NE
Rose Siding, Pa., 28, SE
Rose, Pa., 28, SE
Rosebank, N.Y., 53, SE
Rosebud, Pa., 44, NE
Rosebud, W.Va., 89, NW
Roseby Rock, W.Va., 71, NE
Rosedale, Md., 96, NE
Rosedale, N.J., 83, SW
Rosedale, N.Y., 54, SE
Rosedale, Pa., 81, NE
Rosegarden, Pa., 62, SE
Roseland, N.J., 53, NW
Roselle Park, N.J., 53, SW
Roselle, N.J., 53, SW
Rosemont, Pa., 66, SW
Rosemont, W.Va., 89, NE

Rosemount, Ohio, 25, SW
Rosenhayn, N.J., 99, NE
Roses Mill, Va., 141, SW
Roseville Ave., N.J., 53, NE
Roseville Ave., N.J., 53A
Roseville Siding, N.J., 52, NE
Roslyn, N.Y., 54, NE
Roslyn, Pa., 66, SE
Rosney, Va., 142, SW
Ross Colliery, Pa., 73, NE
Ross Common, Pa., 51, NW
Ross Run Jct., Pa., 15, SW
Ross, Va., 129, NW
Rossburg, N.Y., 5, NE
Rossiter (B&O; NYC; PRR), Pa., 44, NW
Rossland, Ky., 162, NW
Rosslyn, Pa., 56A
Rossmore, W.Va., 135, NW
Rossmoyne, Pa., 63, SW
Rosston, Pa., 42, SE
Rossville, Md., 97, NW
Rossyln (PRR; W&OD), D.C., 112, NW
Rostraver, Pa., 57, SW
Roth, Va., 149, SW
Rothsville, Pa., 64, SW
Roulette, Pa., 17, NE
Round Bay, Md., 96, SE
Round Bottom, Va., 150, NE
Round Hill, Va., 94, SW
Round Top (B&O), W.Va., 76, SE
Round Top (WM), Md., 76, SE
Round Top, Pa., 19, SW
Round Top, Pa., 78, NE
Roup, Pa., 57A
Rouseville, Pa., 27, NE
Rouzer, W.Va., 121, NE
Rovnar, Pa., 59, NW
Rowe, Ky., 148, NW
Rowe's Run, Pa., 57, SW
Rowena, Pa., 59, SW
Rowlands, Pa., 36, NE
Rowlandville, Md., 80, SE
Rowlesburg (B&O; R&S), W.Va., 90, NE
Rowley, Pa., 56, NE
Roxana, Ky., 147, SW
Roxburg, N.J., 51, NE
Roxbury, Md., 77, SE
Roxbury, Va., 158, NE
Roxton, Pa., 67, SW
Roy, Pa., 63, SE

Royal Branch Jct., Pa., 57, SW
Royal Oak, Md., 114, NE
Royal, Pa., 73, NW
Royal, W.Va., 136, NE
Royce (CNJ; LV), N.J., 52, SE
Royer, Pa., 60, NW
Royersford, Pa., 65, SE
Roys, N.J., 37, SE
RS&G Jct., W.Va., 103, NW
Ruffsdale, Pa., 57, SE
Rugby, Va., 127, SW
Rulon Road, N.J., 82, NW
Rum Jct., W.Va., 135, NW
Rumbaugh, Pa., 58, SW
Rumer, W.Va., 103, SW
Rummel, Pa., 59, SW
Rummerfield, Pa., 21, SW
Runnemeade, N.J., 82, NE
Rupert Jct., W.Va., 137, NE
Rupert, Pa., 49, NW
Rupert, W.Va., 137, NE
Rural Bridge, Pa., 42, SW
Rural Retreat, Va., 167, NW
Rush Run (PRR; W&LE), Ohio, 55, NE
Rush, Ky., 118, NW
Rushland, Pa., 66, NE
Russell City, Pa., 16, SW
Russell, Pa., 15, NE
Russell, W.Va., 73, SW
Russellton, Pa., 42, SW
Russellville, W.Va., 122, SW
Rustburg, Va., 154, NE
Ruth, W.Va., 150, NW
Ruthby, Del., 81, SE
Rutherford, N.J., 53, NE
Rutherford, Pa., 63, NW
Rutherford, Va., 126, NW
Rutherford, W.Va., 87, SE
Rutherglen, Va., 144, NW
Ruthford, Pa., 59, NE
Rutland, Ohio, 85, SE
Ruxton, Md., 96, NE
Ryan, Va., 158, SW
Rydal, Pa., 66, SE
Ryde, Pa., 61, NW
Rye (NH; NYW&B), N.Y., 54, NE
Ryers, Pa., 66, SE

S&M Jct., Pa., 73, NW
Sabaraton, W.Va., 73, SW
Sabillasville, Md., 78, SW
Sabine, W.Va., 135, SE

Sabinsville, Pa., 18, NE
Sabot, Va., 143, SE
Sabula (B&O; PRR), Pa., 29, SE
Sacco, Pa., 36, NW
Saegers, Pa., 33, SW
Saegersville, Pa., 50, SE
Saegertown, Pa., 13, SE
Safe Harbor, Pa., 80, NW
Sagamore, Pa., 43, NE
Sagamore, Pa., 58, SW
Sago, W.Va., 106, NE
Sagon, Pa., 49A
Sailors Snug Harbor, N.Y., 53, SE
Saint Brides, Va., 177, SE
Salamanca (Erie), N.Y., 4, SE
Salamanca (PRR), N.Y., 4, SE
Salem (N&W; VGN), Va., 152, NE
Salem, N.J., 82, SW
Salem, Ohio, 40, NW
Salem, Pa., 26, NW
Salem, Pa., 29, SE
Salem, W.Va., 88, NE
Salesville, Ohio, 70, NW
Salford, Pa., 66, NW
Salida, Pa., 57, NW
Salina, Pa., 43, SW
Salineville, Ohio, 40, SW
Salisbury (PRR), Md., 132, NE
Salisbury Jct., Pa., 74, NE
Salisbury Mills, N.Y., 38, NE
Salisbury, N.Y., 54, SE
Salisbury, Va., 139, SW
Salisbury, W.Va., 106, SW
Salix, Pa., 59, NE
Salona, Pa., 32, SW
Salt Rock, W.Va., 119, NE
Saltillo, Pa., 60, SE
Saltpetre, Va., 139, SE
Saltpetre, W.Va., 118, SE
Saltsburg, Pa., 43, SW
Saltsburg, Pa., 58, NW
Saltville, Va., 166, NW
Sam, Va., 150, SW
Sample Run, Pa., 43, SE
Sample, Pa., 42, SW
Sanatoga, Pa., 65, SE
Sand Creek, W.Va., 119, SE
Sand Fork, W.Va., 121, NE
Sand Lick Branch Jct., W.Va., 89, NE
Sand Patch (WM), Pa., 75, NW
Sand Run Jct., W.Va., 91, SW
Sand Run, W.Va., 106, NE

Sand, Pa., 46, NE
Sanderson, W.Va., 121, NW
Sands Eddy, Pa., 51, NE
Sands, W.Va., 134, SE
Sandusky, N.Y., 5, NW
Sandy Creek, Pa., 57, NW
Sandy Huff, W.Va., 149, NW
Sandy Level, Va., 170, NE
Sandy Lick, Pa., 42, SE
Sandy Ridge, Pa., 45, NE
Sandy Run, Pa., 35, SW
Sandy Run, Pa., 45, SW
Sandy Run, Pa., 60, SE
Sandy Summit, W.Va., 103, NE
Sandy Valley, Pa., 29, SW
Sandy, Pa., 27, NW
Sandy, W.Va., 121, NW
Sandyville, W.Va., 103, NE
Sanitaria Springs, N.Y., 10, SW
Sanitorium, Md., 78, SW
Santee, Pa., 51, SW
Sarah Ann, W.Va., 135, SW
Sarah Furnace, Pa., 27, SE
Sarahsville, Ohio, 70, NW
Saratoga, Ohio, 40, NE
Sarita, W.Va., 136, NW
Sarver, Pa., 42, SE
Satterfield Jct., Pa., 34, NW
Sattes, W.Va., 120, NW
Saunders Range, Md., 96, SE
Saunders, W.Va., 135, NE
Savage Branch, Ky., 118, NE
Savage Factory, Md., 96, SW
Savage, Md., 96, SW
Savan, Pa., 44, NW
Savona (DL&W; Erie), N.Y., 7, NE
Saxe, Va., 172, NE
Saxton, Pa., 60, SE
Saybrook (NKP; NYC), Ohio, 12, NW
Saylorsburg Jct., Pa., 51, NW
Saylorsburg, Pa., 51, NW
Sayre, Pa., 20, NE
Sayreville Jct., N.J., 68, NW
Sayreville, N.J., 68, NW
SB, W.Va., 55, NE
SC Booth, Pa., 50C
SC Jct., Pa., 20, NE
SC Tower, Ohio, 70, NW
Scaggs, W.Va., 135, SW
Scale Jct., Pa., 50, NW
Scale Siding, Pa., 35, SW

Scalp Level, Pa., 59, SW
Scarboro, Md., 133, SW
Scarborough, N.Y., 39, SW
Scarbro, W.Va., 136, NE
Scarsdale, N.Y., 39, SW
Scarsdale, N.Y., 54, NW
Scary, W.Va., 120, NW
Scenery Hill, Pa., 56, SE
Schell, W.Va., 91, NW
Schenley, Pa., 42, SE
Schick, Ohio, 55, SW
Schickshinny, Pa., 34, SE
Schoenersville, Pa., 51, SW
Scholes, N.Y., 6, NW
School House, Pa., 46, NW
Schrader, W.Va., 121, NW
Schraders, Pa., 46, SE
Schuman Run Jct., Pa., 59, NW
Schuyler, Pa., 33, SE
Schuyler, Va., 141, NE
Schuylkill Haven (PRR; RDG), Pa., 49, SE
Schwenksville, Pa., 66, NW
Scio (Erie; W&B), N.Y., 6, SW
Scotch Valley, Pa., 49, NW
Scotia, Pa., 46, NW
Scotland, Pa., 77, NE
Scotland, Va., 159, SW
Scott Haven, Pa., 57, NW
Scott, Pa., 56, NE
Scott, W.Va., 120, NW
Scottdale (B&O; PRR), Pa., 57, SE
Scottdale Jct., Pa., 57, SE
Scottglen, Pa., 58, NE
Scotti Jct., W.Va., 123, NW
Scottsburg, Va., 172, NW
Scottsville, Va., 142, NW
Scranton (CNJ; NYO&W), Pa., 35C
Scranton (D&H), Pa., 35C
Scranton (DL&W), Pa., 35C
Scranton (Erie), Pa., 35C
Scranton (L&WV), Pa., 35C
Scranton Shops, Pa., 35C
Scranton-Mulberry St., Pa., 35C
Sea Cliff, N.Y., 54, NE
Sea Girt, N.J., 68, SE
Sea Isle City, N.J., 100, SE
Sea Isle Jct., N.J., 100, SW
Seaboard, Va., 149, SW
Seabright, N.J., 69, NW
Seabrook, Md., 113, NW

Seabrook, N.J., 82, SE
Seacoast, Va., 158, SW
Seaford, Del., 115, SE
Seaman, W.Va., 104, NW
Seanor, Pa., 59, SW
Seaside Heights, N.J., 84, NE
Seaside Park, N.J., 84, NE
Seaside, N.Y., 54, SW
Seaview, N.J., 100, NE
Sebrell, Va., 175, NE
Secane, Pa., 82, NW
Secaucus Yard, N.J., 53B
Secaucus, N.J., 53B
Seco, Ky., 147, SE
Secoal, W.Va., 135, NW
Security (B&O; WM), Md., 77, SE
Security Jct., Md., 77, SE
Sedgwick Ave., N.Y., 54, NW
Sedgwick, Pa., 66, SE
Sedley, Va., 176, NW
See, W.Va., 118, SE
Seebert, W.Va., 123, SE
Seeley Creek, N.Y., 8, SW
Seeleyville, Pa., 23, SW
Seiple, Pa., 50, SE
Seitzville, Pa., 79, NE
Selden, Va., 142, SE
Selinsgrove Jct., Pa., 48, NW
Selinsgrove, Pa., 48, NW
Sell, Pa., 78, NE
Sellersville, Pa., 66, NW
Selma, Va., 139, NW
Seminary, Va., 112, NE
Semmel Siding, Pa., 50, NE
Semmel, Pa., 50, NE
Semora, N.C., 171, SE
Seneca, Va., 154, SE
Seneca, W.Va., 73, SW
Senecaville, Ohio, 70, NW
Sergeant, Pa., 16, SW
Sergent, Ky., 147, SW
Seth, W.Va., 120, SE
Seven Pines, Pa., 62, NW
Seven Stars, Pa., 78, NW
Seven-Mile Ford, Va., 166, NE
Severn, Md., 96, SE
Severn, N.C., 175, SE
Severna Park, Md., 96, SE
Severnside, Md., 97, SW
Sewall's Point, Va., 177, NW
Seward, Pa., 58, NE
Sewaren, N.J., 53, SW
Sewell, Md., 97, NW

Slate Hill, N.J., 38, NW
Slate Hill, Pa., 80, SW
Slate Jct., Pa., 50, NE
Slate Run, Pa., 31, NE
Slate Valley, Pa., 50, NE
Slate, W.Va., 87, SW
Slatefield, Pa., 51, NW
Slatington (LNE; LV) Pa., 50, NE
Sledds Point, Md., 96, SE
Sleepy Creek, W.Va., 76, SE
Slickville, Pa., 57, NE
Sligo, Pa., 28, SW
Sloatsburg, N.Y., 38, SE
Smalley, N.J., 83, NE
Smethport (B&O; PRR; PS&N),
    Pa., 17, NW
Smicksburg, Pa., 43, NE
Smiley, Ky., 163, SW
Smith Summit, Pa., 29, SW
Smith, Pa., 58, NE
Smith, Pa., 80, NW
Smith's Ferry, Pa., 40, SE
Smith's Landing, N.J., 100, NE
Smith's Mills, N.Y., 3, NE
Smith's Run, Pa., 16, SE
Smithboro (Erie; LV), N.Y., 9, SW
Smithburg, W.Va., 88, NE
Smithdale, Pa., 57, SW
Smithers, W.Va., 121, SW
Smithfield St., Pa., 56A
Smithfield, Ohio, 55, NE
Smithfield, Pa., 73, NW
Smiths, Pa., 79, NW
Smithsburg, Md., 77, SE
Smithton (B&O; P&LE), Pa., 57,
    SE
Smithville, N.J., 83, NE
Smithville, Pa., 80, NW
Smock, Pa., 57, SW
Smock, Pa., 73, NW
Smoke Run, Pa., 45, NW
Smokeless, W.Va., 150, NW
Smoketown, Md., 77, SE
Smoky Ordinary, Va., 174, NE
Smyrna, Del., 98, NE
Smyser, Pa., 79, NE
Snodes, Ohio, 40, NW
Snow Flake, W.Va., 137, SE
Snow Fork Jct., Ohio, 85, NE
Snow Hill, Md., 133, SW
Snow Hill, W.Va., 120, NE
Snow Shoe (NYC; PRR), Pa., 31,
    SW

Snow Shoe Int., Pa., 46, NW
Snowden, Pa., 57, NW
Snowden, Va., 140, SW
Snyder, N.Y., 9, SW
Snyder, Va., 125, SE
Snyders, Pa., 50, SW
Snydersburg, Pa., 28, NW
Snydertown (PRR; RDG), Pa., 48,
    NE
Sodom, Pa., 57C
Soldier Run, Pa., 29, SW
Solitude, Va., 139, SE
Somerdale, N.J., 82, NE
Somerfeld, Pa., 74, NW
Somers Point, N.J., 100, NE
Somerset, N.J., 67, NW
Somerset, Pa., 58, SE
Somerset, Va., 127, SE
Somerton, Pa., 66, SE
Somerville, N.J., 52, SE
Sonestown, Pa., 33, NE
Sophia, W.Va., 136A, SW
Soudan, Va., 172, SE
Souderton, Pa., 66, NW
Soughton, Pa., 58, SE
Sours Mills, Pa., 51, NW
South Alberta, Va., 174, NW
South Altoona, Pa., 60, NW
South Amboy (NY&LB; PRR),
    N.J., 68, NW
South Amboy (RR), N.J., 68, NW
South Anna, Va., 144, NW
South Beach, N.Y., 53, SE
South Bellwood, Va., 158, NW
South Boston (N&W; SOU), Va.,
    172, SW
South Bound Brook, N.J., 53, SW
South Branch, Pa., 21, SW
South Bridgeton, N.J., 99, NE
South Brownsville, Pa., 57, SW
South Buckannon, W.Va., 106, NE
South Burgess, Va., 157, SE
South Camden, N.J., 82B
South Carbon, W.Va., 121, SW
South Carnegie, Pa., 56A
South Charleston, W.Va., 120, NE
South Chester, Va., 158, NW
South Clarksville, Va., 172, SE
South Clearfield, Pa., 30, SW
South Cochran, Va., 174, NW
South Costen, Md., 132, SE
South Danville, Pa., 48, NE
South Dayton, N.Y., 3, NE

South Dennis, N.J., 100, SW
South Dewitt, Va., 157, SE
South Dinwiddie, Va., 157, SE
South Duquesne, Pa., 57, NW
South Elizabeth, N.J., 53, SE
South End (PRR; RF&P), Va.,
    112A
South Fayette, W.Va., 121, SE
South Fork, Pa., 59, NW
South Gap, Va., 150, SE
South Glassboro, N.J., 82, SE
South Gloucester, N.J., 82, NE
South Greensburg, Pa., 57, NE
South Heights, Pa., 41, SE
South Hill, Va., 173, SE
South Lakewood, N.J., 68, SW
South Lynch, Va., 158, NW
South Malden, W.Va., 120, NE
South McKenney, Va., 174, NE
South Modena, Pa., 81, NW
South Montrose, Pa., 22, NW
South Mountain, Pa., 64, NE
South Norfolk, Va., 177, NW
South Nyack, N.Y., 39, SW
South Orange (C&O), Va., 127,
    SE
South Orange, N.J., 53, SW
South Orangeburg, N.J., 37, SE
South Patterson, N.J., 53, NE
South Pemberton, N.J., 83, NE
South Penn Jct., Pa., 77, NE
South Petersburg, Va., 158, SW
South Philadelphia, Pa., 82, NE
South Plainfield, N.J., 53, SW
South Point, Ohio, 118, NE
South Rawlings, Va., 174, NW
South Richmond, Va., 158, NW
South River, N.J., 68, NW
South River, Va., 140, NW
South Ruffner, W.Va., 120, NE
South Rush Run, W.Va., 136, NE
South Ryan, Va., 158, SW
South Scranton, Pa., 35C
South Seaville, N.J., 100, SW
South Side Jct., W.Va., 136, NE
South Side Wye, Del., 81A
South St., N.J., 53A
South Sulger, Pa., 28, SE
South Unadilla, N.Y., 11, NW
South Vandalia, N.Y., 4, SE
South Vineland, N.J., 99, NE
South Warfield, Va., 174, NW
South Westville, N.J., 82, NE

South Williamsport, Pa., 33, SW
South Witmer Branch, Pa., 44,
    SE
South Woodstown, N.J., 82, SW
South Yard S. End, Va., 158, NW
South Yard, Va., 144A
Southampton, Pa., 66, SE
Southfields, N.Y., 38, SE
Southport Jct., N.Y., 8, SW
Southport, N.Y., 8, SW
Southside, Pa., 80, NW
Southview, Pa., 56, NW
Southwood, Del., 81, NE
Sowash, Pa., 58, NW
Spangler (NYC; PRR), Pa., 44,
    SW
Spangler, W.Va., 106, SE
Spanglers Mill, W.Va., 136A, SE
Sparkill, N.Y., 39, SW
Sparks, Md., 79, SE
Sparrowbush, N.Y., 37, NE
Sparrows Point Jct., Md., 96A
Sparrows Point, Md., 97, SW
Sparta Jct. (L&HR), N.J., 37, SE
Sparta, N.J., 37, SE
Spartansburg, Pa., 14, NE
Speeceville, Pa., 63, NW
Speedway, Pa., 73, NE
Speedwell, Va., 167, NE
Speers Ferry (CC&O; SOU), Va.,
    164, SE
Spelter, W.Va., 89, NW
Spencer, N.Y., 8, SE
Spencer, N.Y., 9, SW
Spencer, Va., 169, SE
Spencer, W.Va., 104, NW
Spielman, Md., 77, SW
Spitler, Va., 110, SW
Splash Dam, Pa., 34, NE
Splashdam, Va., 148, SW
Splint, Ky., 163, NE
Spotswood, N.J., 68, NW
Spottswood, Va., 140, NE
Spout Spring, Va., 155, NW
Sprague, W.Va., 136A, NE
Sprankle Mills, Pa., 28, SE
Spray Beach, N.J., 84, SE
Spray, N.C., 170, SW
Spread, W.Va., 121, NE
Sprigg, W.Va., 134, SE
Spring City, Pa., 65, SE
Spring Creek, Pa., 14, NE
Spring Creek, Va., 125, NE

Spring Creek, W.Va., 138, NW
Spring Fork, W.Va., 120, NE
Spring Gap, Md., 75, SE
Spring Garden St., Pa., 82A
Spring Grove (PRR; WM), Pa., 79, NW
Spring Grove, Va., 159, SW
Spring Hill, W.Va., 120, NE
Spring Lake, N.J., 68, SE
Spring Mill (PRR; RDG), Pa., 66A
Spring Mount, Pa., 66, NW
Spring St., N.J., 53, SE
Spring Valley, N.Y., 38, SE
Springboro (B&LE; PRR), Pa., 13, NW
Springdale Cemetery, Conn., 39, SE
Springdale, Conn., 39, SE
Springdale, N.J., 83, NW
Springdale, Pa., 42, SW
Springdale, W.Va., 137, NW
Springdell, Pa., 81, NW
Springfield Garden Jct., Pa., 49, SE
Springfield Gardens, N.Y., 54, SW
Springfield, Md., 113, NW
Springfield, N.J., 53, SW
Springfield, Pa., 13, NW
Springfield, Va., 112, NE
Springfield, W.Va., 92, NE
Springton, W.Va., 150, NE
Springtown, N.J., 51, SE
Springtown, Pa., 51, SW
Springvale, Pa., 79, NE
Springville, Pa., 22, SW
Springwood, Va., 139, SE
Sproul, Pa., 60, NW
Sproul, W.Va., 120, NW
Spruce Grove, Pa., 80, NE
Spruce Hill, Pa., 62, NW
Spruce Low Gap, W.Va., 122, NE
Spruce Valley, W.Va., 135, NW
Spruce, Pa., 30, SE
Spur Jct., W.Va., 149, SE
Spuyten Duyvil, N.Y., 54, NW
Squire, W.Va., 149, SE
St. Albans, N.Y., 54D
St. Albans, W.Va., 120, NW
St. Asaph, Va., 112, NE
St. Benedict, Pa., 44, SE
St. Charles (L&N; SOU), Ky., 163, NE

St. Charles, Pa., 43, NW
St. Charles, Va., 164, NW
St. Clair (PRR; RDG), Pa., 49, SE
St. Clair, Va., 150, SW
St. Clairsville, Ohio, 55, SW
St. Davids, Pa., 66, SW
St. Denis, Md., 96, SE
St. George, N.Y., 53, SE
St. Georges, Md., 96, NW
St. Helena, Md., 96A
St. James, Md., 77, SW
St. Joe, Pa., 42, NW
St. Johns Park, N.Y., 53, SE
St. Joseph's Academy, Md., 78, SW
St. Joseph's, N.Y., 24, SE
St. Just, Va., 128, NW
St. Leonard, Pa., 67, SW
St. Martin's, Pa., 66, SE
St. Martins, Md., 133, NE
St. Mary's (PS&N; PRR), Pa., 29, NE
St. Mary's, W.Va., 87, SE
St. Michael, Pa., 59, NE
St. Michaels, Md., 114, NE
St. Nicholas, Pa., 49, NE
St. Paul (CC&O; N&W), Va., 165, NW
St. Peters, Pa., 65, SE
St. Petersburg, Pa., 27, SE
St. Vincent, Pa., 58, NW
St. Xavier, Pa., 58, NW
Staffordville, N.J., 84, SW
Stahl Point, Md., 96, SE
Staley, Va., 167, NW
Stambaugh (B&O; PRR), Pa., 73, NE
Stamford, Conn., 39, SE
Stanaford, W.Va., 136, NE
Stanard, N.Y., 6, SW
Standard Jct., Pa., 42, NW
Standard, W.Va., 121, SW
Standing Stone, Pa., 21, SW
Standing Stone, W.Va., 87, SW
Stanhope, Ohio, 25, NE
Stanhope, Pa., 49, SW
Stanley Jct., Pa., 29, SW
Stanley, Pa., 29, SW
Stanley, Va., 109, SE
Stanleyville, Ohio, 87, NW
Stanton (B&O; PRR), Del., 81, SE

Stanton, N.J., 52, SW
Stanton, Pa., 28, SE
Stanwick Ave. (Moorestown), N.J., 83, NW
Stapleton, N.Y., 53, SE
Stapleton, Va., 155, NW
Star City, W.Va., 73, SW
Star Rock, Pa., 80, NW
Starbrick (NYC; PRR), Pa., 15, NE
Starford, Pa., 44, SW
Stark, W.Va., 135, NE
Starkey, Ohio, 70, NE
Starkey, Va., 153, SW
Starlight, Pa., 16, SE
Starlight, Pa., 23, NW
Starners, Pa., 62, SE
Starnes, Va., 164, SE
Starr, Ohio, 85, NW
Startzman, Md., 77, SW
Starucca, Pa., 23, NW
State Asylum, Md., 96, NE
State College, Pa., 46, NW
State Farm, Va., 143, SW
State Line, Ohio, 12, NE
State Line, Pa., 2, SW
State Line, Pa., 75, SW
State Road, Del., 81, SE
State Road, Ohio, 25, NW
State School, N.Y., 38, NW
State St., N.J., 82B
Staunton (CHW; C&O), Va., 125, SE
Steamburg, N.Y., 4, SW
Stearnes, Va., 142, SE
Steelton (PRR; RDG), Pa., 63, SW
Steiner, W.Va., 107, NW
Steins, Pa., 49, SW
Stell, Pa., 63, NW
Stella, Va., 169, SE
Stelton, N.J., 53, SW
Stemmer's Run, Md., 97, NW
Stemphleytown, Va., 126, NW
Stenton, Pa., 66, SE
Stephens City, Va., 93, SE
Stephens, Va., 164, NE
Stephenson, Va., 93, SE
Stepney, Md., 97, NE
Sterling Forest, N.J., 38, SW
Sterling Run, Pa., 30, NE
Sterling, Va., 95, SW
Sterling, W.Va., 136A, SE

Sterlington (Erie; SMR), N.Y., 38, SE
Sterrett, Va., 140, NW
Steuben, Pa., 51, SW
Steubenville (PRR; W&LE), Ohio, 55, NE
Stevens Point, Pa., 22, NE
Stevens, N.J., 67, SW
Stevens, Pa., 48, NE
Stevens, Pa., 64, SE
Stevenson, Md., 96, NE
Stevenson, Pa., 21, SW
Stevensville, Md., 114, NW
Stewart Manor, N.Y., 54, SE
Stewart, Ohio, 86, NW
Stewart, W.Va., 86, SE
Stewarton (B&O; WM), Pa., 74, NW
Stewarts, Pa., 50, NW
Stewartstown, Pa., 79, NE
Stewartsville, N.J., 51, SE
Stewartsville, Ohio, 55, SW
Stewartsville, Va., 153, SW
Steyer, W.Va., 91, NW
Stickney, W.Va., 135, NE
Stickneys, N.Y., 7, NW
Stier, Pa., 51, NE
Still Pond, Md., 97, NE
Stillwater, N.J., 37, SW
Stillwater, Pa., 34, SW
Stillwell, W.Va., 123, SE
Stirling, N.J., 53, SW
Stirrat, W.Va., 134, SE
Stock Yards, Pa., 57A
Stockdale, Pa., 57, SW
Stockertown, Pa., 51, NW
Stockholm, N.J., 37, SE
Stockley, Del., 116, SW
Stockton, N.J., 67, NW
Stockton, Pa., 50, NW
Stockton, Va., 170, SE
Stoke Park, Pa., 51, SW
Stokes, Va., 142, SE
Stokesdale, Pa., 19, NW
Stokesland, Va., 171, SW
Stokesville, Va., 125, NE
Stollings, W.Va., 135, NW
Stone Branch, W.Va., 134, NE
Stone City, Ohio, 70, NW
Stone Cliff, W.Va., 136, NE
Stone Coal Jct., W.Va., 136, SW
Stone Coal, W.Va., 134, NW

Stone Glen, Pa., 63, NW
Stone Harbor, N.J., 100, SW
Stone House, W.Va., 89, NE
Stone Mountain, Va., 153, SE
Stone, Ky., 134, SW
Stone, Pa., 19, SW
Stoneboro (NYC; PRR), Pa., 26, NE
Stonebreaker, Md., 94, NE
Stonega, Va., 164, NW
Stoneham (PRR; TIV), Pa., 15, NE
Stonemont, Pa., 49, SE
Stones Branch, Ky., 148, NW
Stonewall, W.Va., 136, NE
Stoney, Pa., 63, NW
Stony Bottom, W.Va., 124, NW
Stony Creek (RDG), Pa., 49, SE
Stony Creek, Va., 175, NW
Stony Ford, N.Y., 38, NW
Stony Fork Jct., Ky., 162, SE
Stony Point, N.Y., 39, SW
Stony Point, Pa., 13, SW
Stony Run, Md., 96, SE
Stoops Ferry, Pa., 41, SE
Storage, Pa., 41, SE
Storrs Jct., Pa., 35D
Stotesbury (C&O; VGN), W.Va., 136A, SW
Stoutsburg, N.J., 67, NE
Stover, Pa., 45, SE
Stoverdale, Pa., 63, NE
Stowe (PRR; RDG), Pa., 65, SE
Stowe, Ohio, 87, NW
Stoyestown, Pa., 59, SW
Strafford, Pa., 66, SW
Straight Creek, Ky., 162, NE
Straiton Ave., N.Y., 54, SW
Strange Creek, W.Va., 105, SW
Strangford, Pa., 58, NE
Strasburg Jct. (B&O), Va., 110, NW
Strasburg, Pa., 80, NE
Strasburg, Va., 110, NW
Stratford, N.J., 82, NE
Strathmore, N.J., 100, SE
Strathmore, Va., 142, SW
Strattonville, Pa., 28, SW
Strawberry Ridge, Pa., 33, SE
Strawbridge, Pa., 33, NE
Street, Md., 80, SW
Streets Run Branch SW, Pa., 57A
Stric, Va., 148, NE

Strickler, Pa., 63, SE
Stringer (W&LE), Ohio, 55, SE
Stroudsburg (DL&W; NYS&W), Pa., 51, NE
Stroudsburg Jct., Pa., 51, NE
Strounds, Ohio, 25, NW
Struble, Pa., 46, NW
Strum, Pa., 73, NW
Struthers (P&LE; PRR), Ohio, 25, SE
Struthers, Pa., 15, NE
Stuart, Va., 169, SW
Stuart's Draft, Va., 125, SE
Studa, Pa., 56, NW
Stull, Pa., 34, NE
Stull, Va., 139, SW
Stumptown, Va., 166, NW
Sturgeon, Pa., 56, NE
Sturgisson, W.Va., 73, SW
Succasunna, N.J., 52, NE
Sudan, Pa., 57, SW
Sudbrook, Md., 96, NE
Sudlersville, Md., 98, SW
Suedberg, Pa., 49, SW
Suffern, N.Y., 38, SE
Suffern, N.Y., 38, SE
Suffolk (ACL; A&D; NS; N&W; SAL; VGN), Va., 176, SE
Sugar Creek Jct., W.Va., 136, NE
Sugar Creek, Pa., 27, NW
Sugar Grove, Va., 167, NW
Sugar Hill, Pa., 29, SW
Sugar Loaf, N.Y., 38, NW
Sugar Run, Pa., 16, NW
Suider, W.Va., 90, NE
Suiter, Va., 150, SE
Sulphur Mines, Va., 128, SW
Sulphur Springs, Pa., 62, NE
Summerdale, N.Y., 2, SE
Summerdale, Pa., 66, SE
Summerfield, Md., 96, NE
Summerfield, Ohio, 70, NW
Summerhill, Pa., 59, NW
Summerlee, W.Va., 121, SE
Summerville (LEF&C; PRR), Pa., 28, SE
Summit–Young's Gap, N.Y., 24, NE
Summit (DL&W; RV), N.J., 53, SW
Summit (PRR; W&NB), Pa., 34, SW
Summit Grove, Pa., 79, SE

Summit Hill, Pa., 50C
Summit Park, N.Y., 38, SE
Summit Point, W.Va., 94, NW
Summit Siding, Pa., 56, SE
Summit, Ky., 118, NE
Summit, N.Y., 5, SE
Summit, N.Y., 6, SW
Summit, N.Y., 9, NW
Summit, N.Y., 10, NE
Summit, Ohio, 85, NW
Summit, Pa., 35, NE
Summit, Pa., 42, NW
Summit, Pa., 45, NE
Summit, Pa., 49, SE
Summit, Pa., 49B
Summit, Pa., 51, NW
Summit, Pa., 58, SE
Summit, Pa., 64, NW
Summit, Va., 129, SW
Summit, Va., 149, SE
Summit, Va., 157, NE
Summit, Va., 167, NW
Summit, W.Va., 106, SW
Summit, W.Va., 124, NW
Summit, W.Va., 136A, NW
Summit, W.Va., 150, NE
Summitville, Ohio, 40, SW
Sumnerville, Pa., 13, NW
Sun Flower, W.Va., 104, NW
Sunbright, Va., 164, SW
Sunbury (PRR; RDG), Pa., 48, NW
Suncrest, W.Va., 106, SW
Sunny Side, Va., 156, NE
Sunnybrook, Pa., 66, SE
Sunnyside Jct., N.Y., 54B
Sunnyside, Md., 96, SW
Sunset Hills, Va., 112, NW
Superior, Pa., 56A
Superior, W.Va., 149, NE
Suplee (PRR; RDG), Pa., 65, SW
Surbaugh, W.Va., 122, SW
Surber, Va., 139, SW
Surry Court House, Va., 159, SW
Surveyor (C&O; VGN), W.Va., 136A, NW
Surveyor, Pa., 30, SW
Surveyor, W.Va., 136A, NW
Suscon, Pa., 35, NE
Susquehanna Jct., Pa., 44, NW
Susquehanna Transfer (NNJ; NYS&W), N.J., 53B
Susquehanna, Pa., 22, NE

Susquehanna, Pa., 80, SW
Sussex (LNE; NYS&W), N.J., 37, SE
Sussex Jct., N.J., 37, SE
Sutherland, Va., 157, SE
Sutherland, Va., 164, NE
Sutherlin, Va., 171, SE
Sutton, Ky., 148, NW
Sutton, Md., 96A
Sutton, Pa., 28, SE
Sutton, W.Va., 105, SE
SW Cabin, W.Va., 135, NW
Swain, N.J., 100, SW
Swains (CNYW; Erie; PS&N), N.Y., 6, NW
Swan Creek, Md., 80, SE
Swandale, W.Va., 122, NW
Swanson, Pa., 74, NE
Swanton, W.Va., 91, NE
Swanville (NKP), Pa., 1, SE
Swanville (NYC), Pa., 1, SE
Swart, Pa., 72, NW
Swarthmore, Pa., 82, NW
Swartswood Jct., N.J., 37, SW
Swartswood, N.J., 37, SW
Swartswood, N.Y., 8, SE
Swatara Jct., Pa., 49, SW
Swatara, Pa., 63, NE
Swedeland, Pa., 66A
Swedesboro, N.J., 82, SW
Swedesford Road, Pa., 65, SE, 65A
Sweet Briar, Va., 140, SE
Sweet Hall, Va., 145, SW
Swengel, Pa., 47, NE
Swimley, Va., 93, SE
Swiss (NYC), W.Va., 121, SE
Swiss Jct. (NF&G), W.Va., 121, SE
Swissvale, Pa., 57A
Switch Box No. 6, Del., 81A
Switzer, W.Va., 135, NW
Swope, Va., 125, SE
Swords Creek, Va., 149, SW
Swoyer, Pa., 50, SE
Syberton, Pa., 44, SE
Sycamore, Pa., 72, NE
Sycamore, Va., 154, SW
Sycamore, W.Va., 92, SW
Syfert, Pa., 65, NW
Sygan (PRR; P&WV), Pa., 56, NE
Sykesville (B&O; B&S), Pa., 29, SW

Sykesville, Md., 96, NW
Syler, Pa., 45, NW
Sylmar, Md., 80, SE
Sylvan, Pa., 62, NE
Sylvatus, Va., 168, NW
Syosset, N.Y., 54, NE

Taber, Va., 154, SE
Table Rock, Pa., 78, NE
Tablers, Va., 93, NE
Tabor, Pa., 66, SE
Tacoma (INT; N&W), Va., 164, NE
Tacoma, Ohio, 70, NE
Tacony, Pa., 66, SE
Tad, W.Va., 121, NW
Tainters, Pa., 16, NE
Tait, Pa., 28, SE
Takoma Park, D.C., 112, NE
Talcott, W.Va., 137, SW
Tallmans, N.Y., 38, SE
Tally Ho, Pa., 16, NE
Tamanend (CNJ; RDG), Pa., 49D
Tamanend, Pa., 49D
Tamaqua, Pa., 50, NW
Tamcliff, W.Va., 135, SW
Tamroy, W.Va., 136, NE
Taneytown, Md., 78, SE
Tangas (PRR), Pa., 31, SE
Tangascootac Br. Jct., Pa., 32, SW
Tank, Pa., 34, SE
Tannersville, Pa., 36, SW
Tannery, Md., 79, SW
Taplin, W.Va., 135, NW
Tappan (NNJ; NYC), N.Y., 39, SW
Tarentum, Pa., 42, SW
Tarr, Pa., 57, SE
Tarrtown, Pa., 42, NE
Tarrytown Heights, N.Y., 39, SW
Tarrytown, N.Y., 39, SW
Tasley, Va., 146, SE
Tatesville, Pa., 60, SW
Taylor & Gould, W.Va., 106, NE
Taylor Jct., Pa., 35C
Taylor Yard, Pa., 35C
Taylor, Pa., 35C
Taylor, Va., 153, SW
Taylor, W.Va., 92, SW
Taylor's Valley, Va., 166, SE
Taylorstown, Pa., 56, SW
Taylorville, Va., 144, NW
Tazewell, Va., 149, SE
Teaneck, N.J., 53, NE

Tearing Run Jct., Pa., 43, SE
Teas, Va., 167, NW
Teays, W.Va., 120, NW
Teegarden, Ohio, 40, NW
Teejay, Ky., 162, NE
Tekram, W.Va., 134, NE
Telford, Pa., 66, NW

Tellesburg, Ohio, 55, SW
Teman, Va., 143, NE
Temple (PRR; RDG), Pa., 65, NW
Templeton, Pa., 43, NW
Ten Mile, W.Va., 106, NE
Tenafly, N.J., 54, NW
Tenbury, W.Va., 55, NE
Tennant, N.J., 68, NW
Tennant, W.Va., 72, SE
Tennerton, W.Va., 106, NE
Terminal Jct. (B&O), W.Va., 55A
Terminal Jct. (B&O; PRR), Ohio, 55A
Terminus, N.Y., 54, SE
Terra Alta, W.Va., 90, NE
Terra Cotta, D.C., 113, NW
Terral, Pa., 41, NW
Terry Jct., W.Va., 136, NE
Terry, W.Va., 136, NE
Texas, Md., 96, NE
Thacker Mines, W.Va., 134, SE
Thacker, W.Va., 134, SE
Thaxton, Va., 153, NE
Thayer, W.Va., 136, NE
The Dock, W.Va., 123, SE
The Hunt, Pa., 82, NW
The Plains, Va., 111, NW
The Raunt, N.Y., 54, SW
Thelma, Va., 127, SE
Thiells, N.Y., 38, SE
Third St. (Wellsville), Ohio, 40, SE
Thomas Mill, Pa., 44, SE
Thomas, Pa., 43, SE
Thomas, Pa., 56, SE
Thomas, Va., 148, NE
Thomas, Va., 158, NW
Thomas, W.Va., 91, SW
Thomaston, Pa., 49, SW
Thomasville, Pa., 79, NW
Thompson Run (PRR), Pa., 41, NW
Thompson, Del., 81, SW
Thompson, Md., 114, SE
Thompson, Ohio, 40, SE

Thompson, Pa., 22, NE
Thompson, Pa., 57B
Thompsontown, Pa., 47, SE
Thompsonville, Pa., 56, NE
Thorn Hill, Ohio, 25, SE
Thornburg, Pa., 56A
Thorncliff, Va., 143, SW
Thorndale, Pa., 81, NE
Thornton Jct. (NKP; PRR), Pa., 13, NW
Thornton, W.Va., 90, NW
Thornwood, N.Y., 39, SW
Thorny Creek, W.Va., 123, NE
Thorofare, N.J., 82, NE
Thorofare, Va., 111, NE
Thorpe, W.Va., 149, NE
Three Bridges (CNJ; LV), N.J., 52, SW
Three Mile, W.Va., 121, NW
Three Springs, Pa., 61, SW
Thrifton, D.C., 112, NW
Thurmond, W.Va., 136, NE
Thurmont (PECo.; WM), Md., 78, SW
Tiadaghton, Pa., 19, SW
Tice, W.Va., 72, SE
Ticoal, W.Va., 122, NE
Tidal Jct., Pa., 43, NW
Tide, Md., 96A
Tidedale, Pa., 43, SE
Tidioute, Pa., 15, SW
Tierney, W.Va., 136, SW
Tiffany, Pa., 22, NW
Tigheview, W.Va., 107, NW
Tilly Foster, N.Y., 39, NE
Tiltonville, Ohio, 55, NE
Timbar, W.Va., 135, SW
Timber Ridge, Va., 140, NW
Timberville, Va., 109, SW
Timblin, Pa., 43, NE
Timonium, Md., 96, NE
Tin Bridge, Pa., 20, NW
Tin Bridge, Va., 151, SW
Tinsley, Ky., 162, NW
Tioga (Erie; NYC), Pa., 19, NE
Tioga Center (Erie; LV), N.Y., 9, SW
Tioga Jct., Pa., 19, NE
Tioga, Pa., 82A
Tioga, W.Va., 122, NE
Tiona (PRR; TIV), Pa., 15, SE
Tionesta (PRR; S&T), Pa., 28, NW

Tip Top, N.Y., 6, SW
Tip Top, Va., 150, SW
Tippicanoe, Pa., 57, SW
Tiprell, Tenn., 162, SE
Tipton, Pa., 45, SW
Titusville (NYC; PRR), Pa., 14, SE
Titusville, N.J., 67, NW
Toano, Va., 159, NW
Tobaccoville, Va., 156, NE
Toby No. 3, Pa., 29, NE
Tobyhanna, Pa., 36, SW
Tod, Pa., 44, SE
Toland, Pa., 62, SE
Toler Spur, W.Va., 135, SE
Toler, Ky., 134, SW
Tolleys (C&O; VGN), W.Va., 136A, NW
Tollgate, W.Va., 88, NW
Tom's Brook, Va., 110, NW
Tom's River (CNJ; PRR), N.J., 84, NE
Tome, Md., 80, SE
Tomhicken (LV; PRR), Pa., 49, NE
Tomkins Cove, N.Y., 39, NW
Tomkins, Pa., 19, NE
Tomkinsville, N.Y., 53, SE
Tomlin, N.J., 82, NW
Toms Creek, Va., 165, NW
Tookland, Va., 148, NE
Top Siding, W.Va., 137, NW
Topton, Pa., 50, SE
Torch Hill, Ohio, 86, SE
Torchlight, Ky., 118, SE
Toronto, Ohio, 55, NE
Torpedo, Pa., 15, NW
Torrance, Pa., 58, NE
Torresdale, N.J., 67, SW
Toshes, Va., 171, NW
Tottenville, N.Y., 53, SW
Totz, Ky., 163, NE
Toughkenamon, Pa., 81, NW
Towaco, N.J., 53, NW
Towanda, Pa., 21, NW
Towanda-Washington St., Pa., 21, NW
Tower City, Pa., 48, SE
Tower Hill, N.Y., 39, SW
Tower Hill, N.Y., 53, SE
Towers, Va., 148, NW
Town Creek, Md., 75, SE
Towners (NH; NYC), N.Y., 39, NE

Townley, N.J., 53, SE
Townsbury, N.J., 52, NW
Townsend Inlet, N.J., 100, SE
Townsend, Del., 98, NE
Townsend, Va., 161, SW
Towson Heights, Md., 96, NE
Towson, Md., 96, NE
Towson, W.Va., 90, NW
Tracy's Switch, W.Va., 123, NW
Trade City, Pa., 43, NE
Trafford, Pa., 57, NW
Tralee, W.Va., 136, SW
Trammel, Va., 148, SW
Tranquility, N.J., 52, NW
Transfer (Erie), Pa., 26, NW
Transfer (PRR), Pa., 26, NW
Transfer, Pa., 57, SE
Transit Bridge, N.Y., 5, NE
Trap Rock, Pa., 65, SW
Trappe, Md., 114, SE
Traymore, Pa., 66, SE
Trees Mills, Pa., 57, NE
Trego, Va., 174, SE
Treichler (CNJ; LV), Pa., 50, SE
Tremley, N.J., 53, SE
Tremont Jct., Pa., 49, SW
Tremont, N.Y., 54, NW
Tremont, Pa., 49, SW
Tremont, Pa., 57, SW
Trent, Pa., 74, NW
Trenton (PRR; RDG), N.J., 67, SW
Tresslars, Pa., 36, NW
Treveskyn, Pa., 56, NE
Trevilian, Va., 127, SE
Trevortown, Pa., 48, NE
Trevose, Pa., 67, SW
Trexler, Pa., 50, SW
Trexlertown, Pa., 50, SE
Tridelphia, W.Va., 55, SE
Trimble, Ohio, 85, NE
Trindle Spring, Pa., 62, SE
Triplett, W.Va., 107, SW
Tripoli, Pa., 50, SW
Tripp, W.Va., 134, NW
Trosper, Ky., 162, NW
Trotter, Pa., 57, SE
Trout Brook, N.Y., 23, NE
Trout Dale, Va., 167, SW
Trout Run, Pa., 32, NE
Troutville, Va., 153, NW
Trowbridge, Pa., 20, NW
Troy, Pa., 20, NW

Troy, Va., 142, NE
Trucksville, Pa., 35, NW
Truemans, Pa., 15, SE
Trunkeyville, Pa., 15, SW
Truxall, Pa., 42, SE
Tryonville, Pa., 14, SW
Trythall, Pa., 65, SE
Tub Run, W.Va., 90, SE
Tuckahoe, Md., 115, NW
Tuckahoe, N.J., 100, NW
Tuckahoe, N.Y., 54, NW
Tuckahoe, Va., 143, SE
Tuckahoe, W.Va., 138, NW
Tuckerdale, N.C., 166, SE
Tuckerton, N.J., 84, SW
Tuckerton, Pa., 65, NW
Tudor Terrace, N.J., 100, NE
Tuggle, Va., 156, NW
Tullytown, N.J., 67, SW
Tulpehocken, Pa., 66, SE
Tunkhannock (LV), Pa., 22, SW
Tunnel Hill, Ohio, 70, NE
Tunnel Jct., Pa., 50C
Tunnel Siding, W.Va., 121, SW
Tunnel Siding, W.Va., 136A, SE
Tunnel, N.Y., 10, SE
Tunnel, Ohio, 86, NE
Tunnel, Pa., 44, SE
Tunnel, W.Va., 107, NW
Tunnel, W.Va., 136, NE
Tunnelton (B&O; WVN), W.Va., 90, NE
Tunnelton, Pa., 58, NW
Tunstall, Va., 144, SE
Tuquon, Pa., 80, NW
Turbotville, Pa., 33, SW
Turkey Knob, W.Va., 136, NE
Turkey, Pa., 27, SE
Turner Douglas, W.Va., 91, NW
Turner, Md., 96, SE
Turnpike, Md., 96, NW
Turnpike, Pa., 79, NE
Turrell's Mill, Pa., 34, NW
Turtle Creek, Pa., 57, NW
Turtle Point, Pa., 17, NW
Tuscarora, Md., 95, SW
Tuscarora, Pa., 47, SW
Tuscarora, Pa., 49, NE
Tusten, N.Y., 23, SE
Tuxedo, N.Y., 38, SE
Tweed, Del., 81, SW
Tweedale, Pa., 80, NE
Twin Rocks, Pa., 59, NW

Two Mile, W.Va., 120, NE
Tye River (SOU; VBR), Va., 141, SW
Tygart Jct., W.Va., 89, SE
Tyler (B&O; PS&N; PRR), Pa., 29, SE
Tyler, Va., 143, NE
Tylerdale, Pa., 56, SW
Tylersburg, Pa., 28, NW
Tyree, Va., 154, NE
Tyrell, Ohio, 25, NE
Tyrone, Pa., 45, SE

U Jct., Pa., 57B
Udel, Pa., 57, SE
Uffington, W.Va., 73, SW
Uhlers, Pa., 51, NE
Ulmer, Pa., 49, SE
Ulster, Pa., 20, NE
Ulysses (CPA; NYC), Pa., 18, NW
UN, Pa., 59, NE
Unadilla, N.Y., 11, NW
Unamis, Pa., 74, SW
Uneeda, W.Va., 120, SW
Union Beach, N.J., 68, NE
Union Bridge (PRR; WM), Md., 78, SE
Union City (Erie; PRR), Pa., 14, NW
Union Furnace, Ohio, 85, NW
Union Furnace, Pa., 45, SE
Union Hall St., N.Y., 54D
Union Hall, Va., 170, NE
Union Jct., Pa., 35B
Union Level, Va., 173, SE
Union Mills, Pa., 46, SE
Union No. 1 Jct., Md., 75A
Uniondale, Pa., 23, SW
Uniontown (B&O; PRR), Pa., 73, NE
Unionvale, Ohio, 55, NW
Unionville, N.Y., 37, NE
Unionville, Pa., 46, NW
Unionville, Va., 128, NW
United, Pa., 58, SW
United, W.Va., 136, NW
Unity, Pa., 57, NW
Universal, Pa., 57, NW
University Heights, N.Y., 54, NW
University, D.C., 113, NW
Upland, Pa., 82, NW

Upper Falls, W.Va., 120, NW
Upper Lehigh Jct., Pa., 35, SW
Upper Lehigh, Pa., 35, SW
Upper Mill, Pa., 62, SE
Upper Montclair, N.J., 53, NE
Upper Switch, N.Y., 9, NW
Upsall, Pa., 66, SE
Upton, N.J., 83, NE
Ursina, Pa., 74, NW
Ury, W.Va., 136, SW
Utahville, Pa., 45, SW
Utica, Pa., 27, NW
Utley, Ohio, 86, NW

Vabrook (N&W; VGN), Va., 155, SW
Vail, Pa., 45, SE
Vail's Mill, Va., 166, SW
Vails Gate Jct., N.Y., 38, NE
Vails, N.J., 51, NE
Valcourt, Pa., 26, SE
Vale, Md., 80, SW
Valencia, Pa., 42, SW
Valhalla, N.Y., 39, SW
Valier, Pa., 43, NE
Valley Bend, W.Va., 107, NW
Valley Camp, W.Va., 55, SE
Valley Cottage, N.Y., 39, SW
Valley Creek, Tenn., 162, SW
Valley Falls, Pa., 66, SE
Valley Falls, W.Va., 89, NE
Valley Forge, Pa., 66, SW
Valley Grove, W.Va., 55, SE
Valley Head, W.Va., 106, SE
Valley Jct., N.Y., 37, NE
Valley Jct., Pa., 79, NW
Valley Store, Pa., 65, SE
Valley Stream, N.Y., 54, SE
Valley, N.J., 51, SE
Valley, Ohio, 85, NE
Valley, Pa., 81, NW
Valley, Pa., 82A
Vallonia, Ohio, 71, NW
Van Bibber, Md., 97, NW
Van Buren (NYC), N.Y., 3, NW
Van Buren (PRR), N.Y., 3, NW
Van Cortlandt, N.Y., 54, NW
Van Emman, Pa., 56, NE
Van Etten Jct., N.Y., 8, SE
Van Etten, N.Y., 8, SE
Van Fleet, N.Y., 7, SW
Van Nest Shops, N.Y., 54, NW
Van Nest, N.Y., 54, NW

Van Nostrand Ave., N.J., 53B
Van Ortner, Pa., 45, SW
Van Sant, Va., 148, SE
Van Voorhis, Pa., 57, SW
Van Voorhis, W.Va., 73, SW
Van Zandt, Pa., 73, SW
Van, Pa., 27, NE
Van, Pa., 45, NW
Vance Mill Jct., Pa., 73, NE
Vance, Tenn., 165, SE
Vancluse, Va., 93, SE
Vandalia, N.Y., 4, SE
Vandergrift, Pa., 42, SE
Vandeveer Park, N.Y., 54, SW
Vandyke, Del., 98, NE
Vandyke, Pa., 47, SW
Vanetta, W.Va., 121, SE
Vang Jct., Pa., 59, SW
Vanport, Pa., 41, SW
Vanscoyoc, Pa., 45, SE
Vardo, Md., 77, SE
Varilla, Ky., 162, SE
Varna, N.Y., 9, NW
Varner, W.Va., 103, NE
Vaughn, Va., 110, SW
Veasey, W.Va., 136, NE
Venango, Pa., 13, NE
Venice, Pa., 56, NE
Venters, Ky., 148, NW
Ventnor, N.J., 101, NW
Venus, W.Va., 149, NE
Vera Cruz, Pa., 51, SW
Verda, Ky., 163, NE
Verdon, Va., 143, NE
Verdun, W.Va., 134, NE
Vern, Md., 96, SE
Verner, W.Va., 135, SW
Vernon, N.J., 38, SW
Verona, N.J., 53, NW
Verona, Pa., 42, SW
Verona, Va., 125, SE
Verroy, N.J., 52, SW
Versailles, Pa., 57, NW
Vesta Siding, Pa., 57, SW
Vestal, N.Y., 9, SE
Vesuvius, Va., 140, NE
Veto, W.Va., 71, SW
Vexit Jct., Pa., 50, NW
Viacova, W.Va., 136A, SE
Viaduct, Pa., 28, SE
Viaduct, Pa., 35, NE
Viaduct, Pa., 45, NE
Vicker, Va., 152, SW

Vicksburg, Pa., 48, NW
Vicksville, Va., 175, NE
Victor, W.Va., 121, NW
Victoria, Pa., 74, NW
Victoria, Va., 173, NE
Victory, Pa., 26, NW
Vienna, Md., 132, NW
Vienna, N.J., 52, NW
Vienna, Pa., 56, SW
Vienna, Va., 112, NW
Vienna, W.Va., 86, NE
Villa Nova, W.Va., 105, SW
Villamont, Va., 153, NW
Villanova, Pa., 66, SW
Vimy, W.Va., 134, SE
Vincent, Ky., 118, NW
Vincent, Ohio, 86, NE
Vincentown, N.J., 83, NE
Vineland (CNJ; P-RSL), N.J., 99,
   NE
Vinemont, Pa., 64, NE
Vineyard, Pa., 61, NW
Vinita, Va., 143, SE
Vinton, Ohio, 102, NW
Vinton, Va., 153, NW
Viola, Del., 98, SE
Viola, Pa., 45, NW
Violet, W.Va., 123, SE
Virgie, Ky., 147, NE
Virgilina, N.C., 172, SW
Virginia Beach, Va., 177, NE
Virginia City, Va., 165, NW
Virginia Lane, Md., 75, SW
Virginia Manor, Va., 140, SW
Virginia Mills, Pa., 78, NW
Virginia Mineral Springs, Va.,
   138, SE
Virginia, W.Va., 55, NE
Virginville, Pa., 50, SW
Virso (N&W; SOU; VGN), Va.,
   156, SW
Virwest, W.Va., 136, SW
Vivian, W.Va., 149, NE
Vliettown, N.J., 52, SE
Volant, Pa., 26, SW
Volga, W.Va., 89, SE
Vosburg, Pa., 21, SE
Vowinckel, Pa., 28, NE
Vreeland Ave., N.J., 53, NE
VS Jct., Pa., 35D
Vulcan, Pa., 49, NE
Vulcan, W.Va., 134, SE
Vulcanite, N.J., 51, SE

W&N Jct., Pa., 65, NW
W&W Jct. (W&W), Va., 93, SE
W, Ohio, 55, NW
W, Pa., 59, NW
W.Va. & Pittsburgh Jct., W.Va.,
   89, NW
W.Va. No. 2, W.Va., 136, NW
WA-2, N.J., 53A
WA-3, N.J., 53A
WA-5, N.J., 53A
WA-6, N.J., 53A
Wabun, Va., 152, NE
Wacomah, W.Va., 136, SW
Waddle, Pa., 46, NW
Wadesville, Va., 93, SE
Wagner, Pa., 47, SW
Wagner's, Pa., 35, SE
Wagners Crossroads, Md., 77, SE
Wago Jct., Pa., 63, SE
Wagontown, Pa., 65, SW
Wahneta, Pa., 62, NE
Wainville, W.Va., 122, NE
Waitville, W.Va., 152, NW
Wakefield, Md., 78, SE
Wakefield, N.Y., 54, NW
Wakefield, Va., 176, NW
Walbert, Pa., 50, SE
Walbrook Jct., Md., 96A
Walbrook, Md., 96A
Waldorf, Md., 113, SW
Waldwick, N.J., 38, SE
Walford (P&LE; PRR), Pa., 41,
   NW
Walgrove, W.Va., 121, NW
Walker, Va., 159, NW
Walker, W.Va., 87, SW
Walker, W.Va., 107, NE
Walker's Mill, Pa., 56, NE
Walkerford, Va., 141, SW
Walkersville, Md., 95, NW
Walkersville, W.Va., 106, NW
Walkton, Pa., 50, NE
Wall Rope Works, N.J., 67, SW
Wallace (DL&W; Erie), N.Y., 7,
   NW
Wallace Jct., Pa., 1, SW
Wallace, Pa., 57, SW
Wallace, Va., 165, SE
Wallace, W.Va., 89, NW
Wallaceton (NYC; PRR), Pa., 45,
   NW
Wallingford, Pa., 82, NW
Wallins, Ky., 163, NW

Wallman, W.Va., 91, NW
Wallner, Pa., 50, SE
Wallopsburg, Pa., 43, NE
Wallsend, Ky., 162, NE
Walnut Bend, Pa., 27, NE
Walnut Hill, Pa., 66, SE
Walnut Hill, Pa., 73, NW
Walnut St., N.J., 53, NE
Walnut St., Pa., 46, SE
Walnut, Pa., 32, SE
Walnut, W.Va., 123, NE
Walnutport, Pa., 50, NE
Walsall, Pa., 59, NW
Walson Jct., Pa., 44, NW
Walson, Md., 132, NE
Walters, Va., 176, NW
Waltersburg, Pa., 73, NW
Walton, N.Y., 11, SE
Walton, Pa., 18, NW
Walton, Pa., 30, SW
Walton, Va., 151, SE
Walton-Bridge St., N.Y., 11, SE
Waltonville, Pa., 63, SE
Wanamaker, Pa., 50, SW
Wanaque-Midvale, N.J., 38, SW
Wanda, W.Va., 135, NW
Wandin, Pa., 44, SW
Wangaum, Pa., 36, NE
Wanless, W.Va., 124, NW
Wantage, N.J., 37, NE
Wantagh, N.Y., 54, SE
Wapocomo, W.Va., 92, NE
Wapwallopen, Pa., 34, SE
War Eagle, W.Va., 135, SW
War, W.Va., 149, NE
Ward, Md., 95, SE
Ward, W.Va., 121, SW
Wardensville, W.Va., 92, SE
Wardwell, Ohio, 25, SW
Waretown Jct., N.J., 84, NE
Waretown, N.J., 84, NE
Warfield, Va., 174, NW
Waring, Md., 95, SE
Warminster, Va., 141, SE
Warner, Ohio, 70, SW
Warnock, Ohio, 55, SW
Warren (Erie; PRR), Ohio, 25,
   SW
Warren (NYC; PRR), Pa., 15, NE
Warren St. (Trenton), N.J., 67,
   SW
Warren, Ky., 162, NW
Warren, Va., 141, NE

West Julian, Pa., 46, NW
West Junior, W.Va., 107, NW
West Lake Jct., Pa., 36, NW
West Lebanon, Pa., 43, SW
West Liberty, Pa., 56A
West Line, Pa., 16, NW
West Marietta (O&LK Jct.), Ohio, 87, NW
West Merchantville, N.J., 82B
West Middlesex (Erie; PRR), Pa., 26, SW
West Middlesex, Pa., 56, NW
West Milton, Pa., 33, SW
West Monessen, Pa., 57, SW
West Monterey, Pa., 27, SE
West Moorestown, N.J., 83, NW
West Musgrove, Pa., 43, NW
West Nanticoke, Pa., 34, SE
West New Kensington, Pa., 42, SW
West Newton (B&O; P&LE), Pa., 57, SW
West Newton (P&LE), Pa., 57, SW
West Norfolk, Va., 177, NW, 177A
West Norwood, N.J., 39, SW
West Norwood, N.J., 54, NW
West Notch, N.Y., 5, SE
West Nyack, N.Y., 39, SW
West Oakland, N.J., 38, SW
West Ocean City, Md., 133, NE
West Orange, N.J., 53, NE
West Overton, Pa., 57, SE
West Penn Crossing, Pa., 73, NE
West Penn, Pa., 50, SW
West Perrysburg, N.Y., 3, NE
West Philadelphia Yard, Pa., 82A
West Pilot, Md., 80, SE
West Pittsburgh (B&O; P&LE; WAL), Pa., 41, NW
West Pittston, Pa., 35B
West Point, N.Y., 39, NW
West Point, Pa., 66, SW
West Point, Va., 145, SW
West Port Marion, Pa., 73, SW
West Portal, N.J., 51, SE
West Powhatan, Ohio, 71, NW
West Raleigh, W.Va., 136A, NE
West Rochester, Pa., 41, SW
West Rock, Md., 80, SE
West Romney, W.Va., 92, NW
West Salisbury, Pa., 74, NE
West Shore Line Switch, N.Y., 53, SE

West Side Ave., N.J., 53B
West Springfield, Pa., 13, NW
West Street, N.Y., 54, NE
West Street, Ohio, 12, NW
West Tarentum, Pa., 42, SW
West Taylorstown, Pa., 56, SW
West Templeton, Pa., 43, NW
West Trenton, N.J., 67, NW
West Union, Pa., 56, SW
West Union, W.Va., 88, NW
West Valley, N.Y., 4, NE
West Van Voorhis, W.Va., 73, SW
West Vernon, Pa., 13, SW
West View, Va., 142, SE
West Vindex, Md., 91, NE
West Virginia Central Jct., W.Va., 91, NE
West West Jct., Pa., 49, SW
West Wildwood, N.J., 117, NW
West Willow, Pa., 80, NW
West Winfield, Pa., 42, NE
West Yard, Del., 81A
West York (PRR; WM), Pa., 79, NW
West York Passing, Pa., 79, NE
West Yough Transfer, Pa., 57C
West's Crossing, Va., 154, NW
Westbrook, Pa., 80, NE
Westbrookville, N.Y., 24, SE
Westbury, N.Y., 54, NE
Westchester Ave., N.Y., 39, SW
Westchester, N.Y., 54, NW
Westcolang Park, Pa., 23, SE
Westerly, W.Va., 136, NW
Westernport, W.Va., 91, NE
Westfield (NKP), N.Y., 2, NE
Westfield (NYC), N.Y., 2, NE
Westfield, N.J., 53, SW
Westfield, Pa., 18, NE
Westford, Pa., 13, SW
Westham, Va., 143, SE
Westlakes Crossing, Ohio, 25A
Westland, Pa., 56, NW
Westminster, Md., 79, SW
Westmont, N.J., 82B
Westmont, Pa., 64, NW
Westmore, Md., 95, SE
Westmoreland, Pa., 66, SE
Weston, N.Y., 5, SW
Weston, Pa., 20, SE
Weston, W.Va., 89, SW
Weston-Manville, N.J., 52, SE
Westons, N.Y., 5, SW

Westover, Md., 132, SE
Westover, Pa., 44, NE
Westport, Pa., 31, NW
Westtown, N.Y., 37, NE
Westtown, Pa., 81, NE
Westview, N.J., 53, NE
Westville, N.J., 82, NE
Westville, Pa., 29, SW
Westwood Jct., Pa., 49B
Westwood, N.J., 53, NE
Westwood, N.Y., 54, SE
Wetherill Jct., Pa., 49, NE
Wetmore, Pa., 16, SW
Wetzel, Pa., 65, NE
Wevaco, W.Va., 121, SW
Weverton, Md., 94, NE
Weyburn, Va., 127, SE
Weyers Cave, Va., 126, NW
Whaley, Va., 176, SE
Whaleysville, Md., 133, NW
Wharncliffe, W.Va., 135, SW
Wharton (CNJ; DL&W), N.J., 52, NE
Wharton Jct., N.J., 52, NE
Wharton, Pa., 17, SE
Wharton, W.Va., 135, NE
Wheatfield, Pa., 59, NW
Wheatland (Erie; PRR), Pa., 26, SW
Wheatland, Ohio, 25, SE
Wheeler, N.Y., 7, NW
Wheeler, Va., 162, SE
Wheeler, W.Va., 106, SW
Wheelerville, Pa., 20, SW
Wheeling (B&O; PRR; W&LE), W.Va., 55, SE, 55A
Wheelwright Jct., Ky., 147, NE
Whigville, Ohio, 70, NW
Whippany, N.J., 53, NW
Whipple, Ohio, 70, SW
Whitaker, W.Va., 121, SW
Whitco, Ky., 147, SW
Whitcomb, W.Va., 138, NW
White Bear, Pa., 65, SW
White Bridge, Pa., 74, NW
White Deer, Pa., 33, SW
White Gravel, Pa., 16, NW
White Hall, Md., 79, SE
White Haven (CNJ; LV), Pa., 35, SW
White House (CNJ; P&NJ), N.J., 52, SW
White House, Va., 144, SE

White Marsh, Md., 97, NW
White Marsh, Pa., 66, SE
White Mill, Pa., 44, SW
White Mills, Pa., 23, SE
White Oak Jct., W.Va., 136, NE
White Oak, Pa., 64, SW
White Oak, Va., 171, NW
White Plains North, N.Y., 39, SW
White Plains, N.Y., 39, SW
White Post, Va., 93, SE
White Rock, Pa., 80, NE
White Spring, Pa., 63, NW
White Stick, W.Va., 136, NE
White Sulphur Springs (C&O; WSH), W.Va., 138, NW
White Valley, Pa., 57, NE
White, Pa., 58, NW
White, W.Va., 134, SE
Whiteford, Md., 80, SW
Whitehead, N.J., 99, NE
Whitehorne, Va., 151, SE
Whiteland, Pa., 65, SE
Whitesboro, N.J., 100, SW
Whitesburg, Ky., 147, SW
Whitesett Jct., Pa., 57, SW
Whitestone Landing, N.Y., 54, NW
Whitestone, N.Y., 54, NW
Whitesville, N.Y., 6, SW
Whitetop, Va., 166, SE
Whitewood, Va., 149, SW
Whiting, W.Va., 107, SW
Whitings (CNJ; PRR; TUCK), N.J., 84, NW
Whitlock, Va., 127, SE
Whitman Jct., W.Va., 134, NE
Whitman, W.Va., 134, NE
Whitmer, W.Va., 107, NE
Whitney Point, N.Y., 10, NW
Whitney, Pa., 58, NW
Whittaker, Ky., 147, SE
Whittens Mill, Va., 150, SW
Whittle, Va., 171, NW
Whyte, W.Va., 107, NW
Wick, Ohio, 12, SE
Wickatunk, N.J., 68, NW
Wickham, Va., 144, NW
Wickliffe, Ohio, 25, SE
Wiconisco, Pa., 48, SE
Wide Water, Va., 129, NW
Widen, W.Va., 122, NW
Widnoon, Pa., 43, NW
Wierwood, Va., 161, NW

Wiggins, W.Va., 137, SW
Wigton, Pa., 45, NE
Wigul, W.Va., 136A, SE
Wilawanna, Pa., 20, NE
Wilber, W.Va., 135, NW
Wilburth, N.J., 67, NW
Wilco, W.Va., 149, NE
Wilcox, N.Y., 38, NW
Wilcox, Pa., 16, SE
Wilda, Va., 125, SE
Wildel, W.Va., 107, SW
Wilder, Va., 148, SE
Wildwood (P-RSL; RDG-AC), N.J., 117, NW
Wildwood Crest, N.J., 117, NW
Wildwood Gardens, N.J., 100, SW
Wildwood Jct. (P-RSL; WJ&S), N.J., 100, SW
Wildwood Springs, Pa., 59, NE
Wildwood, Pa., 42, SW
Wildwood, Va., 142, NW
Wiley, W.Va., 136A, SE
Wilfred, Ky., 163, SW
Wilhoit, Ky., 163, NW
Wilkes-Barre (CNJ), Pa., 35A
Wilkes-Barre (L&WV), Pa., 35A
Wilkes-Barre (LV; PRR), Pa., 35A
Wilkins, N.J., 83, NW
Wilkinsburg, Pa., 57A
Willabet, W.Va., 136, SE
Willada, Pa., 56, NE
Willard Branch Jct., W.Va., 89, NW
Willard, Ky., 118, SW
Willards, Md., 133, NW
William Penn Colliery, Pa., 49C
William Tod, Ohio, 25A
William, W.Va., 91, SW
Williams Ave., N.J., 53, NE
Williams Bridge, N.Y., 54, NW
Williams Creek, Ky., 118, NW
Williams Grove, Pa., 62, SE
Williams No. 1 Mine, W.Va., 107, NW
Williams River Branch Switch, W.Va., 122, NE
Williams River Mine, W.Va., 123, NW
Williams Valley Jct., Pa., 48, SE
Williams, Pa., 66, SE
Williamsburg, Md., 115, SW
Williamsburg, Pa., 60, NE
Williamsburg, Va., 159, NE

Williamsfield, Ohio, 12, SE
Williamson School, Pa., 82, NW
Williamson, Pa., 77, NW
Williamson, W.Va., 134, SW
Williamsport (PECo; WM), Md., 77, SW
Williamsport (PRR; RDG), Pa., 32, SE
Williamsport, Md., 77, SW
Williamstown (PRR; RDG), Pa., 48, SE
Williamstown Jct., N.J., 83, NW
Williamstown, N.J., 83, SW
Williamstown, W.Va., 87, NW
Willis Branch, W.Va., 136, NW
Willock, Pa., 57, NW
Willoughby, Md., 114, NE
Willow Ave., N.J., 53B
Willow Farm, Ohio, 87, NW
Willow Grove, Ohio, 55, SW
Willow Grove, Pa., 66, SE
Willow Grove, W.Va., 103, NW
Willow Island, W.Va., 87, NW
Willow Springs, Pa., 34, SW
Wills Hollow, W.Va., 121, NW
Wills, W.Va., 151, NW
Willseyville (DL&W; LV), N.Y., 9, NW
Wilmer, Pa., 65, SE
Wilmerding, Pa., 57, NW
Wilmington (B&O; PRR; RDG), Del., 81A
Wilmington Jct., Pa., 26, SW
Wilmington North, Del., 81A
Wilmington Shops, Del., 81A
Wilmington South, Del., 81A
Wilmore, Pa., 59, NE
Wilmore, W.Va., 149, NW
Wilmoth, W.Va., 90, SW
Wilpen, Pa., 58, NE
Wilsmere, Del., 81A
Wilson Creek, Pa., 74, NE
Wilson Run, Pa., 44, NE
Wilson Siding, W.Va., 106, NW
Wilson, Pa., 28, SW
Wilson, Pa., 57, NW
Wilson, Va., 157, SW
Wilson, W.Va., 91, NW
Wilson, W.Va., 119, NW
Wilson, W.Va., 120, NE
Wilsonburg, W.Va., 89, NW
Wilsondale, W.Va., 134, NW
Wilton, N.J., 83, NW

Wimmers, Pa., 36, NW
Winans, Md., 96, SE
Winbourne, Pa., 45, NE
Winchester (B&O; PRR), Va., 93, SE
Wind Gap (DL&W; LNE), Pa., 51, NW
Wind Gap Jct., Pa., 51, NW
Windber Jct., Pa., 59, SW
Winding Gulf, W.Va., 136A, SE
Windsor Shades, Va., 159, NW
Windsor, N.J., 67, SE
Windsor, N.Y., 10, SE
Windsor, Va., 176, NE
Winfall, Va., 154, SE
Winfield Jct., Pa., 42, SE
Winfield, Pa., 48, NW
Wingert, Pa., 62, SE
Wingerton, Pa., 77, SE
Wingina, Va., 141, SE
Winifrede Jct. (C&O; WNFR), W.Va., 120, SE
Winifrede, W.Va., 120, SE
Winona, W.Va., 89, NE
Winright, Ky., 148, NW
Winslow Jct. (CNJ; P-RSL), N.J., 83, SW
Winslow, Ky., 118, NE
Winston, Va., 127, NE
Winter, Ky., 147, SW
Winterarthur, Del., 81, NE
Winterburn, W.Va., 107, SW
Winterham, Va., 157, NW
Winterpock, Va., 157, NE
Winthrop, Pa., 58, NW
Winton (D&H), Pa., 35D
Winton (NYO&W), Pa., 35D
Winton Jct., Pa., 35D
Wirtz, Va., 153, SW
Wishaw, Pa., 29, SW
Wisner, N.Y., 38, NW
Wissahickon, Pa., 82A, 66B
Wissinoming, Pa., 66, SE
Wistar, Va., 143, SE
Wister, Pa., 66, SE
Witcher, W.Va., 120, SE
Witmer, Pa., 64, SE
Witte, Pa., 51, SW
WK, Ohio, 55, NW
WN Tower, W.Va., 122, NE
Wohlville, Pa., 41, NE
Wolf Creek, W.Va., 137, SE
Wolf Pit, Ky., 148, NW

Wolf Run, N.Y., 4, SW
Wolf Summit, W.Va., 89, NW
Wolf Trap, Va., 172, SW
Wolfert, N.J., 82, NW
Wolford, Va., 149, NW
Wolfsburg, Pa., 59, SE
Wolverton, Pa., 48, NE
Womelsdorf, Pa., 64, NE
Wood Jct., Pa., 50A
Wood, Pa., 31, SE
Wood, Va., 164, NE
Woodberry, Md., 96, NE
Woodbine Jct., N.J., 100, SW
Woodbine, Md., 95, NE
Woodbine, N.J., 100, SW
Woodbine, N.Y., 38, SE
Woodbine, Pa., 80, NW
Woodbine, W.Va., 122, NE
Woodbourne, Pa., 67, SW
Woodbridge, N.J., 53, SW
Woodbrook, Md., 96, NE
Woodbury Heights, N.J., 82, NE
Woodbury, N.J., 82, NE
Woodbury, N.Y., 38, NE
Woodcliff Lake, N.J., 38, SE
Woodcrest, N.J., 82, NE
Wooddale, Del., 81, NE
Woodensburg, Md., 79, SW
Woodfern, N.J., 52, SW
Woodford, Va., 129, SW
Woodhaven Jct., N.Y., 54, SW
Woodland (NYC; PRR), Pa., 45, NW
Woodland, W.Va., 71, NW
Woodlands, N.Y., 39, SW
Woodlane, Pa., 66A
Woodlawn, N.Y., 54, NW
Woodlawn, Va., 144, SE
Woodman, Ky., 134, SE
Woodman, W.Va., 138, NW
Woodmansie, N.J., 84, NW
Woodmere, N.Y., 54, SE
Woodmont Club, Md., 76, SW
Woodmont, Pa., 66, SE
Woodridge, N.Y., 24, SE
Woodrow, Pa., 56, NW
Woodruff, W.Va., 71, NE
Woodruff's Gap, N.J., 37, SE
Woods Run, Pa., 56A
Woods, Pa., 27, NW
Woods, W.Va., 103, SW
Woodsboro, Md., 78, SW
Woodsfield, Ohio, 70, NE

Woodside, Del., 98, SE
Woodside, Md., 112, NE
Woodside, N.J., 53, NE
Woodside, N.Y., 54, SW
Woodson, Va., 140, NE
Woodstock, Md., 96, NW
Woodstock, Pa., 77, NE
Woodstock, Va., 109, NE
Woodstown, N.J., 82, SW
Woodvale, Pa., 16, SE
Woodville, Pa., 18, NW
Woodville, Pa., 56A
Woodville, Pa., 60, SE
Woodville, W.Va., 120, SW
Woodway, Conn., 39, SE
Woodworth, Ohio, 40, NE
World's Fair, N.Y., 54, NW
Wortendyke, N.J., 53, NE
Worth, Pa., 74, SE
Worthington, N.Y., 39, SW
Worthington, Pa., 42, NE
Worthington, W.Va., 89, NW
Worton, Md., 97, NE
Wren, Pa., 34, SE
Wright, W.Va., 136, NE
Wrights, Pa., 17, SE
Wrights, Va., 153, SW
Wrightstown, N.J., 67, SE
Wrightsville, Pa., 63, SE
Wriston, W.Va., 121, SE
WU, Ohio, 55, NW
Wurno, Va., 151, SE
Wurtemburg, Pa., 41, NW

Wyalusing, Pa., 21, SW
Wyano, Pa., 57, SE
Wyatt, W.Va., 89, NW
Wycoff, N.J., 38, SE
Wycombe, Pa., 66, NE
Wye Mills, Md., 114, NE
Wyebrooke, Pa., 65, SW
Wykagyl, N.Y., 54, NW
Wylandville, Pa., 56, SE
Wylie, Pa., 57, NW
Wylo, W.Va., 135, SW
Wyndal, W.Va., 121, SE
Wyndale, Va., 165, SE
Wyndmoor, Pa., 66, SE
Wynn, Pa., 45, NW
Wynnefield Ave., Pa., 82A
Wynnewood, Pa., 66, SW
Wyoanna, Pa., 35, NW
Wyoming (DL&W; LV), Pa., 35B
Wyoming, Del., 98, SE
Wyomissing Jct., Pa., 65, NW
Wyomissing, Pa., 65, NW
Wysor, Va., 151, SE
Wysox, Pa., 21, NW
Wytheville, Va., 167, NE

Y&N Crossing, Ohio, 25A
Yadkin, Va., 177, NW
Yale, Va., 175, NW
Yanda, Ohio, 25A
Yard Tower, Va., 177A
Yardley, Pa., 67, SW
Yardville, N.J., 67, SE

Yates Crossing, W.Va., 119, NE
Yates, W.Va., 89, NE
Yatesboro, Pa., 43, NW
Yatesville (D&H), Pa., 35B
Yatesville (Erie), Pa., 35B
YD (WM), Md., 77, SE
Yeager, Ky., 147, NE
Yeagertown, Pa., 46, SE
Yeatman, Pa., 81, NW
Yellow Spring, Pa., 63, NE
Yellow Springs, Md., 95, NW
Yellow Sulphur, Va., 152, SW
Yerkes, Pa., 66, SW
Yoe, Pa., 79, NE
Yolyn, W.Va., 135, NW
Yonkers, N.Y., 54, NW
York (M&PA; PRR; WM), Pa., 79, NE
York Farm Jct. (LV; RDG), Pa., 49B
York Haven, Pa., 63, SE
York Jct., Pa., 50A
York Road, Pa., 79, NW
York Run Jct., Pa., 73, NW
York Run, Pa., 73, NW
York, Ohio, 25, NE
York, W.Va., 102, NE
Yorklyn, Del., 81, NE
Yorkship, N.J., 82B
Yorktown Heights, N.Y., 39, NW
Yorktown, N.J., 82, SW
Yorkville, Ohio, 55, NE
Youngdale, Pa., 32, SW

Youngs, Va., 149, SE
Youngstown (B&O; Erie; NYC; P&LE; PRR; Y&S), Ohio, 25, SE, 25A
Youngstown (NYC; P&LE), Ohio, 25A
Youngstown Jct., Pa., 73, NE
Youngsville (NYC; PRR), Pa., 15, NW
Youngwood, Pa., 57, SE
Younts, Pa., 60, SW
Yukon Branch Jct., Pa., 57, SE
Yukon, Pa., 57, SE
Yuma, Va., 164, SE

Zaleski, Ohio, 85, NW
Zaners, Pa., 34, SW
Zehners, Pa., 50, NW
Zelda, Ky., 118, SE
Zelienople, Pa., 41, NE
Zenith, Pa., 34, SE
Zeno, Pa., 41, NE
Zevely, W.Va., 90, NE
Zieglersville, Pa., 66, NW
Zimmerman, Pa., 58, SE
Zion Road, N.J., 100, NE
Zion, Pa., 46, NE
Zion, Va., 142, NE
Zionsville, Pa., 65, NE
Zuni, Va., 176, NW

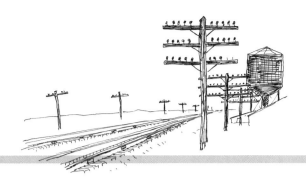

INDEX OF TRACK PANS

On main lines of certain railroads, especially the New York Central and the Pennsylvania, steam locomotives could be resupplied with water without stopping, from water troughs or track pans that were placed between the rails. A scoop would be lowered from the underside of the tender, while over the track pan, and lifted before reaching the end. The railroad operating each track pan is signified by its reporting marks, shown in parentheses.

INDEX OF TUNNELS

The tunnels listed in this index are those that have been identified by name on USGS maps, railroad track charts, or other reliable sources. The reporting marks of the owning railroad are indicated in parentheses.

INDEX OF VIADUCTS

Rivers and valleys were often spanned by large bridges or viaducts, many of which were named. The railroad owning the viaduct is signified by its reporting marks, shown in parentheses.

RICHARD C. CARPENTER was born in Hartford, Connecticut, in 1933 and was raised there and in Wethersfield, Connecticut. He received his bachelor of science degree in history from Boston College, served as a first lieutenant in the U.S. Army artillery, and then received his master's in city and regional planning at the University of Pennsylvania. He was town planner for Wilton, Connecticut, and then, in 1966, joined the South Western Regional Planning Agency of Connecticut as its executive director, retiring in 1999. He continues to serve as a member of the Connecticut Public Transportation Commission. For more than fifty years, Dick Carpenter has closely observed and studied railroad track and signal operating characteristics, collecting in the process invaluable charts, maps, timetables, and other extensive railroad materials. He is a member of the National Railway Historical Society and the Irish Railway Record Society, a corresponding member of the Signalling Record Society, and a member of the Historical and Technical Organizations of nine railroads. He resides in East Norwalk, Connecticut, where he keeps his sailboat, *Phoebe Snow*, named for the symbol of the Delaware, Lackawanna & Western Railroad.